Abstract and Concrete Categories
The Joy of Cats

Jiří Adámek
Institute of Theoretical Computer Science
Technical University of Braunschweig, Germany

Horst Herrlich
Department of Mathematics
University of Bremen, Germany

George E. Strecker
Department of Mathematics
Kansas State University
Manhattan, Kansas USA

Dover Publications, Inc.
Mineola, New York

The authors are grateful for any improvements, corrections, and remarks, and can be reached at the addresses

Jiří Adámek, email: `adamek@iti.cs.tu-bs.de`

Horst Herrlich, email: `horst.herrlich@t-online.de`

George E. Strecker, email: `strecker@math.ksu.edu`

All corrections will be awarded, besides eternal gratefulness, with a piece of delicious cake! You can claim your cake at the KatMAT Seminar, University of Bremen, at any Tuesday (during terms).

The newest edition of the file of the present book can be downloaded from

`http://katmat.math.uni-bremen.de/acc`

Copyright

Copyright © 1990, 2004 by Jiří Adámek, Horst Herrlich, and George E. Strecker
All rights reserved.

Bibliographical Note

This Dover edition, first published in 2009, is an unabridged republication of the revised and corrected version posted on the Internet by the authors in 2004 of the work originally published in 1990 by John Wiley & Sons, Inc., New York.

Library of Congress Cataloging-in-Publication Data

Adámek Jiří, ing.
 Abstract and concrete categories : the joy of cats / J. Adámek, H. Herrlich, and G. E. Strecker. — Dover ed.
  p. cm.
 "Unabridged republication of the revised and corrected version posted on the Internet by the authors in 2004 of the work originally published in 1990 by John Wiley & Sons, Inc., New York."
 Includes bibliographical references and index.
 ISBN-13: 978-0-486-46934-8
 ISBN-10: 0-486-46934-4
 1. Categories (Mathematics) I. Herrlich, Horst. II. Strecker, George E. III. Title.
QA169.A3199 2009
512' 62—dc22

2009010561

Manufactured in the United States by LSC Communications
46934405   2019
www.doverpublications.com

Dedicated to
Bernhard Banaschewski

PREFACE to the 2004 EDITION

Abstract and Concrete Categories was published by John Wiley & Sons, Inc., in 1990, and after several reprints, the book has been sold out and unavailable for several years. We now present an improved and corrected version. This was made possible due to the return of copyright to the authors, and due to many hours of hard work and the exceptional skill of Christoph Schubert, to whom we wish to express our profound gratitude. The illustrations of Edward Gorey are unfortunately missing in the current version (for copyright reasons), but fortunately additional original illustrations by Marcel Erné, to whom additional special thanks of the authors belong, counterbalance the loss.

Besides the acknowledgements appearing at the end of the original preface, we wish to thank all those who have helped to eliminate mistakes that survived the first printing of the text, particularly H. Bargenda, J. Jürgens, W. Meyer, L. Schröder, A. M. Torkabud, and O. Wyler.

January 12, 2004

J. A., H. H., and G. E. S.

PREFACE

Sciences have a natural tendency toward diversification and specialization. In particular, contemporary mathematics consists of many different branches and is intimately related to various other fields. Each of these branches and fields is growing rapidly and is itself diversifying. Fortunately, however, there is a considerable amount of common ground — similar ideas, concepts, and constructions. These provide a basis for a general theory of structures.

The purpose of this book is to present the fundamental concepts and results of such a theory, expressed in the language of category theory — hence, as a particular branch of mathematics itself. It is designed to be used both as a textbook for beginners and as a reference source. Furthermore, it is aimed toward those interested in a general theory of structures, whether they be students or researchers, and also toward those interested in using such a general theory to help with organization and clarification within a special field. The only formal prerequisite for the reader is an elementary knowledge of set theory. However, an additional acquaintance with some algebra, topology, or computer science will be helpful, since concepts and results in the text are illustrated by various examples from these fields.

One of the primary distinguishing features of the book is its emphasis on *concrete categories*. Recent developments in category theory have shown this approach to be particularly useful. Whereas most terminology relating to abstract categories has been standardized for some time, a large number of concepts concerning concrete categories has been developed more recently. One of the purposes of the book is to provide a reference that may help to achieve standardized terminology in this realm. Another feature that distinguishes the text is the systematic treatment of *factorization structures*, which gives a new unifying perspective to many earlier concepts and results and summarizes recent developments not yet published in other books.

The text is organized and written in a "pedagogical style", rather than in a highly economical one. Thus, in order to make the flow of topics self-motivating, new concepts are introduced gradually, by moving from special cases to the more general ones, rather than in the opposite direction. For example,

- equalizers (§7) and products (§10) precede limits (§11),
- factorizations are introduced first for single morphisms (§14), then for sources (§15), and finally for functor-structured sources (§17),
- the important concept of adjoints (§18) comes as a common culmination of three separate paths: 1. via the notions of reflections (§4 and §16) and of free objects (§8), 2. via limits (§11), and 3. via factorization structures for functors (§17).

Each categorical notion is accompanied by many examples — for motivation as well as clarification. Detailed verifications for examples are usually left to the reader as implied exercises. It is not expected that every example will be familiar to or have relevance for each reader. Thus, it is recommended that examples that are unfamiliar should be skipped, especially on the first reading. Furthermore, we encourage those who are working through the text to carry along their favorite category and to keep in mind a "global exercise" of determining how each new concept specializes in that particular setting. The exercises that appear at the end of each section have been designed both as an aid in understanding the material, e.g., by demonstrating that certain hypotheses are needed in various results, and as a vehicle to extend the theory in different directions. They vary widely in their difficulty. Those of greater difficulty are typically embellished with an asterisk (∗).

The book is organized into seven chapters that represent natural "clusters" of topics, and it is intended that these be covered sequentially. The first five chapters contain the basic theory, and the last two contain more recent research results in the realm of concrete categories, cartesian closed categories, and quasitopoi. To facilitate references, each chapter is divided into sections that are numbered sequentially throughout the book, and all items within a given section are numbered sequentially throughout it. We use the symbol □ to indicate either the end of a proof or that there is a proof that is sufficiently straightforward that it is left as an exercise for the reader. The symbol D means that a proof of the dual result has already been given. Symbols such as A 4.19 are used to indicate that no proof is given, since a proof can be obtained by analogy to the one referenced (i.e., to item 19 in Section 4). Two tables of symbols appear at the end of the text. One contains a list (in alphabetical order) of the abbreviated names for special categories that are dealt with in the text. The other contains a list (in order of appearance in the text) of special mathematical symbols that are used. The bibliography contains only books and monographs. However, each section of the text ends with a (chronologically ordered) list of suggestions for further reading. These lists are designed to aid those readers with a particular interest in a given section to "strike out on their own" and they often contain material that can be used to solve the more difficult exercises. They are intended as merely a sampling, and (in view of the vast literature) there has been no attempt to make them complete[1] or to provide detailed historical notes.

Acknowledgements

We are grateful for financial support from each of our "home universities" and from the Natural Sciences and Engineering Research Council of Canada, the National Academies of Sciences of Czechoslovakia and the United States, the U.S. National Science Foundation, and the U.S. Office of Naval Research. We particularly appreciate that such support made it possible for us to meet on several occasions to work together on the manuscript.

Our special thanks go to Marcel Erné for several original illustrations that have been incorporated in the text and to Volker Kühn for his efforts on a frontispiece that the

---

[1] Indeed, although some could serve as a suggested reading for more than one section, none appears in more than one.

Publisher decided not to use. We also express our special thanks to Reta McDermott for her expert typesetting, to Jürgen Koslowski for his valuable TEXnical assistance, and to Y. Liu for assistance in typesetting diagrams. We were also assisted by D. Bressler and Y. Liu in compiling the index and by G. Feldmann in transferring electronic files between Manhattan and Bremen. We are especially grateful to J. Koslowski for carefully analyzing the entire manuscript, to P. Vopěnka for fruitful discussions concerning the mathematical foundations, and to M. Erné, H.L. Bentley, D. Bressler, H. Andréka, I. Nemeti, I. Sain, J. Kincaid, and B. Schröder, each of whom has read parts of earlier versions of the manuscript, has made suggestions for improvements, and has helped to eliminate mistakes. Naturally, none of the remaining mistakes can be attributed to any of those mentioned above, nor can such be blamed on any single author — it is always the fault of the other two.

srohtua eht

Contents

Preface to the 2004 Editon   3

Preface   4

0 Introduction   9
    1   Motivation 11
    2   Foundations 13

I Categories, Functors, and Natural Transformations   19
    3   Categories and functors 21
    4   Subcategories 48
    5   Concrete categories and concrete functors 61
    6   Natural transformations 83

II Objects and Morphisms   99
    7   Objects and morphisms in abstract categories 101
    8   Objects and morphisms in concrete categories 132
    9   Injective objects and essential embeddings 152

III Sources and Sinks   167
    10   Sources and sinks 169
    11   Limits and colimits 193
    12   Completeness and cocompleteness 211
    13   Functors and limits 223

IV Factorization Structures   237
    14   Factorization structures for morphisms 239
    15   Factorization structures for sources 257
    16   E-reflective subcategories 275
    17   Factorization structures for functors 290

V Adjoints and Monads   303
    18   Adjoint functors 305
    19   Adjoint situations 315
    20   Monads 325

VI Topological and Algebraic Categories — 357
 21   Topological categories 359
 22   Topological structure theorems 382
 23   Algebraic categories 389
 24   Algebraic structure theorems 408
 25   Topologically algebraic categories 417
 26   Topologically algebraic structure theorems 429

VII Cartesian Closedness and Partial Morphisms — 435
 27   Cartesian closed categories 437
 28   Partial morphisms, quasitopoi, and topological universes 451

Bibliography — 469

Tables — 473
Functors and morphisms: Preservation properties 473
Functors and morphisms: Reflection properties 473
Functors and limits 474
Functors and colimits 474
Stability properties of special epimorphisms 474

Table of Categories — 475

Table of Symbols — 480

Index — 484

Chapter 0

INTRODUCTION

> There's a tiresome young man in Bay Shore.
> When his fiancée cried, 'I adore
> The beautiful sea',
> He replied, 'I agree,
> It's pretty, but what is it for?'
>
> Morris Bishop

Chapter 0

INTRODUCTION

1 Motivation

Why study categories? Some reasons are these:

1.1 ABUNDANCE

Categories abound in mathematics and in related fields such as computer science. Such entities as sets, vector spaces, groups, topological spaces, Banach spaces, manifolds, ordered sets, automata, languages, etc., all naturally give rise to categories.

1.2 INSIGHT INTO SIMILAR CONSTRUCTIONS

Constructions with similar properties occur in completely different mathematical fields. For example,

(1) *"products"* for vector spaces, groups, topological spaces, Banach spaces, automata, etc.,

(2) *"free objects"* for vector spaces, groups, rings, topological spaces, Banach spaces, etc.,

(3) *"reflective improvements"* of certain objects, e.g., completions of partially ordered sets and of metric spaces, Čech-Stone compactifications of topological spaces, symmetrizations of relations, abelianizations of groups, Bohr compactifications of topological groups, minimalizations of reachable acceptors, etc.

Category theory provides the means to investigate such constructions simultaneously.

1.3 USE AS A LANGUAGE

Category theory provides a language to describe precisely many *similar phenomena* that occur in different mathematical fields. For example,

(1) Each finite dimensional vector space is isomorphic to its dual and hence also to its second dual. The second correspondence is considered "natural", but the first is not. Category theory allows one to precisely make the distinction via the notion of *natural isomorphism*.

(2) Topological spaces can be defined in many different ways, e.g., via open sets, via closed sets, via neighborhoods, via convergent filters, and via closure operations. Why do these definitions describe "essentially the same" objects? Category theory provides an answer via the notion of *concrete isomorphism*.

(3) *Initial structures*, *final structures*, and *factorization structures* occur in many different situations. Category theory allows one to formulate and investigate such concepts with an appropriate degree of generality.

1.4 CONVENIENT SYMBOLISM

Categorists have developed a symbolism that allows one quickly to *visualize* quite complicated facts by means of *diagrams*.

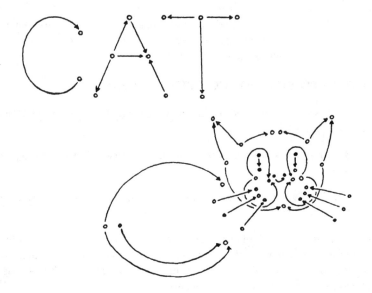

1.5 TRANSPORTATION OF PROBLEMS

Category theory provides a vehicle that allows one to *transport* problems from one area of mathematics (via suitable *functors*) to another area, where solutions are sometimes easier. For example, algebraic topology can be described as an investigation of topological problems (via suitable functors) by algebraic methods.

1.6 DUALITY

The concept of category is well balanced, which allows an economical and useful *duality*. Thus in category theory the "two for the price of one" principle holds: every concept is two concepts, and every result is two results.

The reasons given above show that familiarity with category theory will help those who are confronted with a new field to detect analogies and connections to familiar fields, to organize the new field appropriately, and to separate the general concepts, problems and results from the special ones, which deserve special investigations. Categorical knowledge thus helps to direct and to organize one's thoughts.

2 Foundations

Before delving into categories per se, we need to briefly discuss some foundational aspects. In §1 we have seen that in category theory we are confronted with extremely large collections such as "all sets", "all vector spaces", "all topological spaces", or "all automata". The reader with some set-theoretical background knows that these entities cannot be regarded as sets. For instance, if $\mathcal{U}$ were the set of all sets, then the subset $A = \{x \,|\, x \in \mathcal{U} \text{ and } x \notin x\}$ of $\mathcal{U}$ would have the property that $A \in A$ if and only if $A \notin A$ (Russell's paradox). Someone working, for example, in algebra, topology, or computer science usually isn't (and needn't be) bothered with such set-theoretical difficulties. But it is essential that those who work in category theory be able to deal with "collections" like those mentioned above. To do so requires some foundational restrictions. Nevertheless, certain naturally arising categorical constructions should not be outlawed simply because of the foundational safeguards. Hence, what is needed is a foundation that, on the one hand, is sufficiently flexible so as not to unduly inhibit categorical inquiry and, on the other hand, is sufficiently rigid to give reasonable assurance that the resulting theory is consistent, i.e., does not lead to contradictions. We also require that the foundation be sufficiently close to those foundational systems that are used by most mathematicians. Below we provide a brief outline of the features such a foundation should have.

The basic concepts that we need are those of "sets" and "classes". On a few occasions we will need to go beyond these and also use "conglomerates".

2.1 SETS

Sets can be thought of as the usual sets of intuitive set theory (or of some axiomatic set theory). In particular, we require that the following constructions can be performed with sets.

(1) For each set X and each "property" P, we can form the set $\{x \in X \,|\, P(x)\}$ of all members of X that have the property P.

(2) For each set X, we can form the set $\mathcal{P}(X)$ of all subsets of X (called the **power set** of X).

(3) For any sets X and Y, we can form the following sets:

   (a) the set $\{X, Y\}$ whose members are exactly X and Y,

   (b) the **(ordered) pair** (X, Y) with first coordinate X and second coordinate Y, [likewise for n-tuples of sets, for any natural number $n > 2$],

   (c) the **union** $X \cup Y = \{x \,|\, x \in X \text{ or } x \in Y\}$,

   (d) the **intersection** $X \cap Y = \{x \,|\, x \in X \text{ and } x \in Y\}$,

   (e) the **cartesian product** $X \times Y = \{(x, y) \,|\, x \in X \text{ and } y \in Y\}$,

   (f) the **relative complement** $X \setminus Y = \{x \,|\, x \in X \text{ and } x \notin Y\}$,

(g) the set Y^X of all functions[2] $f : X \to Y$ from X to Y.

(4) For any set I and any family[3] $(X_i)_{i \in I}$ of sets, we can form the following sets:

(a) the **image** $\{X_i \mid i \in I\}$ of the indexing function,

(b) the **union** $\bigcup_{i \in I} X_i = \{x \mid x \in X_i \text{ for some } i \in I\}$,

(c) the **intersection** $\bigcap_{i \in I} X_i = \{x \mid x \in X_i \text{ for all } i \in I\}$, provided that $I \neq \emptyset$,

(d) the **cartesian product** $\prod_{i \in I} X_i = \{f : I \to \bigcup_{i \in I} X_i \mid f(i) \in X_i \text{ for each } i \in I\}$,

(e) the **disjoint union** $\biguplus_{i \in I} X_i = \bigcup_{i \in I} (X_i \times \{i\})$.

(5) We can form the following sets:

$\mathbb{N}$ of all natural numbers,

$\mathbb{Z}$ of all integers,

$\mathbb{Q}$ of all rational numbers,

$\mathbb{R}$ of all real numbers, and

$\mathbb{C}$ of all complex numbers.

The above requirements imply that each topological space is a set. [It is a pair (X, τ), where X is its (underlying) set and τ is a topology (that is the set of all open subsets of X); i.e., $\tau \in \mathcal{P}(\mathcal{P}(X))$.] Analogously, each vector space and each automaton is a set. However, by means of the above constructions, we cannot form "the *set* of all sets", or "the *set* of all vector spaces", etc.

2.2 CLASSES

The concept of "class" has been created to deal with "large collections of sets". In particular, we require that:

(1) the members of each class are sets,

(2) for every "property" P one can form **the class of all sets with property** P.

Hence there is the largest class: the class of all sets, called the **universe** and denoted by $\mathcal{U}$. Classes are precisely the subcollections of $\mathcal{U}$. Thus, given classes A and B, one may form such classes as $A \cup B$, $A \cap B$, and $A \times B$. Because of this, there is no problem in defining functions between classes, equivalence relations on classes, etc. A **family**[4] $(A_i)_{i \in I}$ **of sets** is a function $A : I \to \mathcal{U}$ (sending $i \in I$ to $A(i) = A_i$). In particular, if I is a set, then $(A_i)_{i \in I}$ is said to be **set-indexed** [cf. 2.1(4)].

For convenience we require further

---

[2] A **function** with domain X and codomain Y is a triple (X, f, Y), where $f \subseteq X \times Y$ is a relation such that for each $x \in X$ there exists a unique $y \in Y$ with $(x, y) \in f$ [notation: $y = f(x)$ or $x \mapsto f(x)$]. Functions are denoted by $f : X \to Y$ or $X \xrightarrow{f} Y$. Given functions $X \xrightarrow{f} Y$ and $Y \xrightarrow{g} Z$, the **composite** function $X \xrightarrow{g \circ f} Z$ is defined by $x \mapsto g(f(x))$.

[3] For a formal definition of families of sets see 2.2(2).

[4] One should be aware that a family and its image are different entities and that, moreover, a family is not determined by its image for essentially the same reason that a sequence (i.e., an $\mathbb{N}$-indexed family) is not determined by its set of values. A family $(A_i)_{i \in I}$ is sometimes denoted by $(A_i)_I$.

(3) if $X_1, X_2, \ldots, X_n$ are classes, then so is the n-tuple $(X_1, X_2, \ldots, X_n)$, and

(4) every set is a class (equivalently: every member of a set is a set).

Hence sets are special classes. Classes that are not sets are called **proper classes**. They cannot be members of any class. Because of this, Russell's paradox now translates into the harmless statement that the class of all sets that are not members of themselves is a proper class. Also the universe $\mathcal{U}$, the class of all vector spaces, the class of all topological spaces, and the class of all automata are proper classes.

Notice that in this setting condition 2.1(4)(a) above gives us the *Axiom of Replacement*:

(5) there is no surjection from a set to a proper class.

This means that each set must have "fewer" elements than any proper class.

Therefore sets are also called **small classes**, and proper classes are called **large classes**. This distinction between "large" and "small" turns out to be crucial for many categorical considerations.[5]

The framework of sets and classes described so far suffices for defining and investigating such entities as the category of sets, the category of vector spaces, the category of topological spaces, the category of automata, functors between these categories, and natural transformations between such functors. Thus for most of this book we need not go beyond this stage. Therefore we advise the beginner to skip from here, go directly to §3, and return to this section only when the need arises.

The limitations of the framework described above become apparent when we try to perform certain constructions with categories; e.g., when forming "extensions" of categories or when forming categories that have categories or functors as objects. Since members of classes must be sets and $\mathcal{U}$ is not a set, we can't even form a class $\{\mathcal{U}\}$ whose only member is $\mathcal{U}$, much less a class whose members are all the subclasses of $\mathcal{U}$ or all functions from $\mathcal{U}$ to $\mathcal{U}$. In order to deal effectively with such "collections" we need a further level of generality:

2.3 CONGLOMERATES

The concept of "conglomerate" has been created to deal with "collections of classes". In particular, we require that:

(1) every class is a conglomerate,

(2) for every "property" P, one can form the conglomerate of all classes with property P,

(3) conglomerates are closed under analogues of the usual set-theoretic constructions outlined above (2.1); i.e., they are closed under the formation of pairs, unions, products (of conglomerate-indexed families), etc.

---

[5]See, for example, Remark 10.33.

Thus we can form the conglomerate of all classes as well as such entities as functions between conglomerates and families of conglomerates.

Furthermore, we require

(4) the *Axiom of Choice for Conglomerates*; namely for each surjection between conglomerates $f : X \to Y$, there is an injection $g : Y \to X$ with $f \circ g = id_Y$.

In other words, every equivalence relation on a conglomerate has a system of representatives. Notice that this Axiom of Choice implies an *Axiom of Choice for Classes* and also the familiar *Axiom of Choice for Sets*.

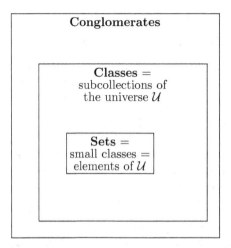

The hierarchy of "collections"

A conglomerate X is said to be **codable** by a conglomerate Y provided that there exists a surjection $Y \to X$ (equivalently: provided that there exists an injection $X \to Y$). Conglomerates that are codable by a class (resp. by a set) are called **legitimate** (resp. **small**) and will sometimes be treated like classes (resp. sets). For example, $\{\mathcal{U}\}$ is a small conglomerate, and $\mathcal{U} \cup \{\mathcal{U}\}$ is a legitimate one. Conglomerates that are not legitimate are called **illegitimate**. For example, $\mathcal{P}(\mathcal{U})$ is an illegitimate conglomerate.

Since our main interest lies with such categories as the category of all sets, the category of all vector spaces, the category of all topological spaces, the category of all automata, and possible "extensions" of these, no need arises to consider any "collections" beyond the level of conglomerates, such as the entity of "all conglomerates".

For a set-theoretic model of the above foundation, see e.g., the Appendix of the monograph of Herrlich and Strecker (see Bibliography), where, in view of the requirement 2.2(3), the familiar Kuratowski definition of an ordered pair $(A, B) = \{\{A\}, \{A, B\}\}$ needs to be replaced by a more suitable one; e.g., by $(A, B) = \{\{\{a\}, \{a, 0\}\} \mid a \in A\} \cup \{\{\{b\}, \{b, 1\}\} \mid b \in B\}$.

Suggestions for Further Reading

Lawvere, F. W. The category of categories as a foundation for mathematics. *Proceedings of the Conference on Categorical Algebra* (La Jolla, 1965), Springer, Berlin–Heidelberg–New York, 1966, 1–20.

Mac Lane, S. One universe as a foundation for category theory. *Springer Lect. Notes Math.* **106** (1969): 192–200.

Feferman, S. Set-theoretical foundations of category theory. *Springer Lect. Notes Math.* **106** (1969): 201–247.

Bénabou, J. Fibred categories and the foundations of naive category theory. *J. Symbolic Logic* **50** (1985): 10–37.

Chapter I

CATEGORIES, FUNCTORS, AND NATURAL TRANSFORMATIONS

In this chapter we introduce the most fundamental concepts of category theory, as well as some examples that we will find to be useful in the remainder of the text.

3 Categories and functors

CATEGORIES

Before stating the formal definition of category, we recall some of the motivating examples from §1. The notion of category should be sufficiently broad that it encompasses

(1) the class of all sets and functions between them,

(2) the class of all vector spaces and linear transformations between them,

(3) the class of all groups and homomorphisms between them,

(4) the class of all topological spaces and continuous functions between them, and

(5) the class of all automata and simulations between them.

3.1 DEFINITION

A **category** is a quadruple $\mathbf{A} = (\mathcal{O}, \hom, id, \circ)$ consisting of

(1) a class $\mathcal{O}$, whose members are called **A-objects**,

(2) for each pair (A, B) of **A**-objects, a set $\hom(A, B)$, whose members are called **A-morphisms from A to B** — [the statement "$f \in \hom(A, B)$" is expressed more graphically[6] by using arrows; e.g., by statements such as "$f: A \to B$ is a morphism" or "$A \xrightarrow{f} B$ is a morphism"],

(3) for each **A**-object A, a morphism $A \xrightarrow{id_A} A$, called the **A-identity** on A,

(4) a **composition law** associating with each **A**-morphism $A \xrightarrow{f} B$ and each **A**-morphism $B \xrightarrow{g} C$ an **A**-morphism $A \xrightarrow{g \circ f} C$, called the **composite** of f and g,

subject to the following conditions:

(a) composition is associative; i.e., for morphisms $A \xrightarrow{f} B$, $B \xrightarrow{g} C$, and $C \xrightarrow{h} D$, the equation $h \circ (g \circ f) = (h \circ g) \circ f$ holds,

(b) **A**-identities act as identities with respect to composition; i.e., for **A**-morphisms $A \xrightarrow{f} B$, we have $id_B \circ f = f$ and $f \circ id_A = f$,

(c) the sets $\hom(A, B)$ are pairwise disjoint.

3.2 REMARKS

If $\mathbf{A} = (\mathcal{O}, \hom, id, \circ)$ is a category, then

(1) The class $\mathcal{O}$ of **A**-objects is usually denoted by $Ob(\mathbf{A})$.

---

[6]Notice that although we use the same notation $f: A \to B$ for a function from A to B (2.1) and for a morphism from A to B, morphisms are not necessarily functions (see Examples 3.3(4) below).

(2) The class of all **A**-morphisms (denoted by $Mor(\mathbf{A})$) is defined to be the union of all the sets $\hom(A,B)$ in **A**.

(3) If $A \xrightarrow{f} B$ is an **A**-morphism, we call A the **domain** of f [and denote it by $dom(f)$] and call B the **codomain** of f [and denote it by $cod(f)$]. Observe that condition (c) guarantees that each **A**-morphism has a *unique* domain and a *unique* codomain. However, this condition is given for technical convenience only, because whenever all other conditions are satisfied, it is easy to "force" condition (c) by simply replacing each morphism f in $\hom(A,B)$ by a triple (A,f,B) (as we did when defining functions in 2.1). For this reason, when verifying that an entity is a category, we will disregard condition (c).

(4) The composition, $\circ$, is a partial binary operation on the class $Mor(\mathbf{A})$. For a pair (f,g) of morphisms, $f \circ g$ is defined if and only if the domain of f and the codomain of g coincide.

(5) If more than one category is involved, subscripts may be used (as in $\hom_{\mathbf{A}}(A,B)$) for clarification.

3.3 EXAMPLES

(1) The category **Set** whose object class is the class of all sets; $\hom(A,B)$ is the set of all functions from A to B, id_A is the identity function on A, and $\circ$ is the usual composition of functions.

(2) The following **constructs**; i.e., categories of structured sets and structure-preserving functions between them ($\circ$ will always be the composition of functions and id_A will always be the identity function on A):

 (a) **Vec** with objects all real vector spaces and morphisms all linear transformations between them.

 (b) **Grp** with objects all groups and morphisms all homomorphisms between them.

 (c) **Top** with objects all topological spaces and morphisms all continuous functions between them.

 (d) **Rel** with objects all pairs (X,ρ), where X is a set and ρ is a (binary) relation on X. Morphisms $f : (X,\rho) \to (Y,\sigma)$ are relation-preserving maps; i.e., maps $f : X \to Y$ such that $x \rho x'$ implies $f(x) \sigma f(x')$.

 (e) **Alg**(Ω) with objects all Ω-algebras and morphisms all Ω-**homomorphisms** between them. Here $\Omega = (n_i)_{i \in I}$ is a family of natural numbers n_i, indexed by a set I. An Ω-algebra is a pair $(X,(\omega_i)_{i \in I})$ consisting of a set X and a family of functions $\omega_i : X^{n_i} \to X$, called n_i-**ary operations** on X. An Ω-homomorphism $f : (X,(\omega_i)_{i \in I}) \to (\hat{X},(\hat{\omega}_i)_{i \in I})$ is a function $f : X \to \hat{X}$ for which the diagram

$$\begin{array}{ccc} X^{n_i} & \xrightarrow{f^{n_i}} & \hat{X}^{n_i} \\ \omega_i \downarrow & & \downarrow \hat{\omega}_i \\ X & \xrightarrow{f} & \hat{X} \end{array}$$

commutes (i.e., $f \circ \omega_i = \hat{\omega}_i \circ f^{n_i}$) for each $i \in I$. In case $n_i = 1$ for each $i \in I$, the symbol $\Sigma = (n_i)_{i \in I}$ is usually used instead of Ω.

(f) **Σ-Seq** with objects all (deterministic, sequential) **Σ-acceptors**, where Σ is a finite set of input symbols. Objects are quadruples (Q, δ, q_0, F), where Q is a finite set of states, $\delta : \Sigma \times Q \to Q$ is a transition map, $q_0 \in Q$ is the initial state, and $F \subseteq Q$ is the set of final states.
A morphism $f : (Q, \delta, q_0, F) \to (Q', \delta', q_0', F')$ (called a **simulation**) is a function $f : Q \to Q'$ that preserves

  (i) transitions, i.e., $\delta'(\sigma, f(q)) = f(\delta(\sigma, q))$,

  (ii) the initial state, i.e., $f(q_0) = q_0'$, and

  (iii) the final states, i.e., $f[F] \subseteq F'$.

(3) For constructs, it is often clear what the morphisms should be once the objects are defined. However, this is not always the case. For instance:

  (a) there are at least three natural constructs each having as objects all metric spaces; namely,

  Met with morphisms all non-expansive maps (= contractions),[7]

  Met$_u$ with morphisms all uniformly continuous maps,

  Met$_c$ with morphisms all continuous maps.

  (b) there are at least two natural constructs each having as objects all Banach spaces; namely,

  Ban with morphisms all linear contractions,[8]

  Ban$_b$ with morphisms all bounded linear maps (= continuous linear maps = uniformly continuous linear maps).

(4) The following categories where the objects and morphisms are *not* structured sets and structure-preserving functions:

  (a) **Mat** with objects all natural numbers, and for which $\hom(m, n)$ is the set of all real $m \times n$ matrices, $id_n : n \to n$ is the unit diagonal $n \times n$ matrix, and composition of matrices is defined by $A \circ B = BA$, where BA denotes the usual multiplication of matrices.

  (b) **Aut** with objects all (deterministic, sequential, Moore) **automata**. Objects are sextuples $(Q, \Sigma, Y, \delta, q_0, y)$, where Q is the set of states, Σ and Y are the sets of input symbols and output symbols, respectively, $\delta : \Sigma \times Q \to Q$ is the transition map, $q_0 \in Q$ is the initial state, and $y : Q \to Y$ is the output map. Morphisms from an automaton $(Q, \Sigma, Y, \delta, q_0, y)$ to an automaton $(Q', \Sigma', Y', \delta', q_0', y')$ are triples (f_Q, f_Σ, f_Y) of functions $f_Q : Q \to Q'$, $f_\Sigma : \Sigma \to \Sigma'$, and $f_Y : Y \to Y'$ satisfying the following conditions:

---

[7] A function $f : (X, d) \to (Y, \hat{d})$ is called **non-expansive** (or a **contraction**) provided that $\hat{d}(f(a), f(b)) \leq d(a, b)$, for all $a, b \in X$.

[8] For Banach spaces the distance between a and b is given by $\| a - b \|$.

(i) preservation of transition: $\delta'(f_\Sigma(\sigma), f_Q(q)) = f_Q(\delta(\sigma, q))$,

(ii) preservation of outputs: $f_Y(y(q)) = y'(f_Q(q))$,

(iii) preservation of initial state: $f_Q(q_0) = q'_0$.

(c) **Classes as categories:**

Every class X gives rise to a category $C(X) = (\mathcal{O}, \text{hom}, id, \circ)$ — the objects of which are the members of X, and whose only morphisms are identities — as follows:

$$\mathcal{O} = X, \quad \text{hom}(x, y) = \begin{cases} \emptyset & \text{if } x \neq y, \\ \{x\} & \text{if } x = y, \end{cases} \quad id_x = x, \quad \text{and} \quad x \circ x = x.$$

$C(\emptyset)$ is called the **empty category**. $C(\{0\})$ is called the **terminal category** and is denoted by **1**.

(d) Preordered classes as categories:

Every preordered class, i.e., every pair $(X, \leq)$ with X a class and $\leq$ a reflexive and transitive relation on X, gives rise to a category $C(X, \leq) = (\mathcal{O}, \text{hom}, id, \circ)$ — the objects of which are the members of X — as follows:

$$\mathcal{O} = X, \quad \text{hom}(x, y) = \begin{cases} \{(x, y)\} & \text{if } x \leq y, \\ \emptyset & \text{otherwise,} \end{cases} \quad id_x = (x, x),$$

and $(y, z) \circ (x, y) = (x, z)$.

(e) Monoids as categories:

Every monoid $(M, \bullet, e)$, i.e., every semigroup $(M, \bullet)$ with unit, e, gives rise to a category $C(M, \bullet, e) = (\mathcal{O}, \text{hom}, id, \circ)$ — with only one object — as follows:

$$\mathcal{O} = \{M\}, \quad \text{hom}(M, M) = M, \quad id_M = e, \quad \text{and} \quad y \circ x = y \bullet x.$$

(f) **Set×Set** is the category that has as objects all pairs of sets (A, B), as morphisms from (A, B) to (A', B') all pairs of functions (f, g) with $A \xrightarrow{f} A'$ and $B \xrightarrow{g} B'$, identities given by $id_{(A,B)} = (id_A, id_B)$, and composition defined by

$$(f_2, g_2) \circ (f_1, g_1) = (f_2 \circ f_1, g_2 \circ g_1).$$

Similarly, for any categories **A** and **B** one can form $\mathbf{A} \times \mathbf{B}$, or, more generally, for finitely many categories $\mathbf{C}_1, \mathbf{C}_2, \ldots, \mathbf{C}_n$, one can form the **product category** $\mathbf{C}_1 \times \mathbf{C}_2 \times \cdots \times \mathbf{C}_n$.

3.4 REMARKS

(1) In the cases of classes, preordered classes, and monoids, for notational convenience we will sometimes not distinguish between them and the categories they determine in the sense of Examples 3.3(4)(c), (d), and (e) above. Thus, we might speak of a preordered class $(X, \leq)$ or of a monoid $(M, \bullet, e)$ as a category.

(2) Morphisms in a category will usually be denoted by lowercase letters, with uppercase letters reserved for objects. The morphism $h = g \circ f$ will sometimes be denoted by $A \xrightarrow{f} B \xrightarrow{g} C$ or by saying that the triangle

$$\begin{array}{ccc} A & \xrightarrow{f} & B \\ & {}_h\searrow & \downarrow{g} \\ & & C \end{array}$$

commutes. Similarly, the statement that the square

$$\begin{array}{ccc} A & \xrightarrow{f} & B \\ {}_h\downarrow & & \downarrow{g} \\ C & \xrightarrow{k} & D \end{array}$$

commutes means that $g \circ f = k \circ h$.

(3) The order of writing the compositions may seem backwards. However, it comes from the fact that in many of the familiar examples (e.g., in all constructs) the composition law is the composition of functions.

(4) Notice that because of the associativity of composition, the notation $A \xrightarrow{f} B \xrightarrow{g} C \xrightarrow{h} D$ is unambiguous.

THE DUALITY PRINCIPLE

3.5 DEFINITION
For any category $\mathbf{A} = (\mathcal{O}, \hom_{\mathbf{A}}, id, \circ)$ the **dual** (or **opposite**) **category of A** is the category $\mathbf{A}^{\mathrm{op}} = (\mathcal{O}, \hom_{\mathbf{A}^{\mathrm{op}}}, id, \circ^{\mathrm{op}})$, where $\hom_{\mathbf{A}^{\mathrm{op}}}(A, B) = \hom_{\mathbf{A}}(B, A)$ and $f \circ^{\mathrm{op}} g = g \circ f$. (Thus $\mathbf{A}$ and $\mathbf{A}^{\mathrm{op}}$ have the same objects and, except for their direction, the same morphisms.)

3.6 EXAMPLES
(1) If $\mathbf{A} = (X, \leq)$ is a preordered class, considered as a category [3.3(4)(d)], then $\mathbf{A}^{\mathrm{op}} = (X, \geq)$.

(2) If $\mathbf{A} = (M, \bullet, e)$ is a monoid, considered as a category [3.3(4)(e)], then $\mathbf{A}^{\mathrm{op}} = (M, \hat{\bullet}, e)$, where $a \mathbin{\hat{\bullet}} b = b \bullet a$.

3.7 REMARK
Because of the way dual categories are defined, every statement $\mathcal{S}_{\mathbf{A}^{\mathrm{op}}}(X)$ concerning an object X in the category $\mathbf{A}^{\mathrm{op}}$ can be translated into a logically equivalent statement $\mathcal{S}_{\mathbf{A}}^{\mathrm{op}}(X)$ concerning the object X in the category $\mathbf{A}$. This observation allows one to

associate (in two steps) with every property $\mathcal{P}$ concerning objects in categories, a **dual property** concerning objects in categories, as demonstrated by the following example:

Consider the property of objects X in **A**:

$\mathcal{P}_{\mathbf{A}}(X) \equiv$ *For any **A**-object A there exists exactly one **A**-morphism $f : A \to X$.*

Step 1: In $\mathcal{P}_{\mathbf{A}}(X)$ replace all occurrences of **A** by $\mathbf{A}^{\text{op}}$, thus obtaining the property

$\mathcal{P}_{\mathbf{A}^{\text{op}}}(X) \equiv$ *For any $\mathbf{A}^{\text{op}}$-object A there exists exactly one $\mathbf{A}^{\text{op}}$-morphism $f : A \to X$.*

Step 2: Translate $\mathcal{P}_{\mathbf{A}^{\text{op}}}(X)$ into the logically equivalent statement

$\mathcal{P}_{\mathbf{A}}^{\text{op}}(X) \equiv$ *For any **A**-object A there exists exactly one **A**-morphism $f : X \to A$.*

Observe that, roughly speaking, $\mathcal{P}_{\mathbf{A}}^{\text{op}}(X)$ is obtained from $\mathcal{P}_{\mathbf{A}}(X)$ by reversing the direction of each arrow and the order in which morphisms are composed. Naturally, in general, $\mathcal{P}_{\mathbf{A}}^{\text{op}}(X)$ is not equivalent to $\mathcal{P}_{\mathbf{A}}(X)$. For example, the above property $\mathcal{P}_{\mathbf{Set}}(X)$ holds if and only if X is a singleton set, whereas the dual property $\mathcal{P}_{\mathbf{Set}}^{\text{op}}(X)$ holds if and only if X is the empty set.

In a similar manner any property about morphisms[9] in categories gives rise to a dual property concerning morphisms in categories, as demonstrated by the following example:

Consider the property of morphisms $A \xrightarrow{f} B$ in **A**:

$\mathcal{Q}_{\mathbf{A}}(f) \equiv$ *There exists an **A**-morphism $B \xrightarrow{g} A$ with $A \xrightarrow{f} B \xrightarrow{g} A = A \xrightarrow{id_A} A$ (i.e., $g \circ f = id_A$) in **A**.*

Step 1: Replace in $\mathcal{Q}_{\mathbf{A}}(f)$ all occurrences of **A** by $\mathbf{A}^{\text{op}}$, thus obtaining the property

$\mathcal{Q}_{\mathbf{A}^{\text{op}}}(f) \equiv$ *There exists an $\mathbf{A}^{\text{op}}$-morphism $B \xrightarrow{g} A$ with $A \xrightarrow{f} B \xrightarrow{g} A = A \xrightarrow{id_A} A$ (i.e., $g \circ f = id_A$) in $\mathbf{A}^{\text{op}}$.*

Step 2: Translate $\mathcal{Q}_{\mathbf{A}^{\text{op}}}(f)$ into the logically equivalent statement

$\mathcal{Q}_{\mathbf{A}}^{\text{op}}(f) \equiv$ *There exists an **A**-morphism $A \xrightarrow{g} B$ with $A \xrightarrow{g} B \xrightarrow{f} A = A \xrightarrow{id_A} A$ (i.e., $f \circ g = id_A$) in **A**.*

For example, the above property $\mathcal{Q}_{\mathbf{Set}}(f)$ holds if and only if f is an injective function with nonempty domain or is the identity on the empty set, whereas the dual property $\mathcal{Q}_{\mathbf{Set}}^{\text{op}}(f)$ holds if and only if f is a surjective function.

More complex properties $\mathcal{P}_{\mathbf{A}}(A, B, \ldots, f, g, \ldots)$ that involve objects $A, B, \ldots$ and morphisms $f, g, \ldots$ in a category **A** can be dualized in a similar way.

If $\mathcal{P} = \mathcal{P}_{\mathbf{A}}(A, B, \ldots, f, g, \ldots)$ holds for all **A**-objects $A, B, \ldots$ and all **A**-morphisms $f, g, \ldots$, then we say that **A** has the property $\mathcal{P}$ or that $\mathcal{P}(\mathbf{A})$ holds.

---

[9] Observe that if a property concerns morphisms $A \xrightarrow{f} B$, then its dual concerns morphisms $B \xrightarrow{f} A$. In particular if a property concerns $dom(f)$, then its dual concerns $cod(f)$.

The **Duality Principle for Categories** states

> Whenever a property $\mathcal{P}$ holds for all categories,
> then the property $\mathcal{P}^{op}$ holds for all categories.

The proof of this (extremely useful) principle follows immediately from the facts that for all categories **A** and properties $\mathcal{P}$

(1) $(\mathbf{A}^{op})^{op} = \mathbf{A}$, and

(2) $\mathcal{P}^{op}(\mathbf{A})$ holds if and only if $\mathcal{P}(\mathbf{A}^{op})$ holds.

For example, consider the property $\mathcal{R} = \mathcal{R}_{\mathbf{A}}(f) \equiv \textit{if } \mathcal{P}_{\mathbf{A}}(dom(f)), \textit{ then } \mathcal{Q}_{\mathbf{A}}(f)$, where $\mathcal{P}$ and $\mathcal{Q}$ are the properties defined above. One can easily show that $\mathcal{R}(\mathbf{A})$ holds for all categories **A**, so that by the Duality Principle $\mathcal{R}^{op}(\mathbf{A})$ holds for all categories **A**, where[10] $\mathcal{R}^{op}_{\mathbf{A}}(f) \equiv \textit{if } \mathcal{P}^{op}_{\mathbf{A}}(cod(f)) \textit{ then } \mathcal{Q}^{op}_{\mathbf{A}}(f)$.

The duality principle

Because of this principle, each result in category theory has two equivalent formulations (which at first glance might seem to be quite different). However, only one of them needs to be proved, since the other one follows by virtue of the Duality Principle.

Often the dual concept $\mathcal{P}^{op}$ of a concept $\mathcal{P}$ is denoted by "*co-$\mathcal{P}$*" (cf. equalizers and coequalizers (7.51 and 7.68), wellpowered and co-wellpowered (7.82 and 7.87), products and coproducts (10.19 and 10.63), etc.). A concept $\mathcal{P}$ is called **self-dual** if $\mathcal{P} = \mathcal{P}^{op}$. An example[11] of a self-dual concept is that of "identity morphism".

---

[10] See footnote 4.
[11] "Live dual, laud evil" — E.H.

Formulation of the duals of definitions and results will be an implied exercise throughout the remainder of the book. However, we find that it is sometimes instructive to provide such formulations. When we do so for results, we usually will conclude with the symbol $\boxed{\text{D}}$ to indicate that the dual result has been stated and proved at an earlier point, so that (by the Duality Principle) no proof is needed.

ISOMORPHISMS

3.8 DEFINITION
A morphism $f : A \to B$ in a category[12] is called an **isomorphism** provided that there exists a morphism $g : B \to A$ with $g \circ f = id_A$ and $f \circ g = id_B$. Such a morphism g is called an **inverse** of f.

3.9 REMARK
It is clear from the above definition that the statement "f is an isomorphism" is self-dual; i.e., f is an isomorphism in **A** if and only if f is an isomorphism in $\mathbf{A}^{op}$.

3.10 PROPOSITION
If $f : A \to B$, $g : B \to A$ and $h : B \to A$ are morphisms such that $g \circ f = id_A$ and $f \circ h = id_B$, then $g = h$.

Proof: $h = id_A \circ h = (g \circ f) \circ h = g \circ (f \circ h) = g \circ id_B = g$. $\square$

3.11 COROLLARY
If g_1 and g_2 are inverses of a morphism f, then $g_1 = g_2$. $\square$

3.12 REMARK
Due to the above corollary we may speak of *the* inverse of an isomorphism f. It will be denoted by f^{-1}.

3.13 EXAMPLES
(1) Every identity id_A is an isomorphism and $id_A^{-1} = id_A$.

(2) In **Set** the isomorphisms are precisely the bijective maps, in **Vec** they are precisely the linear isomorphisms, in **Grp** they are precisely the group-theoretic isomorphisms, in **Top** they are precisely the homeomorphisms, and in **Rel** they are precisely the relational isomorphisms. Observe that in all of these cases every isomorphism is a bijective morphism, but that the converse to this statement, namely, "every bijective morphism is an isomorphism", is true for **Set**, **Vec**, and **Grp**, but not for **Rel** or **Top**.

---

[12]From now on, when making a definition or stating a result that is valid for any category, we will not name the category. Also whenever we speak about morphisms or objects without specifying a category, we usually mean that they belong to the same category. When more than one category is involved and confusion may occur, we will use a hyphenated notation, such as **A**-identity or **A**-isomorphism.

(3) In **Ban**$_b$[3.3(3)] the isomorphisms are precisely the linear homeomorphisms, whereas in **Ban** isomorphisms are precisely the norm-preserving linear bijections.

(4) In **Mat** the isomorphisms are precisely the regular matrices; i.e., the square matrices with nonzero determinant.

(5) A morphism (f_Q, f_Σ, f_Y) in **Aut** is an isomorphism if and only if each of the maps $f_Q, f_\Sigma,$ and f_Y is bijective.

(6) In a monoid, considered as a category, every morphism is an isomorphism if and only if the monoid is a group.

3.14 PROPOSITION

(1) If $A \xrightarrow{f} B$ is an isomorphism, then so is $B \xrightarrow{f^{-1}} A$ and $(f^{-1})^{-1} = f$.

(2) If $A \xrightarrow{f} B$ and $B \xrightarrow{g} C$ are isomorphisms, then so is $A \xrightarrow{g \circ f} C$, and $(g \circ f)^{-1} = f^{-1} \circ g^{-1}$.

Proof:

(1). Immediate from the definitions of inverse and isomorphism (3.8).

(2). By associativity and the definition of inverse, we have: $(g \circ f) \circ (f^{-1} \circ g^{-1}) = g \circ (f \circ f^{-1}) \circ g^{-1} = g \circ id_B \circ g^{-1} = g \circ g^{-1} = id_C$. Similarly, $(f^{-1} \circ g^{-1}) \circ (g \circ f) = id_A$. □

3.15 DEFINITION

Objects A and B in a category are said to be **isomorphic** provided that there is an isomorphism $f : A \to B$.

3.16 REMARK

For any category, **A**, "is isomorphic to" clearly yields an equivalence relation on $Ob(\mathbf{A})$. [Reflexivity follows from the fact that identities are isomorphisms, and symmetry and transitivity are immediate from the proposition above.] Isomorphic objects are frequently regarded as being "essentially" the same.

FUNCTORS

In category theory it is the morphisms, rather than the objects, that have the primary role. Indeed, we will see that it is even possible to define "category" without using the notion of objects at all (3.53). Now, we take a more global viewpoint and consider categories themselves as structured objects. The "morphisms" between them that preserve their structure are called *functors*.

3.17 DEFINITION

If **A** and **B** are categories, then a **functor** F from **A** to **B** is a function that assigns to each **A**-object A a **B**-object $F(A)$, and to each **A**-morphism $A \xrightarrow{f} A'$ a **B**-morphism $F(A) \xrightarrow{F(f)} F(A')$, in such a way that

(1) F preserves composition; i.e., $F(f \circ g) = F(f) \circ F(g)$ whenever $f \circ g$ is defined, and

(2) F preserves identity morphisms; i.e., $F(id_A) = id_{F(A)}$ for each **A**-object A.

3.18 NOTATION
Functors F from **A** to **B** will be denoted by $F : \mathbf{A} \to \mathbf{B}$ or $\mathbf{A} \xrightarrow{F} \mathbf{B}$. We frequently use the simplified notations FA and Ff rather than $F(A)$ and $F(f)$. Indeed, we sometimes denote the action on both objects and morphisms by

$$F(A \xrightarrow{f} B) = FA \xrightarrow{Ff} FB.$$

3.19 REMARK
Notice that a functor $F : \mathbf{A} \to \mathbf{B}$ is technically a family of functions; one from $Ob(\mathbf{A})$ to $Ob(\mathbf{B})$, and for each pair (A, A') of **A**-objects, one from $\hom(A, A')$ to $\hom(FA, FA')$. Since functors preserve identity morphisms and since there is a bijective correspondence between the class of objects and the class of identity morphisms in any category, the object-part of a functor actually is determined by the morphism-parts. Indeed, we will see later that if we choose the "object-free" definition of category (3.53), then a functor between categories can be defined simply as a function between their morphism classes that preserves identities and composition (3.55).

3.20 EXAMPLES
(1) For any category **A**, there is the **identity functor** $id_\mathbf{A} : \mathbf{A} \to \mathbf{A}$ defined by

$$id_\mathbf{A}(A \xrightarrow{f} B) = A \xrightarrow{f} B.$$

(2) For any categories **A** and **B** and any **B**-object B, there is the **constant functor** $C_B : \mathbf{A} \to \mathbf{B}$ with value B, defined by

$$C_B(A \xrightarrow{f} A') = B \xrightarrow{id_B} B.$$

(3) For any of the constructs **A** mentioned above [3.3(2)(3)] there is the **forgetful functor** (or **underlying functor**) $U : \mathbf{A} \to \mathbf{Set}$, where in each case $U(A)$ is the underlying set of A, and $U(f) = f$ is the underlying function of the morphism f.

(4) For any category **A** and any **A**-object A, there is the **covariant hom-functor** $\hom(A, -) : \mathbf{A} \to \mathbf{Set}$, defined by

$$\hom(A, -)(B \xrightarrow{f} C) = \hom(A, B) \xrightarrow{\hom(A,f)} \hom(A, C)$$

where $\hom(A, f)(g) = f \circ g$.

(5) For any category **A** and any **A**-object A, there is the **contravariant**[13] **hom-functor** $\hom(-, A) : \mathbf{A}^{\mathrm{op}} \to \mathbf{Set}$ defined on any $\mathbf{A}^{\mathrm{op}}$-morphism[14] $B \xrightarrow{f} C$ by

$$\hom(-, A)(B \xrightarrow{f} C) = \hom_{\mathbf{A}}(B, A) \xrightarrow{\hom(f, A)} \hom_{\mathbf{A}}(C, A)$$

with $\hom(f, A)(g) = g \circ f$, where the composition is the one in **A**.

A forgetful functor

(6) If **A** and **B** are monoids considered as categories [3.3(4)(e)], then functors from **A** to **B** are essentially just monoid homomorphisms from **A** to **B**.

(7) If **A** and **B** are preordered sets considered as categories [3.3(4)(d)], then functors from **A** and **B** are essentially just order-preserving maps from **A** to **B**.

(8) The **covariant power-set functor** $\mathcal{P} : \mathbf{Set} \to \mathbf{Set}$ is defined by

$$\mathcal{P}(A \xrightarrow{f} B) = \mathcal{P}A \xrightarrow{\mathcal{P}f} \mathcal{P}B$$

where $\mathcal{P}A$ is the power-set of A; i.e., the set of all subsets of A; and for each $X \subseteq A$, $\mathcal{P}f(X)$ is the image $f[X]$ of X under f.

---

[13] Functors are sometimes called *covariant functors*. A **contravariant functor** from **A** to **B** means a functor from $\mathbf{A}^{\mathrm{op}}$ to **B**.

[14] Recall that $\mathbf{A}^{\mathrm{op}}$-objects are precisely the same as **A**-objects, and $B \xrightarrow{f} C$ is an $\mathbf{A}^{\mathrm{op}}$-morphism means that $C \xrightarrow{f} B$ is an **A**-morphism.

(9) The **contravariant**[8] **power-set functor** $\mathcal{Q}: \mathbf{Set}^{\mathrm{op}} \to \mathbf{Set}$ is defined by

$$\mathcal{Q}(A \xrightarrow{f} B) = \mathcal{Q}A \xrightarrow{\mathcal{Q}f} \mathcal{Q}B$$

where $\mathcal{Q}A$ is the power-set of A, and for each $X \subseteq A$, $\mathcal{Q}f(X)$ is the preimage $f^{-1}[X]$ of X under the function $f: B \to A$.

(10) For any positive integer n the **nth power functor** $S^n: \mathbf{Set} \to \mathbf{Set}$ is given by

$$S^n(X \xrightarrow{f} Y) = X^n \xrightarrow{f^n} Y^n,$$

where $f^n(x_1, \ldots, x_n) = (f(x_1), \ldots, f(x_n))$.

(11) The **Stone-functor** $S: \mathbf{Top}^{\mathrm{op}} \to \mathbf{Boo}$ (where **Boo** is the construct of boolean algebras and boolean homomorphisms) assigns to each topological space the boolean algebra of its clopen subsets, and for any continuous map $X \xrightarrow{f} Y$; i.e., for any morphism $Y \xrightarrow{f} X$ in $\mathbf{Top}^{\mathrm{op}}$, $Sf: S(Y) \to S(X)$ is given by $Sf(Z) = f^{-1}[Z]$.

(12) The **duality functor for vector spaces** $(\hat{\ }): \mathbf{Vec}^{\mathrm{op}} \to \mathbf{Vec}$ associates with any vector space V its dual $\hat{V}$ (i.e., the vector space hom$(V, \mathbb{R})$ with operations defined pointwise) and with any $\mathbf{Vec}^{\mathrm{op}}$-morphism $V \xrightarrow{f} W$, i.e., any linear map $W \xrightarrow{f} V$, the morphism $\hat{f}: \hat{V} \to \hat{W}$, defined by $\hat{f}(g) = g \circ f$.

(13) If $M = (M, \bullet, e)$ is a monoid, then functors from M (regarded as a one-object category) into **Set** are essentially just M-**actions**; i.e., pairs $(X, *)$, where X is a set and $*$ is a map from $M \times X$ to X such that $e * x = x$ and $(m \bullet \hat{m}) * x = m * (\hat{m} * x)$. [Associate with any such M-action $(X, *)$ the functor $F: M \to \mathbf{Set}$, defined by $F(M \xrightarrow{m} M) = X \xrightarrow{F(m)} X$, where $F(m)(x) = m * x$.]

PROPERTIES OF FUNCTORS

3.21 PROPOSITION
All functors $F: \mathbf{A} \to \mathbf{B}$ *preserve isomorphisms*; i.e., whenever $A \xrightarrow{k} A'$ is an $\mathbf{A}$-isomorphism, then $F(k)$ is a $\mathbf{B}$-isomorphism.

Proof: $F(k) \circ F(k^{-1}) = F(k \circ k^{-1}) = F(id_{A'}) = id_{FA'}$. Similarly, $F(k^{-1}) \circ F(k) = id_{FA}$. □

3.22 REMARKS
(1) Although the above proposition has a trivial proof, it has interesting consequences. In particular, it can be used to show that certain objects in a category are *not* isomorphic. For example, the fundamental group functor can be used to prove that certain topological spaces are not homeomorphic by showing that their fundamental groups are not isomorphic.

(2) Even though all functors preserve isomorphisms, they need not **reflect isomorphisms** (in the sense that if $F(k)$ is an isomorphism, then k must be an isomorphism). For example, consider the forgetful functor $U : \mathbf{Top} \to \mathbf{Set}$. The identity function from the set of real numbers, with the discrete topology, to $\mathbb{R}$, with its usual topology, is not a homeomorphism (i.e., isomorphism in **Top**), although its underlying function is an identity, and thus is an isomorphism in **Set**.

3.23 PROPOSITION
If $F : \mathbf{A} \to \mathbf{B}$ and $G : \mathbf{B} \to \mathbf{C}$ are functors, then the **composite** $G \circ F : \mathbf{A} \to \mathbf{C}$ defined by
$$(G \circ F)(A \xrightarrow{f} A') = G(FA) \xrightarrow{G(Ff)} G(FA')$$
is a functor.[15] □

3.24 DEFINITION
(1) A functor $F : \mathbf{A} \to \mathbf{B}$ is called an **isomorphism** provided that there is a functor $G : \mathbf{B} \to \mathbf{A}$ such that $G \circ F = id_\mathbf{A}$ and $F \circ G = id_\mathbf{B}$.

(2) The categories $\mathbf{A}$ and $\mathbf{B}$ are said to be **isomorphic** provided that there is an isomorphism $F : \mathbf{A} \to \mathbf{B}$.

3.25 REMARKS
(1) Obviously the functor G in the above definition is uniquely determined by F. It will be denoted by F^{-1}.

(2) Clearly, "is isomorphic to" is an equivalence relation on the conglomerate of all categories. Isomorphic categories are considered to be essentially the same.

3.26 EXAMPLES
(1) For any pair of classes (X, Y), the categories $C(X)$ and $C(Y)$ [3.3(4)(c)] are isomorphic if and only if there exists a bijection from X to Y. A category is isomorphic to a category of the form $C(X)$ if and only if each of its morphisms is an identity. Such categories are called **discrete**.

(2) For any pair $((X, \leq), (Y, \leq))$ of preordered classes, the categories $C(X, \leq)$ and $C(Y, \leq)$ [3.3(4)(d)] are isomorphic if and only if $(X, \leq)$ and $(Y, \leq)$ are order-isomorphic. A category is isomorphic to a category of the form $C(X, \leq)$ if and only if for each pair of objects (A, B), $\hom(A, B)$ has at most one member. Such categories are called **thin**.

(3) For any pair (M, N) of monoids, the categories $C(M)$ and $C(N)$ [3.3(4)(e)] are isomorphic if and only if M and N are isomorphic monoids. A category is isomorphic to a category of the form $C(M)$ if and only if it has precisely one object.

---

[15] Occasionally, for typographical efficiency, we will use juxtaposition to denote composition of functors, i.e., GF rather than $G \circ F$. When F is an **endofunctor**, i.e., when its domain and codomain are the same, we may even use F^2 or F^3 to denote $F \circ F$ or $F \circ F \circ F$, respectively.

(4) The construct **Boo** of boolean algebras [3.20(11)] is isomorphic to the construct **BooRng** of boolean rings[16] and ring homomorphisms.

(5) For any commutative ring R let R-**Mod** (resp. **Mod**-R) denote the construct of left (resp. right) R-modules, and module homomorphisms.[17] Then:

  (a) R-**Mod** is isomorphic to **Mod**-R, for any ring R.

  (b) If $\mathbb{Z}$ denotes the ring of integers, then $\mathbb{Z}$-**Mod** is isomorphic to the construct **Ab** of abelian groups and group homomorphisms.

(6) For any monoid, M, let M-**Act** be the category of all M-actions [3.20(13)] and action homomorphisms [$f(m * x) = m * f(x)$]. If Σ^* is the free monoid of all words over Σ, then Σ^*-**Act** is isomorphic to **Alg**(Σ) [3.3(2)(e)].

3.27 DEFINITION
Let $F : \mathbf{A} \to \mathbf{B}$ be a functor.

(1) F is called an **embedding** provided that F is injective on morphisms.

(2) F is called **faithful** provided that all the hom-set restrictions

$$F : \hom_{\mathbf{A}}(A, A') \to \hom_{\mathbf{B}}(FA, FA')$$

are injective.

(3) F is called **full** provided that all hom-set restrictions are surjective.

(4) F is called **amnestic** provided that an **A**-isomorphism f is an identity whenever Ff is an identity.

3.28 REMARK
Notice that a functor is:

(1) an embedding if and only if it is faithful and injective on objects, and

(2) an isomorphism if and only if it is full, faithful, and bijective on objects.

3.29 EXAMPLES
(1) The forgetful functor $U : \mathbf{Vec} \to \mathbf{Set}$ is faithful and amnestic, but is neither full nor an embedding. This is the case for all of the constructs mentioned above (except, of course, for **Set** itself).

(2) The covariant power-set functor $\mathcal{P} : \mathbf{Set} \to \mathbf{Set}$ and the contravariant power-set functor $\mathcal{Q} : \mathbf{Set}^{\mathrm{op}} \to \mathbf{Set}$ [3.20(8)(9)] are both embeddings that are not full.

---

[16] A **boolean ring** is a ring, with unit, in which each element is idempotent with respect to multiplication.

[17] Notice that for the ring $\mathbb{R}$ of real numbers, $\mathbb{R}$-**Mod** = **Vec**.

(3) The functor $U : \mathbf{Met_c} \to \mathbf{Top}$ defined by

$$U((X,d) \xrightarrow{f} (X',d')) = (X, \tau_d) \xrightarrow{f} (X', \tau_{d'})$$

(where τ_d denotes the topology induced on X by the metric d) is full and faithful, but not an embedding.

(4) For any category $\mathbf{A}$, the unique functor from $\mathbf{A}$ to $\mathbf{1}[3.3(4)(c)]$ is faithful if and only if $\mathbf{A}$ is thin.

(5) The **discrete space functor** $D : \mathbf{Set} \to \mathbf{Top}$ defined by

$$D(X \xrightarrow{f} Y) = (X, \delta_X) \xrightarrow{f} (Y, \delta_Y)$$

(where δ_Z denotes the discrete topology on the set Z) is a full embedding.

(6) The **indiscrete space functor** $N : \mathbf{Set} \to \mathbf{Top}$ defined by

$$N(X \xrightarrow{f} Y) = (X, \iota_X) \xrightarrow{f} (Y, \iota_Y)$$

(where ι_Z denotes the indiscrete topology on the set Z) is a full embedding.

3.30 PROPOSITION

Let $F : \mathbf{A} \to \mathbf{B}$ and $G : \mathbf{B} \to \mathbf{C}$ be functors.

(1) If F and G are both isomorphisms (resp. embeddings, faithful, or full), then so is $G \circ F$.

(2) If $G \circ F$ is an embedding (resp. faithful), then so is F.

(3) If F is surjective on objects and $G \circ F$ is full, then G is full.

3.31 PROPOSITION

If $F : \mathbf{A} \to \mathbf{B}$ is a full, faithful functor, then for every $\mathbf{B}$-morphism $f : FA \to FA'$, there exists a unique $\mathbf{A}$-morphism $g : A \to A'$ with $Fg = f$.

Furthermore, g is an $\mathbf{A}$-isomorphism if and only if f is a $\mathbf{B}$-isomorphism.

Proof: The morphism exists by fullness, and it is unique by faithfulness. Since by Proposition 3.21 functors preserve isomorphisms, f is an isomorphism if g is. If $f : FA \to FA'$ is a $\mathbf{B}$-isomorphism, let $g' : A' \to A$ be the unique $\mathbf{A}$-morphism with $F(g') = f^{-1}$. Then $F(g' \circ g) = Fg' \circ Fg = f^{-1} \circ f = id_{FA} = F(id_A)$, so that by faithfulness $g' \circ g = id_A$. Likewise $g \circ g' = id_{A'}$. Hence g is an isomorphism. □

3.32 COROLLARY

Functors $F : \mathbf{A} \to \mathbf{B}$ that are full and faithful reflect isomorphisms; i.e., whenever g is an $\mathbf{A}$-morphism such that $F(g)$ is a $\mathbf{B}$-isomorphism, then g is an $\mathbf{A}$-isomorphism. □

Recall that isomorphic categories are considered as being essentially the same. This concept of "sameness" is very restrictive. The following slightly weaker and more flexible notion of "essential sameness" called *equivalence of categories* is much more frequently satisfied. It will turn out that equivalent categories have the same behavior with respect to all interesting categorical properties.

3.33 DEFINITION

(1) A functor $F : \mathbf{A} \to \mathbf{B}$ is called an **equivalence** provided that it is full, faithful, and **isomorphism-dense** in the sense that for any **B**-object B there exists some **A**-object A such that $F(A)$ is isomorphic to B.

(2) Categories **A** and **B** are called **equivalent** provided that there is an equivalence from **A** to **B**.

3.34 REMARK

It is shown in Proposition 3.36 that "is equivalent to" is an equivalence relation on the conglomerate of all categories.

3.35 EXAMPLES

(1) Each isomorphism between categories is an equivalence. Hence isomorphic categories are equivalent.

(2) The category **Mat** [3.3(4)(a)] is equivalent to the construct of finite-dimensional vector spaces (and linear transformations), but is not isomorphic to it. The fact that there is no isomorphism can be deduced from the observation that in **Mat** different objects cannot be isomorphic. An equivalence is given by the functor that assigns to each natural number $n \in Ob(\mathbf{Mat})$ the vector space $\mathbb{R}^n$ and to each $n \times m$ matrix $A \in Mor(\mathbf{Mat})$ the linear map from $\mathbb{R}^n$ to $\mathbb{R}^m$ that assigns to each $(x_1, x_2, \ldots, x_n) \in \mathbb{R}^n$ the $1 \times m$ matrix $[x_1 x_2 \ldots x_n] A$ (given by matrix multiplication) considered as an m-tuple in $\mathbb{R}^m$.

(3) The constructs $\mathbf{Met_c}$ of metric spaces and continuous maps and $\mathbf{Top_m}$ of metrizable topological spaces and continuous maps are equivalent. The functor that associates with each metric space its induced topological space is an equivalence that is not an isomorphism.

(4) Posets,[18] considered as categories, are equivalent if and only if they are isomorphic. However, preordered sets, considered as categories, can be equivalent without being isomorphic (cf. Exercise 3H).

(5) The category of all minimal acceptors (i.e., those with a minimum number of states for accepting the given language), as a full subcategory of Σ-**Seq**, is equivalent to the poset of all recognizable languages (ordered by inclusion and considered as a thin category). In fact, for two minimal acceptors A and A', there exists at most one simulation $A \to A'$, and such a simulation exists if and only if A' accepts each word accepted by A.

---

[18] A **poset** (or **partially ordered set**) is a pair $(X, \leq)$ that consists of a set X and a transitive, reflexive, and antisymmetric relation $\leq$ on X.

3.36 PROPOSITION

(1) *If* $\mathbf{A} \xrightarrow{F} \mathbf{B}$ *is an equivalence, then there exists an equivalence* $\mathbf{B} \xrightarrow{G} \mathbf{A}$.

(2) *If* $\mathbf{A} \xrightarrow{F} \mathbf{B}$ *and* $\mathbf{B} \xrightarrow{H} \mathbf{C}$ *are equivalences, then so is* $\mathbf{A} \xrightarrow{H \circ F} \mathbf{C}$.

Proof:
(1) For each object B of $\mathbf{B}$, choose an object $G(B)$ of $\mathbf{A}$ and a $\mathbf{B}$-isomorphism $\varepsilon_B : F(G(B)) \to B$. Since F is full and faithful, for each $\mathbf{B}$-morphism $g : B \to B'$ there is a unique $\mathbf{A}$-morphism $G(g) : G(B) \to G(B')$ with

$$F(G(g)) = \varepsilon_{B'}^{-1} \circ g \circ \varepsilon_B : F(G(B)) \to F(G(B')).$$

Hence $G(g)$ is the unique $\mathbf{A}$-morphism for which the diagram

$$\begin{array}{ccc} F(G(B)) & \xrightarrow{F(G(g))} & F(G(B')) \\ {\scriptstyle \varepsilon_B}\downarrow & & \downarrow{\scriptstyle \varepsilon_{B'}} \\ B & \xrightarrow{g} & B' \end{array} \qquad (*)$$

commutes. That G preserves identities follows immediately from the uniqueness requirement in the above diagram. That G preserves composition follows from the uniqueness, the commutativity of the diagram

$$\begin{array}{ccccc} F(G(B)) & \xrightarrow{F(G(g))} & F(G(B')) & \xrightarrow{F(G(h))} & F(G(B'')) \\ {\scriptstyle \varepsilon_B}\downarrow & & \downarrow{\scriptstyle \varepsilon_{B'}} & & \downarrow{\scriptstyle \varepsilon_{B''}} \\ B & \xrightarrow{g} & B' & \xrightarrow{h} & B'' \end{array}$$

and the fact that F preserves composition. Thus G is a functor. G is full because for each $\mathbf{A}$-morphism $f : G(B) \to G(B')$, the morphism $\varepsilon_{B'} \circ F(f) \circ \varepsilon_B^{-1} : B \to B'$ (which we denote by g) has the property that $g \circ \varepsilon_B = \varepsilon_{B'} \circ F(f)$, and this implies [by uniqueness for $(*)$] that $f = G(g)$. G is faithful since given $B \underset{g_2}{\overset{g_1}{\rightrightarrows}} B'$ with $G(g_1) = G(g_2) = f$, an application of $(*)$ yields

$$g_1 = \varepsilon_{B'} \circ F(G(g_1)) \circ \varepsilon_B^{-1} = \varepsilon_{B'} \circ F(f) \circ \varepsilon_B^{-1} = \varepsilon_{B'} \circ F(G(g_2)) \circ \varepsilon_B^{-1} = g_2.$$

Finally, G is isomorphism-dense because in view of Proposition 3.31 for each $\mathbf{A}$-object A, the $\mathbf{B}$-isomorphism $\varepsilon_{FA} : F(G(FA)) \to FA$ is the image of some $\mathbf{A}$-isomorphism $GFA \to A$.

(2) By Proposition 3.30 it suffices to show that $H \circ F$ is isomorphism-dense. Given a $\mathbf{C}$-object C, the fact that both F and H are isomorphism-dense gives a $\mathbf{B}$-object B, an isomorphism $h : H(B) \to C$, and an $\mathbf{A}$-object A, with an isomorphism $k : F(A) \to B$. Thus $h \circ H(k) : (H \circ F)(A) \to C$ is an isomorphism. $\square$

3.37 REMARK
The concept of equivalence is especially useful when duality is involved. There are numerous examples of pairs of familiar categories where each category is equivalent to the dual of the other.

3.38 DEFINITION
Categories **A** and **B** are called **dually equivalent** provided that $\mathbf{A}^{\mathrm{op}}$ and **B** are equivalent.

3.39 EXAMPLES
(1) The construct **Boo** of boolean algebras is dually equivalent to the construct **BooSpa** of boolean spaces (i.e., to the construct of zero-dimensional compact Hausdorff spaces and continuous maps). An equivalence can be obtained by associating with each boolean space its boolean algebra of clopen subsets (Stone Duality).

(2) The category of finite-dimensional real vector spaces is dually equivalent to itself. An equivalence can be obtained by associating with each finite-dimensional vector space its dual space [cf. 3.20(12)].

(3) **Set** is dually equivalent to the category of complete atomic boolean algebras and complete boolean homomorphisms. An equivalence can be obtained by associating with each set its power-set, considered as a complete atomic boolean algebra.

(4) The category of compact Hausdorff abelian groups is dually equivalent to **Ab**. An equivalence can be obtained by associating with each compact Hausdorff abelian group G its group of characters $\hom(G, \mathbb{R}/\mathbb{Z})$ (Pontrjagin Duality).

(5) The category of locally compact abelian groups is dually equivalent to itself. An equivalence can be obtained as in (4) above.

(6) The category **HComp** of compact Hausdorff spaces (and continuous functions) is dually equivalent to the category of C^*-algebras and algebra homomorphisms. An equivalence can be obtained by associating with each compact Hausdorff space X the C^*-algebra $C(X, \mathbb{C})$ of complex-valued continuous functions (Gelfand-Naimark Duality).

3.40 REMARK
Recall that we have formulated a duality principle related to objects, morphisms, and categories (3.7). We now extend this to functors; i.e., we introduce for any functor $F : \mathbf{A} \to \mathbf{B}$, the concept of its dual functor that can be used to formulate the duals of categorical statements involving functors. The reader may guess as to what the dual of a categorical statement involving a functor might be. Two common mistakes are to either do too little (by dualizing just one of the categories **A** or **B**) or to do too much (by even reversing the arrow representing the functor F). In either of these cases, one generally does *not* obtain a new functor. The proper dual concept is the following:

3.41 DEFINITION
Given a functor $F : \mathbf{A} \to \mathbf{B}$, the **dual** (or **opposite**) functor $F^{\mathrm{op}} : \mathbf{A}^{\mathrm{op}} \to \mathbf{B}^{\mathrm{op}}$ is the functor defined by
$$F^{\mathrm{op}}(A \xrightarrow{f} A') = FA \xrightarrow{Ff} FA'.$$

3.42 REMARK
Obviously $(F^{\mathrm{op}})^{\mathrm{op}} = F$. To form the dual of a categorical statement that involves functors, make the same statement, but with each category and each functor replaced by its dual. Then translate this back into a statement about the original categories and functors.

3.43 PROPOSITION
Each of the following properties of functors is self-dual: "isomorphism", "embedding", "faithful", "full", "isomorphism-dense", and "equivalence". □

CATEGORIES OF CATEGORIES

We have seen above that functors act as morphisms between categories; they are closed under composition, which is associative (since it is just the composition of functions between classes) and the identity functors act as identities with respect to the composition. Because of this, one is tempted to consider the "category of all categories". However, there are two difficulties that arise when we try to *form* this entity. First, the "category of all categories" would have objects such as **Vec** and **Top**, which are proper classes, so that since proper classes cannot be elements of classes, the conglomerate of all objects would not be a class (thus violating condition 3.1(1) in the definition of category). Second, given any categories $\mathbf{A}$ and $\mathbf{B}$, it is not generally true that the conglomerate of all functors from $\mathbf{A}$ to $\mathbf{B}$ forms a set. This violates condition 3.1(2) in the definition of category. However, if we restrict our attention to categories that are *sets*, then both problems are eliminated.

3.44 DEFINITION
A category $\mathbf{A}$ is said to be **small** provided that its class of objects, $Ob(\mathbf{A})$, is a set. Otherwise it is called **large**.

3.45 REMARK
Notice that when $Ob(\mathbf{A})$ is a set, then $Mor(\mathbf{A})$ must be a set, so that the category $\mathbf{A} = (Ob(\mathbf{A}), \hom, id, \circ)$ must *also* be a set (cf. Exercise 3M).

3.46 EXAMPLES
Mat is small; so are all preordered sets considered as categories, and all monoids considered as categories. However, **Mon**, the category of all monoids and monoid homomorphisms between them, is not small.

3.47 DEFINITION
The category **Cat** of small categories has as objects all small categories, as morphisms from **A** to **B** all functors from **A** to **B**, as identities the identity functors, and as composition the usual composition of functors.

3.48 REMARKS
(1) That **Cat** is indeed a category follows immediately from the facts that
   (a) since each small category is a set, the conglomerate of all small categories is a class, and
   (b) for each pair $(\mathbf{A}, \mathbf{B})$ of small categories, the conglomerate of all functors from **A** to **B** is a set.

(2) **Cat** itself is not small. In fact, there are full embeddings from each of the constructs **Set** and **Mon** into **Cat** (3.3(4)(c) and 3I).

Since (because of size restrictions) we can't form the category of all categories, and other naturally occurring entities that we will want to investigate later, we introduce the concept of "quasicategories". This is done by freeing the concept of category from its set-theoretical restrictions:

3.49 DEFINITION
A **quasicategory** is a quadruple $\mathbf{A} = (\mathcal{O}, \text{hom}, id, \circ)$ defined in the same way as a category except that the restrictions that $\mathcal{O}$ be a class and that each conglomerate $\text{hom}(A, B)$ be a set are removed. Namely,

(1) $\mathcal{O}$ is a conglomerate, the members of which are called objects,

(2) for each pair (A, B) of objects, $\text{hom}(A, B)$ is a conglomerate called the conglomerate of all morphisms from A to B (with $f \in \text{hom}(A, B)$ denoted by $f : A \to B$),

(3) for each object A, $id_A : A \to A$ is called the identity morphism on A,

(4) for each pair of morphisms $(f : A \to B, g : B \to C)$ there is a composite morphism $g \circ f : A \to C$,

subject to the following conditions:

(a) composition is associative,

(b) identity morphisms act as identities with respect to composition,

(c) the conglomerates $\text{hom}(A, B)$ are pairwise disjoint.

3.50 DEFINITION
The quasicategory[19] **CAT** of all categories has as objects all categories, as morphisms from **A** to **B** all functors from **A** to **B**, as identities the identity functors, and as composition the usual composition of functors.

---
[19]Frequently proper quasicategories (see 3.51(2)) will be denoted by all capital letters to distinguish them from categories.

3.51 REMARK

(1) Clearly each category is a quasicategory.

(2) **CAT** is a proper quasicategory in the sense that it is not a category. [Notice that hom(**Set**, **Set**) is not a set.]

(3) Virtually every categorical concept has a natural analogue or interpretation for quasicategories. The names for such quasicategorical concepts will be the same as those of their categorical analogues. Thus we have, for example, the notions of functor between quasicategories, equivalence of quasicategories, discrete and thin quasicategories, etc. Because the main object of our study is categories, most notions will only be specifically formulated for categories. However, we will freely make use of implied quasicategorical analogues, especially when it allows clearer or more convenient expression. For example, at this point it is clear that an isomorphism between categories (3.24) is precisely the same as an isomorphism in **CAT** (3.8). Not every categorical concept has a reasonable quasicategorical interpretation. An outstanding example of this is the fact that quasicategories in general lack hom-functors into **Set**.

(4) Dealing with quasicategories and forming **CAT** gives us the possibility of applying category theory to itself. There are advantages to doing this (some of which are indicated above) as well as certain dangers. One danger is the tendency to want to form something like the "quasicategory of all quasicategories". However, to do so causes a Russell-like paradox that cannot be salvaged within our foundational system, as outlined in §2. Because our main interest is in categories, as opposed to quasicategories, we will never need to consider such an entity as the "quasicategory of all quasicategories".

OBJECT-FREE DEFINITION OF CATEGORIES

Because of the bijection between the class of objects and the class of identity morphisms in any category (given by $A \mapsto id_A$) and the fact that identities in a category can be characterized by their behavior with respect to composition, it is possible to obtain an "object-free" definition of category. This definition, given below, is formally simpler than the original one and is "essentially" equivalent to it (3.55). The reason for choosing the definition given in 3.1 is that it is more closely associated with standard examples of categories.

3.52 DEFINITION

(1) A **partial binary algebra** is a pair $(X, *)$ consisting of a class X and a partial binary operation $*$ on X; i.e., a binary operation defined on a subclass of $X \times X$. [The value of $*(x, y)$ is denoted by $x * y$.]

(2) If $(X, *)$ is a partial binary algebra, then an element u of X is called a **unit** of $(X, *)$ provided that $x * u = x$ whenever $x * u$ is defined, and $u * y = y$ whenever $u * y$ is defined.

3.53 DEFINITION

An **object-free category** is a partial binary algebra $\mathbf{C} = (M, \circ)$, where the members of M are called **morphisms**, that satisfies the following conditions:

(1) *Matching Condition*: For morphisms f, g, and h, the following conditions are equivalent:

   (a) $g \circ f$ and $h \circ g$ are defined,

   (b) $h \circ (g \circ f)$ is defined, and

   (c) $(h \circ g) \circ f$ is defined.

(2) *Associativity Condition*: If morphisms f, g, and h satisfy the matching conditions, then $h \circ (g \circ f) = (h \circ g) \circ f$.

(3) *Unit Existence Condition*: For every morphism f there exist units u_C and u_D of $(M, \circ)$ such that $u_C \circ f$ and $f \circ u_D$ are defined.

(4) *Smallness Condition*: For any pair of units (u_1, u_2) of $(M, \circ)$ the class $\hom(u_1, u_2) = \{f \in M \mid f \circ u_1 \text{ and } u_2 \circ f \text{ are defined}\}$ is a set.

3.54 PROPOSITION

If $\mathbf{A}$ is a category, then

(1) $(Mor(\mathbf{A}), \circ)$ is an object-free category, and

(2) an $\mathbf{A}$-morphism is an $\mathbf{A}$-identity if and only if it is a unit of $(Mor(\mathbf{A}), \circ)$.

Proof: $(Mor(\mathbf{A}), \circ)$ is clearly a partial binary algebra, where $f \circ g$ is defined if and only if the domain of f is the codomain of g. Thus each $\mathbf{A}$-identity is a unit. If $A \xrightarrow{u} B$ is a unit in $(Mor(\mathbf{A}), \circ)$, then $u = u \circ id_A = id_A$, where the first equality holds since id_A is a $\mathbf{A}$-identity and the second one holds since u is a unit. Thus (2) is established. From this, (1) is immediate. □

3.55 REMARK

We now have two versions of the concept of category, the "standard" one (3.1), which is more intuitive and is more easily associated with familiar examples, and the "object-free" one (3.53), which is more succinctly stated and so, in many cases, more convenient to use. Next we will see that these two concepts are equivalent. Proposition 3.54 shows that with every category we can associate an object-free category. Even though this correspondence is neither injective nor surjective, it provides an *equivalence* between the "standard" and the "object-free" definitions of category. This claim can be made precise as follows:

(1) One can define functors between object-free categories to be functions between their classes of morphisms that preserve both units (= identities) and composition.

(2) Parallel to the definition of the quasicategory **CAT** of all categories one can define the quasicategory $\mathbf{CAT}_{of}$ of all object-free categories and functors between them.

(3) The correspondence from Proposition 3.54 is the object part of a functor between the quasicategories **CAT** and **CAT**$_{of}$ that can be shown to be an equivalence in the sense of Definition 3.33.

In this sense the two concepts of category are essentially the same; i.e., essentially the same "category theory" will result if one proceeds from either of the two formulations.

EXERCISES

3A. Graphs of Categories

A **graph** is a quadruple (V, E, d, c) consisting of a set V (of vertices), a set E (of (directed) edges), and functions $d, c : E \to V$ (giving the domain and codomain of an edge). A **large graph** is the same concept except that V and E are allowed to be classes. The **graph** $G(\mathbf{A})$ **of a category** $\mathbf{A}$ is the obvious large graph with $V = Ob(\mathbf{A})$ and $E = Mor(\mathbf{A})$.

(a) Verify that a thin category is determined up to isomorphism by its graph.

(b) Find two non-isomorphic categories with the same graph.

(c) Determine which of the following graphs are of the form $G(\mathbf{A})$ for some category $\mathbf{A}$ (where vertices and identity edges are indicated by dots, and non-identity edges are indicated by arrows).

(d) Show that for each of the following graphs G there exists up to isomorphism precisely one category $\mathbf{A}$ with $G(\mathbf{A}) = G$.

(e) The **free category** generated by a graph (V, E, d, c) is the category $\mathbf{A}$ with $Ob(\mathbf{A}) = V$, $Mor(\mathbf{A}) = $ all paths ($=$ all finite sequences in E in which the domain of each edge is the codomain of the preceding one), composition is the obvious composition of paths, and identity morphisms are the empty paths. Verify that $\mathbf{A}$ is indeed a category.

3B. Pointed Categories

(a) Show that there is a category whose objects are all pairs of the form (A, a), where A is a set and $a \in A$ and
$$\hom((A, a), (B, b)) = \{\, f \mid f : A \to B \text{ and } f(a) = b \,\}.$$
This is called the **category of pointed sets** (and base-point-preserving functions). It is denoted by **pSet**.

(b) Show that there is a faithful functor $U : \mathbf{pSet} \to \mathbf{Set}$ with the property that $U(f) = f$ for each base-point-preserving function, f. Does U reflect isomorphisms?

(c) Prove that **Set** and **pSet** are not equivalent.

(d) Mimic the above construction of the category of pointed sets to obtain the categories **pTop** of **pointed topological spaces** and **pGrp** of **pointed groups**. Determine whether or not **Top** and **pTop** (resp. **Grp** and **pGrp**) are equivalent or isomorphic.

3C. Alternative Definition of Category

Define a **category of type 2** to be a quintuple $\mathbf{A} = (\mathcal{O}, \mathcal{M}, dom, cod, \circ)$ consisting of

(1) a class $\mathcal{O}$, of $\mathbf{A}$-objects,

(2) a class $\mathcal{M}$, of $\mathbf{A}$-morphisms,

(3) functions $dom : \mathcal{M} \to \mathcal{O}$ and $cod : \mathcal{M} \to \mathcal{O}$, assigning to each morphism its domain and codomain, and

(4) a function $\circ$ from $D = \{(f, g) \mid f, g \in \mathcal{M} \text{ and } dom(f) = cod(g)\}$ to $\mathcal{M}$ [with $\circ(f, g)$ written $f \circ g$],

subject to the following conditions:

(a) If $(f, g) \in D$, then $dom(f \circ g) = dom(g)$ and $cod(f \circ g) = cod(f)$.

(b) If (f, g) and (h, f) belong to D, then $h \circ (f \circ g) = (h \circ f) \circ g$.

(c) For each $A \in \mathcal{O}$ there exists a morphism e such that $dom(e) = A = cod(e)$ and
   (1) $f \circ e = f$ whenever $(f, e) \in D$, and
   (2) $e \circ g = g$ whenever $(e, g) \in D$.

(d) For any $(A, B) \in \mathcal{O} \times \mathcal{O}$, the class
$$\{ f \mid f \in \mathcal{M}, \ dom(f) = A, \ \text{and} \ cod(f) = B \}$$
is a set.

Compare the definition of category of type 2 with that of category (3.1) and object-free category (3.53) and determine in which sense these definitions can be considered "equivalent".

If functors between categories of type 2 are defined to be functions between their morphism classes that preserve identities and composition, show that the quasicategory $\mathbf{CAT}_2$ of categories of type 2 and functors between them is equivalent to $\mathbf{CAT}$ and to $\mathbf{CAT}_{of}$.

3D. Identities

(a) Show that whenever $\mathbf{A} = (\mathcal{O}, \text{hom}, id, \circ)$ and $\mathbf{A}^* = (\mathcal{O}, \text{hom}, id^*, \circ)$ are categories, then $\mathbf{A} = \mathbf{A}^*$.

(b) A functor F is said to **reflect identities** provided that if $F(k)$ is an identity then k must be an identity. Show that every isomorphism between categories reflects identities, but that not every equivalence does.

3E. Duality

(a) Show that none of the following categories is dually equivalent to itself: **Set**, **Vec** (but compare 3.39(2)), **Grp**, **Top**, **Rel**, **Pos**.

(b) Determine whether or not the category of finite sets is dually equivalent to itself.

(c) Establish the following consequence of the duality principle:

> If $\mathcal{S}$ is a categorical statement, then $\mathcal{S}$ holds for all categories satisfying property $\mathcal{P}$ if and only if $\mathcal{S}^{op}$ holds for all categories satisfying $\mathcal{P}^{op}$.

3F. Isomorphisms

(a) Describe the isomorphisms in each of the categories **Met**, **Met$_u$**, and **Met$_c$**.

(b) Show that a monoid is a group if and only if, considered as a category, each of its morphisms is an isomorphism.

3G. Functors

(a) Consider the category $\mathbf{A}$ with exactly one object A, with $\text{hom}(A, A) = \{a, b\}$, and with composition defined by $a \circ a = a$ and $a \circ b = b \circ a = b \circ b = b$. Consider further the category $\mathbf{B}$ with exactly one object B, with $\text{hom}(B, B) = \{b\}$, and with $b \circ b = b$. Let F be defined by $F(B) = A$, $F(b) = b$. Is $F : \mathbf{B} \to \mathbf{A}$ a functor?

(b) Show that a category $\mathbf{A}$ is thin (resp. empty) if and only if every functor with domain $\mathbf{A}$ is faithful (resp. full).

3H. Equivalences

(a) Show that none of the following categories is equivalent to any of the others: **Set**, **Vec**, **Grp**, **Top**, **Rel**, **Pos**.

(b) Show that two posets (resp. monoids), considered as categories, are equivalent if and only if they are isomorphic.

(c) Show that, considered as categories, every preordered set is equivalent to a poset, and that a preordered set is isomorphic to a poset if and only if it is a poset itself.

(d) Show that an equivalence is an embedding if and only if it reflects identities.

(e) Prove that categories $\mathbf{A}$ and $\mathbf{B}$ are isomorphic if and only if there exists an equivalence $E : \mathbf{A} \to \mathbf{B}$ such that for each $\mathbf{A}$-object A the number of isomorphic copies of A in $\mathbf{A}$ coincides with the number of isomorphic copies of $E(A)$ in $\mathbf{B}$.

3I.
Show that there is a full embedding from the category **Mon** of monoids into **Cat**.

3J.
Show that there is a covariant power-set functor $\mathcal{R} : \mathbf{Set} \to \mathbf{Set}$ that is different from the one given in Example 3.20(8): such that on objects A, $\mathcal{R}A$ is the power set of A, and on functions $A \xrightarrow{f} B$, $(\mathcal{R}f)(X) = \{\, b \in B \mid f^{-1}(b) \subseteq X \,\}$.

3K. Comma Categories
If $\mathbf{A} \xrightarrow{F} \mathbf{C}$ and $\mathbf{B} \xrightarrow{G} \mathbf{C}$ are functors, then the **comma category** $(F \downarrow G)$ is the category whose objects are triples (A, f, B) with $A \in Ob(\mathbf{A})$, $B \in Ob(\mathbf{B})$, and $FA \xrightarrow{f} GB \in \mathrm{Mor}(\mathbf{C})$; whose morphisms from (A, f, B) to (A', f', B') are pairs (a, b) with $A \xrightarrow{a} A' \in \mathrm{Mor}(\mathbf{A})$, and $B \xrightarrow{b} B' \in \mathrm{Mor}(\mathbf{B})$ such that the square

$$\begin{array}{ccc} FA & \xrightarrow{Fa} & FA' \\ {\scriptstyle f}\downarrow & & \downarrow{\scriptstyle f'} \\ GB & \xrightarrow[Gb]{} & GB' \end{array}$$

commutes; whose identities are $id_{(A,f,B)} = (id_A, id_B)$; and whose composition is defined componentwise, i.e., by $(\bar{a}, \bar{b}) \circ (a, b) = (\bar{a} \circ a, \bar{b} \circ b)$.

(a) Verify that $(F \downarrow G)$ is indeed a category.

(b) For any category $\mathbf{A}$ and any $\mathbf{A}$-object K, denote by $C_K : \mathbf{1} \to \mathbf{A}$ the constant functor with value K. Give explicit descriptions of

- the comma category $(id_\mathbf{A} \downarrow C_K)$, also denoted by $(\mathbf{A} \downarrow K)$ and called the **category of objects over** K,
- the comma category $(C_K \downarrow id_\mathbf{A})$, also denoted by $(K \downarrow \mathbf{A})$ and called the **category of objects under** K,
- the comma category $(id_\mathbf{A} \downarrow id_\mathbf{A})$, also denoted by $\mathbf{A}^\mathbf{2}$ and called the **arrow category of A**.

(c) Show that $(\mathbf{A} \downarrow K)^{\mathrm{op}} = (K \downarrow \mathbf{A}^{\mathrm{op}})$.

(d) Show that if P is a singleton set, then $(P \downarrow \mathbf{Set})$ is isomorphic with **pSet** (cf. Exercise 3B).

3L. Quasicategories as Objects
Show that one may not form the "quasicategory of all quasicategories". [Hint: Russell's paradox].

3M. Small Categories
Let $\mathbf{A}$ be a category whose object-class is a set. Show that (a) – (d) below are true. [Hint: see 2.1.]

(a) $Mor(\mathbf{A}) = \bigcup \{hom(A, B) \mid (A, B) \in Ob(\mathbf{A}) \times Ob(\mathbf{A})\}$ is a set.
(b) $hom : Ob(\mathbf{A}) \times Ob(\mathbf{A}) \to \mathcal{P}(Mor(\mathbf{A}))$ is a set.
(c) $\circ \subseteq Mor(\mathbf{A}) \times Mor(\mathbf{A}) \times Mor(\mathbf{A})$ is a set.
(d) $\mathbf{A}$ is a set.

3N. Decompositions of Functors

Let $\mathbf{A} \xrightarrow{F} \mathbf{B}$ be a functor. Show that

(a) There exist functors $\mathbf{A} \xrightarrow{G} \mathbf{C}$ and $\mathbf{C} \xrightarrow{H} \mathbf{B}$ with the following properties:
  (1) $F = H \circ G$,
  (2) G is full and bijective on objects,
  (3) H is faithful,
  (4) whenever $\mathbf{A} \xrightarrow{F} \mathbf{B} = \mathbf{A} \xrightarrow{\overline{G}} \overline{\mathbf{C}} \xrightarrow{\overline{H}} \mathbf{B}$ and $\overline{H}$ is faithful, then there exists a unique functor $\mathbf{C} \xrightarrow{K} \overline{\mathbf{C}}$, such that the diagram

$$\begin{array}{ccc} \mathbf{A} & \xrightarrow{G} & \mathbf{C} \\ \overline{G} \downarrow & \swarrow K & \downarrow H \\ \overline{\mathbf{C}} & \xrightarrow{\overline{H}} & \mathbf{B} \end{array} \qquad (*)$$

commutes, i.e., $\overline{G} = K \circ G$ and $H = \overline{H} \circ K$.

(b) There exist functors $\mathbf{A} \xrightarrow{G} \mathbf{C}$ and $\mathbf{C} \xrightarrow{H} \mathbf{B}$ with the following properties:
  (1) $F = H \circ G$,
  (2) G is bijective on objects,
  (3) H is full and faithful,
  (4) whenever $\mathbf{A} \xrightarrow{F} \mathbf{B} = \mathbf{A} \xrightarrow{\overline{G}} \overline{\mathbf{C}} \xrightarrow{\overline{H}} \mathbf{B}$ and $\overline{H}$ is full and faithful, then there exists a unique functor $\mathbf{C} \xrightarrow{K} \overline{\mathbf{C}}$ such that the diagram $(*)$ above commutes.

4 Subcategories

In §3 we have seen several instances where one category is included in another (**Ab** in **Grp**, **HComp** in **Top**, etc.). In this section we investigate this phenomenon more thoroughly.

4.1 DEFINITION
(1) A category **A** is said to be a **subcategory** of a category **B** provided that the following conditions are satisfied:

  (a) $Ob(\mathbf{A}) \subseteq Ob(\mathbf{B})$,

  (b) for each $A, A' \in Ob(\mathbf{A})$, $\hom_{\mathbf{A}}(A, A') \subseteq \hom_{\mathbf{B}}(A, A')$,

  (c) for each **A**-object A, the **B**-identity on A is the **A**-identity on A,

  (d) the composition law in **A** is the restriction of the composition law in **B** to the morphisms of **A**.

(2) **A** is called a **full subcategory** of **B** if, in addition to the above, for each $A, A' \in Ob(\mathbf{A})$, $\hom_{\mathbf{A}}(A, A') = \hom_{\mathbf{B}}(A, A')$.

4.2 REMARKS
(1) Because of the nature of full subcategories, a full subcategory of a category **B** can be specified by merely specifying its object class within **B**.

(2) Notice that conditions (a), (b), and (d) of part (1) of the definition do *not* imply (c). (See Exercise 4A.)

(3) If $F: \mathbf{A} \to \mathbf{B}$ is a full functor or is injective on objects, then the image of **A** under F is a subcategory of **B**. However, for arbitrary functors $F: \mathbf{A} \to \mathbf{B}$, the image of **A** under F need not be a subcategory of **B**. Consider, e.g., a functor from $\begin{smallmatrix}\bullet \to \bullet \\ \bullet \to \bullet\end{smallmatrix}$ to $\begin{smallmatrix}\bullet \to \bullet \\ \searrow \downarrow \\ \bullet\end{smallmatrix}$.

4.3 EXAMPLES
(1) For any category **A**, the empty category and **A** itself are full subcategories of **A**.

(2) The class of all Hausdorff spaces specifies the full subcategory **Haus** of **Top**; likewise, all Tychonoff spaces (i.e., completely regular T_1 spaces) yields a full subcategory **Tych** of **Haus**; and **HComp** is a full subcategory of **Tych**.

(3) The class of all preordered sets (i.e., all sets supplied with a reflexive and transitive relation) determines a full subcategory **Prost** of **Rel**. The class of all partially ordered sets (i.e., all sets supplied with a reflexive, transitive and antisymmetric relation) determines a full subcategory **Pos** of **Prost**. The category **Lat** that consists of all lattices (i.e., all partially ordered sets for which each pair of elements has a

meet and a join) together with all lattice homomorphisms (i.e., all maps preserving meets and joins of pairs) is a nonfull subcategory of **Pos**. The category **JCPos** of complete lattices and join-preserving maps is a nonfull subcategory of **Pos**. The category **CLat** of complete lattices and meet- and join-preserving maps is a nonfull subcategory of **JCPos** that has the same objects as **JCPos**.

(4) The category **Grp** of groups is a full subcategory of the construct **Mon** that consists of all monoids (i.e., semigroups with unit) and monoid homomorphisms (i.e., unit-preserving semigroup homomorphisms). **Mon** is a nonfull subcategory of the construct **Sgr** of all semigroups and semigroup homomorphisms.

(5) **Ban** is a nonfull subcategory of **Ban**$_b$ that has the same class of objects.

(6) The subcategories of a monoid M, considered as a category, are precisely the empty category and the submonoids of M.

4.4 REMARK

For every subcategory **A** of a category **B** there is a naturally associated **inclusion functor** $E : \mathbf{A} \hookrightarrow \mathbf{B}$. Moreover, each such inclusion is

(1) an embedding;

(2) a full functor if and only if **A** is a full subcategory of **B**.

As the next proposition shows, inclusions of subcategories are (up to isomorphism) precisely the embedding functors and (up to equivalence) precisely the faithful functors.

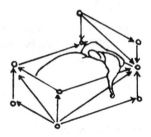

A full embedding

4.5 PROPOSITION

(1) *A functor $F : \mathbf{A} \to \mathbf{B}$ is a (full) embedding if and only if there exists a (full) subcategory $\mathbf{C}$ of $\mathbf{B}$ with inclusion functor $E : \mathbf{C} \to \mathbf{B}$ and an isomorphism $G : \mathbf{A} \to \mathbf{C}$ with $F = E \circ G$.*

(2) *A functor $F : \mathbf{A} \to \mathbf{B}$ is faithful if and only if there exist embeddings $E_1 : \mathbf{D} \to \mathbf{B}$ and $E_2 : \mathbf{A} \to \mathbf{C}$ and an equivalence $G : \mathbf{C} \to \mathbf{D}$ such that the diagram*

$$\begin{array}{ccc} \mathbf{A} & \xrightarrow{F} & \mathbf{B} \\ {\scriptstyle E_2}\downarrow & & \uparrow{\scriptstyle E_1} \\ \mathbf{C} & \xrightarrow{G} & \mathbf{D} \end{array}$$

commutes.

Proof:

(1). One direction is immediate and the other is a consequence of the fact that (full) embeddings are closed under composition.

(2). One direction follows from the compositive nature of faithful functors. To show, conversely, that every faithful functor can be decomposed as stated, let $E_1 : \mathbf{D} \to \mathbf{B}$ be the inclusion of the full subcategory $\mathbf{D}$ of $\mathbf{B}$ that has as objects all images (under F) of $\mathbf{A}$-objects. Let $\mathbf{C}$ be the category with $Ob(\mathbf{C}) = Ob(\mathbf{A})$, with

$$\hom_{\mathbf{C}}(A, A') = \hom_{\mathbf{B}}(FA, FA'),$$

and with identities and composition defined as in $\mathbf{B}$. $\mathbf{C}$ is easily seen to be a category. Now define functors $E_2 : \mathbf{A} \to \mathbf{C}$ and $G : \mathbf{C} \to \mathbf{D}$ by

$$E_2(A \xrightarrow{f} A') = A \xrightarrow{Ff} A' \quad \text{and} \quad G(C \xrightarrow{g} C') = FC \xrightarrow{g} FC'.$$

Then E_2 is an embedding, G is an equivalence, and $F = E_1 \circ G \circ E_2$. □

4.6 DEFINITION

A category $\mathbf{A}$ is said to be **fully embeddable** into $\mathbf{B}$ provided that there exists a full embedding $\mathbf{A} \to \mathbf{B}$, or, equivalently, provided that $\mathbf{A}$ is isomorphic to a full subcategory of $\mathbf{B}$.

4.7 EXAMPLES

Although it is far from easy to prove (see the monograph Pultr-Trnková in the References), each category of the form $\mathbf{Alg}(\Omega)$ is fully embeddable into each of the following constructs:

(a) **Sgr**,

(b) **Rel**,

(c) $\mathbf{Alg}(1,1)$, i.e., the construct of unary algebras on two operations.

Under an additional set-theoretical hypothesis (the non-existence of measurable cardinals), *every* construct is fully embeddable into **Sgr** (or **Rel** or $\mathbf{Alg}(1,1)$).

4.8 REMARK

Because full subcategories are determined by their object classes, they are often regarded as "properties of objects". Since most of the interesting properties P satisfy the condition that whenever an object A has property P then every object isomorphic to A also has P, we often require that full subcategories have the property (defined below) of being *isomorphism-closed*.

4.9 DEFINITION

A full subcategory $\mathbf{A}$ of a category $\mathbf{B}$ is called

(1) **isomorphism-closed** provided that every **B**-object that is isomorphic to some **A**-object is itself an **A**-object,

(2) **isomorphism-dense** provided that every **B**-object is isomorphic to some **A**-object.

4.10 REMARK
If **A** is a full subcategory of **B**, then the following conditions are equivalent:

(1) **A** is an isomorphism-dense subcategory of **B**,

(2) the inclusion functor $\mathbf{A} \hookrightarrow \mathbf{B}$ is isomorphism-dense,

(3) the inclusion functor $\mathbf{A} \hookrightarrow \mathbf{B}$ is an equivalence.

4.11 EXAMPLE
The full subcategory of **Set** with the single object $\mathbb{N}$ (of natural numbers) is neither isomorphism-closed nor isomorphism-dense in **Set**. It is equivalent to the isomorphism-closed full subcategory of **Set** consisting of all countable infinite sets.

There are instances when one wishes to consider full subcategories in which different objects cannot be isomorphic:

4.12 DEFINITION
A **skeleton** of a category is a full, isomorphism-dense subcategory in which no two distinct objects are isomorphic.

4.13 EXAMPLES
(1) The full subcategory of all cardinal numbers is a skeleton for **Set**.

(2) The full subcategory determined by the powers $\mathbb{R}^m$, where m runs through all cardinal numbers, is a skeleton for **Vec**.

4.14 PROPOSITION
(1) Every category has a skeleton.

(2) Any two skeletons of a category are isomorphic.

(3) Any skeleton of a category $\mathbf{C}$ is equivalent to $\mathbf{C}$.

Proof:
(1). This follows from the Axiom of Choice [2.3(4)] applied to the equivalence relation "is isomorphic to" on the class of objects of the category.

(2). Let **A** and **B** be skeletons of **C**. Then each **A**-object A is isomorphic in **C** to a unique **B**-object. Denote the latter by $F(A)$ and choose for each **A**-object A a **C**-isomorphism $f_A : A \to F(A)$. Then the functor $F : \mathbf{A} \to \mathbf{B}$ defined by:

$$F(A \xrightarrow{h} A') = FA \xrightarrow{f_A^{-1}} A \xrightarrow{h} A' \xrightarrow{f_{A'}} FA'$$

is an isomorphism.

(3). The inclusion of a skeleton of **C** into **C** is an equivalence. (See 4.10(2).) □

4.15 COROLLARY
Two categories are equivalent if and only if they have isomorphic skeletons. □

REFLECTIVE AND COREFLECTIVE SUBCATEGORIES

4.16 DEFINITION
Let **A** be a subcategory of **B**, and let B be a **B**-object.

(1) An **A-reflection** (or **A-reflection arrow**) for B is a **B**-morphism $B \xrightarrow{r} A$ from B to an **A**-object A with the following universal property:

for any **B**-morphism $B \xrightarrow{f} A'$ from B into some **A**-object A', there exists a unique **A**-morphism $f' : A \to A'$ such that the triangle

commutes.

By an "abuse of language" an object A is called an **A**-reflection for B provided that there exists an **A**-reflection $B \xrightarrow{r} A$ for B with codomain A.

(2) **A** is called a **reflective subcategory** of **B** provided that each **B**-object has an **A**-reflection.

4.17 EXAMPLES
Several familiar constructions in mathematics such as certain "completions", certain formations of "quotients", and certain "modifications" of structures can be regarded in a natural way as reflections. Here we list a few such examples of reflective subcategories **A** of categories **B**. In every example except the last two, **A** is a full subcategory of **B**.

A. Modifications of the Structure

(1) Making a relation symmetric: **B** = **Rel**, **A** = **Sym**, the full subcategory of symmetric relations. $(X, \rho) \xrightarrow{id_X} (X, \rho \cup \rho^{-1})$ is an **A**-reflection[20] for (X, ρ).

(2) Making a topological space completely regular: **B** = **Top**, **A** = the full subcategory of completely regular (not necessarily T_1) spaces. If (X, τ) is a topological space, then the collection of all cozero sets[21] in (X, τ) is a base for a topology τ_c on X. $(X, \tau) \xrightarrow{id_X} (X, \tau_c)$ is an **A**-reflection for (X, τ).

---
[20] $\rho^{-1} = \{(y, x) \mid (x, y) \in \rho\}$.
[21] A is called a **cozero set** in (X, τ) provided that there is a continuous map $f : (X, \tau) \to \mathbb{R}$ with $A = X - f^{-1}[\{0\}]$.

Sec. 4]   Subcategories   53

B. Improving Objects by Forming Quotients

(3) Making a preordered set partially ordered: **B** = **Prost** (preordered sets and order-preserving maps), **A** = **Pos**. If $(X, \leq)$ is a preordered set, define an equivalence relation $\approx$ on X by: $x \approx y \Leftrightarrow (x \leq y$ and $y \leq x)$. Let $p : X \to X/\approx$ be the canonical map. Then $(X, \leq) \xrightarrow{p} (X/\approx, (p \times p)[\leq])$ is an **A**-reflection for $(X, \leq)$.

(4) Making a group abelian: **B** = **Grp**, **A** = **Ab**. Let G be a group and let G' be the commutator subgroup of G. Then the canonical map $G \to G/G'$ is an **A**-reflection for G.

(5) Making a topological space T_0: **B** = **Top**, **A** = **Top**$_0$, the full subcategory consisting of all T_0 spaces. Let X be a topological space, and let $\approx$ be the equivalence relation on X, given by: $x \approx y \Leftrightarrow$ the closure of $\{x\}$ = the closure of $\{y\}$. Then the canonical map $X \to X/\approx$ is an **A**-reflection for X.

(6) Making an abelian group torsion-free: **B** = **Ab**, **A** = **TfAb**, the full subcategory of torsion-free abelian groups. Let G be an abelian group and let TG be the torsion subgroup of G. Then the canonical map $G \to G/TG$ is an **A**-reflection for G.

(7) Making a reachable acceptor minimal: **B** = the full subcategory of Σ-**Seq** consisting of all **reachable acceptors** (i.e., those for which each state can be reached from the initial one by an input word), **A** = the full subcategory of **B** consisting of all **minimal acceptors** (i.e., those reachable acceptors with the property that no two different states are **observably equivalent**. The observability equivalence $\approx$ on a reachable acceptor B is given by: $q \approx q'$ provided that whenever the initial state of B is changed to q, the resulting acceptor recognizes the same language as it does when the initial state is changed to q'). Then the canonical map $B \to B/\approx$, where $B/\approx$ is the induced acceptor on the set of all $\approx$-equivalence classes of states of B, is an **A**-reflection for B.

C. Completions

(8) **B** = **Met**$_u$(metric spaces and uniformly continuous maps) or **Met**(metric spaces and non-expansive maps), **A** = the full subcategory of complete metric spaces. In either case the metric completion $(X, d) \hookrightarrow (X^*, d^*)$ is an **A**-reflection for (X, d).

(9) **B** = **Tych**, **A** = **HComp**. If X is a Tychonoff space, then the Čech-Stone compactification $X \hookrightarrow \beta X$ is an **A**-reflection for X.

(10) **B** = **JPos** (posets and join-preserving maps), **A** = **JCPos**. Let $(B, \leq)$ be a poset, and let B^* be the collection of all subsets S of B that satisfy

   (a) S is an **lower-set**, i.e., $b \in S$ and $b' \leq b \Rightarrow b' \in S$,

   (b) S is closed under the formation of all existing joins, i.e., if $A \subseteq S$ and b is the join of A in $(B, \leq)$, then $b \in S$.

   Then $(B^*, \subseteq)$ is a complete lattice, and the function

   $$(B, \leq) \to (B^*, \subseteq), \quad \text{defined by} \quad b \mapsto \{x \in B \mid x \leq b\}$$

is an **A**-reflection for $(B, \leq)$. Note that these **A**-reflections differ, in general, from the Mac Neille completions. For example, an **A**-reflection for a 3-element discrete poset is an 8-element complete lattice, whereas the Mac Neille completion is a 5-element one.

(11) **B** = **Pos**, **A** = the nonfull (!) subcategory **JCPos**. Let $(B, \leq)$ be a poset and let $\tilde{B}$ be the collection of all lower-sets S of B [cf. (10)(a)]. Then $(\tilde{B}, \subseteq)$ is a complete lattice and the function $(B, \leq) \to (\tilde{B}, \subseteq)$ defined as in (10) is an **A**-reflection.

(12) **B** = **Sgr**, **A** = the nonfull (!) subcategory **Mon**. If $(X, \bullet)$ is a semigroup, then the extension $(X, \bullet) \hookrightarrow (X \cup \{e\}, \hat{\bullet}, e)$, obtained by adding a unit element $e \notin X$ of the operation $\hat{\bullet}$, is an **A**-reflection for $(X, \bullet)$.

4.18 REMARK

Observe that in the last two examples given above, the reflection arrows are *never* surjective. This fact may seem surprising since in each of the other examples each **A**-reflection arrow for an **A**-object is an isomorphism. It is not surprising, however, in view of Proposition 4.20.

4.19 PROPOSITION

Reflections are essentially unique, i.e.,

*(1) if $B \xrightarrow{r} A$ and $B \xrightarrow{\hat{r}} \hat{A}$ are **A**-reflections for B, then there exists an **A**-isomorphism $k : A \to \hat{A}$ such that the triangle*

commutes,

*(2) if $B \xrightarrow{r} A$ is an **A**-reflection for B and $A \xrightarrow{k} \hat{A}$ is an **A**-isomorphism, then $B \xrightarrow{k \circ r} \hat{A}$ is an **A**-reflection for B.*

Proof:

(1). The existence of a morphism k with $\hat{r} = k \circ r$ follows from the definition of reflection and the fact that $\hat{A}$ is an **A**-object. Similarly there is a morphism $\hat{k}$ with $r = \hat{k} \circ \hat{r}$. Now $(\hat{k} \circ k) \circ r = r = id_A \circ r$, so that by the uniqueness requirement in the definition of reflection, $\hat{k} \circ k = id_A$. Analogously, one can see that $k \circ \hat{k} = id_{\hat{A}}$, so that k is an isomorphism.

(2). Obvious. □

4.20 PROPOSITION

*If **A** is a reflective subcategory of **B**, then the following conditions are equivalent:*

*(1) **A** is a full subcategory of **B**,*

(2) *for each* **A**-*object* A, $A \xrightarrow{id_A} A$ *is an* **A**-*reflection,*

(3) *for each* **A**-*object* A, **A**-*reflection arrows* $A \xrightarrow{r_A} A^*$ *are* **A**-*isomorphisms,*

(4) *for each* **A**-*object* A, **A**-*reflection arrows* $A \xrightarrow{r_A} A^*$ *are* **A**-*morphisms.*

Proof: That (1) $\Rightarrow$ (2) $\Rightarrow$ (3) $\Rightarrow$ (4) is clear. To see that (4) $\Rightarrow$ (1), let $A \xrightarrow{f} A'$ be a **B**-morphism between **A**-objects. By the definition of reflection there is an **A**-morphism $A^* \xrightarrow{\overline{f}} A'$ with $f = \overline{f} \circ r_A$. Thus f is the composite of **A**-morphisms, and so must be an **A**-morphism. □

4.21 REMARK
There exist nonfull reflective subcategories **A** of **B** such that every **A**-reflection arrow is a **B**-isomorphism; e.g., let $\mathbf{A} = \bullet \to \bullet$ and $\mathbf{B} = \bullet \rightleftarrows \bullet$.

4.22 PROPOSITION
Let **A** *be a reflective subcategory of* **B**, *and for each* **B**-*object* B *let* $r_B : B \to A_B$ *be an* **A**-*reflection arrow. Then there exists a unique functor* $R : \mathbf{B} \to \mathbf{A}$ *such that the following conditions are satisfied:*

(1) $R(B) = A_B$ *for each* **B**-*object* B,

(2) *for each* **B**-*morphism* $f : B \to B'$, *the diagram*

$$\begin{array}{ccc} B & \xrightarrow{r_B} & R(B) \\ f \downarrow & & \downarrow R(f) \\ B' & \xrightarrow{r_{B'}} & R(B') \end{array}$$

commutes.

Proof: There exists a unique function R on objects satisfying (1) and (by the definition of **A**-reflection arrows) a unique function R on morphisms satisfying (2). It remains to be shown that R is a functor, i.e., that R preserves identities and composition. The first fact follows from the commutativity of the diagram

$$\begin{array}{ccc} B & \xrightarrow{r_B} & RB \\ id_B \downarrow & & \downarrow id_{RB} \\ B & \xrightarrow{r_B} & RB \end{array}$$

and the second one from the commutativity of the diagram

$$\begin{array}{ccc} B & \xrightarrow{r_B} & RB \\ g \circ f \downarrow & & \downarrow Rg \circ Rf \\ B'' & \xrightarrow{r_{B''}} & RB'' \end{array}$$

obtained by pasting together the corresponding diagrams for $B \xrightarrow{f} B'$ and $B' \xrightarrow{g} B''$. □

4.23 DEFINITION
A functor $R : \mathbf{B} \to \mathbf{A}$ constructed according to the above proposition is called a **reflector for A**.

4.24 REMARK
If **A** is a reflective subcategory of **B**, then a reflector for **A** depends upon the choice of the reflection arrows. Hence, there are usually many different reflectors for **A**. In Proposition 6.7 we will see that any two such reflectors are essentially the same; i.e., they will turn out to be "naturally isomorphic".

The dual of the concept *reflective subcategory* is *coreflective subcategory*. That is, **A** is a coreflective subcategory of **B** if and only if $\mathbf{A}^{\mathrm{op}}$ is a reflective subcategory of $\mathbf{B}^{\mathrm{op}}$. Although each of the above statements *is* an adequate definition, to aid the reader we provide a detailed dual formulation that doesn't involve the dual categories. We occasionally provide such explicit dual formulations although, strictly speaking, to do so is redundant.

4.25 DEFINITION
Let **A** be a subcategory of **B** and let B be a **B**-object.

(1) An **A-coreflection** (or **A-coreflection arrow**) for B is a **B**-morphism $A \xrightarrow{c} B$ from an **A**-object A to B with the following universal property:

for any **B**-morphism $A' \xrightarrow{f} B$ from some **A**-object A' to B there exists a unique **A**-morphism $f' : A' \to A$ such that the triangle

commutes.

By an "abuse of language" an object A is called an **A**-coreflection for B provided that there exists an **A**-coreflection $A \xrightarrow{c} B$ for B with domain A.

(2) **A** is called a **coreflective subcategory** of **B** provided that each **B**-object has an **A**-coreflection.

4.26 EXAMPLES
Several mathematical constructions such as certain "modifications" of structures and certain "selections" of convenient subobjects can be regarded in a natural way as coreflections. Here we list a few examples of full coreflective subcategories **A** of categories **B**.

A. Modifications of the Structure

(1) Making a relation symmetric: **B** = **Rel**, **A** = **Sym**. $(X, \rho \cap \rho^{-1}) \xrightarrow{id_X} (X, \rho)$ is an **A**-coreflection for (X, ρ).

This example shows one of the rare instances where a subcategory is simultaneously reflective and coreflective [cf. 4.17(1)].

(2) Making a topological space sequential: **B**=**Top**, **A**= the full subcategory of sequential spaces (i.e., spaces in which every sequentially closed set is closed). If (X, τ) is a topological space, then $\tau' = \{A \subseteq X \mid X \setminus A \text{ is sequentially closed in } (X, \tau)\}$ is a topology on X and $(X, \tau') \xrightarrow{id_X} (X, \tau)$ is an **A**-coreflection for (X, τ).

B. Sorting Out Convenient Subobjects

(3) Making an abelian group a torsion group: **B** = **Ab**, **A** = the full subcategory of abelian torsion groups. For any abelian group G the canonical embedding $TG \hookrightarrow G$ of the torsion-subgroup TG of G into G is an **A**-coreflection for G.

(4) Making a sequential acceptor reachable: **B** = Σ-**Seq**, **A** = the full subcategory of reachable acceptors. For any acceptor A the canonical embedding $RA \hookrightarrow A$ of the acceptor RA, formed by removing from the state set of A all states that cannot be reached from the initial state, is an **A**-coreflection for A. Thus a minimalization of sequential acceptors is obtained in two steps: first the coreflection, which yields a reachable acceptor, and then the reflection, which gives the minimal quotient of the reachable part [cf. 4.17(7)].

4.27 PROPOSITION

If **A** *is a coreflective subcategory of* **B** *and for each* **B***-object* B, $A_B \xrightarrow{c_B} B$ *is an* **A***-coreflection arrow, then there exists a unique functor* $C : \mathbf{B} \to \mathbf{A}$ *(called a* **coreflector** *for* **A***) such that the following conditions are satisfied:*

(1) $C(B) = A_B$ *for each* **B***-object* B,

(2) for each **B***-morphism* $f : B \to B'$ *the diagram*

$$\begin{array}{ccc} C(B) & \xrightarrow{c_B} & B \\ C(f) \downarrow & & \downarrow f \\ C(B') & \xrightarrow{c_{B'}} & B' \end{array}$$

commutes. $\boxed{D}$ [22]

Suggestions for Further Reading

Herrlich, H. *Topologische Reflexionen und Coreflexionen.* Lecture Notes in Mathematics, Vol. 78, Springer, Berlin–Heidelberg–New York, 1968.

---

[22]See Remark 3.7.

Banaschewski, B., and E. Nelson. Completions of partially ordered sets. *SIAM J. Comp.* **11** (1982): 521–528.

Kelly, G. M. On the ordered set of reflective subcategories. *Bull. Austral. Math. Soc.* **36** (1987): 137–152.

Tholen, W. Reflective subcategories. *Topol. Appl.* **27** (1987): 201–212.

EXERCISES

4A. Subcategories and Identities

Consider the categories **A** and **B** described in Exercise 3G(a). If $A = B$, then is **B** a subcategory of **A**?

4B. Isomorphism-Closed Subcategories

A (not necessarily full) subcategory **A** of a category **B** is called **isomorphism-closed** provided that every **B**-isomorphism with domain in **A** belongs[23] to **A**. Show that every subcategory **A** of **B** can be embedded into a smallest isomorphism-closed subcategory **A**′ of **B** that contains **A**. The inclusion functor $\mathbf{A} \hookrightarrow \mathbf{A}'$ is an equivalence iff all **B**-isomorphisms between **A**-objects belong to **A**.

4C. Full Subcategories

(a) Show that a category is discrete if and only if each of its subcategories is full.

(b) Show that in a poset, considered as a category,

- every subcategory is isomorphism-closed,
- every (co)reflective subcategory is full.

4D. Reflective Subcategories of Special Categories

*(a) Show that **Set** has

- precisely three full, isomorphism-closed, reflective subcategories,
- precisely two full, isomorphism-closed, coreflective subcategories,
- infinitely many reflective subcategories,
- infinitely many coreflective subcategories.

*(b) Show that **HComp** has precisely two full, isomorphism-closed, coreflective subcategories.

(c) Show that a full subcategory **A** of a poset **B**, considered as category, is reflective in **B** if and only if for each element b of **B** the set $\{a \in \mathbf{A} \mid b \leq a\}$ has a smallest element.

---

[23] Observe that if a morphism belongs to **A**, so does its domain and its codomain.

(d) Consider the poset of natural numbers as a category, **B**. Verify that a subset A of **B**, considered as full subcategory of **B**, is

- reflective in **B** if and only if A is infinite,
- coreflective in **B** if and only if $0 \in A$.

(e) Show that no finite monoid, considered as a category, has a proper reflective subcategory. However, if **A** is the monoid of all maps from $\mathbb{N}$ into $\mathbb{N}$, considered as a category, then the subcategory of **A**, consisting of all maps $f: \mathbb{N} \to \mathbb{N}$ with $f(0) = 0$, is reflective in **A**.

(f) Show that the (nonfull) subcategory **A** of **Pos**, consisting of those posets for which every nonempty subset has a meet, and those morphisms that preserve meets of nonempty subsets, is reflective in **Pos**. For a poset A, $A \xrightarrow{id} A$ is an **A**-reflection if and only if A is well-ordered.

(g) Verify that the category $\bullet \rightrightarrows \bullet$ has no proper reflective subcategory.

4E. Subcategories That Are Simultaneously Reflective and Coreflective

(a) Show that in **Rel** the full subcategory of symmetric relations is both reflective and coreflective. What about the full subcategory of reflexive relations, ..., or of transitive relations?

(b) Verify that in **Alg**(1), the category of unary algebras on one operation, all idempotent algebras form a full subcategory that is both reflective and coreflective.

(c) Show that the poset of natural numbers, considered as a category, has infinitely many isomorphism-closed subcategories that are simultaneously reflective and coreflective. [Cf. 4D(c).]

*(d) Show that neither **Set** nor **Top** has a proper isomorphism-closed full subcategory that is both reflective and coreflective. What about nonfull isomorphism-closed subcategories of **Set**?

4F. Intersections of Reflective Subcategories

(a) Show that in the poset of natural numbers, considered as a category **B**, the following hold:

- the intersection of any nonempty family of coreflective subcategories of **B** is coreflective in **B**,
- every full subcategory of **B** is an intersection of two reflective full subcategories of **B**.

* (b) Let **BiTop** be the construct of bitopological spaces (i.e., triples (X, τ_1, τ_2), where τ_1 and τ_2 are topologies on X) and bicontinuous maps (i.e., maps that are separately continuous with respect to the first topologies and with respect to the second topologies). Verify that the full subcategory **B**$_1$ that consists of all bitopological spaces with τ_1 compact Hausdorff is reflective in **BiTop**. By symmetry the full

subcategory $\mathbf{B}_2$ that consists of all bitopological spaces with τ_2 compact Hausdorff is also reflective in **BiTop**. Show, however, that **BiComp** $= \mathbf{B}_1 \cap \mathbf{B}_2$ is not reflective in **BiTop**. [Hint: A space (X, τ_1, τ_2), where X is infinite and τ_1 and τ_2 are discrete, has no reflection arrow.]

* (c) Let **PsTop** be the construct of pseudotopological spaces (i.e., pairs (X, α), where X is a set and α is a function that assigns to each ultrafilter on X a subset of X, in such a way that for each point $x \in X$, $x \in \alpha(\dot{x})$ [where $\dot{x}$ is the ultrafilter of all supersets of $\{x\}$]) and continuous maps (i.e., functions $(X, \alpha) \xrightarrow{f} (Y, \beta)$ having the property that for each $x \in \alpha(\mathcal{F})$ it follows that $f(x) \in \beta(f(\mathcal{F}))$, where $f(\mathcal{F}) = \{M \subseteq Y \mid f^{-1}[M] \in \mathcal{F}\}$). Verify that the full subcategory of **PsTop** consisting of all compact Hausdorff spaces, i.e., all spaces (X, α) with the property that each $\alpha(\mathcal{F})$ is a singleton set, is not reflective even though it is an intersection of a class of full reflective subcategories. [Hint: Let $\mathbf{B}_k$ be the subcategory that consists of all (X, α) such that $\alpha(\mathcal{F})$ is a singleton whenever some member of $\mathcal{F}$ is a set of cardinality less than k.]

4G. Subcategories of Subcategories

Let **A** be a subcategory of **B** and **B** be a subcategory of **C**. Prove that

(a) **A** is a subcategory of **C**,

(b) if **A** is reflective in **B** and **B** is reflective in **C**, then **A** is reflective in **C**,

(c) if **A** is reflective in **C** and **B** is a full subcategory of **C**, then **A** is reflective in **B**,

(d) if **A** is reflective in **C**, then **A** need not be reflective in **B**. [Construct an example.]

4H. Reflectors

(a) Let **A** be a reflective subcategory of **B**. Show that if **A** is full in **B**, then there exists a reflector $R : \mathbf{B} \to \mathbf{A}$ with $R \circ E \circ R = R$. Does the converse hold?

(b) Let **A** be the full subcategory of **Vec** consisting of all one-element vector spaces. Verify that **A** is simultaneously reflective and coreflective in **Vec**, and every functor from **Vec** to **A** is simultaneously a reflector and a coreflector.

4I. Skeletons

Given two skeletons of a category **A**, show that there is an isomorphism $\mathbf{A} \to \mathbf{A}$ that restricts to an isomorphism between the skeletons.

* ## 4J. Universal Categories

Does there exist a category **A** such that every category can be fully embedded into **A**?

4K. Full Embeddability

Show that

(a) **Prost** is fully embeddable into **Top** as a coreflective subcategory,

(b) **Alg**(1) is fully embeddable into **Rel** as a reflective subcategory,

(c) Σ-**Seq** is fully embeddable into **Aut**.

5 Concrete categories and concrete functors

As we have seen, many familiar categories such as **Vec** and **Top** are constructs (i.e., categories of structured sets and structure-preserving functions between them). If we regard such constructs as purely *abstract* categories (as we have done so far), some valuable information (concerning underlying sets of objects and underlying functions of morphisms) is lost. Fortunately, category theory enables us to retain this information by providing a means for a formal definition of construct — a construct being a pair $(\mathbf{A}, U)$ consisting of a category $\mathbf{A}$ and a faithful functor $U : \mathbf{A} \to \mathbf{Set}$. A careful analysis reveals that, for instance, many of the interesting properties of the constructs of vector spaces and topological spaces are not properties of the corresponding abstract categories **Vec** and **Top** but rather of the corresponding constructs $(\mathbf{Vec}, U)$ and $(\mathbf{Top}, V)$, where U and V denote the obvious underlying functors. In fact, they are often properties of just the underlying functors U and V. For example, the facts that the construct of vector spaces is "algebraic" and the construct of topological spaces is "topological" are very conveniently expressed by specific properties of the underlying functors, rather than by properties of the abstract categories **Vec** and **Top** (see Chapter VI).

If, instead of vector spaces or topological spaces, we want to investigate objects with more complex structures, e.g., topological vector spaces, we may supply the corresponding abstract category **TopVec** with forgetful functors not only into **Set** (forgetting both the topology and the linear structure), but also into **Top** (just forgetting the linear structure) or into **Vec** (just forgetting the topology). Hence we can consider **TopVec** as a concrete category over **Set** or over **Top** or over **Vec**. This leads to the concept of *concrete categories* over a category $\mathbf{X}$ as pairs $(\mathbf{A}, U)$ consisting of a category $\mathbf{A}$ and a faithful functor $U : \mathbf{A} \to \mathbf{X}$. In this way we may decompose complex structures into simpler ones, or, conversely, construct more complex structures out of simpler ones by composing forgetful functors. The concept of concrete categories over arbitrary base categories provides a suitable language to carry out such investigations.

5.1 DEFINITION

(1) Let $\mathbf{X}$ be a category. A **concrete category** over $\mathbf{X}$ is a pair $(\mathbf{A}, U)$, where $\mathbf{A}$ is a category and $U : \mathbf{A} \to \mathbf{X}$ is a faithful functor. Sometimes U is called the **forgetful** (or **underlying**) **functor** of the concrete category and $\mathbf{X}$ is called the **base category** for $(\mathbf{A}, U)$.

(2) A concrete category over **Set** is called a **construct**.

5.2 EXAMPLES

(1) Every category $\mathbf{A}$ can be regarded via the identity functor as a concrete category $(\mathbf{A}, id_\mathbf{A})$ over itself.

(2) If **A** is a construct in the sense of Example 3.3(2) and U is its naturally associated underlying functor [3.20(3)], then the pair $(\mathbf{A}, U)$ is a construct in the sense of Definition 5.1. Frequently we will denote the construct $(\mathbf{A}, U)$ (by an "abuse of notation") by **A** alone. For example, the abstract category of vector spaces and the construct of vector spaces will both be denoted by **Vec**. It will always be clear from the context which of these entities is being considered.

(3) **Ban** [3.3(3)(b)] can be considered as a construct in two natural ways:

   (a) via the obvious forgetful functor $U :$ **Ban** $\to$ **Set**,

   (b) via the less obvious but useful "unit ball" functor $O :$ **Ban** $\to$ **Set**, where $O(X) = \{\, x \in X \mid \; \| x \| \leq 1 \,\}$ and $O(f)$ is the corresponding restriction of $U(f)$.

   [It will turn out that, as constructs, $(\mathbf{Ban}, O)$ has more pleasant properties than $(\mathbf{Ban}, U)$.]

(4) Whenever $(\mathbf{A}, U)$ is a construct, then $(\mathbf{A}^{\mathrm{op}}, \mathcal{Q} \circ U^{\mathrm{op}})$ is also a construct (where $\mathcal{Q} : \mathbf{Set}^{\mathrm{op}} \to \mathbf{Set}$ is the contravariant power-set functor [cf. 3.20(9)]). In particular, $(\mathbf{Set}^{\mathrm{op}}, \mathcal{Q})$ is a construct.

(5) The category **TopVec** of topological vector spaces and continuous linear transformations can be regarded naturally via the obvious forgetful functors as

   (a) an abstract category,

   (b) a construct,

   (c) a concrete category over **Top**,

   (d) a concrete category over **Vec**.

   A similar situation occurs for **TopGrp**, the category of topological groups (and continuous homomorphisms).

(6) The category **Aut** of deterministic sequential Moore automata [3.3(4)(b)] can naturally be considered as a concrete category over **Set** $\times$ **Set** $\times$ **Set** [cf. 3.3(4)(f)]. This is the case since for any object $(Q, \Sigma, Y, \delta, q_0, y)$ of **Aut**, each of Q, Σ, and Y is a set and **Aut**-morphisms (i.e., simulations) are triples of functions that carry Q, Σ, and Y into the corresponding sets of the codomain.

(7) Since for any category **A** the unique functor $\mathbf{A} \to \mathbf{1}$ is faithful if and only if **A** is thin, the concrete categories over **1** are essentially just the thin categories; i.e., the preordered classes [3.26(2)].

5.3 REMARK

We adopt the following conventions:

(1) Since faithful functors are injective on hom-sets, we usually assume [in view of 3.2(3)] that $\hom_{\mathbf{A}}(A, B)$ is a subset of $\hom_{\mathbf{X}}(UA, UB)$ for each pair (A, B) of **A**-objects. This familiar convention allows one to express the property that "for **A**-objects A and

B and an **X**-morphism $UA \xrightarrow{f} UB$ there exists a (necessarily unique) **A**-morphism $A \to B$ with $U(A \to B) = UA \xrightarrow{f} UB$" much more succinctly, by stating

"$UA \xrightarrow{f} UB$ is an **A**-morphism (from A to B)".[24]

Observe, however, that since U doesn't need to be injective on objects, the expression

"$UA \xrightarrow{id_X} UB$ is an **A**-morphism (from A to B)"

does not imply that $A = B$ or that $id_X = id_A$, although it does imply that $UA = UB = X$. (This is the case, for example, for the morphism in **Pos** from $(\{0,1\}, =)$ to $(\{0,1\}, \leq)$ that is the identity function on $\{0,1\}$.) To avoid possible confusion in such a circumstance, we call an **A**-morphism $A \xrightarrow{f} B$ **identity-carried** if $Uf = id_X$.

(2) Sometimes we will write simply **A** for the concrete category $(\mathbf{A}, U)$ over **X**, when U is clear from the context; e.g., in the example above. In these cases the underlying object of an **A**-object A will sometimes be denoted by $|A|$; i.e., "$|\ |$" will serve as a standard notation for underlying functors.

(3) If P is a property of categories (or of functors), then we will say that a concrete category $(\mathbf{A}, U)$ **has property** P provided that **A** (or U) has property P.

FIBRES IN CONCRETE CATEGORIES

5.4 DEFINITION

Let $(\mathbf{A}, U)$ be a concrete category over **X**.

(1) The **fibre** of an **X**-object X is the preordered class consisting of all **A**-objects A with $U(A) = X$ ordered by:

(2) $A \leq B$ if and only if $id_X : UA \to UB$ is an **A**-morphism.

(3) **A**-objects A and B are said to be **equivalent** provided that $A \leq B$ and $B \leq A$.

(4) $(\mathbf{A}, U)$ is said to be **amnestic** provided that its fibres are partially ordered classes; i.e., no two different **A**-objects are equivalent.

(5) $(\mathbf{A}, U)$ is said to be **fibre-small** provided that each of its fibres is small, i.e., a preordered set.

5.5 EXAMPLES

If **A** is a construct in the sense of 3.3(2)(3) and X is a set, then the **A**-fibre of X is (up to order-isomorphism) the class of all **A**-structures on X, ordered appropriately, e.g.,

---

[24] Observe that analogues of this expression are frequently used in "concrete situations", e.g., in saying that a certain function between vector spaces "is linear" or that a certain function between topological spaces "is continuous", etc.

(1) if **A** = **Rel**, then for relations ρ_1 and ρ_2 on X

$$\rho_1 \leq \rho_2 \quad \Leftrightarrow \quad \rho_1 \subseteq \rho_2,$$

(2) if **A** = **Met**, then for metrics d_1 and d_2 on X

$$d_1 \leq d_2 \quad \Leftrightarrow \quad \forall (x,y) \in X \times X, \ d_2(x,y) \leq d_1(x,y),$$

(3) if **A** = Σ-**Seq**, then for acceptors (Q, δ, q_0, F) and (Q, δ', q_0', F') on a set Q

$$(\delta, q_0, F) \leq (\delta', q_0', F') \quad \Leftrightarrow \quad \delta = \delta', \ q_0 = q_0', \text{ and } F \subseteq F',$$

(4) if **A** = **Top**, then for topologies τ_1 and τ_2 on X

$$\tau_1 \leq \tau_2 \quad \Leftrightarrow \quad \tau_2 \subseteq \tau_1,$$

(5) if **A** = **Vec**, then for vector space structures ν_1 and ν_2 on X

$$\nu_1 \leq \nu_2 \quad \Leftrightarrow \quad \nu_1 = \nu_2.$$

Observe that for **Top** the fibres are complete lattices (with smallest element the discrete topology and largest element the indiscrete topology), whereas for **Vec** the fibres are ordered by the equality relation, i.e., no two different vector space structures on the same set are related to each other. These properties of the fibres of **Top** (resp. **Vec**) are typical for "topological" (resp. "algebraic") constructs. See Chapter VI.

5.6 REMARKS
A concrete category $(\mathbf{A}, U)$ is amnestic if and only if the functor U is amnestic (cf. 3.27(4)). Most of the familiar concrete categories are both amnestic and fibre-small. However:

(1) No proper class, considered as a concrete category over **1**, is fibre-small. Fibres of the quasiconstruct[25] of quasitopological spaces[26] are not even preordered classes, but rather are proper conglomerates. The latter statement follows from the fact that for each set X with at least two points, the fibre of X cannot be put in bijective correspondence with any subclass of the universe $\mathcal{U}$.

(2) The constructs **Ban**$_b$ and **Met**$_c$ are not amnestic. In Proposition 5.33 we will see that each concrete category $(\mathbf{A}, U)$ over **X** has an associated "amnestic modification" that in most respects behaves like $(\mathbf{A}, U)$.

---

[25] A **quasiconstruct** is a pair $(\mathbf{A}, U)$, where **A** is a quasicategory and U is a faithful functor from U to **Set**.
[26] See, e.g., E. Spanier, Quasi-topologies, *Duke Math. J.* **30** (1963): 1–14.

5.7 DEFINITION
A concrete category is called

(1) **fibre-complete** provided that its fibres are (possibly large) complete lattices,[27]

(2) **fibre-discrete** provided that its fibres are ordered by equality.

5.8 PROPOSITION
A concrete category $(\mathbf{A}, U)$ *over* $\mathbf{X}$ *is fibre-discrete if and only if* U **reflects identities** *(i.e., if* $U(k)$ *is an* $\mathbf{X}$*-identity, then* k *must be an* $\mathbf{A}$*-identity).* □

CONCRETE FUNCTORS

Just as functors can be considered as the natural "morphisms" between abstract categories, and are a useful tool for investigating them, concrete functors are the natural "morphisms" between concrete categories.

5.9 DEFINITION
If $(\mathbf{A}, U)$ and $(\mathbf{B}, V)$ are concrete categories over $\mathbf{X}$, then a **concrete functor from** $(\mathbf{A}, U)$ **to** $(\mathbf{B}, V)$ is a functor $F: \mathbf{A} \to \mathbf{B}$ with $U = V \circ F$. We denote such a functor by $F: (\mathbf{A}, U) \to (\mathbf{B}, V)$.

5.10 PROPOSITION
(1) Every concrete functor is faithful.

(2) Every concrete functor is completely determined by its values on objects.

(3) Objects that are identified by a full concrete functor are equivalent.

(4) Every full concrete functor with amnestic domain is an embedding.

Proof: Let $F: (\mathbf{A}, U) \to (\mathbf{B}, V)$ be a concrete functor.

(1). This follows from Proposition 3.30(2).

(2). Suppose that $G: (\mathbf{A}, U) \to (\mathbf{B}, V)$ is a concrete functor with $G(A) = F(A)$ for each $\mathbf{A}$-object A. Then for any $\mathbf{A}$-morphism $A \xrightarrow{f} A'$ we have the $\mathbf{B}$-morphisms

$$GA = FA \underset{Gf}{\overset{Ff}{\rightrightarrows}} FA' = GA'$$

with $V(Ff) = U(f) = V(Gf)$. Since V is faithful, $Ff = Gf$. Hence $F = G$.

(3). Let A and A' be $\mathbf{A}$-objects with $FA = FA'$ ($= B$). Then, by using Proposition 3.31, $id_B: FA \to FA'$ can be lifted to an $\mathbf{A}$-isomorphism $g: A \to A'$. Hence A and A' are equivalent.

---

[27] A partially ordered class $(X, \leq)$ is called a **large complete lattice** provided that every subclass of X has a join and a meet.

(4). F is full, faithful, and injective on objects; hence it is an embedding. □

5.11 EXAMPLES
Because of part (2) of the above proposition, when specifying concrete functors, we need only describe how they map objects.

(1) The forgetful functor **Rng** → **Ab** that "forgets" multiplication is a concrete functor from the category of rings (and ring homomorphisms) to the category of abelian groups, where both categories are considered as constructs. In a similar way, the forgetful functors **Met** → **Top** (that assigns to each metric space (X, d) the topological space determined by the distance function d) and **TopVec** → **Vec** (that assigns to each topological vector space its underlying vector space) are concrete functors.

(2) There are many functors from **Set** to itself, but there is only one concrete functor between the construct (**Set**, $id_{\mathbf{Set}}$) and itself — namely, the identity functor.

(3) The "discrete-space functor" and the "indiscrete-space functor" [3.29(5),(6)] are examples of concrete functors from **Set** to **Top**.

(4) Any functor between concrete categories over **1** is already concrete.

5.12 REMARK
A **concrete isomorphism** $F : (\mathbf{A}, U) \to (\mathbf{B}, V)$ between concrete categories over **X** is a concrete functor that is an isomorphism of categories. All of the isomorphisms given in Examples 3.26(2), (4), (5), and (6) are concrete. That such a concrete isomorphism exists means, informally, that each structure in **A**, i.e., each object A of **A**, can be completely substituted by a structure in **B**, namely $F(A)$ (keeping, of course, the same morphisms). For example, the standard descriptions of topological spaces by means of

- neighborhoods,
- open sets,
- closure operators, or
- convergent filters,

give technically different constructs, all of which are concretely isomorphic. This is why the differences between the various descriptions are regarded as inessential and we can in good conscience call each of them "**Top**". The concept of concretely isomorphic concrete categories gives rise to an equivalence relation that is stronger than the relation of isomorphism of categories. For example, assuming that no measurable cardinals exist, **Top** (and, indeed, any construct) can be thought of as being isomorphic to a full subcategory of **Rel** (cf. 4.7). However, **Top** is not concretely isomorphic to such a subcategory, because there are more topologies on $\mathbb{N}$ (namely, $2^{2^{\aleph_0}}$) than there are binary relations on $\mathbb{N}$ (namely, $2^{\aleph_0}$).

5.13 REMARK

If $F : (\mathbf{A}, U) \to (\mathbf{B}, V)$ is a concrete isomorphism, then its inverse $F^{-1} : \mathbf{B} \to \mathbf{A}$ is concrete from $(\mathbf{B}, V)$ to $(\mathbf{A}, U)$. Unfortunately, the corresponding result does not hold for concrete equivalences (i.e., concrete functors that are equivalences) since if $F : (\mathbf{A}, U) \to (\mathbf{B}, V)$ is a concrete equivalence, then it may happen that there is no concrete equivalence from $(\mathbf{B}, V)$ to $(\mathbf{A}, U)$ even though there are equivalences from $\mathbf{B}$ to $\mathbf{A}$; cf. Proposition 3.36. For example, the embedding of the skeleton of cardinal numbers into **Set** is such a concrete equivalence of constructs that is not invertible. Thus, even though it makes sense to say that two concrete categories over $\mathbf{X}$ are concretely isomorphic, it makes little sense to say that they are concretely equivalent since the relation between concrete categories of being concretely equivalent is *not* symmetric.

The following proposition provides the basis for forming the quasicategory of all concrete categories over a given base category.

5.14 PROPOSITION

(1) The identity functor on a concrete category is a concrete isomorphism.

(2) Any composite of concrete functors over $\mathbf{X}$ is a concrete functor over $\mathbf{X}$. □

5.15 DEFINITION

The quasicategory that has as objects all concrete categories over $\mathbf{X}$ and as morphisms all concrete functors between them is denoted by $\mathbf{CAT}(\mathbf{X})$. In particular, $\mathbf{CONST} = \mathbf{CAT}(\mathbf{Set})$ is the quasicategory of all constructs.

5.16 REMARK

Notice that even when $\mathbf{X}$ is small, $\mathbf{CAT}(\mathbf{X})$ is never an actual category (unless $\mathbf{X}$ is the empty category). $\mathbf{CAT}(\mathbf{1})$ is essentially the quasicategory of all preordered classes and order-preserving functions.

5.17 EXAMPLE

There is a natural embedding $F : \mathbf{Mon}^{\mathrm{op}} \to \mathbf{CONST}$ that assigns to any monoid M the category $M\text{-}\mathbf{Act}$. If $f : M \to N$ is a monoid homomorphism, then $F(f) : N\text{-}\mathbf{Act} \to M\text{-}\mathbf{Act}$ is the concrete functor that assigns to any N-action $(X, *_N)$ the M-action $(X, *_M)$, where $m *_M x$ is defined to be $f(m) *_N x$.

5.18 DEFINITION

If F and G are both concrete functors from $(\mathbf{A}, U)$ to $(\mathbf{B}, V)$, then F is **finer than** G (or G is **coarser than** F), denoted by: $F \leq G$, provided that $F(A) \leq G(A)$ for each $\mathbf{A}$-object A [5.4(1)].

5.19 EXAMPLES

(1) For order-preserving functions considered as concrete functors over **1**, $f \leq g$ if and only if this relation holds pointwise.

(2) Among the concrete functors from **Set** to **Top** there is a finest one (the discrete-space functor), and a coarsest one (the indiscrete-space functor).

DUALITY FOR CONCRETE CATEGORIES

5.20 REMARK
For every concrete category $(\mathbf{A}, U)$ over $\mathbf{X}$, its dual $(\mathbf{A}^{\mathrm{op}}, U^{\mathrm{op}})$ is a concrete category over $\mathbf{X}^{\mathrm{op}}$. Moreover, for every concrete functor $F : (\mathbf{A}, U) \to (\mathbf{B}, V)$ over $\mathbf{X}$ its dual functor $F^{\mathrm{op}} : (\mathbf{A}^{\mathrm{op}}, U^{\mathrm{op}}) \to (\mathbf{B}^{\mathrm{op}}, V^{\mathrm{op}})$ is a concrete functor over $\mathbf{X}^{\mathrm{op}}$. Thus, there is a duality principle for concrete categories and concrete functors. Observe, however, that unless $\mathbf{X} = \mathbf{X}^{\mathrm{op}}$ there is *no* duality for concrete categories over a fixed base category $\mathbf{X}$. In particular, we don't have a duality principle for constructs. However, since $\mathbf{1} = \mathbf{1}^{\mathrm{op}}$, there is a duality principle for concrete categories over $\mathbf{1}$ (i.e., for preordered classes). This is indeed a familiar duality.

CONCRETE SUBCATEGORIES

5.21 CONVENTION
If $(\mathbf{B}, U)$ is a concrete category over $\mathbf{X}$ and $\mathbf{A}$ is a subcategory of $\mathbf{B}$ with inclusion $E : \mathbf{A} \hookrightarrow \mathbf{B}$, then $\mathbf{A}$ will often be regarded (via the functor $U \circ E$) as a concrete category $(\mathbf{A}, U \circ E)$ over $\mathbf{X}$. In such cases we will call $(\mathbf{A}, U \circ E)$ a **concrete subcategory** of $(\mathbf{B}, U)$. In the case that the base category is **Set**, we will call $(\mathbf{A}, U \circ E)$ a **subconstruct** of $(\mathbf{B}, U)$.

5.22 DEFINITION
(1) A concrete subcategory $(\mathbf{A}, U)$ of $(\mathbf{B}, V)$ is called **concretely reflective** in $(\mathbf{B}, V)$ (or a **reflective modification** of $(\mathbf{B}, V)$) provided that for each **B**-object there exists an identity-carried **A**-reflection arrow.

(2) Reflectors induced by identity-carried reflection arrows are called **concrete reflectors**.[28]

DUAL NOTIONS: **Concretely coreflective subcategory** (or **coreflective modification**); **concrete coreflectors**.

5.23 EXAMPLES
(1) The construct of symmetric relations is simultaneously a concretely reflective and a concretely coreflective subconstruct of the construct **Rel** of (binary) relations [cf. 4.17(1) and 4.26(1)].

(2) The construct of completely regular spaces is a concretely reflective subconstruct of **Top** [cf. 4.17(2)].

---
[28]CAUTION: Concrete functors that are reflectors need not be concrete reflectors. See 5.23(6).

(3) The construct of sequential topological spaces is a concretely coreflective subconstruct of **Top** [cf. 4.26(2)].

(4) None of the reflective subconstructs described in Examples 4.17(3)–(12) is concretely reflective.

(5) None of the coreflective subconstructs described in Examples 4.26(3) and (4) is concretely coreflective.

(6) Let **X** be a category consisting of a single object X and two morphisms id_X and s with $s \circ s = id_X$. Let **A** be the concrete category over **X**, consisting of two objects A_0 and A_1 and the morphism sets

$$\hom_{\mathbf{A}}(A_i, A_j) = \begin{cases} \{id_X\} & \text{if } i = j \\ \{s\} & \text{if } i \neq j. \end{cases}$$

Consider **A** as a concretely reflective subcategory of itself. Then $id_{\mathbf{A}} \colon \mathbf{A} \to \mathbf{A}$ is a concrete reflector, and the concrete functor $R \colon \mathbf{A} \to \mathbf{A}$, defined by $R(A_i) = A_{1-i}$, is a reflector that is not a concrete reflector.

5.24 PROPOSITION
Every concretely reflective subcategory of an amnestic concrete category is a full subcategory.

Proof: Let $(\mathbf{A}, U)$ be a concretely reflective subcategory of an amnestic $(\mathbf{B}, V)$, let A be an **A**-object, and let $r \colon A \to A^*$ be an identity-carried **A**-reflection arrow for A. We wish to show that $r = id_A$ so that Proposition 4.20 can be applied. By reflectivity there exists a unique **A**-morphism $s \colon A^* \to A$ such that the diagram

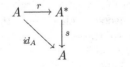

commutes.

Since r is identity-carried, $V(r) = id_{VA}$. Since also $V(id_A) = id_{VA}$, we conclude that $V(s) = id_{VA}$ as well. Faithfulness of V gives us $r \circ s = id_{A^*}$. Hence r is a **B**-isomorphism with $V(r) = id_{VA}$. Amnesticity of $(\mathbf{B}, V)$ yields $r = id_A$. Therefore, by Proposition 4.20, **A** is full in **B**. □

5.25 REMARK
A concretely reflective subcategory of a nonamnestic concrete category need not be full, as Example 4.21 shows. (The categories given there should be considered as concrete categories over **1**).

5.26 PROPOSITION
*For a concrete full subcategory $(\mathbf{A}, U)$ of a concrete category $(\mathbf{B}, V)$ over **X**, with inclusion functor $E \colon (\mathbf{A}, U) \hookrightarrow (\mathbf{B}, V)$, the following are equivalent:*

(1) $(\mathbf{A}, U)$ is concretely reflective in $(\mathbf{B}, V)$,

(2) there exists a concrete functor $R: (\mathbf{B}, V) \to (\mathbf{A}, U)$ that is a reflector with $R \circ E = id_{\mathbf{A}}$ and $id_{\mathbf{B}} \leq E \circ R$,[29]

(3) there exists a concrete functor $R: (\mathbf{B}, V) \to (\mathbf{A}, U)$ with $R \circ E \leq id_{\mathbf{A}}$ and $id_{\mathbf{B}} \leq E \circ R$.

Proof: (1) $\Rightarrow$ (2). For each **B**-object B choose a reflection arrow $B \xrightarrow{r_B} A_B$ such that r_B is identity-carried, and in case that B is an **A**-object $B \xrightarrow{r_B} A_B = B \xrightarrow{id_B} B$ (4.20). The associated reflector (4.23) has the properties required in (2).

(2) $\Rightarrow$ (3). Immediate.

(3) $\Rightarrow$ (1). If R is a concrete functor that satisfies the requirements of (3), then for each **B**-object B, we have $B \leq RB$; i.e., there is an identity-carried **B**-morphism $B \xrightarrow{r_B} RB$. To show that r_B is an **A**-reflection arrow for B, let A be an **A**-object and $B \xrightarrow{f} A$ be a **B**-morphism. Since R is a concrete functor, $RB \xrightarrow{f} RA$ is an **A**-morphism. In view of the fact that $RA \leq A$, this implies that $RB \xrightarrow{f} A$ is an **A**-morphism. Hence it is the unique **A**-morphism that makes the triangle

commute. □

5.27 REMARK

By the above proposition every full concretely reflective subcategory has a reflector that is a concrete functor. The converse holds (Proposition 5.31) for isomorphism-closed full concrete subcategories of transportable concrete categories, defined below. However, for nonfull subcategories [cf. Exercise 5E(d)] or non-transportable categories [cf. Exercise 5E(c)] the converse need not hold.

TRANSPORTABILITY

We have seen that many of the familiar constructs have the extra property of being amnestic. Another frequently encountered convenient property is that of transportability.

5.28 DEFINITION

A concrete category $(\mathbf{A}, U)$ over $\mathbf{X}$ is said to be **(uniquely) transportable** provided that for every **A**-object A and every **X**-isomorphism $UA \xrightarrow{k} X$ there exists a (unique) **A**-object B with $UB = X$ such that $A \xrightarrow{k} B$ is an **A**-isomorphism.

---
[29] Observe that $R \circ E = id_{\mathbf{A}}$ just means that $RA = A$ for each **A**-object A and that $id_{\mathbf{B}} \leq E \circ R$ just means that $B \leq RB$ for each **B**-object B [5.4(1)].

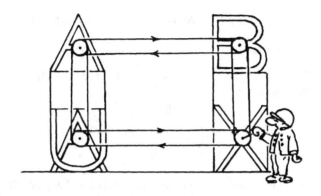

Transportability

5.29 PROPOSITION

A concrete category is uniquely transportable if and only if it is transportable and amnestic.

Proof: Let $(\mathbf{A}, U)$ be uniquely transportable and let $f : A \to A'$ be an **A**-isomorphism such that $Uf = id_X$. Since id_A is also an **A**-isomorphism with domain A and $U(id_A) = id_X$, by uniqueness it follows that $f = id_A$.

Conversely, if $(\mathbf{A}, U)$ is transportable and amnestic, A is an **A**-object, and $A \xrightarrow{k_1} A_1$ and $A \xrightarrow{k_2} A_2$ are **A**-isomorphisms with $Uk_1 = Uk_2$, then $A_1 \xrightarrow{k_2 \circ k_1^{-1}} A_2$ is an identity-carried **A**-isomorphism. Hence, by amnesticity $k_2 \circ k_1^{-1} = id_{A_1}$, so that $k_1 = k_2$. □

5.30 EXAMPLES

(1) For a preordered class $(A, \leq)$ considered as a concrete category over **1** [cf. 5.2(7)], the following conditions are equivalent:

   (a) $(A, \leq)$ is uniquely transportable,

   (b) $(A, \leq)$ is amnestic,

   (c) $\leq$ is antisymmetric, i.e., $\leq$ is a partial-order on A.

(2) Familiar constructs such as **Vec**, **Rel**, **Grp**, and **Top** are uniquely transportable.

(3) The skeleton of **Vec** formed by all spaces $\mathbb{R}^m$ [see 4.13(2)] is amnestic, but not transportable.

5.31 PROPOSITION

If $(\mathbf{A}, U)$ is an isomorphism-closed full concrete subcategory of a transportable concrete category $(\mathbf{B}, V)$ over $\mathbf{X}$, then the following are equivalent:

   1. $(\mathbf{A}, U)$ is concretely reflective in $(\mathbf{B}, V)$,

2. there exists a reflector $R : \mathbf{B} \to \mathbf{A}$ that is concrete from $(\mathbf{B}, V)$ to $(\mathbf{A}, U)$.

Proof: (1) $\Rightarrow$ (2). Immediate (cf. 5.26).

(2) $\Rightarrow$ (1). Let R be as in (2) and for each **B**-object B, let $B \xrightarrow{r_B} RB$ be an **A**-reflection arrow. Then the diagram

$$\begin{array}{ccc} B & \xrightarrow{r_B} & RB \\ {\scriptstyle r_B}\downarrow & & \downarrow{\scriptstyle Rr_B} \\ RB & \xrightarrow{r_{RB}} & R(RB) \end{array}$$

commutes (cf. 4.22). This implies (by uniqueness in the definition of reflection arrows) that $Rr_B = r_{RB}$. Since RB is an **A**-object and **A** is full in **B**, r_{RB} is an **A**-isomorphism (4.20). Hence $U(RB \xrightarrow{r_{RB}=Rr_B} R(RB))$ is an **X**-isomorphism. By the concreteness of R, $V(B \xrightarrow{r_B} RB) = U(RB \xrightarrow{Rr_B} R(RB))$. Hence $VRB \xrightarrow{r_B^{-1}} VB$ is an **X**-isomorphism. By the transportability of $(\mathbf{B}, V)$ there exists an isomorphism $RB \xrightarrow{r_B^{-1}} \hat{B}$ in **B**, so that since **A** is isomorphism-closed in **B**, $\hat{B}$ is an **A**-object. Consequently, the composite $B \xrightarrow{r_B} RB \xrightarrow{r_B^{-1}} \hat{B}$ is an identity-carried **A**-reflection arrow for B. □

5.32 REMARK
For concretely reflective subcategories there usually exist (even in the uniquely transportable cases) several reflectors that are concrete functors [see Example 5.23(6)]. However, in the amnestic case there exists only one *concrete* reflector.

Although transportability and amnesticity are quite useful and convenient properties for concrete categories, they are not especially strong ones. The next propositions show that if a concrete category fails to have one of them, it can be replaced via slight modifications by concrete categories that do have them.

5.33 PROPOSITION
For every concrete category $(\mathbf{A}, U)$ over **X**, there exists an amnestic concrete category $(\mathbf{B}, V)$ over **X** that has the following properties:

(1) there exists an (injective) concrete equivalence $E : (\mathbf{B}, V) \to (\mathbf{A}, U)$,

(2) there exists a surjective concrete equivalence $P : (\mathbf{A}, U) \to (\mathbf{B}, V)$.

Moreover, if $(\mathbf{A}, U)$ is transportable, then so is $(\mathbf{B}, V)$.

Proof: The relation for **A**-objects, given by "A is equivalent to $\hat{A}$ if and only if $A \leq \hat{A}$ and $\hat{A} \leq A$", is an equivalence relation on the class of **A**-objects. If **B** is a full subcategory of **A** that contains, as objects, precisely one member from each equivalence class, $E : \mathbf{B} \hookrightarrow \mathbf{A}$ is the inclusion, and $V = U \circ E$, then $(\mathbf{B}, V)$ is an amnestic concrete category and $E : (\mathbf{B}, V) \to (\mathbf{A}, U)$ satisfies condition (1).

If $(\mathbf{B}, V)$ is constructed as above, then the unique concrete functor $(\mathbf{A}, U) \xrightarrow{P} (\mathbf{B}, V)$ that sends each $\mathbf{A}$-object A to its unique equivalent $\mathbf{B}$-object satisfies condition (2).

That transportability is not destroyed is obvious from the construction. □

5.34 REMARK

The construction of $(\mathbf{A}, U) \xrightarrow{P} (\mathbf{B}, V)$, given above, is unique in the sense that for every surjective concrete equivalence $(\mathbf{A}, U) \xrightarrow{\hat{P}} (\hat{\mathbf{B}}, \hat{V})$ with $(\hat{\mathbf{B}}, \hat{V})$ amnestic, there exists a concrete isomorphism $K : (\mathbf{B}, V) \to (\hat{\mathbf{B}}, \hat{V})$ such that $\hat{P} = K \circ P$. In view of this the above $(\mathbf{B}, V)$ can be called "the" **amnestic modification** of $(\mathbf{A}, U)$. With respect to almost every interesting categorical property[30] a concrete category is indistinguishable from its amnestic modification. If one takes the view that in non-amnestic categories there are "too many objects floating around in the fibres", then the amnestic modifications can be considered to be more natural than the concrete categories that give rise to them. For an example of this phenomenon, notice that the amnestic modification of the construct $\mathbf{Met_c}$ of metric spaces and continuous maps is[31] the construct $\mathbf{Top_m}$ of metrizable topological spaces and continuous maps.

5.35 LEMMA

For every concrete category $(\mathbf{A}, U)$ over $\mathbf{X}$, there exists a transportable concrete category $(\mathbf{B}, V)$ over $\mathbf{X}$ and a concrete equivalence $E : (\mathbf{A}, U) \to (\mathbf{B}, V)$.

Proof: Define $\mathbf{B}$ as follows: Each $\mathbf{B}$-object is a triple (A, a, X), with $A \in Ob(\mathbf{A})$, $X \in Ob(\mathbf{X})$, and $a : UA \to X$ an $\mathbf{X}$-isomorphism.

$$\hom_{\mathbf{B}}((A, a, X), (\hat{A}, \hat{a}, \hat{X})) = \hom_{\mathbf{A}}(A, \hat{A}).$$

Identities and composition are as in $\mathbf{A}$. Define $V : \mathbf{B} \to \mathbf{X}$ by

$$V((A, a, X) \xrightarrow{f} (\hat{A}, \hat{a}, \hat{X})) = X \xrightarrow{\hat{a} \circ Uf \circ a^{-1}} \hat{X};$$

i.e., such that the square

$$\begin{array}{ccc} UA & \xrightarrow{a} & X \\ {\scriptstyle Uf}\downarrow & & \downarrow{\scriptstyle Vf} \\ U\hat{A} & \xrightarrow{\hat{a}} & \hat{X} \end{array}$$

commutes.

Clearly, V is a faithful functor, and if $V(A, a, X) \xrightarrow{f} Y$ is an $\mathbf{X}$-isomorphism, then $id_A : (A, a, X) \to (A, f \circ a, Y)$ is a $\mathbf{B}$-isomorphism with $V(id_A) = f$. Hence $(\mathbf{B}, V)$ is transportable. $E : (\mathbf{A}, U) \to (\mathbf{B}, V)$, defined by: $E(A) = (A, id_{UA}, UA)$, is a concrete equivalence. □

---

[30] Amnesticity obviously is not one of them.
[31] with respect to the canonical forgetful functor $\mathbf{Met_c} \to \mathbf{Top_m}$

5.36 PROPOSITION

For every concrete category $(\mathbf{A}, U)$ *over* $\mathbf{X}$ *there exists a uniquely transportable concrete category* $(\mathbf{B}, V)$ *over* $\mathbf{X}$ *and a concrete equivalence* $E : (\mathbf{A}, U) \to (\mathbf{B}, V)$ *that is uniquely determined*[32] *up to concrete isomorphism.*

Proof: According to Lemma 5.35 there is a transportable concrete category $(\mathbf{C}, W)$ over $\mathbf{X}$ and a concrete equivalence $F : (\mathbf{A}, U) \to (\mathbf{C}, W)$. By Proposition 5.33 there is an amnestic concrete category $(\mathbf{B}, V)$ over $\mathbf{X}$ and a surjective concrete equivalence $P : (\mathbf{C}, W) \to (\mathbf{B}, V)$. Consequently, $E = P \circ F : (\mathbf{A}, U) \to (\mathbf{B}, V)$ is a concrete equivalence, and by Proposition 5.33 $(\mathbf{B}, V)$ is uniquely transportable.

To show the essential uniqueness, let $(\mathbf{A}, U) \xrightarrow{\overline{E}} (\overline{\mathbf{B}}, \overline{V})$ be a concrete equivalence, with $(\overline{\mathbf{B}}, \overline{V})$ uniquely transportable. For each $\mathbf{B}$-object B choose an $\mathbf{A}$-object A_B and a $\mathbf{B}$-isomorphism $EA_B \xrightarrow{h_B} B$ such that whenever $B \in E[Ob(\mathbf{A})]$ then h_B is a $\mathbf{B}$-identity, $EA_B \xrightarrow{h_B} B = B \xrightarrow{id_B} B$. Then $V(EA_B \xrightarrow{h_B} B) = UA_B \xrightarrow{h_B} VB = \overline{V}(\overline{E}A_B) \xrightarrow{h_B} VB$ is an $\mathbf{X}$-isomorphism. Since $(\overline{\mathbf{B}}, \overline{V})$ is uniquely transportable, there is a unique $\overline{\mathbf{B}}$-isomorphism $\overline{E}A_B \to C_B$ with $\overline{V}(\overline{E}A_B \to C_B) = \overline{V}(\overline{E}A_B) \xrightarrow{h_B} VB$. A simple computation establishes that there exists a unique full concrete functor $H : (\mathbf{B}, V) \to (\overline{\mathbf{B}}, \overline{V})$ with $HB = C_B$ for each $\mathbf{B}$-object B. By Proposition 5.10, H is a concrete embedding. To show that $\overline{E} = H \circ E$, let A be an $\mathbf{A}$-object and let $B = EA$. By our construction $EA_B \xrightarrow{h_B} B$ is a $\mathbf{B}$-identity, which implies that $EA_B = EA$. Hence, by Proposition 5.10, A_B and A are equivalent $\mathbf{A}$-objects. Consequently, $\overline{E}A_B$ and $\overline{E}A$ are equivalent $\overline{\mathbf{B}}$-objects. Since $(\overline{\mathbf{B}}, \overline{V})$ is amnestic, this implies that $\overline{E}A_B = \overline{E}A$. Hence $\overline{E}A_B \xrightarrow{id} \overline{E}A$ is a $\overline{\mathbf{B}}$-isomorphism with

$$\overline{V}(\overline{E}A_B \xrightarrow{id} \overline{E}A) = \overline{V}(\overline{E}A_B) \xrightarrow{h_B} VB,$$

so that by the definition of C_B we have $C_B = \overline{E}A$; i.e., $(H \circ E)A = HB = C_B = \overline{E}A$.

It remains to be shown that H is surjective. Let C be a $\overline{\mathbf{B}}$-object. Since $\overline{E}$ is isomorphism-dense, there exists some $\mathbf{A}$-object A and some $\overline{\mathbf{B}}$-isomorphism $\overline{E}A \xrightarrow{h} C$. Then $\overline{V}(\overline{E}A \xrightarrow{h} C) = UA \xrightarrow{h} \overline{V}C = V(EA) \xrightarrow{h} \overline{V}C$ is an $\mathbf{X}$-isomorphism, so that by the unique transportability of $(\mathbf{B}, V)$, there exists a unique $\mathbf{B}$-isomorphism $EA \to B$, with $V(EA \to B) = V(EA) \xrightarrow{h} \overline{V}C$. We claim that $HB = C$. To see this, consider the chosen $\mathbf{B}$-isomorphism $EA_B \xrightarrow{h_B} B$. Since $EA \xrightarrow{h} B$ is a $\mathbf{B}$-isomorphism, $EA_B \xrightarrow{h^{-1} \circ h_B} EA$ is a $\mathbf{B}$-isomorphism. By the fullness and faithfulness of E, this implies that $A_B \xrightarrow{h^{-1} \circ h_B} A$ is an $\mathbf{A}$-isomorphism (3.31), and hence $\overline{E}A_B \xrightarrow{h^{-1} \circ h_B} \overline{E}A$ is a $\overline{\mathbf{B}}$-isomorphism. Consequently,

$$\overline{E}A_B \xrightarrow{h_B} C = \overline{E}A_B \xrightarrow{h^{-1} \circ h_B} \overline{E}A \xrightarrow{h} C$$

---

[32] That is, whenever $(A, U) \xrightarrow{\overline{E}} (\overline{B}, \overline{V})$ is a concrete equivalence to a uniquely transportable concrete category over $\mathbf{X}$, then there exists a concrete isomorphism $H : (\mathbf{B}, V) \to (\overline{\mathbf{B}}, \overline{V})$ such that $\overline{E} = H \circ E$.

is a $\overline{\mathbf{B}}$-isomorphism with

$$\overline{V}(\overline{E}A_B \xrightarrow{h_B} C) = \overline{V}(\overline{E}A_B) \xrightarrow{h_B} \overline{V}C = VB,$$

which, by the definition of HB, implies that $HB = C$. Thus $H : (\mathbf{B}, V) \to (\overline{\mathbf{B}}, \overline{V})$ is a concrete isomorphism with $\overline{E} = H \circ E$. □

FUNCTORS "INDUCING" CONCRETE CATEGORIES

Many familiar constructs of an "algebraic" or "topological" nature have natural descriptions that can be accomplished in two steps. The first step, which will be formalized below (separately for algebraic and topological cases), consists of defining algebraic (resp. topological) categories by means of certain functors. The second step consists of singling out full, concrete subcategories by imposing certain axioms on the objects (cf. 16.11 and 22.1).

5.37 DEFINITION
Let $T : \mathbf{X} \to \mathbf{X}$ be a functor. $\mathbf{Alg}(T)$ is the concrete category over $\mathbf{X}$, the objects of which (called T-**algebras**) are pairs (X, h) with X an $\mathbf{X}$-object and $h : T(X) \to X$ an $\mathbf{X}$-morphism. Morphisms $f : (X, h) \to (X', h')$ (called T-**homomorphisms**) are $\mathbf{X}$-morphisms $f : X \to X'$ such that the diagram

$$\begin{array}{ccc} T(X) & \xrightarrow{h} & X \\ T(f) \downarrow & & \downarrow f \\ T(X') & \xrightarrow{h'} & X' \end{array}$$

commutes. The underlying functor to $\mathbf{X}$ is given by: $\left|(X, h) \xrightarrow{f} (X', h')\right| = X \xrightarrow{f} X'$.

5.38 EXAMPLES
(1) Consider the squaring functor [3.20(10)] $S^2 : \mathbf{Set} \to \mathbf{Set}$. The construct $\mathbf{Alg}(S^2)$ is obviously the construct of **binary algebras** $\mathbf{Alg}(\Omega_0)$, where Ω_0 consists of the singleton natural number 2 [3.3(2)(e)]. Observe that such familiar constructs as $\mathbf{Sgr}$, $\mathbf{Mon}$, $\mathbf{Grp}$, and $\mathbf{Ab}$ can be considered as subcategories of $\mathbf{Alg}(S^2)$. [Later we shall see that each $\mathbf{Alg}(\Omega)$ is isomorphic to $\mathbf{Alg}(T)$ for some T (10U).]

(2) A unary Σ-algebra [3.3(2)(e)] $(X, \sigma(-))_{\sigma \in \Sigma}$ gives rise to a function $X \times \Sigma \xrightarrow{h} X$ that maps (x, σ) to $\sigma(x)$. Define a functor $(- \times \Sigma) : \mathbf{Set} \to \mathbf{Set}$ as follows. On objects: $(- \times \Sigma)(X) = X \times \Sigma$, and on morphisms: $(- \times \Sigma)(f) = f \times id_\Sigma$. Then the passage from $(X, \sigma(-))_{\sigma \in \Sigma}$ to (X, h) defines a concrete isomorphism

$$F : \mathbf{Alg}(\Sigma) \to \mathbf{Alg}(- \times \Sigma)$$

that is the identity function on morphisms.

(3) Let $\hat{S}^2$: **Pos** → **Pos** be the squaring functor for **Pos**; i.e., $\hat{S}^2(X, \leq) = (X \times X, \leq)$, where the order relation on $X \times X$ is defined coordinatewise, and $\hat{S}^2(f) = f \times f$, for any order-preserving map f. Then the concrete category **Alg**$(\hat{S}^2)$ over **Pos** is concretely isomorphic to the category of ordered binary algebras, considered as a concrete category over **Pos**.

(4) For any category **X** and **X**-object, X, the constant functor C_X : **X** → **X** [3.20(2)] yields the category **Alg**(C_X) which is the comma category $(X \downarrow \mathbf{X})$ of objects under X (see Exercise 3K).

5.39 PROPOSITION
Each concrete category of the form **Alg**(T) *is fibre-discrete.* □

5.40 DEFINITION
Let T : **X** → **Set** be a functor. **Spa**(T) is the concrete category over **X**, the objects of which (called T-**spaces**) are pairs (X, α) with $\alpha \subseteq T(X)$. Morphisms $(X, \alpha) \xrightarrow{f} (Y, \beta)$ (called T-**maps**) are **X**-morphisms f : $X \to Y$ such that $T(f)[\alpha] \subseteq \beta$. The underlying functor to **X** is given by: $|(X, \alpha) \xrightarrow{f} (Y, \beta)| = X \xrightarrow{f} Y$. Concrete categories of the form **Spa**(T) are called **functor-structured categories**.

5.41 EXAMPLES
(1) Consider the squaring functor S^2 : **Set** → **Set** [3.20(10)]. Then the construct **Spa**(S^2) is equal to the construct **Rel**. Observe that such familiar constructs as **Prost** and **Pos** can be considered as full subcategories of **Spa**(S^2).

(2) Any functor T : **1** → **Set** consists of choosing a distinguished set A, and so, for this case, **Spa**(T) is just the power-set of A regarded as a poset or as a concrete category over **1**.

5.42 PROPOSITION
Each concrete category of the form **Spa**(T) *is fibre-complete.* □

5.43 REMARK
We will see later (22.3) that every small-fibred "topological" category $(\mathbf{A}, U)$ over **X** is isomorphic to a full concrete subcategory of **Spa**$(\hat{T})$ for a suitable $\hat{T}$: **X** → **Set**. A more natural representation of $(\mathbf{A}, U)$ is often obtained in terms of a full concrete subcategory of $(\mathbf{Spa}(T))^{\mathrm{op}}$ considered as a concrete category over $\mathbf{X} = (\mathbf{X}^{\mathrm{op}})^{\mathrm{op}}$ for some functor T : $\mathbf{X}^{\mathrm{op}}$ → **Set**. Concrete categories of the form $(\mathbf{Spa}(T))^{\mathrm{op}}$ are called **functor-costructured categories**.

5.44 EXAMPLE
Consider the contravariant power-set functor $\mathcal{Q}$: **Set**^{op} → **Set**. Then the construct $(\mathbf{Spa}(\mathcal{Q}))^{\mathrm{op}}$ has as objects all pairs (X, τ) consisting of a set X and a set τ of subsets of X, where a morphism f : $(X, \tau) \to (Y, \sigma)$ is a function f : $X \to Y$ such that $\mathcal{Q}f[\sigma] \subseteq \tau$. In particular, **Top** is a full subconstruct of $(\mathbf{Spa}(\mathcal{Q}))^{\mathrm{op}}$.

5.45 REMARK

Concrete categories of the form **Spa**(T) can be very large. In fact, every construct **A** can be fully embedded in

(a) **Spa**($\mathcal{P}$), where $\mathcal{P}$ is the covariant power-set functor;

(b) (**Spa**($\mathcal{Q}$))^{op}, where $\mathcal{Q}$ is the contravariant power-set functor;

(c) the concrete category of topological spaces and open, continuous functions.

These and related results can be found in the monograph Pultr-Trnková (see References). However, the corresponding embeddings usually will not be concrete.

Suggestions for Further Reading

Katětov, M. Allgemeine Stetigkeitsstrukturen. *Proceedings of the International Congress of Mathematicians* (Stockholm, 1962), Inst. Mittag-Leffler, Djursholm, 1963, 473–479.

Isbell, J. R. Two set-theoretical theorems in categories. *Fund. Math.* **53** (1963): 43–49.

Barr, M. Coequalizers and free triples. *Math. Z.* **116** (1970): 307–322.

Ehresmann, A., and C. Ehresmann. Categories of sketched structures. *Cahiers Topol. Geom. Diff.* **13** (1972): 105–214.

Kučera, L., and A. Pultr. On a mechanism of defining morphisms in concrete categories. *Cahiers Topol. Geom. Diff.* **13** (1972): 397–410.

Freyd, P. J. Concreteness. *J. Pure Appl. Algebra* **3** (1973): 171–191.

Rosický, J., and M. Sekanina. Realizations of topologies by set systems. *Colloquium on Topology* (Keszthely, 1972), *Colloq. Math. Soc. János Bolyai* **8**, North Holland, Amsterdam, 1974, 535–555.

Koubek, V. Each concrete category has a representation by T_2 paracompact topological spaces. *Comment. Math. Univ. Carolinae* **15** (1976): 655–664.

Porst, H.-E., and W. Tholen Concrete Dualities. *Category Theory at Work* (eds. H. Herrlich and H.-E. Porst). Heldermann Verlag 1991, 111–136.

EXERCISES

5A. Fibres

(a) Show that the category **TopGrp**, as a concrete category over **Top**, is fibre-discrete and, as a concrete category over **Grp**, is fibre-complete. What about the construct **TopGrp**? What about the constructs **HComp** and **Comp** (of compact spaces)?

(b) Which concrete categories have the property that each fibre is a singleton?

5B. Non-concrete Isomorphisms

Show that there is an illegitimate conglomerate of isomorphisms of **Set** onto itself, although the only concrete isomorphism of the construct **Set** onto itself is *id*.

5C. Categories of Concrete Categories

Verify that for each small category **X** there is a category whose objects are all fibre-small concrete categories over **X** and whose morphisms are all concrete functors. Nevertheless, show that the conglomerate of all fibre-small constructs is illegitimate. [In fact, the conglomerate of all full subcategories of **Set** is illegitimate.]

5D. Concrete Functors Between Constructs

(a) Show that there is precisely one concrete functor from **Rel** to **Set** (the forgetful functor) and that there are precisely three concrete functors from **Set** to **Rel**, the discrete functor, the diagonal functor ($FX = (X, \Delta)$, where $\Delta = \{(x,x) \mid x \in X\}$), and the indiscrete functor.

(b) Show that there is precisely one concrete functor from **Set** to **Pos**.

(c) Show that there is no concrete functor from **Set** to **Vec**. Generalize this to other algebraic constructs, and to Σ-**Seq**.

*(d) Show that there are precisely two concrete functors from **Set** to **Top**, but a proper class of concrete functors from **Top** into itself. [Hint: For each cardinal number α let $F_\alpha : $ **Top** $\to$ **Top** assign to each topological space X the space $F_\alpha X$ generated by intersections of α-indexed families of X-open sets.]

5E. Concrete Reflections

(a) Show that no proper subconstruct of **Grp** is concretely reflective (or coreflective). Generalize this to all fibre-discrete concrete categories.

(b) Show that the full subconstruct of **Rel** formed by all objects (X, ρ) without proper cycles (i.e., if $x_1 \rho x_2$, $x_2 \rho x_3$, $\cdots$, $x_{n-1} \rho x_n$, and $x_1 = x_n$, then $x_1 = x_2 = \cdots = x_n$) is reflective, but not concretely reflective.

(c) Show that a reflective, isomorphism-closed, full concrete subcategory $(\mathbf{A}, U)$ of a (non-transportable) concrete category $(\mathbf{B}, V)$ over **X** with a reflector functor $(\mathbf{B}, V) \xrightarrow{R} (\mathbf{A}, U)$ that is concrete need not be concretely reflective. [Hint: Consider the additive monoid of integers as base category **X**; let **B** be the category with $Ob(\mathbf{B}) = \mathbb{N}$ and $\hom_{\mathbf{B}}(n, m) = \begin{cases} \emptyset, & \text{if } m = 0 < n \\ \{(n,m)\}, & \text{otherwise} \end{cases}$; let the functor $\mathbf{B} \xrightarrow{V} \mathbf{X}$ be defined by $V(n,m) = n - m$; and let $(\mathbf{A}, U)$ be the full concrete subcategory of $(\mathbf{B}, V)$ obtained by deleting the object 0.] Cf. Proposition 5.31.

*(d) Show that a reflective (nonfull) concrete subcategory $(\mathbf{A}, U)$ of a transportable concrete category $(\mathbf{B}, V)$ over **X** with a reflector $(\mathbf{B}, V) \xrightarrow{R} (\mathbf{A}, U)$ that is a concrete functor need not be concretely reflective. [Hint: Consider the compositive monoid of all order-preserving endomorphisms of the set $\mathbb{N}$, with its natural order, as base

category **X**, and let $(\mathbf{A}, U)$ (resp. $(\mathbf{B}, V)$) be the concrete category over **X** with objects all pairs $(\mathbb{N}, n)$, with $n \in \mathbb{N}$, and whose morphisms $(\mathbb{N}, n) \xrightarrow{f} (\mathbb{N}, m)$ are all **X**-morphisms $\mathbb{N} \xrightarrow{f} \mathbb{N}$ that satisfy $f(n+p) = m+p$ for all $p \in \mathbb{N}$ (resp. $f(n+p) = m+p$ for all $p \geq 1$ and $f(n) \leq m$).] Cf. Proposition 5.31.

5F. Amnestic Modification

(a) Describe the amnestic modification of the construct $\mathbf{Met_u}$ of metric spaces and uniformly continuous maps.

(b) What are the amnestic modifications for concrete categories over **1**?

5G. Categories of *T*-Algebras

(a) Describe $\mathbf{Alg}(T)$ for the constant functor $T: \mathbf{Set} \to \mathbf{Set}$

   (1) with value $\emptyset$,

   (2) with value $1 = \{0\}$.

(b) Find a functor $T: \mathbf{Set} \to \mathbf{Set}$ such that $\mathbf{Alg}(T)$ is concretely isomorphic to the category of commutative binary algebras (i.e., algebras with a binary operation $\cdot$ satisfying $x \cdot y = y \cdot x$).

(c) Describe $\mathbf{Alg}(id)$.

(d) Let $T: \mathbf{Pos} \to \mathbf{Pos}$ be the functor that assigns to each poset $(X, \leq)$ the discretely ordered set, and is defined on morphisms by: $Tf = f$. Describe $\mathbf{Alg}(T)$.

5H. Categories of *T*-Spaces

(a) Describe $\mathbf{Spa}(T)$ for the constant functor $T: \mathbf{Set} \to \mathbf{Set}$

   (1) with value $\emptyset$,

   (2) with value $1 = \{0\}$.

(b) Find a functor $T: \mathbf{Set} \to \mathbf{Set}$ such that $\mathbf{Spa}(T)$ is concretely isomorphic to the construct of symmetric binary relations (i.e., if x is related to y, then y must be related to x).

(c) Show that there is no functor $T: \mathbf{Set} \to \mathbf{Set}$ such that $\mathbf{Spa}(T)$ is concretely isomorphic to the construct of reflexive binary relations.

5I. Cat as a Concrete Category

Cat can be viewed in various ways as a concrete category. Depending on the chosen definition of category (cf. 3.1, 3.53 and 3C) and on the chosen forgetful functor, various situations arise:

(a) Show that by assuming any of the two definitions of categories 3.1 or 3C, **Cat** can be considered as a concrete category over $\mathbf{Set} \times \mathbf{Set}$ via the functor $U: \mathbf{Cat} \to \mathbf{Set} \times \mathbf{Set}$, defined by

$$U(\mathbf{A} \xrightarrow{F} \mathbf{B}) = (Ob(\mathbf{A}) \xrightarrow{F_O} Ob(\mathbf{B}), Mor(\mathbf{A}) \xrightarrow{F_M} Mor(\mathbf{B})),$$

(where F_O is the restriction of F to objects and F_M is its restriction to morphisms), and that in both cases U is uniquely transportable.

(b) Show that by assuming any of the three definitions of categories, **Cat** can be considered as construct via the functor $U : \mathbf{Cat} \to \mathbf{Set}$, defined by $U(\mathbf{A} \xrightarrow{F} \mathbf{B}) = Mor(\mathbf{A}) \xrightarrow{F_M} Mor(\mathbf{B})$. In the cases of Definition 3.1 and of Exercise 3C, U is transportable, but not amnestic; and in the case of Definition 3.53, U is amnestic and transportable. [Thus, when considered as a construct, $\mathbf{Cat}_{of}$ is usually preferred to **Cat**.]

5J. Concretizable Categories

A category **A** is called **concretizable over X** provided that there exists a faithful functor from **A** to **X**. Verify that

(a) **A** is concretizable over **1** if and only if **A** is thin.

(b) $\mathbf{Set}^{\mathrm{op}}$ is concretizable over **Set**.

(c) If **A** is concretizable over **Set**, then so is $\mathbf{A}^{\mathrm{op}}$.

* (d) There exist categories that are not concretizable over **Set**, e.g., the category **hTop**, whose objects are topological spaces and whose morphisms are homotopy equivalence classes of continuous maps.

(e) **A** is concretizable over **Set** if and only if **A** is embeddable into **Set**.

(f) There exist categories **A** that are not concretizable over $\mathbf{A}^{\mathrm{op}}$.

* 5K. Subconstructs That Are Simultaneously Reflective and Coreflective Modifications

Show that the construct **Rel** has precisely six subconstructs that are simultaneously reflective and coreflective modifications and that the five proper ones can be described by the following implications:

(1) $x\rho y \Rightarrow y\rho x$ (symmetry),

(2) $x\rho y \Rightarrow x\rho x$,

(3) $x\rho y \Rightarrow y\rho y$,

(4) $x\rho y \Rightarrow (x\rho x \text{ and } y\rho y)$,

(5) $x\rho y \Rightarrow (x\rho x \text{ and } y\rho x)$.

5L. The Constructs $\mathbf{Top}_0^{\mathrm{op}}$ and Fram

(a) If $\mathbf{Top}_0$ denotes the construct of T_0 topological spaces, show that $\mathbf{Top}_0^{\mathrm{op}}$ can be considered as a construct via the functor $L : \mathbf{Top}_0^{\mathrm{op}} \to \mathbf{Set}$ that forgets **not** the "structure" but rather the underlying set; i.e., for a continuous function $(Y,\sigma) \xrightarrow{f} (X,\tau)$, $L((X,\tau) \xrightarrow{f} (Y,\sigma)) = \tau \xrightarrow{Lf} \sigma$, where $Lf(A) = f^{-1}[A]$.

(b) Let **Fram** be the construct whose objects are **frames**, i.e., complete lattices that satisfy the law $a \wedge \bigvee_{i \in I} b_i = \bigvee_{i \in I} (a \wedge b_i)$, and whose morphisms are **frame-homomorphisms**, i.e., functions that preserve finite meets and arbitrary joins. Show that there is a unique concrete functor $\mathbf{Top}_0^{\mathrm{op}} \xrightarrow{T} \mathbf{Fram}$ over **Set** that sends each topological space to its set of open subsets, ordered by inclusion.

[**Loc**=**Fram**^{op} is called the category of locales. The functor $T^{\mathrm{op}} : \mathbf{Top}_0 \to \mathbf{Loc}$ is a nonfull coreflective embedding, and its restriction to sober spaces is a full coreflective embedding].

5M. HComp^{op} as a Construct

Show that **HComp**^{op} can be considered as a construct via the contravariant hom-functor $\hom(-,[0,1]) : \mathbf{HComp}^{\mathrm{op}} \to \mathbf{Set}$.

5N. Concrete Isomorphisms

(a) Show that the following constructs are concretely isomorphic:

 (1) **MCPos** (complete lattices and meet-preserving maps),

 (2) **JCPos**,

 (3) the full subconstruct of $\mathbf{Alg}(\mathcal{P})$ consisting of those $\mathcal{P}$-algebras (X, h) that satisfy the following two conditions:

   (i) $h(\{x\}) = x$ for each $x \in X$,

   (ii) $h(\cup \mathcal{A}) = h(\{h(A) \mid A \in \mathcal{A}\})$ for each $\mathcal{A} \subseteq \mathcal{P} X$.

(b) Show that the three constructs defined below are pairwise concretely isomorphic and therefore provide three different approaches to topology.

 (1) Approach via closure; the construct **Clos**:

   Objects are **closure spaces**; i.e., pairs (X, cl) with $cl : \mathcal{P}(X) \to \mathcal{P}(X)$ a function that satisfies:

   (i) $cl(\emptyset) = \emptyset$,

   (ii) $A \subseteq cl(A)$ for each $A \in \mathcal{P}(X)$,

   (iii) $cl(A \cup B) = cl(A) \cup cl(B)$ for each $A, B \subseteq X$.

   Morphisms are **closure-preserving maps**; i.e., functions $(X, cl) \xrightarrow{f} (X', cl')$ such that $f[cl(A)] \subseteq cl'(f[A])$ for each $A \subseteq X$.

 (2) Approach via convergence; the construct **PrTop**:

   Objects are **pretopological spaces**; i.e., pairs $(X, conv)$, where $conv$ is the relation that shows which filters converge to which points of X, subject to the following conditions:

   (i) for every $x \in X$ the fixed ultrafilter at x ($\dot{x}$) converges to x,

   (ii) if a filter converges to x, then so does every finer filter,

(iii) if each member of a family of filters converges to x, then so does the intersection of the family.

Morphisms are **convergence-preserving maps**; i.e., functions $(X, conv) \xrightarrow{f} (X', conv')$ such that whenever $\mathcal{F}$ converges to x, then the filter generated by $f[\mathcal{F}]$ converges to $f(x)$.

(3) Approach via neighborhoods; the construct **Neigh**:

Objects are **neighborhood spaces**; i.e., pairs $(X, \mathcal{N})$, where $\mathcal{N}$ associates with any $x \in X$ a filter $\mathcal{N}(x)$, the neighborhood filter of x, subject to the condition that if U is a neighborhood of x then x is a member of U.

Morphisms are all functions $(X, \mathcal{N}) \xrightarrow{f} (X', \mathcal{N}')$ such that whenever U is a neighborhood of $f(x)$, then $f^{-1}[U]$ is a neighborhood of x.

5O. Realizations

Full concrete embeddings are called **realizations**. Show that

(a) There are precisely two realizations from **Pos** to **Rel**.

* (b) There are precisely two realizations from **Top** to **Spa**$(\mathcal{Q})^{\mathrm{op}}$.

(c) There is a realization from **Alg**(1) to **Rel**.

(d) There is a realization from **Sgr** to **Spa**(S^3).

(e) For each Ω, there is a realization from **Alg**(Ω) to some functor-structured category.

(f) There is a realization from **Prost** to **Top**.

(g) There is no realization from **Rel** to **Sgr**.

(h) There is a proper class of realizations from **Tych** to **Spa**$(\mathcal{Q})^{\mathrm{op}}$.

(i) There are at least two realizations from **Haus** to **Spa**$(\mathcal{Q})^{\mathrm{op}}$. [It is unknown whether there are more than two.]

(j) There is a no fibre-small construct $(\mathbf{A}, U)$ such that every fibre-small construct has a realization to $(\mathbf{A}, U)$.

* (k) There is a construct $(\mathbf{A}, U)$ such that every construct has a realization to $(\mathbf{A}, U)$.

5P. Amnesticity

Show that a concrete category $(\mathbf{A}, U)$ over $\mathbf{X}$ is amnestic if and only if each $\mathbf{A}$-isomorphism f is an $\mathbf{A}$-identity whenever Uf is an $\mathbf{X}$-identity.

6 Natural transformations

Let V be a finite-dimensional real vector space and let $\hat{V}$ be its dual (i.e., the set of all linear functionals $V \to \mathbb{R}$ with vector-space operations defined pointwise). V and $\hat{V}$ are known to be isomorphic. Hence V and its second dual $\hat{\hat{V}}$ are isomorphic as well. However, there is a fundamental difference between these two situations. There is a "natural" isomorphism $\tau : V \to \hat{\hat{V}}$ which to every vector x assigns the "evaluate at x" functional $\tau(x) : \hat{V} \to \mathbb{R}$. But there is no "natural" isomorphism between V and $\hat{V}$. This section provides a formal definition for the intuitive notion of "natural isomorphism" and (more generally) of "natural transformation".

6.1 DEFINITION
Let $F, G : \mathbf{A} \to \mathbf{B}$ be functors. A **natural transformation** τ from F to G (denoted by $\tau : F \to G$ or $F \xrightarrow{\tau} G$) is a function that assigns to each **A**-object A a **B**-morphism $\tau_A : FA \to GA$ in such a way that the following **naturality condition** holds: for each **A**-morphism $A \xrightarrow{f} A'$, the square

$$\begin{array}{ccc} FA & \xrightarrow{\tau_A} & GA \\ {\scriptstyle Ff}\downarrow & & \downarrow{\scriptstyle Gf} \\ FA' & \xrightarrow{\tau_{A'}} & GA' \end{array}$$

commutes.

6.2 EXAMPLES
(1) Let **A** be a reflective subcategory of **B** with inclusion functor E, let $B \xrightarrow{r_B} RB$ be an **A**-reflection arrow for each **B**-object B, and let $R : \mathbf{B} \to \mathbf{A}$ be the associated reflector (4.23). Then $r = (r_B)_{B \in Ob(\mathbf{B})}$ is a natural transformation: $id_\mathbf{B} \xrightarrow{r} E \circ R$.

(2) Let $U : \mathbf{Grp} \to \mathbf{Set}$ be the forgetful functor, and let $S : \mathbf{Grp} \to \mathbf{Set}$ be the "squaring-functor", defined by $S(G \xrightarrow{f} H) = G^2 \xrightarrow{f^2} H^2$. For each group G, its multiplication is a function $\tau_G : G^2 \to G$. The family $\tau = (\tau_G)$ is a natural transformation from S to U. The naturality condition simply means that $f(x \cdot y) = f(x) \cdot f(y)$ for any group homomorphism $G \xrightarrow{f} H$ and any $x, y \in G$. Thus "multiplication" in groups can be regarded as a natural transformation. Likewise, for any type of algebras, each of the defining operations can be considered as a natural transformation between suitable functors.

(3) Let $(\hat{\ }) : \mathbf{Vec} \to \mathbf{Vec}$ be the **second-dual functor for vector spaces** defined by

$$\mathbf{Vec} \xrightarrow{(\hat{\ })} \mathbf{Vec} = (\mathbf{Vec}^{op})^{op} \xrightarrow{(\hat{\ })^{op}} \mathbf{Vec}^{op} \xrightarrow{(\hat{\ })} \mathbf{Vec},$$

where $(\hat{\ })^{op}$ is the dual of the duality functor for vector spaces (cf. 3.20(12) and 3.41), and let $id_{\mathbf{Vec}}$ be the identity functor on **Vec**. Then the linear transformations $\tau_V : V \to \hat{V}$, defined by $(\tau_V(x))(f) = f(x)$, yield a natural transformation $id_{\mathbf{Vec}} \xrightarrow{\tau} (\hat{\ })$.

(4) The assignment of the Hurewicz homomorphism $\pi_n(X) \to H_n(X)$ to each topological space X is a natural transformation from the nth homotopy functor $\pi_n : \mathbf{Top} \to \mathbf{Grp}$ to the nth homology functor $H_n : \mathbf{Top} \to \mathbf{Grp}$.

(5) If $B \xrightarrow{f} C$ is an **A**-morphism, then

$$\hom_{\mathbf{A}}(C, -) \xrightarrow{\tau_f} \hom_{\mathbf{A}}(B, -),$$

defined by $\tau_f(g) = g \circ f$, and

$$\hom_{\mathbf{A}}(-, B) \xrightarrow{\sigma_f} \hom_{\mathbf{A}}(-, C),$$

defined by $\sigma_f(g) = f \circ g$, are natural transformations.

(6) Let $U : \Sigma\text{-}\mathbf{Seq} \to \mathbf{Set}$ be the forgetful functor. For each $\sigma \in \Sigma$, and each acceptor $A = (Q, \delta, q_0, F)$, let $\hat{\sigma}_A$ be the function $\delta(-, \sigma) : Q \to Q$. Then $\hat{\sigma} = (\hat{\sigma}_A) : U \to U$ is a natural transformation.

6.3 DEFINITION

If $G, G' : \mathbf{A} \to \mathbf{B}$ are functors and $G \xrightarrow{\tau} G'$ is a natural transformation, then

(1) for each functor $F : \mathbf{C} \to \mathbf{A}$, the natural transformation $\tau F : G \circ F \to G' \circ F$ is defined by

$$(\tau F)_C = \tau_{FC},$$

(2) for each functor $H : \mathbf{B} \to \mathbf{D}$, the natural transformation $H\tau : H \circ G \to H \circ G'$ is defined by

$$(H\tau)_A = H(\tau_A).$$

Likewise the natural transformation $G'^{op} \xrightarrow{\tau^{op}} G^{op}$ is defined by

$$\tau_A^{op} = \tau_A.$$

6.4 EXAMPLE

If $S^2 : \mathbf{Set} \to \mathbf{Set}$ is the squaring functor [3.20(10)] and $\Delta : id \to S^2$ is the natural transformation that associates with every set X the diagonal map $\Delta_X : X \to X^2$ given by $x \mapsto (x, x)$, then

(1) $S^2\Delta : S^2 \to S^2 \circ S^2$ is given by $(x, y) \mapsto ((x, x), (y, y))$,

(2) $\Delta S^2 : S^2 \to S^2 \circ S^2$ is given by $(x, y) \mapsto ((x, y), (x, y))$.

NATURAL ISOMORPHISMS

6.5 DEFINITION
Let $F, G : \mathbf{A} \to \mathbf{B}$ be functors.

(1) A natural transformation $F \xrightarrow{\tau} G$ whose components τ_A are isomorphisms is called a **natural isomorphism** from F to G.

More generally, a natural transformation from F to G whose components belong to some specified class M of **B**-morphisms is called an M-**transformation**.

(2) F and G are said to be **naturally isomorphic** (denoted by $F \cong G$) provided that there exists a natural isomorphism from F to G.[33]

6.6 EXAMPLES
(1) For each functor $F : \mathbf{A} \to \mathbf{B}$ we have the **identity natural transformation on F**, $id_F : F \to F$ given by $(id_F)_A = id_{FA}$, which is clearly a natural isomorphism.

(2) Let $F : \mathbf{Set} \to \mathbf{Vec}$ be a functor that assigns to each set X a vector space FX with basis X, and to each function $X \xrightarrow{f} Y$ the unique linear extension $FX \xrightarrow{Ff} FY$ of f. This actually is not a correct definition of a functor, since there are many different vector spaces with the same basis. However, the definition is "correct up to natural isomorphism". Whenever we choose, for each set X, a specific vector space FX with basis X, we do obtain a functor $F : \mathbf{Set} \to \mathbf{Vec}$ (since the above condition determines the action of F on functions uniquely). Furthermore, any two functors that are obtained in this way are naturally isomorphic.

(3) The natural transformation $\tau : id_{\mathbf{Vec}} \to (\hat{})$ from the identity functor on **Vec** to the second-dual functor for vector spaces $(\hat{})$ given in Example 6.2(3) becomes a natural isomorphism when the above functors are restricted to the full subcategory of finite-dimensional vector spaces.

(4) For any 2-element set A, $\hom(A, -)$ is naturally isomorphic to the squaring functor S^2 [3.20(10)] and $\hom(-, A)$ is naturally isomorphic to the contravariant power-set functor $\mathcal{Q}$ [3.20(9)].

(5) If f is a morphism and τ_f and σ_f are the associated natural transformations for the hom-functors [6.2(5)], then the following are equivalent:

  (a) f is an isomorphism,

  (b) σ_f is a natural isomorphism,

  (c) τ_f is a natural isomorphism.

Thus, if A and B are isomorphic objects, then $\hom(A, -)$ and $\hom(B, -)$ are naturally isomorphic functors, and so are $\hom(-, A)$ and $\hom(-, B)$. The converse holds as well (cf. Exercise 6M).

---

[33] Observe that the relation $\cong$ is an equivalence relation on the conglomerate of all functors from $\mathbf{A}$ to $\mathbf{B}$.

6.7 PROPOSITION

If **A** is a reflective subcategory of **B**, then any two reflectors for **A** are naturally isomorphic.

Proof: Let R and S be reflectors for **A** with associated reflection arrows $B \xrightarrow{r_B} RB$ and $B \xrightarrow{s_B} SB$. Then there exist **A**-morphisms $RB \xrightarrow{f_B} SB$ and $SB \xrightarrow{g_B} RB$ such that the diagrams

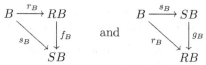

and

commute. Uniqueness in the definition of reflection arrows implies that $g_B \circ f_B = id_{RB}$ and $f_B \circ g_B = id_{SB}$. Hence the f_B's are **A**-isomorphisms. That $(f_B)_{B \in Ob(\mathbf{A})}$ is a natural transformation (hence a natural isomorphism) follows from the fact that for any **B**-morphism $B \xrightarrow{f} B'$ the following diagrams $\boxed{1}$ and $\boxed{2}$ — hence, by the uniqueness property for $B \xrightarrow{r_B} RB$, also $\boxed{3}$ — commute:

$$
\begin{array}{ccc}
B \xrightarrow{s_B} SB & & B \xrightarrow{r_B} RB \xrightarrow{f_B} SB \\
f \downarrow \; \boxed{1} \; \downarrow Sf & = & f \downarrow \; \boxed{2} \; \downarrow Rf \; \boxed{3} \; \downarrow Sf \\
B' \xrightarrow{s_{B'}} SB' & & B' \xrightarrow{r_{B'}} RB' \xrightarrow{f_{B'}} SB'
\end{array}
$$

In Proposition 3.36 we have seen that if $\mathbf{A} \xrightarrow{F} \mathbf{B}$ is an equivalence, then there is an equivalence $\mathbf{B} \xrightarrow{G} \mathbf{A}$. Now that we have the notion of natural isomorphism, we are able to sharpen this result:

6.8 PROPOSITION

A functor $\mathbf{A} \xrightarrow{F} \mathbf{B}$ is an equivalence if and only if there exists a functor $\mathbf{B} \xrightarrow{G} \mathbf{A}$ such that $id_{\mathbf{A}} \cong G \circ F$ and $F \circ G \cong id_{\mathbf{B}}$.

Proof: Let F be an equivalence. By following the proof of Proposition 3.36 we see that there is a functor G and a natural isomorphism $\varepsilon : F \circ G \to id_{\mathbf{B}}$. Now for each **A**-object A, $\varepsilon_{FA}^{-1} : FA \to (F \circ G \circ F)(A)$ is an isomorphism. Since F is full and faithful, Proposition 3.31 implies that there is a unique isomorphism $\eta_A : A \to (G \circ F)(A)$ such that $F(\eta_A) = \varepsilon_{FA}^{-1}$. The naturality of η follows from that of ε^{-1} (cf. Exercise 6L) and the faithfulness of F.

Conversely, let $G : \mathbf{B} \to \mathbf{A}$ be a functor and let $id_{\mathbf{A}} \xrightarrow{\eta} G \circ F$ and $F \circ G \xrightarrow{\varepsilon} id_{\mathbf{B}}$ be natural isomorphisms. Since $F(GB) \xrightarrow{\varepsilon_B} B$ is an isomorphism for any **B**-object B, it follows that F is isomorphism-dense. F is faithful, since for any pair $A \underset{g}{\overset{f}{\rightrightarrows}} A'$ of **B**-morphisms the equality $Ff = Fg$ implies $\eta_{A'} \circ f = GFf \circ \eta_A = GFg \circ \eta_A = \eta_{A'} \circ g$; hence $f = g$. F is full, since for any **B**-morphism $f : FA \to FA'$, $\eta_{A'}^{-1} \circ Gf \circ \eta_A$ is an **A**-morphism $g : A \to A'$ with $Fg = f$. □

6.9 DEFINITION
A functor $F : \mathbf{A} \to \mathbf{Set}$ is called **representable** (by an **A**-object A) provided that F is naturally isomorphic to the hom-functor $\hom(A, -) : \mathbf{A} \to \mathbf{Set}$.

6.10 REMARK
Objects that represent the same functor (or two naturally isomorphic functors) are isomorphic. (Cf. Exercise 6M.) □

6.11 EXAMPLES
(1) Forgetful functors are often representable. For example,

  (a) **Vec** $\to$ **Set** is represented by the vector space $\mathbb{R}$,

  (b) **Grp** $\to$ **Set** is represented by the group of integers $\mathbb{Z}$,

  (c) **Top** $\to$ **Set** is represented by any one-point topological space.

(2) The underlying functor U for the construct **Ban** [5.2(3)] is not representable (see Exercise 10J). However, the faithful unit ball functor $O : \mathbf{Ban} \to \mathbf{Set}$ is represented in the complex case by the Banach space $\mathbb{C}$ of complex numbers.

(3) The forgetful functor for Σ-**Seq** is not representable (since for each (finite) acceptor A there exists a nonempty acceptor B with $\hom(A, B) = \emptyset$).

6.12 REMARK
For constructs $(\mathbf{A}, U)$ the forgetful functor is represented by an object A if and only if A is a *free object* over a singleton set [see Definition 8.22(2)]. This provides many additional examples of representations.

FUNCTOR CATEGORIES

6.13 DEFINITION
If $F, G, H : \mathbf{A} \to \mathbf{B}$ are functors and $F \xrightarrow{\sigma} G$ and $G \xrightarrow{\tau} H$ are natural transformations, then the **composition of natural transformations** $\tau \circ \sigma : F \to H$ is the natural transformation that assigns to each **A**-object A the morphism $\tau_A \circ \sigma_A : F(A) \to H(A)$.

6.14 REMARK
It is obvious that the composition of natural transformations is a natural transformation, that this composition is associative, and that the identity natural transformations act as units.

6.15 DEFINITION
For categories **A** and **B** the **functor quasicategory** $[\mathbf{A}, \mathbf{B}]$ has as objects all functors from **A** to **B**, as morphisms from F to G all natural transformations from F to G, as identities the identity natural transformations, and as composition the composition of natural transformations given above.

6.16 REMARKS

(1) If **A** and **B** are small categories, then [**A**, **B**] is a category. If **A** is small and **B** is large, then [**A**, **B**], though being a proper quasicategory, is isomorphic to a category. Quasicategories that are isomorphic to categories are called **legitimate quasicategories** and are treated as categories. If **A** and **B** are both large, then [**A**, **B**] will generally fail to be isomorphic to a category. Such quasicategories are called **illegitimate**.

(2) A natural transformation between functors from **A** to **B** is a natural isomorphism if and only if it is an isomorphism in [**A**, **B**].

6.17 EXAMPLES

(1) If **A** is a discrete category with one object, then [**A**, **B**] is isomorphic to **B**.

(2) If **A** is a discrete category with two objects, then [**A**, **B**] is isomorphic to **B** × **B**.

(3) If **2** is a category of the form • → • with two objects and one non-identity morphism, then [**2**, **B**] is isomorphic to the arrow category of **B** (cf. Exercise 3K).

(4) If M is a monoid, considered as a category, then [M, **Set**] is isomorphic to the category M-**Act** of M-actions.

6.18 PROPOSITION

*For any functor $F :$ **A** $\to$ **Set**, any **A**-object A and any element $a \in F(A)$, there exists a unique natural transformation $\tau :$ hom$(A, -) \to F$ with $\tau_A(id_A) = a$.*

Proof: Let $\tau_B(f) = (F(f))(a)$. Pointwise evaluations establish that τ is a natural transformation. If $\delta :$ hom$(A, -) \to F$ is such that $\delta_A(id_A) = a$, then by the naturality of δ, $\delta_B(f) = \delta_B(f \circ id_A) = (\delta_B \circ \text{hom}(A, f))(id_A) = (F(f) \circ \delta_A)(id_A) = F(f)(a) = \tau_B(f)$. □

6.19 COROLLARY (YONEDA LEMMA)

*If $F :$ **A** $\to$ **Set** is a functor and A is an **A**-object, then the following function*

$$Y : [\text{hom}(A, -), F] \to F(A) \quad \text{defined by} \quad Y(\sigma) = \sigma_A(id_A),$$

is a bijection (where $[\text{hom}(A, -), F]$ is the set of all natural transformations from $\text{hom}(A, -)$ to F). □

6.20 THEOREM

*For any category **A**, the functor $E :$ **A** $\to$ [**A**^{op}, **Set**], defined by*

$$E(A \xrightarrow{f} B) = \text{hom}(-, A) \xrightarrow{\sigma_f} \text{hom}(-, B),$$

where $\sigma_f(g) = f \circ g$, is a full embedding.

Proof: The described assignment clearly preserves identities and composition. Thus it is a functor. If f and f' are distinct members of hom(A, B), then σ_f and $\sigma_{f'}$ clearly differ on id_A. Hence E is faithful. Fullness follows from Corollary 6.19 with $F = \text{hom}(B, -)$. □

6.21 EXAMPLES

(1) If M is a monoid considered as a category, then the unique object assigned to it by the above embedding is the M^{op}-action on the set M defined by right translations in M.

(2) If **A** is a poset considered as a thin category, then the above embedding is the representation of **A** by all the principal ideals in **A**.

CONCRETE NATURAL TRANSFORMATIONS AND GALOIS CORRESPONDENCES

6.22 REMARK

From now on, when investigating concrete categories, we typically will use the notational conventions described in Remark 5.3. In particular,

(1) we will denote a concrete category $(\mathbf{A}, U)$ over **X** by **A** alone and denote the underlying functor U by $|\ |$,

(2) the expression "$|A| \xrightarrow{f} |B|$ is an **A**-morphism" means that for the **X**-morphism $|A| \xrightarrow{f} |B|$ there exists a (necessarily unique) **A**-morphism $A \to B$, which will also be denoted by f, with $|A \to B| = |A| \xrightarrow{f} |B|$.

6.23 DEFINITION

If **A** and **B** are concrete categories over **X** and $F, G : \mathbf{A} \to \mathbf{B}$ are concrete functors (5.9), then a natural transformation $\tau : F \to G$ is called **concrete** (or **identity-carried**) provided that $|\tau_A| = id_{|A|}$ for each **A**-object A.

6.24 PROPOSITION

If $F, G : \mathbf{A} \to \mathbf{B}$ are concrete functors, then the following are equivalent:

(1) $F \leq G$ (5.18),

(2) there exists a (necessarily unique) concrete natural transformation $\tau : F \to G$. □

6.25 DEFINITION

Let **A** and **B** be concrete categories over **X**. If $G : \mathbf{A} \to \mathbf{B}$ and $F : \mathbf{B} \to \mathbf{A}$ are concrete functors over **X**, then the pair (F, G) is called a **Galois correspondence** (between **A** and **B** over **X**) provided that $F \circ G \leq id_{\mathbf{A}}$ and $id_{\mathbf{B}} \leq G \circ F$.

6.26 EXAMPLES

(1) **Galois isomorphisms**: If $K : \mathbf{A} \to \mathbf{B}$ is a concrete isomorphism, then (K^{-1}, K) is a Galois correspondence, called a **Galois isomorphism**.

(2) **Galois reflections and coreflections**:

(a) If $E : \mathbf{A} \to \mathbf{B}$ is a concrete embedding and $R : \mathbf{B} \to \mathbf{A}$ is a concrete reflector, then (R, E) is a Galois correspondence, called a **Galois reflection**.

(b) If $E : \mathbf{A} \to \mathbf{B}$ is a concrete embedding and $C : \mathbf{B} \to \mathbf{A}$ is a concrete coreflector, then (E, C) is a Galois correspondence, called a **Galois coreflection**.

(3) **Galois correspondences for constructs**:

(a) Let $U : \mathbf{Top} \to \mathbf{Set}$ be the forgetful functor, let $D : \mathbf{Set} \to \mathbf{Top}$ be the discrete functor, and let $N : \mathbf{Set} \to \mathbf{Top}$ be the indiscrete functor [3.29(5) and (6)]. Then (D, U) and (U, N) are both Galois correspondences.

(b) Let $G : \mathbf{Top} \to \mathbf{Rel}$ be the concrete functor (over $\mathbf{Set}$) defined on objects by: $(X, \tau) \mapsto (X, \rho_\tau)$, where $x \rho_\tau y$ if and only if x is in the τ-closure of $\{y\}$. Let $F : \mathbf{Rel} \to \mathbf{Top}$ be the concrete functor defined on objects by $(X, \rho) \mapsto (X, \tau_\rho)$, where A is τ_ρ-closed if and only if A is a **lower-set** for ρ [i.e., $a \in A$ and $a' \rho a \Rightarrow a' \in A$]. Then (F, G) is a Galois correspondence.

(c) If $U : \mathbf{Unif} \to \mathbf{Top}$ is the forgetful functor that assigns to any uniform space the completely regular space underlying it, and $F : \mathbf{Top} \to \mathbf{Unif}$ is the **fine functor** that assigns to any topological space the fine uniform space determined by it, then (F, U) is a Galois correspondence.

(4) **Galois connections**:

(a) Recall that when $\mathbf{X} = \mathbf{1}$, then concrete categories over $\mathbf{X}$ are essentially pre-ordered classes and concrete functors between them are essentially order-preserving functions. Historically, a Galois connection between preordered classes A and B has been defined as a pair (f, g) of order-preserving[34] functions $g : A \to B$ and $f : B \to A$ with the property that for all $a \in A$ and $b \in B$, $f(b) \leq a$ if and only if $b \leq g(a)$. Notice that the latter condition is equivalent to: $(f \circ g)(a) \leq a$ for all $a \in A$ and $b \leq (g \circ f)(b)$ for all $b \in B$. Thus **Galois connections** are just Galois correspondences with the base category $\mathbf{X} = \mathbf{1}$. (See also Proposition 6.28 below.)

(b) Important special cases of Galois connections arise from (binary) relations:

Let ρ be a relation from the set X to the set Y, i.e., $\rho \subseteq X \times Y$. Denote by A the poset of all subsets of X ordered by inverse inclusion and by B the poset of all subsets of Y ordered by inclusion. Then the following maps yield a Galois connection:

$$g : A \to B, \quad \text{defined by} \quad g(S) = \{y \in Y \mid s \rho y \text{ for all } s \in S\}$$

$$f : B \to A, \quad \text{defined by} \quad f(T) = \{x \in X \mid x \rho t \text{ for all } t \in T\}.$$

---

[34] Many of the earlier definitions of Galois connections deal only with the case of partially ordered sets and they usually adopt a "contravariant" formulation as follows: If $g : A \to B$ and $f : B \to A$ are order-reversing functions between posets, then (f, g) is a (contravariant) Galois connection provided that $a \leq f(b)$ if and only if $b \leq g(a)$. Notice that this is the same as the formulation above if the order on A is reversed. The order-preserving version corresponds to what are frequently called residuated-residual pairs or (sometimes) Galois connections of mixed type.

6.27 PROPOSITION

(1) If (F, G) is a Galois correspondence between $\mathbf{A}$ and $\mathbf{B}$ and $(\hat{F}, \hat{G})$ is a Galois correspondence between $\mathbf{B}$ and $\mathbf{C}$, then $(F \circ \hat{F}, \hat{G} \circ G)$ is a Galois correspondence between $\mathbf{A}$ and $\mathbf{C}$ [sometimes denoted by $(\hat{F}, \hat{G}) \circ (F, G)$].

(2) If (F, G) is a Galois correspondence between $\mathbf{A}$ and $\mathbf{B}$ over $\mathbf{X}$, then $(G^{\text{op}}, F^{\text{op}})$ is a Galois correspondence between $\mathbf{B}^{\text{op}}$ and $\mathbf{A}^{\text{op}}$ over $\mathbf{X}^{\text{op}}$.

Proof:
(1). For any $\mathbf{A}$-object A we obtain $\hat{F}\hat{G}(GA) \leq GA$. Application of the functor F yields $F\hat{F}\hat{G}G(A) \leq FGA$. Since $FGA \leq A$, transitivity of $\leq$ yields $(F\hat{F})(\hat{G}G)(A) \leq A$. Likewise $B \leq (\hat{G}G)(F\hat{F})B$ for each $\mathbf{B}$-object B.

(2). Obvious. □

6.28 PROPOSITION

Let $G : \mathbf{A} \to \mathbf{B}$ and $F : \mathbf{B} \to \mathbf{A}$ be concrete functors over $\mathbf{X}$. Then the following are equivalent:

(1) (F, G) is a Galois correspondence,

(2) an $\mathbf{X}$-morphism $|F(B)| \xrightarrow{f} |A|$ is an $\mathbf{A}$-morphism if and only if $|B| \xrightarrow{f} |G(A)|$ is a $\mathbf{B}$-morphism.

Proof:
(1) $\Rightarrow$ (2). If $|FB| \xrightarrow{f} |A|$ is an $\mathbf{A}$-morphism, then by applying G and using $id_{\mathbf{B}} \leq G \circ F$, one has that $|B| \xrightarrow{id_{|B|}} |(G \circ F)(B)| \xrightarrow{f} |GA|$ is a $\mathbf{B}$-morphism. Conversely, the facts that $|B| \xrightarrow{f} |GA|$ is a $\mathbf{B}$-morphism and that $F \circ G \leq id_{\mathbf{A}}$ imply that $|FB| \xrightarrow{f} |(F \circ G)(A)| \xrightarrow{id_{|A|}} |A|$ is an $\mathbf{A}$-morphism.

(2) $\Rightarrow$ (1). Since each $|FB| \xrightarrow{id_{|FB|}} |FB|$ is an $\mathbf{A}$-morphism, from (2) we see that $B \leq (G \circ F)(B)$ for each $\mathbf{B}$-object B. Similarly, $(F \circ G)(A) \leq A$ for each $\mathbf{A}$-object A. Thus (F, G) is a Galois correspondence. □

6.29 PROPOSITION

The functors in a Galois correspondence between amnestic concrete categories determine each other uniquely; in particular, if (F, G) and (F', G) are such Galois correspondences, then $F = F'$.

Proof: Let B be a $\mathbf{B}$-object. If (F, G) is a Galois correspondence, $B \leq (G \circ F)(B)$, so that by the above proposition, if (F', G) is a Galois correspondence, $F'B \leq FB$. Similarly, $FB \leq F'B$, so that by amnesticity, $F = F'$. [Dually, it can be shown that if each of (F, G) and (F, G') are Galois correspondences, then $G = G'$.] □

6.30 PROPOSITION

If (F,G) is a Galois correspondence between amnestic concrete categories, then $G \circ F \circ G = G$ and $F \circ G \circ F = F$.

Proof: Clearly, $id_\mathbf{B} \leq G \circ F$ implies $F = F \circ id_\mathbf{B} \leq F \circ G \circ F$. Similarly, $F \circ G \leq id_\mathbf{A}$ implies $F \circ G \circ F \leq id_\mathbf{A} \circ F = F$, so that by amnesticity $F = F \circ G \circ F$. The other equation holds by duality. □

6.31 COROLLARY

If (F,G) is a Galois correspondence between amnestic concrete categories, then $(G \circ F) \circ (G \circ F) = G \circ F$ and $(F \circ G) \circ (F \circ G) = F \circ G$. □

6.32 COROLLARY

Let $G : \mathbf{A} \to \mathbf{B}$ and $F : \mathbf{B} \to \mathbf{A}$ be concrete functors between amnestic concrete categories such that (F,G) is a Galois correspondence, and let $\mathbf{A}^*$ be the full subcategory of $\mathbf{A}$ with objects: $\{F(B) \mid B \in Ob(\mathbf{B})\}$ and $\mathbf{B}^*$ the full subcategory of $\mathbf{B}$ with objects: $\{G(A) \mid A \in Ob(\mathbf{A})\}$. Then

(1) $\mathbf{A}^*$ is coreflective in $\mathbf{A}$, and $A \in Ob(\mathbf{A}^*)$ if and only if $A = (F \circ G)(A)$.

(2) $\mathbf{B}^*$ is reflective in $\mathbf{B}$, and $B \in Ob(\mathbf{B}^*)$ if and only if $B = (G \circ F)(B)$.

(3) The restrictions of G and F to $\mathbf{A}^*$ and $\mathbf{B}^*$ are concrete isomorphisms, $G^* : \mathbf{A}^* \to \mathbf{B}^*$ and $F^* : \mathbf{B}^* \to \mathbf{A}^*$, that are inverse to each other. □

6.33 EXAMPLES

(1) If (R, E) is a Galois reflection between amnestic concrete categories with $E : \mathbf{A} \to \mathbf{B}$, then $\mathbf{A}^* = \mathbf{B}^* = \mathbf{A}$. Similarly for Galois coreflections.

(2) For the Galois correspondence (D, U) of Example 6.26(3)(a) $\mathbf{Set}^* = \mathbf{Set}$ and $\mathbf{Top}^*$ is the full subcategory of discrete spaces.

(3) For the Galois correspondence (U, N) of Example 6.26(3)(a) $\mathbf{Set}^* = \mathbf{Set}$ and $\mathbf{Top}^*$ is the full subcategory of indiscrete spaces.

(4) For the Galois correspondence (F, G) of Example 6.26(3)(b) $\mathbf{Rel}^* = \mathbf{Prost}$ and $\mathbf{Top}^*$ consists of those topological spaces for which arbitrary intersections of open sets are open.

(5) For the Galois correspondence (F, U) of Example 6.26(3)(c) $\mathbf{Top}^*$ is the full subcategory of completely regular (= uniformizable) spaces, and $\mathbf{Unif}^*$ is the full subcategory of fine uniform spaces.

6.34 PROPOSITION

Let $G : \mathbf{A} \to \mathbf{B}$ and $F : \mathbf{B} \to \mathbf{A}$ be concrete functors between amnestic concrete categories such that (F,G) is a Galois correspondence. Then the following are equivalent:

(1) G is a full embedding,

(2) G is full,

(3) G is injective on objects,

(4) F is surjective on objects,

(5) $F \circ G = id_{\mathbf{A}}$,

(6) up to Galois isomorphism, (F, G) is a Galois reflection; i.e., there exists a Galois reflection (R, E) and a Galois isomorphism (K^{-1}, K) with $(F, G) = (R, E) \circ (K^{-1}, K)$.

Proof:
(1) $\Rightarrow$ (2). Trivial.

(2) $\Rightarrow$ (3). If $GA = GA'$, then by fullness, $A \leq A'$ and $A' \leq A$. Hence, by amnesticity, $A = A'$.

(3) $\Rightarrow$ (4). By Proposition 6.30 for any **A**-object A, we have $(G \circ F \circ G)(A) = G(A)$, so that by (3), $F(GA) = A$.

(4) $\Rightarrow$ (5). For any **A**-object A there exists, by (4), a **B**-object B with $F(B) = A$. Hence, by Proposition 6.30, $(F \circ G)(A) = (F \circ G \circ F)(B) = F(B) = A$.

(5) $\Rightarrow$ (6). Let $\mathbf{B}^*$ be the full subcategory of **B** with objects $\{G(A) \mid A \in Ob(\mathbf{A})\}$; let K be the codomain restriction of G, i.e., $K : \mathbf{A} \to \mathbf{B}^*$; let $E : \mathbf{B}^* \hookrightarrow \mathbf{B}$ be the inclusion; and let $R = K \circ F$. Then by Corollary 6.32 $K^{-1} = F \circ E$ and (K^{-1}, K) is a Galois isomorphism. Clearly, $G = E \circ K$ and $F = K^{-1} \circ R$. It remains to be shown that (R, E) is a Galois reflection; i.e., that R is a concrete reflector. For each **B**-object B, $B \leq (G \circ F)(B)$. But $(G \circ F)(B) = R(B)$. Thus $|B| \xrightarrow{id_{|B|}} |R(B)|$ is a **B**-morphism so that the reflection arrows are identity-carried.

(6) $\Rightarrow$ (1). By amnesticity the embedding E must be full (5.24). Thus $G = E \circ K$ is the composition of full embeddings and so must be one too. $\square$

6.35 DECOMPOSITION THEOREM FOR GALOIS CORRESPONDENCES

Every Galois correspondence (F, G) between amnestic concrete categories is a composite $(F, G) = (R, E_{\mathbf{B}}) \circ (K^{-1}, K) \circ (E_{\mathbf{A}}, C)$ of

(1) a Galois coreflection, $(E_{\mathbf{A}}, C)$,

(2) a Galois isomorphism, (K^{-1}, K) and

(3) a Galois reflection, $(R, E_{\mathbf{B}})$.

Proof: If $G : \mathbf{A} \to \mathbf{B}$ and $F : \mathbf{B} \to \mathbf{A}$ are concrete functors such that (F, G) is a Galois correspondence, then let $\mathbf{A}^*$ and $\mathbf{B}^*$ be the full subcategories of $\mathbf{A}$ and $\mathbf{B}$ determined by the images of F and G, respectively (Corollary 6.32). Let $C : \mathbf{A} \to \mathbf{A}^*$ and $R : \mathbf{B} \to \mathbf{B}^*$ be the codomain restrictions of $F \circ G$ and $G \circ F$, respectively, let $K : \mathbf{A}^* \to \mathbf{B}^*$ and $K^{-1} : \mathbf{B}^* \to \mathbf{A}^*$ be the corresponding restrictions of G and F, and let $E_A : \mathbf{A}^* \to \mathbf{A}$ and $E_B : \mathbf{B}^* \to \mathbf{B}$ be the full embedding of these subcategories. The result then follows immediately from Proposition 6.34 and its dual. □

6.36 REMARK
Notice that the above theorem actually gives a characterization of Galois correspondences between amnestic concrete categories since the composition of Galois correspondences is a Galois correspondence.

Suggestions for Further Reading

Eilenberg, S., and S. Mac Lane. General theory of natural equivalences. *Trans. Amer. Math. Soc.* **58** (1945): 231–294.

Yoneda, N. On the homology theory of modules. *J. Fac. Sci. Tokyo* **7** (1954): 193–227.

Herrlich, H., and M. Hušek. Galois connections categorically. *J. Pure Appl. Algebra* **68** (1990): 165–180.

EXERCISES

6A. Composition of Natural Transformations
Let $F, F' : \mathbf{A} \to \mathbf{B}$ and $G, G' : \mathbf{B} \to \mathbf{C}$ be functors and let $F \xrightarrow{\tau} F'$ and $G \xrightarrow{\sigma} G'$ be natural transformations. Show that

(a) $\sigma F' \circ G\tau = G'\tau \circ \sigma F$. [This natural transformation is called the **star product** of τ and σ and is denoted by $G \circ F \xrightarrow{\sigma * \tau} G' \circ F'$.]

(b) $\sigma F = \sigma * id_F$ and $G\tau = id_G * \tau$.

(c) $id_G * id_F = id_{G \circ F}$.

(d) If $H, H' : \mathbf{C} \to \mathbf{D}$ are functors and $H \xrightarrow{\delta} H'$ is a natural transformation, then

$$\delta * (\sigma * \tau) = (\delta * \sigma) * \tau.$$

(e) If $\mathbf{A} \xrightarrow{F''} \mathbf{B}$ and $\mathbf{B} \xrightarrow{G''} \mathbf{C}$ are functors and $F' \xrightarrow{\tau'} F''$ and $G' \xrightarrow{\sigma'} G''$ are natural transformations, then

$$(\sigma' \circ \sigma) * (\tau' \circ \tau) = (\sigma' * \tau') \circ (\sigma * \tau).$$

(f) If $\mathbf{C} \xrightarrow{H} \mathbf{D}$ is a functor, then $(H\sigma)F = H(\sigma F)$.

(g) If $\mathbf{C} \xrightarrow{H} \mathbf{D}$ is a functor, then $(H \circ G)\tau = H(G\tau)$.

(h) If $\mathbf{D} \xrightarrow{K} \mathbf{A}$ is a functor, then $\sigma(F \circ K) = (\sigma F)K$.

(i) If $\mathbf{B} \xrightarrow{G''} \mathbf{C}$ and $\mathbf{C} \xrightarrow{H} \mathbf{D}$ are functors and $G' \xrightarrow{\sigma'} G''$ is a natural transformation, then $H(\sigma' \circ \sigma)F = (H\sigma'F) \circ (H\sigma F)$.

6B. Counting Natural Transformations

(a) Show that there is precisely one natural transformation $id_{\mathbf{Set}} \to id_{\mathbf{Set}}$.

(b) Let $\mathbf{M}$ be a monoid considered as a category. Show that an element x of M yields a natural transformation $id_{\mathbf{M}} \to id_{\mathbf{M}}$ if and only if $x \circ y = y \circ x$ for each $y \in M$.

(c) How many natural transformations are there from S^2 to $\mathcal{P}$? Cf. 3.20(8) and 3.20(10). [Hint: Observe that S^2 is naturally isomorphic to $\hom(\mathbf{2}, -)$ and use Corollary 6.19.]

6C. Functor-Structured Categories

Let $S, T : \mathbf{X} \to \mathbf{Set}$ be functors. Show that $\mathbf{Spa}(S)$ and $\mathbf{Spa}(T)$ are concretely isomorphic if and only if S and T are naturally isomorphic.

6D. Functors Naturally Isomorphic to $id_{\mathbf{A}}$

Show that

(a) If a functor $\mathbf{A} \xrightarrow{F} \mathbf{A}$ is naturally isomorphic to $id_{\mathbf{A}}$, then F is an equivalence.

(b) A functor $\mathbf{Set} \xrightarrow{F} \mathbf{Set}$ is naturally isomorphic to $id_{\mathbf{Set}}$ if and only if F is an equivalence.

(c) If $\mathbf{G}$ is a group considered as a category, then a functor $\mathbf{G} \xrightarrow{F} \mathbf{G}$ (i.e., a group endomorphism F) is naturally isomorphic to $id_{\mathbf{G}}$ if and only if F is an inner automorphism of $\mathbf{G}$.

(d) If $\mathbf{A}$ is a discrete category, then $id_{\mathbf{A}}$ is the only functor from $\mathbf{A}$ to $\mathbf{A}$ that is naturally isomorphic to $id_{\mathbf{A}}$.

6E. $\mathbf{Ban} \xrightarrow{O} \mathbf{Set}$ and $\mathbf{Ban} \xrightarrow{U} \mathbf{Set}$

Show that O is representable, but U is not. Cf. 5.2(3).

6F. Representability of Power-Set Functors

Show that the contravariant power-set functor $\mathbf{Set}^{\mathrm{op}} \xrightarrow{\mathcal{Q}} \mathbf{Set}$ is representable (by any two-element set), but that the covariant power-set functor $\mathbf{Set} \xrightarrow{\mathcal{P}} \mathbf{Set}$ is not representable.

6G. Maps Induce Galois Connections

Let $A \xrightarrow{f} B$ be a map. Consider the functions $\mathcal{P}A \xrightarrow{\mathcal{P}f} \mathcal{P}B$ and $\mathcal{P}B \xrightarrow{\mathcal{Q}f} \mathcal{P}A$ [cf. 3.20(8) and (9)] as concrete functors between the power-sets $\mathcal{P}A$ and $\mathcal{P}B$, ordered by inclusion, and show that $(\mathcal{P}f, \mathcal{Q}f)$ is a Galois connection.

6H. Legitimate Functor Quasicategories.

(a) Prove that for each small category **A** all quasicategories [**A**, **B**] are legitimate.

(b) Prove that **1** and **∅** are the only categories **B** with the property that each quasicategory [**A**, **B**] is legitimate.

(c) Prove that [**Set**, **Set**] is illegitimate.

6I. Total Categories

A category **A** is called **total** provided that the natural embedding $E : \mathbf{A} \to [\mathbf{A}^{\mathrm{op}}, \mathbf{Set}]$ of Theorem 6.20 maps it onto a reflective subcategory $E[\mathbf{A}]$ of $[\mathbf{A}^{\mathrm{op}}, \mathbf{Set}]$.

(a) Prove that **Set** is total. [Hint: A reflection of $F : \mathbf{Set}^{\mathrm{op}} \to \mathbf{Set}$ is given by $\tau : F \to \hom(-, F(1))$, where $1 = \{0\}$ and τ_A maps $a \in F(A)$ to the function $\tau_A(a) : A \to F(1)$ given by $t \mapsto Ff(a)$ for $f : 1 \to A$, with $f(0) = t$.]

(b) Prove that $\mathbf{Spa}(T)$ is total for each $T : \mathbf{Set} \to \mathbf{Set}$. [Hint: Analogous to (a) with $(1, \emptyset)$ substituted for 1.]

(c) Prove that if **A** is a total category then every full reflective subcategory of **A** is total.

(d) Prove that a category isomorphic to a total category is total. Conclude that **Pos**, **Vec**, **Sgr**, and $\mathbf{Alg}(\Omega)$ are total categories. [Hint: Combine 6I(c) and 4K.]

6J. Yoneda Embedding

Show that the bijective function Y of Corollary 6.19 is "natural in the variables A and F", i.e., define functors $H, G : [\mathbf{A}, \mathbf{Set}] \times \mathbf{A} \to \mathbf{Set}$ on objects by $H(F, A) = F(A)$ and $G(F, A) = [\hom(A, -), F]$ such that Y becomes a natural isomorphism from G to H.

6K. Representable Functors

If $F : \mathbf{A} \to \mathbf{Set}$ is a functor, then a **universal point** of F is a pair (A, a) consisting of an **A**-object A and a point $a \in FA$ with the following (universal) property: for each **A**-object B and each point $b \in FB$ there exists a unique **A**-morphism $f : A \to B$ with $Ff(a) = b$.

(a) Prove that a functor is representable if and only if it has a universal point.

(b) Find a universal point of each of the forgetful functors of **Vec**, **Top**, and **Pos**. Show that the forgetful functor of Σ-**Seq** has no universal point.

(c) Find a universal point of $\mathcal{Q} : \mathbf{Set}^{\mathrm{op}} \to \mathbf{Set}$.

(d) For which sets M does the functor $- \times M : \mathbf{Set} \to \mathbf{Set}$ have a universal point?

(e) Does $\mathcal{P} : \mathbf{Set} \to \mathbf{Set}$ have a universal point?

6L. "Naturally Isomorphic" is an Equivalence Relation

Let F, G, and H be functors from **A** to **B** and let $F \xrightarrow{\sigma} G$ and $G \xrightarrow{\tau} H$ be natural isomorphisms. Show that:

(a) $\sigma^{-1} = (\sigma_A^{-1}) : G \to F$ is a natural isomorphism.

(b) $\tau \circ \sigma : F \to H$ is a natural isomorphism.

6M. Naturally Isomorphic Hom-Functors

Let $\tau : \hom(A, -) \to \hom(B, -)$ be a natural isomorphism. Show that $\tau_A(id_A) : B \to A$ is an isomorphism.

Chapter II

OBJECTS AND MORPHISMS

7 Objects and morphisms in abstract categories

INITIAL AND TERMINAL OBJECTS

7.1 DEFINITION
An object A is said to be an **initial object** provided that for each object B there is exactly one morphism from A to B.

7.2 EXAMPLES
(1) The empty set $\emptyset$ is the unique initial object for **Set**. Likewise, the empty partially ordered set (resp. the empty topological space) is the unique initial object for **Pos** (resp. **Top**).

(2) Every one-element group is an initial object for **Grp**; likewise for **Vec**.

(3) The empty category (i.e., the category with no objects and no morphisms) is the only initial object for **Cat**. It is also the only initial object for the quasicategory **CAT**, and, considered as a concrete category over **X** (via the inclusion), is the only initial object in **CAT(X)**.

(4) For any category of the form **Spa**(T) (Definition 5.40) an object (X, α) is initial if and only if X is an initial object in **X** and $\alpha = \emptyset$.

(5) For any category of the form **Alg**(Ω) [Example 3.3(2)] let $\Omega_n = \{i \in I \mid n_i = n\}$. If $\Omega_0 = \emptyset$, then the unique initial object is the empty algebra. If $\Omega_0 \neq \emptyset$, then an initial object in **Alg**(Ω) is the **term algebra**. Its members are **terms**, defined inductively as follows:

  (a) each element of Ω_0 is a term;

  (b) if $i \in \Omega_n$ and if $\omega_1, \omega_2, \ldots, \omega_n$ are terms, then $i\omega_1\omega_2 \ldots \omega_n$ is a term; and

  (c) all terms are obtained by iterative application of (a) and (b) above.

  Each term algebra operation is concatenation via rule (b) resp. (a).

(6) The ring of integers is an initial object in the construct **Rng** of rings with unit and unitary ring homomorphisms. For a ring R with unit e, the unique homomorphism $f : \mathbb{Z} \to R$ is defined by $f(n) = e + e + \cdots + e$ [n summands], and $f(-n) = -f(n)$.

(7) The two-element boolean algebras are initial objects in **Boo**.

(8) Σ-**Seq** has no initial object.

(9) In a poset considered as a category, an object is an initial object if and only if it is a smallest element.

7.3 PROPOSITION

Initial objects are essentially unique, i.e.,

(1) if A and B are initial objects, then A and B are isomorphic,

(2) if A is an initial object, then so is every object that is isomorphic to A.

Proof:
(1). By definition, there are morphisms $A \xrightarrow{k} B$ and $B \xrightarrow{h} A$. Furthermore, $h \circ k = id_A$ since id_A is the unique morphism from A to A. Analogously, $k \circ h = id_B$. Thus k is an isomorphism.

(2). Let $k : A' \to A$ be an isomorphism. For each object B there is a unique morphism $f : A \to B$. Then $f \circ k : A' \to B$ is a morphism from A' to B. It is unique since if $g : A' \to B$, then $g \circ k^{-1} : A \to B$. So $g \circ k^{-1}$ must be f; i.e., g must be $f \circ k$. □

Next we define terminal objects. They are dual to initial objects; i.e., A is terminal in **A** if and only if A is initial in $\mathbf{A}^{op}$.

7.4 DEFINITION
An object A is called a **terminal object** provided that for each object B there is exactly one morphism from B to A.

A terminal object

7.5 EXAMPLES

(1) Every singleton set is a terminal object for **Set**.

(2) Frequently for constructs, there is only one structure on the singleton set $\{0\}$, and in these cases the corresponding object is a terminal object. This is the case, for example, in **Vec**, **Pos**, **Grp**, **Top**, and $\mathbf{Cat}_{of}$.

Sec. 7] Objects and morphisms in abstract categories 103

(3) In **Rel** there are two structures on the set $\{0\}$. Of these, the pair $(\{0\}, \{(0,0)\})$ is a terminal object.

(4) For any category of the form **Spa**(T) (Definition 5.40) an object (X, α) is terminal if and only if X is a terminal object object in **X** and $\alpha = T(X)$.

(5) In Σ-**Seq** the acceptor with exactly one state that is both an initial and a final state is a terminal object.

(6) In a poset considered as a category an object is a terminal object if and only if it is a largest element.

(7) In **CAT**(**X**) an object $(\mathbf{A}, U)$ is terminal if and only if $U : \mathbf{A} \to \mathbf{X}$ is an isomorphism. In particular, $(\mathbf{X}, id_\mathbf{X})$ is a terminal object.

7.6 PROPOSITION
Terminal objects are essentially unique. $\boxed{D}$

ZERO OBJECTS

7.7 DEFINITION
An object A is called a **zero object** provided that it is both an initial object and a terminal object.

7.8 REMARK
Notice that since "terminal object" is dual to "initial object", the notion of zero object is self-dual; i.e., A is a zero object in **A** if and only if it is a zero object in $\mathbf{A}^{op}$.

7.9 EXAMPLES
(1) **Set** and **Top** don't have zero objects, but **pSet** and **pTop** (cf. Exercise 3B) do have zero objects — the "singletons".

(2) **Vec**, **Ban**, **Ban**$_b$, **TopVec**, and **Mon** have zero objects, but **Sgr** doesn't.

(3) **Ab** and **Grp** have zero objects, but **Rng** doesn't.

(4) **Pos** and **Cat** don't have zero objects.

SEPARATORS AND COSEPARATORS

7.10 DEFINITION
An object S is called a **separator** provided that whenever $A \underset{g}{\overset{f}{\rightrightarrows}} B$ are distinct morphisms, there exists a morphism $S \xrightarrow{h} A$ such that

$$S \xrightarrow{h} A \xrightarrow{f} B \neq S \xrightarrow{h} A \xrightarrow{g} B.$$

7.11 EXAMPLES

(1) In **Set** the separators are precisely the nonempty sets.

(2) In **Top** (resp. **Pos**) the separators are precisely the nonempty spaces (resp. nonempty posets).

(3) In **Vec** the separators are precisely the nonzero vector spaces.

(4) The group of integers $\mathbb{Z}$ under addition is a separator for **Grp** and for **Ab**. The monoid of natural numbers $\mathbb{N}$ under addition is a separator for **Mon**.

(5) (X, ρ) is a separator in **Rel** if and only if $X \neq \emptyset = \rho$.

7.12 PROPOSITION

An object S of a category **A** *is a separator if and only if* $\hom(S, -) : \mathbf{A} \to \mathbf{Set}$ *is a faithful functor.*

Proof: The faithfulness of $\hom(S, -)$ means, by definition, that given distinct **A**-morphisms $A \xrightarrow[g]{f} B$, their $\hom(S, -)$ images are distinct; i.e., they differ in at least one element $h \in \hom(S, A)$. In other words, $f \circ h \neq g \circ h$, and this is precisely the definition of separator. □

7.13 REMARK

As we will see later, the existence of a separator in **A** often serves as a useful "smallness" condition for **A** that guarantees that there are not "too many" **A**-objects. A slightly weaker condition that serves the same purpose is the existence of a separating *set*.

7.14 DEFINITION

A set $\mathcal{T}$ of objects is called a **separating set** provided that for any pair $A \xrightarrow[g]{f} B$ of distinct morphisms, there exists a morphism $S \xrightarrow{h} A$ with domain S a member of $\mathcal{T}$ such that $f \circ h \neq g \circ h$.

7.15 EXAMPLES

(1) The empty set is a separating set for **A** if and only if **A** is thin.

(2) A one-element set $\{S\}$ is a separating set if and only if S is a separator.

(3) **Set** × **Set** has no separators, but the set consisting of the two objects $(\emptyset, \{0\})$ and $(\{0\}, \emptyset)$ is a separating set.

(4) **Aut** has no separator, but $\{A_1, A_2\}$ is a separating set, where A_1 has states $\{q_0, q_1\}$, with q_0 initial, no input, and output set $\{0, 1, 2\}$, with $y(q_0) = 0$ and $y(q_1) = 1$; and A_2 has states $\{q_i \mid i \in \mathbb{N}\}$, with q_0 initial, one input σ, with $\delta(\sigma, q_i) = q_{i+1}$, and $y(q_i) = i$.

Next we introduce coseparators — the dual concept to separators. From the list of examples one can discern that in familiar categories coseparators are more rare than separators.

7.16 DEFINITION
An object C is a **coseparator** provided that whenever $B \overset{f}{\underset{g}{\rightrightarrows}} A$ are distinct morphisms, there exists a morphism $A \xrightarrow{h} C$ such that

$$B \xrightarrow{f} A \xrightarrow{h} C \neq B \xrightarrow{g} A \xrightarrow{h} C.$$

7.17 PROPOSITION
C is a coseparator for **A** if and only if $\hom(-, C) : \mathbf{A}^{\mathrm{op}} \to \mathbf{Set}$ is faithful. □

7.18 EXAMPLES
(1) In **Set** the coseparators are precisely those sets that have at least two elements. [If C has at least two elements and if $B \overset{f}{\underset{g}{\rightrightarrows}} A$ differ on $b \in B$, let $h : A \to C$ be any function with $h(f(b)) \neq h(g(b))$.]

(2) In **Vec** the coseparators are precisely the nonzero vector spaces.

(3) The coseparators in **Pos** are precisely the nondiscrete posets. To see this, suppose that C has elements $x < y$ and $B \overset{f}{\underset{g}{\rightrightarrows}} A$ differ on $b \in B$, say, $f(b) \not\leq g(b)$. Define $h : A \to C$ by

$$h(a) = \begin{cases} x & \text{if } a \leq g(b) \\ y & \text{if } a \not\leq g(b). \end{cases}$$

(4) The coseparators in **Top** are precisely the non-T_0-spaces. [If C is a non-T_0-space, then C has an indiscrete subspace with at least two points. Using this fact, proceed as in (1) above.]

(5) The coseparators in **Top$_0$** are precisely the non-T_1-spaces. To see this, suppose that C is a T_0-space that is not T_1. Then there exist elements x and y with $y \in c\ell\{x\}$ and $x \notin c\ell\{y\}$. If $B \overset{f}{\underset{g}{\rightrightarrows}} A$ differ on $b \in B$, then there exists an open set W in A that contains precisely one of $f(b)$ and $g(b)$. Define $h : A \to C$ by

$$h(a) = \begin{cases} x & \text{if } a \in W \\ y & \text{if } a \notin W. \end{cases}$$

(6) In a preordered class considered as a category, every element (= object) is simultaneously a separator and a coseparator.

(7) Any two-element boolean algebra is a coseparator for **Boo**. The closed unit interval is a coseparator for **Tych**. The Banach space of complex numbers is a coseparator for both **Ban$_b$** and **Ban** (Hahn-Banach Theorem). The circle group $\mathbb{R}/\mathbb{Z}$ is a coseparator for **Ab** and, considered as a compact group, it is a coseparator for the category of locally compact abelian groups. **Set** is a coseparator for the quasicategory **CAT**. [If $F, G : \mathbf{B} \to \mathbf{A}$, with $F \neq G$, then $\hom(A, -) \circ F \neq \hom(A, -) \circ G$ for some **A**-object A.]

(8) None of the categories **Rng**, **Grp**, **Sgr**, or **Haus** has a coseparator. (This follows for rings from the existence of arbitrarily large fields, for groups and semigroups from the existence of arbitrarily large simple groups, and for Hausdorff spaces from the fact that for each Hausdorff space X there exists a Hausdorff space Y with more than one point such that every continuous map from Y to X is constant.) Indeed, none of these categories even has a coseparating set.

SECTIONS AND RETRACTIONS

7.19 DEFINITION
A morphism $A \xrightarrow{f} B$ is called a **section** provided that there exists some morphism $B \xrightarrow{g} A$ such that $g \circ f = id_A$ (i.e., provided that f has a "left-inverse").

Sections

7.20 EXAMPLES
(1) A morphism in **Set** is a section if and only if it is an injective function and is not the empty function from the empty set to a nonempty set.

(2) In **Vec** the sections are exactly the injective linear transformations.

(3) In **Ab** a homomorphism $f : A \to B$ is a section if and only if it is injective and $f[A]$ is a direct summand of B.

(4) If X and Y are sets (resp. topological spaces) and if $a \in Y$, then the function $f: X \to X \times Y$ defined by $f(x) = (x, a)$ is a section in **Set** (resp. **Top**). [Note that the image of f is a "cross-section" of the product, which is in one-to-one correspondence (resp. homeomorphic) to X. This motivates our use of the word "section" in Definition 7.19.]

(5) The sections described in (4) are just special cases of the following situation: Let $f: X \to Y$ be a morphism in **Set** (resp. **Top, Grp,** R–**Mod**). Consider the graph of f as a subset (resp. subspace, subgroup, submodule) of the product $X \times Y$. Then the embedding of X into $X \times Y$ defined by $x \mapsto (x, f(x))$ is a section in the category in question.

(6) If T is a terminal object, then every morphism with domain T is a section.

(7) In a thin category, sections are precisely the isomorphisms.

7.21 PROPOSITION

(1) If $A \xrightarrow{f} B$ and $B \xrightarrow{g} C$ are sections, then $A \xrightarrow{f} B \xrightarrow{g} C$ is a section.

(2) If $A \xrightarrow{f} B \xrightarrow{g} C$ is a section, then f is a section.

Proof:
(1). Given h with $h \circ f = id_A$ and k with $k \circ g = id_B$, then $(h \circ k) \circ (g \circ f) = id_A$.

(2). Given h with $h \circ (g \circ f) = id_A$, we have $(h \circ g) \circ f = id_A$. $\square$

7.22 PROPOSITION

Every functor **preserves sections** (i.e., if $F: \mathbf{A} \to \mathbf{B}$ is a functor and f is an **A**-section, then $F(f)$ is a **B**-section).

Proof: If $h \circ f = id_A$, then $Fh \circ Ff = F(h \circ f) = F(id_A) = id_{FA}$. $\square$

7.23 PROPOSITION

Every full, faithful functor **reflects sections** (i.e., if $F: \mathbf{A} \to \mathbf{B}$ is full and faithful and $F(f)$ is a **B**-section, then f is an **A**-section).

Proof: Given $h: FB \to FA$ with $h \circ F(f) = id_{FA}$, by fullness there is $k: B \to A$ with $h = F(k)$. Thus $F(k \circ f) = id_{FA} = F(id_A)$, so that by faithfulness $k \circ f = id_A$. $\square$

The dual concept for "section" is "retraction". The name comes from topology, where a subspace Y of a space X is called a *retract* if there is a continuous function $f: X \to Y$ with $f(y) = y$ for each $y \in Y$; so that if $e: Y \to X$ is the inclusion, $f \circ e = id_Y$.

7.24 DEFINITION

A morphism $A \xrightarrow{f} B$ is called a **retraction** provided that there exists some morphism $B \xrightarrow{g} A$ such that $f \circ g = id_B$ (i.e., provided that f has a "right-inverse"). If there exists such a retraction, then B will be called a **retract of** A.

7.25 EXAMPLES

(1) The retractions in **Set** are precisely the surjective functions. [Notice that this statement is equivalent to the *Axiom of Choice* for sets.]

(2) In **Vec** the retractions are exactly the surjective linear transformations.

(3) A morphism f in **Top** is a retraction if and only if there is a topological retraction r and a homeomorphism h such that $f = h \circ r$. In other words, the retractions in **Top** are (up to homeomorphism) exactly the topological retractions.

(4) In any thin category the concepts "retraction", "section", and "isomorphism" agree.

(5) The usual underlying functor $U : \textbf{Top} \to \textbf{Set}$ is a retraction in the quasicategory **CAT**. It has precisely two right inverses, the discrete functor and the indiscrete functor.

(6) In **Ab** retractions are (up to isomorphism) the projection homomorphisms, i.e., $f : A \to B$ is a retraction if and only if there is an abelian group C such that $A \xrightarrow{f} B = A \xrightarrow{h} B \times C \xrightarrow{p} B$, where h is an isomorphism and p is a projection.

7.26 PROPOSITION

For a morphism f the following are equivalent:

(1) f is an isomorphism,

(2) f is a section and a retraction.

Proof: This follows immediately from Proposition 3.10. □

7.27 PROPOSITION

(1) If $A \xrightarrow{f} B$ and $B \xrightarrow{g} C$ are retractions, then $A \xrightarrow{g \circ f} C$ is a retraction.

(2) If $A \xrightarrow{g \circ f} C$ is a retraction, then g is a retraction. $\boxed{D}$

7.28 PROPOSITION

Every functor preserves retractions. $\boxed{D}$

7.29 PROPOSITION

Every full, faithful functor reflects retractions. $\boxed{D}$

7.30 REMARKS

(1) Notice that by combining Proposition 7.23 and its dual (7.29) we get the result (already proved [3.32]) that every full, faithful functor reflects isomorphisms.

(2) Notice also that we did not need to define "sect" as the dual of retract, since "retract" is self-dual.

7.31 PROPOSITION

An isomorphism-closed full reflective subcategory **A** *of a category* **B** *contains with any object A each retract of A in* **B**.

Proof: Let $A \xrightarrow{r} B$ be a **B**-retraction of an **A**-object A. Then there exist a section $s: B \to A$ such that $r \circ s = id_B$ and an **A**-reflection $B \xrightarrow{u} A_B$. By the definition of reflection there is an **A**-morphism $A_B \xrightarrow{t} A$ such that $s = t \circ u$. Thus $(r \circ t) \circ u = r \circ s = id_B$. Hence $u \circ (r \circ t) \circ u = u = id_{A_B} \circ u$, so that by uniqueness $u \circ (r \circ t) = id_{A_B}$. Thus u is an isomorphism, and since **A** is isomorphism-closed in **B**, B must belong to **A**. □

MONOMORPHISMS AND EPIMORPHISMS

7.32 DEFINITION

A morphism $A \xrightarrow{f} B$ is said to be a **monomorphism** provided that for all pairs $C \underset{k}{\overset{h}{\rightrightarrows}} A$ of morphisms such that $f \circ h = f \circ k$, it follows that $h = k$ (i.e., f is "left-cancellable" with respect to composition).

7.33 EXAMPLES

(1) A function is a monomorphism in **Set** if and only if it is injective. [To show that a monomorphism $f: A \to B$ must be injective, take $a, b \in A$ and consider the constant functions $\hat{a}, \hat{b}: \{p\} \to A$.]

(2) For any morphism f in **Vec**, the following are equivalent:

   (a) f is a monomorphism,

   (b) f is a section,

   (c) f is injective.

(3) In many constructs, monomorphisms are precisely those morphisms that have injective underlying functions; e.g., this is the case for **Pos**, **Top**, **Grp**, **Ab**, **Sgr**, **Rng**, **Rel**, and **Alg**(Ω). In fact this is true for any construct with a representable underlying functor (7.37 and 7.38).

(4) In the construct $\{\sigma\}$-**Seq** (whose forgetful functor is not representable), monomorphisms are precisely the injective simulations. [If $f: A \to B$ is not injective, then there exist distinct states x_1 and x_2 with $f(x_1) = f(x_2)$, and there exist n, m, and k such that $\sigma^{n+k} x_1 = \sigma^k x_1$ and $\sigma^{m+k} x_2 = \sigma^k x_2$. Let C be the acceptor obtained from A by adding $k + nm$ nonfinal states $q_1, \ldots, q_{k+nm}$ such that $\sigma q_i = q_{i+1}$ for $i < k + nm$ and $\sigma q_{k+nm} = q_{k+1}$. For $i = 1, 2$, there is a unique simulation $g_i: C \to A$ that is the inclusion on A-states and such that $g_i(q_1) = x_i$. Then $f \circ g_1 = f \circ g_2$.]

(5) In the category **DivAb** of divisible abelian groups and group homomorphisms there are monomorphisms that have non-injective underlying functions. Consider the natural quotient $\mathbb{Q} \to \mathbb{Q}/\mathbb{Z}$, where $\mathbb{Q}$ is the additive group of rational numbers and $\mathbb{Z}$ is the additive group of integers.

(6) There is a monomorphism $X \xrightarrow{f} Y$ in **Top** such that the homotopy class $X \xrightarrow{\tilde{f}} Y$ of f is not a monomorphism in the category **hTop**, whose objects are topological spaces and whose morphisms are homotopy equivalence classes of continuous maps. [Consider the usual embedding of the bounding circle of a disc into the disc.]

(7) The monomorphisms in **CAT** are precisely the embedding functors.

(8) In the category of all fields[35] and homomorphisms between them, every morphism is a monomorphism.

(9) In any thin category every morphism is a monomorphism.

7.34 PROPOSITION

(1) If $A \xrightarrow{f} B$ and $B \xrightarrow{g} C$ are monomorphisms, then $A \xrightarrow{f} B \xrightarrow{g} C$ is a monomorphism.

(2) If $A \xrightarrow{f} B \xrightarrow{g} C$ is a monomorphism, then f is a monomorphism.

Proof: Let $h, k : D \to A$.

(1). $(g \circ f) \circ h = (g \circ f) \circ k \Rightarrow g \circ (f \circ h) = g \circ (f \circ k) \Rightarrow f \circ h = f \circ k \Rightarrow h = k$.

(2). $f \circ h = f \circ k \Rightarrow (g \circ f) \circ h = (g \circ f) \circ k \Rightarrow h = k$. □

7.35 PROPOSITION

Every section is a monomorphism.

Proof: Suppose that $g \circ f = id$ and $f \circ h = f \circ k$. Then $h = g \circ f \circ h = g \circ f \circ k = k$. □

7.36 PROPOSITION

For any morphism f the following are equivalent:

(1) f is an isomorphism,

(2) f is a retraction and a monomorphism.

Proof: (1) $\Rightarrow$ (2) is clear from Propositions 7.26 and 7.35. To show that (2) $\Rightarrow$ (1), let f be a monomorphism with $f \circ g = id$. Then $f \circ (g \circ f) = (f \circ g) \circ f = id \circ f = f \circ id$, so that by left-cancellation, $g \circ f = id$. Hence f is an isomorphism. □

7.37 PROPOSITION

(1) *Every representable functor* **preserves monomorphisms**, *i.e., if $F : \mathbf{A} \to \mathbf{Set}$ is representable and if f is a monomorphism in $\mathbf{A}$, then $F(f)$ is a monomorphism in* **Set** *(i.e., an injective function).*

(2) *Every faithful functor* **reflects monomorphisms**, *i.e., if $F : \mathbf{A} \to \mathbf{B}$ is faithful and $F(f)$ is a $\mathbf{B}$-monomorphism, then f is an $\mathbf{A}$-monomorphism.*

---
[35] Recall that in each field $0 \neq 1$.

Proof:

(1). Simple computations show that

   (a) hom-functors $\hom(A, -) : \mathbf{A} \to \mathbf{Set}$ preserve monomorphisms,

   (b) whenever functors F and G are naturally isomorphic and F preserves monomorphisms, then so does G.

(2). Suppose that $f \circ h = f \circ k$. Then $Ff \circ Fh = Ff \circ Fk$ implies that $Fh = Fk$, so that by faithfulness $h = k$. $\square$

7.38 COROLLARY

In any construct all morphisms with injective underlying functions are monomorphisms. When the underlying functor is representable, the monomorphisms are precisely the morphisms with injective underlying functions. $\square$

The categorical dual of "monomorphism" is "epimorphism".

7.39 DEFINITION

A morphism $A \xrightarrow{f} B$ is said to be an **epimorphism** provided that for all pairs $B \underset{k}{\overset{h}{\rightrightarrows}} C$ of morphisms such that $h \circ f = k \circ f$, it follows that $h = k$ (i.e., f is "right-cancellable" with respect to composition).

7.40 EXAMPLES

(1) In both **Set** and **Vec** the following are equivalent for any morphism f:

   (a) f is an epimorphism,

   (b) f is a retraction,

   (c) f is surjective.

[To show that an epimorphism $A \xrightarrow{f} B$ in **Set** is surjective, consider two functions from B to $\{0, 1\}$, one of them mapping every point of B to 0 and the other mapping precisely the points of $f[A]$ to 0. To show that $A \xrightarrow{f} B$ in **Vec** is surjective, use the above idea with the quotient $B/f[A]$ replacing $\{0, 1\}$.]

(2) In a number of constructs the epimorphisms are precisely the morphisms with surjective underlying functions. This is the case, for instance, in **Top**, **Rel**, each $\mathbf{Alg}(\Omega)$, **Lat**, **Σ-Seq**, **Pos**, **Ab**, **Grp**, and **HComp**. [For **Top**, argue as in **Set**, where $\{0, 1\}$ is given the indiscrete topology; similarly for **Rel**. For **Ab** and **HComp**, argue as in **Vec**.] However, this situation occurs less frequently than that of monomorphisms being precisely those morphisms with injective underlying functions. In quite a few familiar constructs, epimorphisms fail to be surjective (see below), and even when they are surjective, the proof may be far from obvious; e.g., for **Grp** see Exercise 7H.

(3) In **Haus** the epimorphisms are precisely the continuous functions with dense images. Also in **Ban**$_b$, and in **Ban** (with either of the two natural forgetful functors) [5.2(3)], the epimorphisms are precisely the morphisms with dense images. In the category of Hausdorff topological groups there exist epimorphisms with non-dense images.

(4) In the category of torsion-free abelian groups, a morphism $A \xrightarrow{f} B$ is an epimorphism if and only if the factor group $B/f[A]$ is a torsion group. Thus, in this category, the inclusion $2\mathbb{Z} \hookrightarrow \mathbb{Z}$ is a non-surjective epimorphism.

(5) In **Rng** and **Sgr** there are epimorphisms that are not surjective; e.g., the usual embedding $\mathbb{Z} \xrightarrow{f} \mathbb{Q}$ of the integers into the rationals is an epimorphism in **Rng** and in **Sgr**. [If h and k are homomorphisms such that $h \circ f = k \circ f$ and if $n/m \in \mathbb{Q}$, then

$$h(n/m) = h(n) \cdot h(1/m) \cdot h(1) = k(n) \cdot h(1/m) \cdot k(1)$$
$$= k(n) \cdot h(1/m) \cdot k(m) \cdot k(1/m) = k(n) \cdot h(1/m) \cdot h(m) \cdot k(1/m)$$
$$= k(n) \cdot h(1) \cdot k(1/m) = k(n) \cdot k(1) \cdot k(1/m) = k(n/m).]$$

(6) In **Cat** there are epimorphisms that are not surjective. Consider the epimorphism $F: \mathbf{A} \to \mathbf{B}$, where $\mathbf{A} = \bullet \xrightarrow{g} \bullet$, $\mathbf{B}$ is the additive monoid of natural numbers, and F is the unique functor from $\mathbf{A}$ and $\mathbf{B}$ with $F(g) = 1$.

(7) There is an epimorphism $X \xrightarrow{f} Y$ in **Top** such that the homotopy class $X \xrightarrow{\tilde{f}} Y$ of f is *not* an epimorphism in the category **hTop**. Consider the covering projection of the real line onto the circle, defined by: $x \mapsto e^{ix}$.

(8) In a thin category, each morphism is an epimorphism.

7.41 PROPOSITION
(1) If $A \xrightarrow{f} B$ and $B \xrightarrow{g} C$ are epimorphisms, then $A \xrightarrow{f} B \xrightarrow{g} C$ is an epimorphism.

(2) If $A \xrightarrow{f} B \xrightarrow{g} C$ is an epimorphism, then g is an epimorphism. □

7.42 PROPOSITION
Every retraction is an epimorphism. □

7.43 PROPOSITION
For a morphism, f, the following are equivalent:

(1) f is an isomorphism.

(2) f is a section and an epimorphism. □

7.44 PROPOSITION
Every faithful functor reflects epimorphisms. □

7.45 COROLLARY
In any construct all morphisms with surjective underlying functions are epimorphisms. □

7.46 REMARK
Although faithful functors reflect epimorphisms and monomorphisms, they need not preserve them (as the above examples show). In fact, even full embeddings may fail to do so.[36] For example, the full embedding $E :$ **Haus** $\hookrightarrow$ **Top** doesn't preserve epimorphisms [7.40(2) and (3)] and so the full embedding $E^{op} :$ **Haus**^{op} $\hookrightarrow$ **Top**^{op} doesn't preserve monomorphisms. However, if such functors are also isomorphism-dense, then they preserve monomorphisms and epimorphisms, as the following shows:

7.47 PROPOSITION
Every equivalence functor preserves and reflects each of the following: monomorphisms, epimorphisms, sections, retractions, and isomorphisms.

Proof: By duality and Propositions 7.22, 7.23, and 7.37, we need only show preservation of monomorphisms. Let $F : \mathbf{A} \to \mathbf{B}$ be an equivalence, let $A' \xrightarrow{f} A$ be an **A**-monomorphism, and let $B \underset{s}{\overset{r}{\rightrightarrows}} FA'$ be morphisms with $Ff \circ r = Ff \circ s$. Since F is isomorphism-dense, there exists an **A**-object A'' and a **B**-isomorphism $FA'' \xrightarrow{k} B$. By fullness there are **A**-morphisms $A'' \underset{s'}{\overset{r'}{\rightrightarrows}} A'$ with $Fr' = r \circ k$ and $Fs' = s \circ k$. Thus

$$F(f \circ r') = Ff \circ Fr' = Ff \circ r \circ k = Ff \circ s \circ k = Ff \circ Fs' = F(f \circ s').$$

Faithfulness and the fact that f is a monomorphism imply that $r' = s'$, from which it follows readily that $r = s$. □

7.48 REMARK
The above is typical of equivalences. They preserve and reflect virtually all properties that are considered to be categorical. In fact, a reasonable way to define a "categorical property" would be as "a property of categories that is preserved and reflected by all equivalences". This is why equivalent categories are considered to be almost as much alike as isomorphic ones.

7.49 DEFINITION
(1) A morphism is called a **bimorphism** provided that it is simultaneously a monomorphism and an epimorphism.

(2) A category is called **balanced** provided that each of its bimorphisms is an isomorphism.

---

[36]However, we will see later [Proposition 18.6 and Example 19.12(1)] that embeddings of *reflective* subcategories must preserve monomorphisms.

7.50 EXAMPLES

(1) **Set**, **Vec**, **Grp**, **Ab**, and **HComp** are balanced categories.

(2) **Rel**, **Pos**, **Top**, **Mon**, **Sgr**, **Rng**, **Cat**, **Ban**, and **Ban**$_b$ are not balanced categories. The inclusion $\mathbb{Z} \hookrightarrow \mathbb{Q}$ is a non-isomorphic bimorphism in **Mon**, **Sgr**, **Rng**, and **Cat**. The function $\ell^\infty \xrightarrow{f} c_0$,[37] defined by $f(x_n) = (\frac{x_n}{n})$, is a non-isomorphic bimorphism in **Ban** and **Ban**$_b$.

REGULAR AND EXTREMAL MONOMORPHISMS

Originally it was believed that monomorphisms would constitute the correct categorical abstraction of the notion "embeddings of substructures" that exists in various constructs. However, in many instances the concept of monomorphism is too weak; e.g., in **Top** monomorphisms are just injective continuous maps and thus need not be embeddings. Below we introduce two stronger notions (and there are several others in current use — see Exercises 7D and 14C) that more frequently correspond with embeddings in categories. However, a satisfactory concept of "embeddings" seems to be possible only in the setting of constructs (see 8.6).

7.51 DEFINITION

Let $A \underset{g}{\overset{f}{\rightrightarrows}} B$ be a pair of morphisms. A morphism $E \xrightarrow{e} A$ is called an **equalizer of** f **and** g provided that the following conditions hold:

(1) $f \circ e = g \circ e$,

(2) for any morphism $e' : E' \to A$ with $f \circ e' = g \circ e'$, there exists a unique morphism $\bar{e} : E' \to E$ such that $e' = e \circ \bar{e}$, i.e., such that the triangle

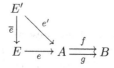

commutes.

7.52 EXAMPLES

(1) Let **A** be one of the categories **Set**, **Vec**, **Pos**, **Top**, or **Grp**. If $A \underset{g}{\overset{f}{\rightrightarrows}} B$ are **A**-morphisms, and if E denotes the set $\{a \in A \mid f(a) = g(a)\}$ considered as a subset (resp. linear subspace, subposet, subspace, subgroup) of A, then the inclusion from E to A is an equalizer of f and g. [If $E' \xrightarrow{e'} A$ is such that $f \circ e' = g \circ e'$, then $\bar{e}$ is the codomain restriction of e'.]

---

[37] ℓ^∞ (resp. c_0) is the classical Banach space of all bounded sequences in the field K (resp. all sequences in K that converge to 0) with the sup-norm.

(2) If **A** is the full subcategory of **Top** that consists of sequential spaces (resp. compactly generated spaces) and if E (as in (1)) is supplied with the coarsest sequential (resp. compactly generated) topology for which the inclusion $E \xrightarrow{e} A$ is continuous, then e is an equalizer of f and g in **A**.

7.53 PROPOSITION

Equalizers are essentially unique; i.e., given $A \underset{g}{\overset{f}{\rightrightarrows}} B$ in a category, then the following hold:

(1) *if each of $E \xrightarrow{e} A$ and $E' \xrightarrow{e'} A$ is an equalizer of f and g, then there is an isomorphism $k : E' \to E$ with $e' = e \circ k$,*

(2) *if $E \xrightarrow{e} A$ is an equalizer of f and g, and if $E' \xrightarrow{k} E$ is an isomorphism, then $E' \xrightarrow{e \circ k} A$ is also an equalizer of f and g.*

Proof:

(1). Since $f \circ e' = g \circ e'$, there is a k with $e' = e \circ k$. Analogously, there is an h with $e = e' \circ h$. Thus $e \circ id = e = e' \circ h = e \circ (k \circ h)$, so that by the uniqueness requirement in the definition of equalizer, $id = k \circ h$. Similarly, $h \circ k = id$, so that k is an isomorphism.

(2). This is clear, since whenever $\bar{e}$ is a unique morphism with $e \circ \bar{e} = e'$, then $k^{-1} \circ \bar{e}$ will be a unique morphism with $(e \circ k) \circ (k^{-1} \circ \bar{e}) = e'$. □

7.54 PROPOSITION

If $E \xrightarrow{e} A$ is an equalizer of $A \underset{g}{\overset{f}{\rightrightarrows}} B$, then the following are equivalent:

(1) $f = g$,

(2) e *is an epimorphism,*

(3) e *is an isomorphism,*

(4) id_A *is an equalizer of f and g.* □

7.55 PROPOSITION

If $A \xrightarrow{g} B \xrightarrow{f} A = A \xrightarrow{id_A} A$, then g is an equalizer of $g \circ f$ and id_B.

Proof: Clearly $(g \circ f) \circ g = g \circ (f \circ g) = g = id_B \circ g$. Given $h : C \to B$ with $(g \circ f) \circ h = id_B \circ h$, then $f \circ h : C \to A$ is the unique morphism k with $h = g \circ k$. □

7.56 DEFINITION

A morphism $E \xrightarrow{e} A$ is called a **regular monomorphism** provided that it is an equalizer of some pair of morphisms.

7.57 REMARKS

(1) It is clear from the uniqueness requirement in the definition of equalizer that regular monomorphisms must be monomorphisms.

(2) One should be aware of the difference between the concepts of equalizer and regular monomorphism. "Equalizer" is defined relative to pairs of morphisms, whereas "regular monomorphism" is absolute. The difference is more than just a technical one. For example, later we will see that it is possible for a functor to preserve regular monomorphisms without preserving equalizers (13.6).

7.58 EXAMPLES

(1) In **Set** the regular monomorphisms are the injective functions, i.e., up to isomorphism, precisely the inclusions of subsets. [If $E \xrightarrow{e} A$ is an inclusion of a subset, consider functions f and g from A to $\{0,1\}$, where f maps every point of A to 1 and g maps precisely the points of E to 1. Then e is an equalizer of f and g.]

(2) In **Top** the regular monomorphisms are, up to isomorphism, precisely the embeddings of subspaces. [If $E \xrightarrow{e} A$ is an inclusion of a subspace, proceed as in (1), and supply the set $\{0,1\}$ with the indiscrete topology.]

(3) In **Haus** the regular monomorphisms are, up to isomorphism, precisely the inclusions of closed subspaces. [If $e : E \to A$ is an embedding of a closed subspace in **Haus**, let $A_1 = A_2 = A$, and let $\tilde{A}$ be the topological sum of A_1 and A_2 with embeddings μ_1 and μ_2. Let $q : \tilde{A} \to B$ be the quotient map that for each $x \in E$ identifies $\mu_1(x)$ and $\mu_2(x)$, but identifies nothing else. Then B is Hausdorff and e is an equalizer of $q \circ \mu_1$ and $q \circ \mu_2$.] Similarly, in **Met$_c$** (resp. **Ban$_b$**) the regular monomorphisms are the topological embeddings of closed (linear) subspaces. In **Met** (resp. **Ban**) the regular monomorphisms are the isometric embeddings of closed (linear) subspaces.

(4) In many "algebraic" categories, e.g., in **Vec**, **Grp**, and **Alg**(Ω), and also in **HComp**, all monomorphisms are regular. [If $e : E \to A$ is a monomorphism in **Vec**, consider the quotient space $B = A/e[E]$, the zero map $f : A \to B$, and the natural quotient map $g : A \to B$. Then e is an equalizer of f and g.] However, in **Sgr** and **Rng** monomorphisms need not be regular, e.g., the inclusion $\mathbb{Z} \hookrightarrow \mathbb{Q}$ is a non-regular monomorphism. [If it were regular, then since it is an epimorphism [7.40(5)] by Corollary 7.63 and Proposition 7.66 it would necessarily be an isomorphism.]

(5) In $\{\sigma\}$-**Seq** the regular monomorphisms are precisely the injective simulations that map each nonfinal state to a nonfinal state.

7.59 PROPOSITION

(1) Every section is a regular monomorphism.

(2) Every regular monomorphism is a monomorphism.

Proof: (1) is immediate from Proposition 7.55 and (2) follows from the uniqueness requirement in the definition of equalizer. □

7.60 REMARK
Neither implication of the previous proposition can be reversed. For example, $\mathbb{Z} \hookrightarrow \mathbb{Q}$ is a monomorphism in **Sgr** that is not regular, and every embedding of a non-connected topological space into a connected one is a regular monomorphism in **Top** that is not a section. Also the embedding of the unit circle into the unit disc is not a section in **Top**. (This is the crucial lemma for Brouwer's fixed-point theorem.)

Next, we introduce another type of monomorphism that is particularly useful because (as is the case with regular monomorphisms) it often coincides with embeddings in constructs, and, since it is defined by intrinsic properties, it is sometimes easier to handle than regular monomorphisms.

7.61 DEFINITION
A monomorphism m is called **extremal** provided that it satisfies the following **extremal condition**: If $m = f \circ e$, where e is an epimorphism, then e must be an isomorphism.

7.62 PROPOSITION
Let $A \xrightarrow{f} B$ and $B \xrightarrow{g} C$ be morphisms.

(1) If f is an extremal monomorphism and g is a regular monomorphism, then $g \circ f$ is an extremal monomorphism.

(2) If $g \circ f$ is an extremal monomorphism, then f is an extremal monomorphism.

(3) If $g \circ f$ is a regular monomorphism and g is a monomorphism, then f is a regular monomorphism.

Proof:
(1). Let g be an equalizer of r and s. Let $g \circ f = h \circ e$, where e is an epimorphism. Then $r \circ h \circ e = r \circ g \circ f = s \circ g \circ f = s \circ h \circ e$ implies that $r \circ h = s \circ h$. Thus there exists a unique morphism k with $h = g \circ k$. Hence $g \circ k \circ e = h \circ e = g \circ f$. Consequently, $k \circ e = f$, which implies that e is an isomorphism.

(2). Immediate.

(3). If $g \circ f$ is an equalizer of r and s, then f is an equalizer of $r \circ g$ and $s \circ g$. □

7.63 COROLLARY
Every regular monomorphism is extremal. □

7.64 REMARK
A composite of extremal monomorphisms may fail to be extremal. However, if all extremal monomorphisms in a category are regular, then, by Proposition 7.62, in that category the class of extremal (= regular) monomorphisms is closed under composition and is cancellable from the left; i.e., its compositive behavior is similar to that of the class of all monomorphisms (cf. Proposition 7.34). This is the case for most of the familiar categories, e.g., for **Set**, **Vec**, **Rel**, **Pos**, **Top**, **Haus**, **Grp**, and **Alg**(Ω). (See also the dual of Proposition 14.14.) However, this is not always the case, as the following example and Exercises 7J and 14I show.

7.65 EXAMPLE

Let **FHaus** denote the full subcategory of **Top** consisting of the functionally Hausdorff spaces.[38] Then a morphism $A \xrightarrow{f} B$ in **FHaus** is

(a) a monomorphism if and only if f is injective,

(b) an extremal monomorphism if and only if f is an embedding such that for every subspace C of B with $f[A] \subsetneq C$, there exists a non-constant morphism $C \xrightarrow{g} \mathbb{R}$ with the restriction $g|f[A]$ constant,

(c) a regular monomorphism if and only if f is an embedding of a subspace of B that is an intersection of zerosets[39] in B (cf. 7J).

7.66 PROPOSITION

For any morphism f the following are equivalent:

(1) f is an isomorphism,

(2) f is an extremal monomorphism and an epimorphism.

Proof: (1) $\Rightarrow$ (2) is immediate from Proposition 7.59, Corollary 7.63, and duality. If f satisfies (2), the trivial factorization $f = id \circ f$ shows that f is an isomorphism. □

7.67 PROPOSITION

*For any category **A**, the following are equivalent:*

*(1) **A** is balanced,*

*(2) in **A** each monomorphism is extremal.*

Proof: (1) $\Rightarrow$ (2) is immediate from Proposition 7.34(2), and (2) $\Rightarrow$ (1) follows from the proposition above (7.66). □

REGULAR AND EXTREMAL EPIMORPHISMS

One of the nice insights that can be gleaned from category theory is that the formation of quotient structures (such as groups of cosets and identification topologies) can be viewed as the dual of the formation of substructures. Analogous to the situation where monomorphisms frequently are too weak to represent substructures, epimorphisms are frequently too weak to represent quotient structures. The notions of regular epimorphism and extremal epimorphism, which are the duals of the notions of regular monomorphism and extremal monomorphism, frequently coincide with natural quotients in abstract categories. However, as is the case with subobjects, a truly satisfactory concept of "quotient" seems to be possible only in the setting of constructs (see 8.10).

---

[38] A topological space A is called a **functionally Hausdorff space** provided that for any two distinct points a and b of A there exists a continuous map from A to the real numbers that has different values at a and b.

[39] Z is called a **zeroset** in B provided that there exists a continuous map from B to the real numbers such that Z is the inverse image of 0.

7.68 DEFINITION

Let $A \underset{g}{\overset{f}{\rightrightarrows}} B$ be a pair of morphisms. A morphism $B \xrightarrow{c} C$ is called a **coequalizer of** f **and** g provided that the following conditions hold:

(1) $c \circ f = c \circ g$,

(2) for any morphism $c' : B \to C'$ with $c' \circ f = c' \circ g$, there exists a unique morphism $\bar{c} : C \to C'$ such that $c' = \bar{c} \circ c$; i.e., such that the triangle

$$A \underset{g}{\overset{f}{\rightrightarrows}} B \xrightarrow{c} C$$

with c' going from B to C' and $\bar{c}$ going from C to C',

commutes.

7.69 EXAMPLES

1. Given two functions $A \underset{g}{\overset{f}{\rightrightarrows}} B$ in **Set**, let $\sim$ be the smallest equivalence relation on B such that $f(a) \sim g(a)$ for all $a \in A$. Then the natural map $q : B \to B/\sim$, which assigns to each $b \in B$ the equivalence class to which b belongs, is a coequalizer of f and g.

2. If $A \underset{g}{\overset{f}{\rightrightarrows}} B$ are continuous functions in **Top**, then the procedure given above yields a coequalizer of f and g, provided that $B/\sim$ is assigned the finest topology that makes q continuous (i.e., the final (or identification) topology).

3. Coequalizers in **Rel** and **Prost** are formed analogously to the way that they are formed in **Top**. In **Pos** they are formed by first forming the coequalizers in **Prost** and then taking the **Pos**-reflection of the resulting preordered set.

4. If $A \underset{g}{\overset{f}{\rightrightarrows}} B$ are **Ab**-morphisms, let B_0 be the subgroup $\{f(a) - g(a) \,|\, a \in A\}$ of B. Then the natural map $q : B \to B/B_0$ to the group of cosets is a coequalizer of f and g.

5. If $A \underset{g}{\overset{f}{\rightrightarrows}} B$ are homomorphisms in **Alg**(Ω), then a coequalizer for them is obtained analogously as in **Set** by using, instead of equivalence relations, the smallest congruence relation $\sim$ such that $f(a) \sim g(a)$ for all $a \in A$.

6. If $A \underset{g}{\overset{f}{\rightrightarrows}} B$ are simulations in Σ-**Seq**, then a coequalizer for them is obtained analogously as in **Alg**(Ω) with final states precisely the congruence classes that contain some final state of B.

7.70 REMARK

Since the concept of coequalizer is dual to that of equalizer, the general results about equalizers can be translated (via the Duality Principle) into results about coequalizers. For example,

(1) Coequalizers are essentially unique.

(2) If $A \xrightarrow{g} B \xrightarrow{f} A = A \xrightarrow{id_A} A$, then f is the coequalizer of $g \circ f$ and id_B.

(3) If $B \xrightarrow{c} C$ is a coequalizer of $A \underset{g}{\overset{f}{\rightrightarrows}} B$, then the following are equivalent:

   (a) $f = g$,
   (b) c is a monomorphism,
   (c) c is an isomorphism,
   (d) id_B is a coequalizer of f and g.

7.71 DEFINITION

A morphism $B \xrightarrow{c} C$ is called a **regular epimorphism** provided that it is a coequalizer of some pair of morphisms.

7.72 EXAMPLES

(1) In **Set** the regular epimorphisms are the surjective functions. [Given $e : A \to B$ surjective, consider the two projections from $D = \{(a, a') \in A \times A \mid e(a) = e(a')\}$ to A.] In fact, in **Set**,

$$\text{Retr} = \text{RegEpi} = \text{Surj} = \text{Epi}.$$

(2) In **Top** the regular epimorphisms are precisely the topological quotient maps; i.e., the surjective continuous maps onto spaces with the final topology. [For such a map e proceed as in **Set** with D considered as a subspace of $A \times A$.] In Σ-**Seq** regular epimorphisms are precisely the surjective simulations that have the property that each final state in the codomain has a final preimage in the domain. Thus, in both of these categories,

$$\text{RegEpi} \subsetneq \text{Surj} = \text{Epi}.$$

(3) In each **Alg**(Ω), **Vec**, **Grp**(cf. 7H), **Ab**, **Lat**, and **HComp**,

$$\text{RegEpi} = \text{Surj} = \text{Epi}.$$

(4) In **Sgr**, **Mon**, and **Rng**,

$$\text{RegEpi} = \text{Surj} \subsetneq \text{Epi}.$$

(5) In the constructs **Cat** and (**Ban**, O) there exist regular epimorphisms that are not surjective. In **Cat**, consider the regular epimorphism $\mathbf{A} \xrightarrow{F} \mathbf{B}$ of Example 7.40(6). In (**Ban**, O), consider the regular epimorphism $e : c_0 \to K$, where K is the field of real (or complex) numbers considered as a Banach space, c_0 is the classical Banach space of all sequences in K converging to 0 with the sup-norm, and $e(x_n) = \sum_{n=1}^{\infty} x_n/2^n$.

7.73 REMARK

Each of the forgetful functors from **Vec**, **Grp**, and **Top**$_1$ (T_1-topological spaces) into **Set** preserves regular epimorphisms. However, none of them preserves coequalizers. (Cf. Remark 13.6.)

7.74 DEFINITION

An epimorphism e is called **extremal** provided that it satisfies the following **extremal condition**:   If $e = m \circ f$, where m is a monomorphism, then m must be an isomorphism.

7.75 PROPOSITION

(1) Every retraction is a regular epimorphism.

(2) Every regular epimorphism is an extremal epimorphism. $\boxed{D}$

7.76 REMARK

(1) In most of the familiar categories the regular epimorphisms and the extremal epimorphisms coincide. See, e.g., Proposition 14.14. However, in **Cat** there exist extremal epimorphisms that are not regular epimorphisms. An example is the composite $G \circ F$ of the regular epimorphism (functor) $F: \mathbf{A} \to \mathbf{B}$ of Example 7.40(6) with the regular epimorphism (functor) $G: \mathbf{B} \to \mathbf{C}$, where $\mathbf{C}$ is the multiplicative submonoid $\{0,1\}$ of $\mathbb{Z}$ (considered as a category), $G(n) = 0$ for $n > 0$, and $G(0) = 1$.

(2) In general, "between" regular epimorphisms and extremal epimorphisms there are several other commonly used types of epimorphisms, the "strong epimorphisms" (the dual of which is introduced in Exercise 14C), the "swell epimorphisms" (introduced in 15A), and the "strict epimorphisms" (introduced in Exercise 7D). Thus we have the following diagram that summarizes the relative strengths of the various notions introduced. Notice that the notions in the boxes frequently coincide (see (1) above), but none of the implications can be reversed in general.

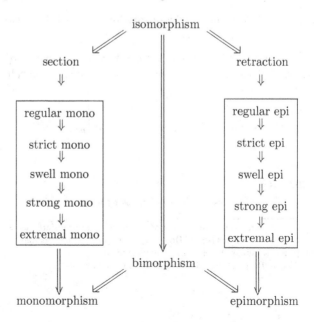

SUBOBJECTS

For most of the familiar constructs, a canonical notion of "subobject" is generally understood. Category theory has made it apparent that "A being a subobject of B" is not only a property of the object A, but also of the naturally associated "inclusion morphism" from A into B. Hence we will define subobjects of B to be pairs (A, m), where A is an object and m is an "inclusion morphism".

Since, as we have seen, for different categories one may need different concepts of morphism[40] to characterize (up to isomorphism) the "inclusion morphisms of subobjects", the concept of subobject will be made dependent upon a chosen class M of morphisms that represents (in each case) such inclusions.

7.77 DEFINITION

Let M be a class of monomorphisms. An **M-subobject** of an object B is a pair (A, m), where $A \xrightarrow{m} B$ belongs to M. In case M consists of all (regular, extremal) monomorphisms, M-subobjects are called (**regular**, **extremal**) **subobjects**.

7.78 REMARK

By the above definition, a subobject of a set B in **Set** is a pair (A, m), where $m : A \to B$ is an injective function. In order for the notion of subobject to correspond more closely to the notion of subset, two subobjects (A, m) and (A', m') should be considered to be essentially the same if $m[A] = m'[A']$. Furthermore a subobject (A, m) of B in **Set** should be considered to be "smaller than" the subobject (A', m') of B provided that $m[A] \subseteq m'[A']$. The following definitions capture these ideas.

7.79 DEFINITION

Let (A, m) and (A', m') be subobjects of B.

(1) (A, m) and (A', m') are called **isomorphic** provided that there exists an isomorphism $h : A \to A'$ with $m = m' \circ h$.

(2) (A, m) is said to be **smaller than** (A', m') — denoted by $(A, m) \leq (A', m')$ — provided that there exists some (necessarily unique) morphism[41] $h : A \to A'$ with $m = m' \circ h$

$$\begin{array}{ccc} A & \xrightarrow{h} & A' \\ & \searrow m & \downarrow m' \\ & & B \end{array}$$

7.80 REMARK

Observe that for the class of all subobjects of a given object

---

[40] For example, in **Rng** the class of monomorphisms is appropriate, whereas the class of regular (= extremal) monomorphisms is not. However, for **Top** the class of regular (= extremal) monomorphisms is appropriate, whereas the class of monomorphisms is not.

[41] Observe that by Proposition 7.34, h must be a monomorphism.

(1) the relation of being isomorphic is an equivalence relation, and
(2) the relation $\leq$ is a preorder; i.e., it is reflexive and transitive. In general, it fails to be antisymmetric, but $(A, m) \leq (A', m')$ and $(A', m') \leq (A, m)$ imply that these two subobjects are isomorphic.

7.81 EXAMPLES

(1) Let B be a set, let $\mathcal{P}B$ be the set of all subsets of B, let $\mathcal{S}B$ be the class of all subobjects of B in **Set**, and let $H : \mathcal{S}B \to \mathcal{P}B$ be the function defined by: $H(A, m) = m[A]$. Then
   (a) (A, m) and (A', m') are isomorphic if and only if $H(A, m) = H(A', m')$, and
   (b) $(A, m) \leq (A', m')$ if and only if $H(A, m) \subseteq H(A', m')$.

   Hence, up to isomorphism, subsets of B correspond bijectively to subobjects of B, where the correspondence is order-preserving.

(2) Similarly, in the categories **Vec**, **Sgr**, **Mon**, **Grp**, **Ab**, **Rng**, **Alg**(Ω), and **HComp**, subobjects correspond to linear subspaces, subsemigroups, submonoids, subgroups, subrings, subalgebras, and compact subspaces, respectively.

(3) Analogously, in **Top**, **Rel**, and **Pos**, regular (= extremal) subobjects correspond to subspaces, subrelations, and subposets, respectively.

7.82 DEFINITION

Let M be a class of monomorphisms of a category **A**.

(1) **A** is called M-**wellpowered** provided that no **A**-object has a proper class of pairwise[42] non-isomorphic M-subobjects.

(2) In case M is the class of all (regular, extremal) monomorphisms, then M-wellpowered is called (**regular, extremally**) **wellpowered**.

7.83 REMARK

Every wellpowered category must be regular wellpowered and extremally wellpowered, but not conversely. [The thin category **Ord** of all ordinal numbers (with its usual order) is wellpowered. **Ord**^{op} is extremally wellpowered, but is not wellpowered. In fact, each of its objects has a proper class of pairwise non-isomorphic subobjects, but only one extremal subobject.] Thus one should be aware that even though the phrase "extremally wellpowered" sounds stronger than "wellpowered", it is actually weaker.

QUOTIENT OBJECTS

Often in mathematics one has a "quotient" or "identification" procedure that maps points to equivalence classes or "collapses" parts of a structure. Frequently, both of

---
[42] "Pairwise" in this context means that any pair consisting of *distinct* members are assumed to be non-isomorphic.

these processes are considered to be essentially the same, and are used interchangeably. For example, collapsing a subgroup of an abelian group to the identity element is understood either as the homomorphism onto the group of cosets or as the homomorphism onto a particular set of representatives of the cosets. The reason that mathematicians don't distinguish these procedures is because they both give rise to essentially the same quotient object — where "quotient object" is the dual concept to that of subobject.

7.84 DEFINITION
Let E be a class of epimorphisms. An E-**quotient object** of an object A is a pair (e, B), where $A \xrightarrow{e} B$ belongs to E. In case E consists of all (regular, extremal) epimorphisms, E-quotient objects are called (**regular**, **extremal**) **quotient objects**.

7.85 DEFINITION
Let (e, B) and (e', B') be quotient objects of A.

(1) (e, B) and (e', B') are called **isomorphic** provided that there exists an isomorphism $h : B \to B'$ with $e' = h \circ e$.

(2) (e, B) is said to be **larger than** (e', B') — denoted by $(e, B) \geq (e', B')$ — provided that there exists some (necessarily unique) morphism $h : B \to B'$ with $e' = h \circ e$.

7.86 EXAMPLES
(1) Let A be a set, let $\mathcal{E}A$ be the set of all equivalence relations on A, let $\mathcal{K}A$ be the set of all quotient objects of A in **Set**, and let $H : \mathcal{K}A \to \mathcal{E}A$ be the function defined by: $H(e, B) = \{(x, y) \in A \times A \,|\, e(x) = e(y)\}$. Then

(a) (e, B) and (e', B') are isomorphic if and only if $H(e, B) = H(e', B')$, and

(b) $(e, B) \geq (e', B')$ if and only if $H(e, B) \subseteq H(e', B')$.

Hence, up to isomorphism, quotient objects of A correspond bijectively to equivalence relations on A, where the correspondence is order-reversing.

(2) Similarly, in the constructs **Top**, **Rel**, and **Pos** the regular (= extremal) quotient objects of an object correspond (up to isomorphism) bijectively to the equivalence relations on its underlying set.

(3) Analogously, in **HComp** the quotient objects (= regular quotient objects) of an object A correspond (up to isomorphism) bijectively to those equivalence relations ρ on the underlying set of A that, considered as subsets of $A \times A$, are closed.

(4) In algebraic constructs such as **Vec**, **Sgr**, **Mon**, **Grp**, **Ab**, **Rng**, and **Alg**(Ω), the regular (= extremal) quotient objects of an object A correspond (up to isomorphism) bijectively to the congruence relations on A; i.e., to the equivalence relations on the underlying set of A that respect the operations.

7.87 DEFINITION

Let E be a class of epimorphisms of a category $\mathbf{A}$.

(1) $\mathbf{A}$ is called E-**co-wellpowered** provided that no $\mathbf{A}$-object has a proper class of pairwise non-isomorphic E-quotient objects.

(2) In case E is the class of all (regular, extremal) epimorphisms, then E-co-wellpowered is called (**regular**, **extremally**) **co-wellpowered**.

7.88 THEOREM

Every construct is regular wellpowered and regular co-wellpowered.

Proof: Let $(\mathbf{A}, U)$ be a construct.

(1). To show that $\mathbf{A}$ is a regular co-wellpowered category, let A be an $\mathbf{A}$-object and let $\{(e_i, B_i) \mid i \in I\}$ be a class of pairwise non-isomorphic regular quotient objects of A. Let $\mathcal{E}A$ be the set of all equivalence relations on UA and let the function $H: I \to \mathcal{E}A$ be defined by: $H(i) = \{(a,b) \in UA \times UA \mid e_i(a) = e_i(b)\}$. To see that H is injective, let $i_1, i_2 \in I$ with $H(i_1) = H(i_2)$. For $k = 1, 2$, let (e_{i_k}, B_{i_k}) be a coequalizer of $A_k \underset{s_k}{\overset{r_k}{\rightrightarrows}} A$. The equality $e_{i_1} \circ r_1 = e_{i_1} \circ s_1$ implies that for each $a \in UA_1$ $(r_1(a), s_1(a)) \in H(i_1) = H(i_2)$. Consequently, $e_{i_2} \circ r_1 = e_{i_2} \circ s_1$, so that by the definition of coequalizer $(e_{i_1}, B_{i_1}) \geq (e_{i_2}, B_{i_2})$. By symmetry, and by using the dual of Remark 7.80(2), we have that (e_{i_1}, B_{i_1}) and (e_{i_2}, B_{i_2}) are isomorphic and hence, by assumption, $i_1 = i_2$. Thus $H: I \to \mathcal{E}A$ is an injective function, so that since $\mathcal{E}A$ is a set, I must be a set.

(2). That $\mathbf{A}$ is regular wellpowered follows immediately from (1) and the fact that $(\mathbf{A}^{op}, \mathcal{Q} \circ U^{op})$ is a construct [5.2(4)]. $\square$

7.89 COROLLARY

Every category with a separator or a coseparator is regular wellpowered and regular co-wellpowered.

Proof: If S is a separator in $\mathbf{A}$, then $(\mathbf{A}, \hom(S, -))$ is a construct. If C is a coseparator in $\mathbf{A}$, then $(\mathbf{A}, \mathcal{Q} \circ \hom(-, C)^{op})$ is a construct, where $\mathcal{Q}: \mathbf{Set}^{op} \to \mathbf{Set}$ is the contravariant power-set functor. $\square$

7.90 EXAMPLES

(1) Fibre-small transportable constructs for which the epimorphisms are precisely the morphisms with surjective underlying functions must be co-wellpowered.

(2) Constructs are frequently co-wellpowered, even when they have non-surjective epimorphisms; e.g., **Sgr**, **Rng**, and **Haus**. However, in these cases establishing co-wellpoweredness is more involved.

(3) The construct of Urysohn spaces[43] is not co-wellpowered. The proof is nontrivial. Another non-co-wellpowered subcategory of **Top** is described in Exercise 7L. Both of these constructs are regular co-wellpowered (since they are constructs). **Ord** is obviously not co-wellpowered.

(4) Constructs need not be extremally co-wellpowered: consider the construct whose objects are all sets and whose morphisms are all identities and all constant functions from sets A to sets B, where the cardinality of B is greater than that of A.

Suggestions for Further Reading

Grothendieck, A. Sur quelque points d'algébre homologique. *Tohoku Math. J.* **9** (1957): 119–221.

Isbell, J. R. Some remarks concerning categories and subspaces. *Canad. J. Math.* **9** (1957): 563–577.

Burgess, W. The meaning of mono and epi in some familiar categories. *Canad. Math. Bull.* **8** (1965): 759–769.

Baron, S. Note on epi in T_0. *Canad. Math. Bull.* **11** (1968): 503–504.

Isbell, J. R. Epimorphisms and dominions IV. *J. London Math. Soc.* **1** (1969): 265–273.

Poguntke, D. Epimorphisms of compact groups are onto. *Proc. Amer. Math. Soc.*, **26** (1970): 503–504.

Pareigis, B., and M. E. Sweedler On generators and cogenerators. *Manuscr. Math.* **2** (1970): 49–66.

Schröder, J. The category of Urysohn spaces is not cowellpowered. *Topol. Appl.* **16** (1983): 237–241.

EXERCISES

7A. Zero Morphisms

A morphism $A \xrightarrow{f} B$ is called a **constant morphism** provided that for any pair $A' \underset{s}{\overset{r}{\rightrightarrows}} A$ of morphisms we have $f \circ r = f \circ s$. Morphisms that are simultaneously constant and coconstant are called **zero morphisms**. Show that

(a) If f is a constant morphism then so is any composite $h \circ f \circ g$.

(b) If f can be factored through a terminal object, then f is constant.

(c) If **A** has a zero object, 0, then an **A**-morphism is a zero-morphism if and only if it can be factored through 0.

---
[43] A topological space is called a **Urysohn space** provided that any two distinct points have disjoint closed neighborhoods.

7B. Pointed Categories

A category is called **pointed** provided that each of its morphism sets $\hom(A, B)$ contains a zero morphism. Show that a category is pointed if and only if it is a full subcategory of a category with a zero object.

7C. Kernels and Normal Monomorphisms

In a pointed category an equalizer of $A \xrightarrow{f} B$ and a zero morphism $A \to B$ is called a **kernel** of f. A morphism is called a **normal monomorphism** provided that it is the kernel of some morphism. Show that

(a) Every normal monomorphism is regular.

(b) In **Grp** the normal monomorphisms are the embeddings of normal subgroups, whereas the regular monomorphisms are the embeddings of subgroups. [Use 7H(a)].

(c) If $A \xrightarrow{f} B \xrightarrow{g} C$ are morphisms in a pointed category and g is a monomorphism, then f is a kernel of g if and only if A is a zero object.

(d) Each nonempty pointed category has a zero object if each of its morphisms has a kernel.

(e) If **A** is a pointed category such that each of its morphisms has a kernel and a cokernel, then the following hold:

(1) A morphism f is a normal monomorphism if and only if it is a kernel of a cokernel of f.

(2) For each object A the preordered class of all non-isomorphic normal subobjects of A is anti-isomorphic with the preordered class of all non-isomorphic normal quotient objects of A.

7D. Strict Monomorphisms

A morphism $A \xrightarrow{f} B$ is called a **strict monomorphism** provided that whenever $A' \xrightarrow{f'} B$ is a morphism with the property that for all morphisms $B \underset{s}{\overset{r}{\rightrightarrows}} C$, $r \circ f = s \circ f$ implies that $r \circ f' = s \circ f'$, then there exists a unique morphism $A' \xrightarrow{\overline{f}} A$ with $f' = f \circ \overline{f}$. Show that:

(a) Every regular monomorphism is a strict monomorphism, but not vice versa.

(b) Every strict monomorphism is an extremal monomorphism, but not vice versa (cf. 14I).

(c) If $A \xrightarrow{f} B$ is an extremal monomorphism and $B \xrightarrow{g} C$ is a strict monomorphism, then $A \xrightarrow{f} B \xrightarrow{g} C$ is an extremal monomorphism.

(d) If $A \xrightarrow{f} B$ is a strict monomorphism and $B \xrightarrow{g} C$ is a section, then $A \xrightarrow{f} B \xrightarrow{g} C$ is a strict monomorphism.

(e) The composite of two strict monomorphisms is an extremal but not necessarily strict monomorphism.

(f) If $A \xrightarrow{f} B \xrightarrow{g} C$ is a strict monomorphism and $B \xrightarrow{g} C$ is a monomorphism, then $A \xrightarrow{f} B$ is a strict monomorphism.

7E. Sections
Show that

(a) Every morphism with a terminal domain is a section.

(b) If $A \xrightarrow{f} B \xrightarrow{g} C$ is a section and $A \xrightarrow{f} B$ is an epimorphism, then $B \xrightarrow{g} C$ is a section.

(c) If $A \xrightarrow{f} B$ is a function and A has at least two elements, then in **Set** the following are equivalent:

   (1) f is an isomorphism,

   (2) there exists precisely one function $B \xrightarrow{g} A$ with $g \circ f = id_A$.

(d) If **A** is a full subcategory of **B**, then an **A**-reflection arrow is a section if and only if it is an isomorphism.

(e) In a construct each section (but not necessarily each regular monomorphism) is injective.

(f) If a natural transformation $\tau = (\tau_A)$ regarded as a morphism in the functor quasi-category $[\mathbf{A}, \mathbf{B}]$ is a section, then each τ_A is a **B**-section (but the converse need not be true).

7F. Full Embeddings of Reflective Subcategories
Let **A** be a full reflective subcategory of **B**. Show that the inclusion functor $\mathbf{A} \hookrightarrow \mathbf{B}$

(a) preserves and reflects monomorphisms, sections, retractions, and isomorphisms,

(b) preserves but need not reflect regular monomorphisms, extremal monomorphisms, and strict monomorphisms,

(c) reflects but need not preserve regular epimorphisms, extremal epimorphisms, strict epimorphisms, and epimorphisms.

7G. Faithful Functors
(a) Show that a faithful functor that reflects extremal epimorphisms must reflect isomorphisms.

(b) Show that a functor $F : \mathbf{A} \to \mathbf{B}$ that **reflects equalizers** (i.e., whenever $E \xrightarrow{e} A$ and $A \underset{g}{\overset{f}{\rightrightarrows}} B$ are **A**-morphisms such that Fe is an equalizer of Ff and Fg, then e is an equalizer of f and g) is faithful.

7H. Epimorphisms for Groups

* (a) Show that if K is a subgroup of the (finite) group H, then there exists a (finite) group G and group homomorphisms $f_1, f_2 : H \to G$ such that

$$K = \{\, h \in H \mid f_1(h) = f_2(h) \,\}.$$

[Hint: Consider the set X obtained from the set $\{\, hK \mid h \in H \,\}$ of all left K-cosets of H by adjoining a single new element $\hat{K}$. Let G be the permutation group of X, and let $\rho : X \to X$ be the permutation that interchanges the elements eK ($= K$) and $\hat{K}$, and leaves all other elements of X fixed.
Define $f_1, f_2 : H \to G$ by

$$f_1(h)(S) = \begin{cases} hh'K & \text{if } S = h'K \\ \hat{K} & \text{if } S = \hat{K} \end{cases}$$

$$f_2(h) = \rho \circ f_1(h) \circ \rho^{-1}.]$$

(b) Show that the epimorphisms in **Grp** are precisely the surjective homomorphisms.

(c) Show that the epimorphisms in the category of finite groups are precisely the surjective homomorphisms.

(d) Show that in the category of commutative cancellative semigroups, $A \xrightarrow{f} B$ is an epimorphism if and only if for each $b \in B$ there exist $a_1, a_2 \in A$ with $f(a_1) + b = f(a_2)$. [Hint: Embed each commutative cancellative semigroup in an abelian group. There the epimorphisms are surjective.]

(e) Show that in the category of torsion-free abelian groups, a morphism $A \xrightarrow{f} B$ is an epimorphism if and only if the factor group $B/f[A]$ is a torsion group.

7I. Epimorphisms in FHaus

(a) Show that in **FHaus** a morphism $A \xrightarrow{f} B$ is an epimorphism if and only if each morphism $B \xrightarrow{g} \mathbb{R}$ is constant whenever $g \circ f$ is constant.

(b) Let B be the space with underlying set $\mathbb{R}$ that has $\tau \cup \{\mathbb{R} \setminus \mathbb{Q}\}$ as a subbase for its topology (where τ is the usual topology on $\mathbb{R}$). Let A be the subspace of B with underlying set $\mathbb{Q}$. Show that the (closed) embedding $A \hookrightarrow B$ is an epimorphism in **FHaus**. Notice that this implies that $A \hookrightarrow B$ is not an extremal monomorphism in **FHaus**.

7J. Regular Monomorphisms in FHaus

In the category of functionally Hausdorff spaces (cf. 7.65) consider the discrete spaces A and B with underlying sets $\{\frac{1}{n} \mid n \in \mathbb{N}^+\}$ and $\{0\} \cup \{\frac{1}{n} \mid n \in \mathbb{N}^+\}$, the space C with underlying set $\mathbb{R}$, having $\tau \cup \{\mathbb{R} \setminus A\}$ as a subbase for its topology (where τ is the usual topology on $\mathbb{R}$), and the space D obtained as a quotient of C by identifying the points 0 and 1.

(a) Show that the natural inclusions $A \xrightarrow{f} B$ and $B \xrightarrow{g} C$ are regular monomorphisms, but that $g \circ f$ is not regular.

(b) Show that if $C \xrightarrow{h} D$ is the natural quotient map, then $h \circ (g \circ f)$ is a regular monomorphism, but that $g \circ f$ is not regular.

7K. Subobjects

Show that it is possible for (A, m) and (A', m') to be non-isomorphic subobjects of an object B even though A and A' are isomorphic.

* 7L. A Non-co-wellpowered Category of Topological Spaces

Let **A** be the full subcategory of **Top** that consists of those topological spaces in which every compact subspace is Hausdorff. Show that

(a) **A** is a reflective subcategory of **Top**, and

(b) **A** is not co-wellpowered.

7M. Coequalizers

Let $A \xrightarrow{e} B$ be an epimorphism and let $B \underset{g}{\overset{f}{\rightrightarrows}} C$ be a pair of morphisms. Show that c is a coequalizer of f and g if and only if c is a coequalizer of $f \circ e$ and $g \circ e$.

7N. Extremal Monomorphisms

Show that the composite of two extremal monomorphisms need not be extremal.

7O. Epi = RegEpi and Mono = RegMono

Show that the above equations hold in any of the following categories: **Set**, **Vec**, **Ab**, **Grp**, **Alg**(Ω), and **HComp**.

* 7P. A Characterization of Monomorphisms by Sections

Let f be a morphism in a category **A**. Show that f is a monomorphism in **A** if and only if **A** can be embedded into a category **B** such that f is a section in **B**.

7Q. Separating Sets and Concretizable Categories

(a) Show that if **A** has a separating set $\{A_i \mid i \in I\}$ and **I** is the discrete category associated with I, then **A** is concretizable over $[\mathbf{I}, \mathbf{Set}]$. (Cf. Proposition 7.12.)

(b) Show that if **A** has a separating set, then **A** is concretizable over **Set**.

(c) Exhibit a construct that has no separating set.

7R. Epi-Transformations and Mono-Transformations

Let $S, T : \mathbf{X} \to \mathbf{Set}$ be functors. Show that

(a) If there exists an Epi-transformation from S to T, then $\mathbf{Spa}(T)$ is concretely isomorphic to a reflective modification of $\mathbf{Spa}(S)$.

(b) If there exists a Mono-transformation from S to T, then $\mathbf{Spa}(S)$ is concretely isomorphic to a coreflective modification of $\mathbf{Spa}(T)$.

7S. Regular Epimorphisms in Cat

(a) Let $\mathbf{A} \xrightarrow{F} \mathbf{B}$ be the functor described in 7.40(6), let $\mathbf{B} \xrightarrow{G} \mathbf{C}$ be the functor described in 7.76(1), let $\mathbf{D}$ be the category $\bullet \underset{g}{\overset{f}{\rightrightarrows}} \bullet$, let $\mathbf{A} \xrightarrow{H} \mathbf{D}$ be the inclusion functor, and let $\mathbf{D} \xrightarrow{K} \mathbf{B}$ be the functor defined by $K(g) = 1$ and $K(f) = 2$. Show that

  (1) F and G are regular epimorphisms in $\mathbf{Cat}$, but $G \circ F$ is not a regular epimorphism,

  (2) $F = K \circ H$ is a regular epimorphism in $\mathbf{Cat}$, but K is not a regular epimorphism,

(b) Consider the functors:

$$A_1 \xrightarrow{f_1} B_1 \qquad A_2 \xrightarrow{f_2} B_2 \qquad A_1 \xrightarrow{f_1} B_1 = A_2 \xrightarrow{f_2} B_2 \qquad A_2 \xrightarrow{f_1} \bullet \xrightarrow{f_2} B_2$$
$$\xrightarrow{F} \qquad \xrightarrow{f} \qquad \xrightarrow{G} \qquad \| \qquad \|$$
$$A_3 \xrightarrow[f_3]{} B_3 \qquad A_3 \xrightarrow[f_3]{} B_3 \qquad A_3 \xrightarrow[f=f_3]{} B_3$$

Show that in $\mathbf{Cat}$:

(1) F is a regular epimorphism,

(2) G is a retraction,

(3) $G \circ F$ is not a regular epimorphism.

[Contrast this with Exercise 10M.]

8 Objects and morphisms in concrete categories

In this section we study special objects and morphisms in concrete categories. It will become apparent that these concepts depend heavily upon the underlying functors, and, hence, can be defined in the context of arbitrary (not necessarily faithful) functors. The notational conventions of Remark 6.22 will be used throughout this section. In particular, forgetful functors will be denoted by | |.

DISCRETE AND INDISCRETE OBJECTS

8.1 DEFINITION
An object A in a concrete category **A** over **X** is called **discrete** whenever, for each object B, every **X**-morphism $|A| \to |B|$ is an **A**-morphism.

8.2 EXAMPLES
(1) In the construct **Pos** the discrete objects are precisely the sets ordered by equality.

(2) In **Cat** the discrete objects are the discrete small categories.

(3) In **Top** the discrete objects are the discrete topological spaces.

(4) Let **PMet** denote the construct of pseudometric spaces[44] and contractions. For each set X, the object (X, d), where d is defined by

$$d(x, y) = \begin{cases} 0, & \text{if } x = y \\ \infty, & \text{if } x \neq y, \end{cases}$$

is a discrete object in **PMet**. In contrast, in the construct **Met** no space with at least two points is discrete.

(5) The constructs **Vec**, **Grp**, and Σ-**Seq** have no discrete objects. For $\Omega \neq \emptyset$, **Alg**(Ω) has no discrete object, with the possible exception of the empty Ω-algebra.

(6) In **TopGrp**, considered as a concrete category over **Grp**, the discrete objects are the topological groups with discrete topology. In **TopGrp**, considered as a concrete category over **Top**, no object is discrete.

---
[44] A pseudometric on a set X is a function $d: X \times X \to [0, \infty]$, such that:
(1) $d(x, x) = 0$ for all $x \in X$,
(2) $d(x, y) = d(y, x)$ for all $x, y \in X$,
(3) $d(x, z) \leq d(x, y) + d(y, z)$ for all $x, y, z \in X$.

(7) In an abstract category considered as concrete over itself (w.r.t. the identity functor) every object is discrete. In a poset A, considered as a concrete category over **1**, an object is discrete if and only if it is the smallest element of A.

Next we consider the concrete dual (5.20) of the concept of discrete object.

8.3 DEFINITION

An object A in a concrete category **A** over **X** is called **indiscrete** whenever, for each object B, every **X**-morphism $|B| \to |A|$ is an **A**-morphism.

8.4 REMARK

Notice that if A is a discrete (resp. indiscrete) object in the concrete category **A**, then A is the smallest (resp. largest) element in the fibre of $|A|$ (cf. 5.4). The converse need not hold, as seen below [8.5(2)].

8.5 EXAMPLES

(1) In the construct **Pos** the only indiscrete objects are the empty poset and the singleton posets; similarly in **Cat**.

(2) In **Top** the indiscrete objects are precisely the indiscrete topological spaces. In the full subcategory **Top**$_1$ of **Top**, consisting of all T_1-spaces, only the empty space and the singleton spaces are indiscrete. However, each fibre has a largest element — the cofinite topology.

(3) In **PMet** (8.2) the indiscrete objects are precisely those for which every pair of points has distance 0. The only indiscrete objects in **Met** are the metric spaces with at most one point.

(4) In **Vec**, **Grp**, and **Mon** only the trivial objects with precisely one element are indiscrete.

(5) In Σ-**Seq** an acceptor is indiscrete if and only if it has only one state and that state is final.

(6) If **TopGrp** is considered as a concrete category over **Grp**, then the indiscrete objects are the topological groups with indiscrete topology. If **TopGrp** is considered as a concrete category over **Top** or **Set**, then the indiscrete objects are trivial.

EMBEDDINGS

In §7 several classes of monomorphisms have been introduced in order to formalize the intuitive concept of "embeddings of subobjects". However, none of these concepts works in all situations. (Recall that monomorphisms fail in **Top** [7.33(3)] and extremal and regular monomorphisms fail in **Sgr** [7.58(4), 7I].) Below, we introduce the concept of *embedding*, which agrees with the intuitive notion of "embeddings of subobjects" in every familiar construct.

8.6 DEFINITION
Let **A** be a concrete category over **X**.

(1) An **A**-morphism $A \xrightarrow{f} B$ is called **initial**[45] provided that for any **A**-object C an **X**-morphism $|C| \xrightarrow{g} |A|$ is an **A**-morphism whenever $|C| \xrightarrow{f \circ g} |B|$ is an **A**-morphism.

(2) An initial morphism $A \xrightarrow{f} B$ that has a monomorphic underlying **X**-morphism $|A| \xrightarrow{f} |B|$ is called an **embedding**.

(3) If $A \xrightarrow{f} B$ is an embedding, then (f, B) is called an **extension** of A and (A, f) is called an **initial subobject** of B.

8.7 PROPOSITION
For any concrete category the following hold:

(1) Each embedding is a monomorphism.

(2) Each section (and in particular each isomorphism) is an embedding.

(3) If the forgetful functor preserves regular monomorphisms, then each regular monomorphism is an embedding.

Proof:
(1). Immediate from Proposition 7.37(2).

(2). Suppose that $A \xrightarrow{s} B$ and $B \xrightarrow{r} A$ are **A**-morphisms with $r \circ s = id_A$. Let $|C| \xrightarrow{g} |A|$ be an **X**-morphism for which $|C| \xrightarrow{s \circ g} |B|$ is an **A**-morphism. Then $g = r \circ (s \circ g)$ is an **A**-morphism. Thus $A \xrightarrow{s} B$ is initial, hence an embedding.

(3). Suppose that $A \xrightarrow{m} B$ is an equalizer of a pair of morphisms (u, v), and let $g : |C| \to |A|$ be an **X**-morphism such that $m \circ g$ is an **A**-morphism. Then since $u \circ (m \circ g) = v \circ (m \circ g)$, there exists an **A**-morphism $k : C \to A$ with $m \circ g = m \circ k$. Since, by assumption, m is a **X**-monomorphism, this implies that $k = g$. Thus $g = k : C \to A$ is an **A**-morphism. Hence m is initial, and so is an embedding. □

8.8 EXAMPLES[46]
(1) If an abstract category **A** is considered to be concrete over itself via the identity functor, then every morphism is initial. In particular,
$$\text{Emb}(\mathbf{A}) = \text{Mono}(\mathbf{A}).$$

(2) In the construct $\mathbf{A} = \mathbf{Top}$ a continuous map $f : (X, \tau) \to (Y, \sigma)$ is initial if and only if τ is the "initial topology" with respect to f and σ, i.e., $\tau = \{ f^{-1}[S] \mid S \in \sigma \}$. Thus embeddings are precisely the "topological embeddings", i.e., homeomorphisms onto subspaces. In particular,
$$\text{RegMono}(\mathbf{A}) = \text{ExtrMono}(\mathbf{A}) = \text{Emb}(\mathbf{A}) \subsetneq \text{Mono}(\mathbf{A}).$$

---
[45] Notwithstanding their names, the concepts of initial object (7.1) and initial morphism are unrelated.
[46] Emb(**A**) is the class of all **A**-embeddings. Similarly, Mono(**A**) is the class of all **A**-monomorphisms, etc. See the Table of Symbols.

(3) In the construct $\mathbf{A} = \mathbf{Haus}$ the initial morphisms are precisely the topological embeddings. Hence

$$\mathrm{ExtrMono}(\mathbf{A}) \subsetneq \mathrm{Init}(\mathbf{A}) = \mathrm{Emb}(\mathbf{A}) \subsetneq \mathrm{Mono}(\mathbf{A}).$$

(4) In the construct $\mathbf{A} = \mathbf{FHaus}$ of functionally Hausdorff spaces the initial morphisms are precisely the topological embeddings. Hence (cf. 7.65, 7I and 7J)

$$\mathrm{RegMono}(\mathbf{A}) \subsetneq \mathrm{ExtrMono}(\mathbf{A}) \subsetneq \mathrm{Init}(\mathbf{A}) = \mathrm{Emb}(\mathbf{A}) \subsetneq \mathrm{Mono}(\mathbf{A}).$$

(5) In some "algebraic" constructs $\mathbf{A}$ such as $\mathbf{Vec}$ and $\mathbf{Grp}$ the initial morphisms coincide with each of the other "reasonable" families of monomorphisms; i.e.,

$$\mathrm{RegMono}(\mathbf{A}) = \mathrm{Init}(\mathbf{A}) = \mathrm{Emb}(\mathbf{A}) = \mathrm{Mono}(\mathbf{A}).$$

However, if $\mathbf{A}$ is any of the constructs $\mathbf{Rng}$, $\mathbf{Ban}_b$, or $(\mathbf{Ban}, O)$, then

$$\mathrm{ExtrMono}(\mathbf{A}) \subsetneq \mathrm{Init}(\mathbf{A}) = \mathrm{Emb}(\mathbf{A}) = \mathrm{Mono}(\mathbf{A}).$$

(6) In the construct $\mathbf{A} = (\mathbf{Ban}, U)$ injective morphisms usually fail to be initial; instead the initial morphisms are precisely the isometric embeddings of (automatically closed) subspaces. Thus

$$\mathrm{ExtrMono}(\mathbf{A}) = \mathrm{Init}(\mathbf{A}) = \mathrm{Emb}(\mathbf{A}) \subsetneq \mathrm{Mono}(\mathbf{A}).$$

(7) In the construct $\mathbf{A} = \mathbf{Met}$ the initial morphisms are precisely the isometric embeddings of (not necessarily closed) subspaces. Thus

$$\mathrm{ExtrMono}(\mathbf{A}) \subsetneq \mathrm{Init}(\mathbf{A}) = \mathrm{Emb}(\mathbf{A}) \subsetneq \mathrm{Mono}(\mathbf{A}).$$

(8) In the constructs $\mathbf{Rel}$ and $\mathbf{Pos}$ a morphism $f : (X, \rho) \to (Y, \sigma)$ is initial if and only if the equivalence: $x \rho y \Leftrightarrow f(x) \sigma f(y)$ holds. Thus in $\mathbf{Pos}$ initial morphisms are precisely the embeddings, and in both cases

$$\mathrm{ExtrMono}(\mathbf{A}) = \mathrm{Emb}(\mathbf{A}) \subsetneq \mathrm{Mono}(\mathbf{A}).$$

(9) Extremal monomorphisms need not be embeddings. In the full subconstruct $\mathbf{A} = \mathbf{DRail}$ of $\mathbf{Top}$ consisting of all discrete spaces and all infinite indiscrete spaces, every injective map with finite discrete domain and indiscrete range is an extremal monomorphism, but (for non-trivial domain) not an embedding. Hence

$$\mathrm{ExtrMono}(\mathbf{A}) \not\subseteq \mathrm{Emb}(\mathbf{A}) = \mathrm{RegMono}(\mathbf{A}) \subsetneq \mathrm{ExtrMono}(\mathbf{A}).$$

(10) In the construct $\mathbf{Cat}$ the embeddings are precisely the embedding functors.

(11) In the construct $\Sigma\text{-}\mathbf{Seq}$ embeddings are those injective simulations $A \xrightarrow{f} B$ for which any state $a \in A$, with $f(a)$ final in B, must itself be final in A.

(12) In a preordered class **A** considered as a concrete category over **1**, initial morphisms and isomorphisms coincide. Thus, if **A** is a nondiscrete poset, then

$$\text{Iso}(\mathbf{A}) = \text{Init}(\mathbf{A}) = \text{Emb}(\mathbf{A}) = \text{ExtrMono}(\mathbf{A}) \subsetneq \text{Mono}(\mathbf{A}).$$

(13) If **TopGrp** is considered as a concrete category over **Grp** (resp. over **Top**), then a morphism $G \xrightarrow{f} \hat{G}$ is initial if and only if G has the initial topology with respect to f (resp. if and only if it is injective).

8.9 PROPOSITION

(1) If $A \xrightarrow{f} B$ and $B \xrightarrow{g} C$ are initial morphisms (resp. embeddings), then $A \xrightarrow{g \circ f} C$ is an initial morphism (resp. an embedding).

(2) If $A \xrightarrow{g \circ f} C$ is an initial morphism (resp. an embedding), then f is initial (resp. an embedding). □

QUOTIENT MORPHISMS

The concepts of *final morphism* and *quotient morphism* are dual to the concepts of initial morphism and embedding, respectively.

8.10 DEFINITION

Let **A** be a concrete category over **X**.

(1) An **A**-morphism $A \xrightarrow{f} B$ is called **final** provided that for any **A**-object C, an **X**-morphism $|B| \xrightarrow{g} |C|$ is an **A**-morphism whenever $|A| \xrightarrow{g \circ f} |C|$ is an **A**-morphism.

(2) A final morphism $A \xrightarrow{f} B$ with epimorphic underlying **X**-morphism $|A| \xrightarrow{f} |B|$ is called a **quotient morphism**.

(3) If $A \xrightarrow{f} B$ is a quotient morphism, then (f, B) is called a **final quotient object** of A.

8.11 EXAMPLES

(1) In the construct **Top** a continuous function $f : (X, \tau) \to (Y, \sigma)$ is final if and only if $\sigma = \{A \subseteq Y \mid f^{-1}[A] \in \tau\}$; i.e., σ is the "final topology" on Y with respect to f and τ. Thus in **Top** the quotient morphisms are the topological quotient maps.

(2) In the construct **Rel** a morphism $f : (X, \rho) \to (Y, \sigma)$ is final if and only if σ is the final relation on Y with respect to f and ρ; i.e., $\sigma = \{(f(x), f(y)) \mid (x, y) \in \rho\}$.

(3) For any of the constructs **Grp**, **Ab**, **Vec**, **Boo**, **Lat**, **HComp**, **BooSp**, **Mon**, **Sgr**, and **Σ-Seq**, we have[47]

$$\text{Final}(\mathbf{A}) = \text{Quot}(\mathbf{A}) = \text{RegEpi}(\mathbf{A}).$$

---

[47] Final(**A**) is the class of all final morphisms in **A**. Similarly, Quot(**A**) is the class of all **A**-quotient morphisms, etc. See the Table of Symbols.

(4) For the constructs **Cat** and (**Ban**, O) there exist regular epimorphisms that are not final. Some examples of such morphisms are the non-surjective regular epimorphisms described in Example 7.72(5).

(5) In the construct **Cat** there exist surjective extremal epimorphisms that are not final. An example of this is the functor $G \circ F : \mathbf{A} \to \mathbf{C}$ described in Remark 7.76(1).

(6) A morphism $(X, d) \xrightarrow{f} (Y, \hat{d})$ is final in **PMet** if and only if

$$\hat{d}(y, y') = \inf \left\{ \sum_{i=1}^{n-1} dist_d(f^{-1}(y_i), f^{-1}(y_{i+1})) \,|\, (y_1, \ldots, y_n) \text{ is a finite sequence in } Y \right.$$
$$\left. \text{ with } y_1 = y \text{ and } y_n = y' \right\}.$$

8.12 PROPOSITION
For any concrete category the following hold:

(1) Each quotient morphism is an epimorphism.

(2) Each retraction (and in particular each isomorphism) is a quotient morphism.

(3) If the forgetful functor preserves regular epimorphisms, then each regular epimorphism is a quotient morphism. □

8.13 PROPOSITION
(1) If $A \xrightarrow{f} B$ and $B \xrightarrow{g} C$ are final morphisms (resp. quotient morphisms), then $A \xrightarrow{g \circ f} C$ is final (resp. a quotient morphism).

(2) If $A \xrightarrow{g \circ f} C$ is a final morphism (resp. a quotient morphism), then g is final (resp. a quotient morphism). □

8.14 PROPOSITION
In a concrete category $\mathbf{A}$ over $\mathbf{X}$, the following conditions are equivalent for each $\mathbf{A}$-morphism f:

(1) f is an $\mathbf{A}$-isomorphism.

(2) f is an initial morphism and an $\mathbf{X}$-isomorphism.

(3) f is a final morphism and an $\mathbf{X}$-isomorphism.

Proof: (1) ⇒ (2) follows from Proposition 3.21 and Proposition 8.7(2).

(2) ⇒ (1). If $A \xrightarrow{f} B$ is an initial $\mathbf{X}$-isomorphism, then $|B| \xrightarrow{f^{-1}} |A| \xrightarrow{f} |B| = |B| \xrightarrow{id_B} |B|$ implies, by initiality, that $|B| \xrightarrow{f^{-1}} |A|$ is an $\mathbf{A}$-morphism. Thus $A \xrightarrow{f} B$ is an $\mathbf{A}$-isomorphism.

(1) ⇔ (3) follows by duality from (1) ⇔ (2). □

STRUCTURED ARROWS

For concrete categories **A** over **X**, we have introduced properties of **A**-objects and of **A**-morphisms. Next we will investigate properties of pairs (f, A) consisting of an **X**-morphism f and a suitably related **A**-object A. First we define three concepts of "generation". Whereas for "topological" constructs the concept of *generation* works better than that of *extremal generation*, and for "algebraic" constructs the concept of extremal generation works better than that of generation, the concept of *concrete generation* works well in either setting.

8.15 DEFINITION
Let **A** be a concrete category over **X**.

(1) A **structured arrow with domain** X is a pair (f, A) consisting of an **A**-object A and an **X**-morphism $X \xrightarrow{f} |A|$. Such a structured arrow will often be denoted by $X \xrightarrow{f} |A|$.

(2) A structured arrow (f, A) is said to be **generating** provided that for any pair of **A**-morphisms $r, s : A \to B$ the equality $r \circ f = s \circ f$ implies that $r = s$.

(3) A generating arrow (f, A) is called **extremally generating** (resp. **concretely generating**) provided that each **A**-monomorphism (resp. **A**-embedding) $m : A' \to A$, through which f factors (i.e., $f = m \circ g$ for some **X**-morphism g), is an **A**-isomorphism.

(4) In a construct, an object A is (**extremally** resp. **concretely**) **generated by a subset** X **of** $|A|$ provided that the inclusion map $X \hookrightarrow |A|$ is (extremally resp. concretely) generating.

8.16 PROPOSITION
In a concrete category **A** *over* **X** *the following hold for each structured arrow* $f : X \to |A|$:

(1) If (f, A) is extremally generating, then (f, A) is concretely generating.

(2) If (f, A) is concretely generating, then (f, A) is generating.

*(3) If $X \xrightarrow{f} |A|$ is an **X**-epimorphism, then (f, A) is generating.*

(4) If $X \xrightarrow{f} |A|$ is an extremal epimorphism in **X**, *and if* $|\ |$ *preserves monomorphisms, then (f, A) is extremally generating.* □

8.17 EXAMPLES[48]
(1) If an abstract category **A** is considered to be concrete over itself via the identity functor, then an **A**-morphism $A \xrightarrow{f} B$, considered as a structured arrow (f, B), is

---

[48] Gen(**A**) is the class of all generating structured arrows in **A**, ExtrGen(**A**) is the class of all extremally generating structured arrows in **A**, ConcGen(**A**) is the class of all concretely generating structured arrows in **A**, etc. See the Table of Symbols.

generating (resp. extremally or concretely generating) if and only if f is an epimorphism (resp. an extremal epimorphism). That is,

$$\text{Gen}(\mathbf{A}) = \text{Epi}(\mathbf{A}) \quad \text{and} \quad \text{ExtrGen}(\mathbf{A}) = \text{ConcGen}(\mathbf{A}) = \text{ExtrEpi}(\mathbf{A}).$$

(2) In **Vec**, **Grp**, **Sgr**, **Rng**, and other algebraic constructs, the concepts of concrete generation and of extremal generation coincide with the familiar (non-categorical) concept of generation.
In the constructs **Sgr** and **Rng** the inclusion map $\mathbb{Z} \hookrightarrow \mathbb{Q}$ is generating, but is not concretely generating [cf. 7.40(5)].

(3) In the construct $\mathbf{A} = \mathbf{Top}$ we have

$$\text{ConcGen}(\mathbf{A}) = \text{Gen}(\mathbf{A}) = \text{Surjective maps, and}$$
$$\text{ExtrGen}(\mathbf{A}) = \text{Surjective maps with discrete codomain.}$$

(4) In the construct $\mathbf{A} = \mathbf{Haus}$ we have

$$\text{Gen}(\mathbf{A}) = \text{Dense maps,}$$
$$\text{ConcGen}(\mathbf{A}) = \text{Surjective maps, and}$$
$$\text{ExtrGen}(\mathbf{A}) = \text{Surjective maps with discrete codomain.}$$

(5) In a partially ordered set $\mathbf{A}$, considered as a concrete category over **1**, every structured arrow is concretely generating. A structured arrow (f, a) is extremally generating if and only if a is a minimal element of $\mathbf{A}$.

(6) In Σ-**Seq**, given an acceptor A, the inclusion map of $\{q_0\}$, where q_0 is the initial state, concretely generates A if and only if A is reachable.

8.18 REMARK
In a concrete category $\mathbf{A}$, if an $\mathbf{A}$-morphism $A \xrightarrow{f} B$ is regarded as a structured arrow (f, B) with domain $|A|$, then the following hold:

(1) $A \xrightarrow{f} B$ is an epimorphism if and only if (f, B) is generating.

(2) If (f, B) is extremally generating and the forgetful functor preserves monomorphisms, then $A \xrightarrow{f} B$ is an extremal epimorphism.

(3) If $A \xrightarrow{f} B$ is an extremal epimorphism, then (f, B) is concretely generating.

8.19 DEFINITION
Let $\mathbf{A}$ be a concrete category over $\mathbf{X}$.

(1) Structured arrows (f, A) and (g, B) in $\mathbf{A}$ with the same domain are said to be **isomorphic** provided that there exists an $\mathbf{A}$-isomorphism $k : A \to B$ with $k \circ f = g$.

(2) **A** is said to be **concretely co-wellpowered** provided that for each **X**-object X any class of pairwise non-isomorphic concretely generating arrows with domain X is a set.

8.20 EXAMPLES
Most of the familiar constructs such as **Vec, Grp, Top, HComp, Pos, Alg(Ω)** for each Ω, and Σ-**Seq** for each Σ are concretely co-wellpowered. The construct with objects all sets and morphisms all identity mappings is fibre-small and co-wellpowered, but not concretely co-wellpowered. The constructs **CLat** and **CBoo** are not concretely co-wellpowered (cf. Exercise 8E). A proper class, ordered by equality and considered as a concrete category over **1**, is co-wellpowered, but is not concretely co-wellpowered.

8.21 PROPOSITION
Each concretely co-wellpowered concrete category is extremally co-wellpowered.

Proof: This follows immediately from the following two facts:
(1) Each extremal quotient (f, B) of A, considered as a structured arrow with domain $|A|$, is concretely generating. Cf. Remark 8.18(3).
(2) Extremal quotients (f_1, B_1) and (f_2, B_2) are isomorphic if and only if the structured arrows (f_1, B_1) and (f_2, B_2) are isomorphic. □

UNIVERSAL ARROWS AND FREE OBJECTS

8.22 DEFINITION
In a concrete category **A** over **X**

(1) a **universal arrow** over an **X**-object X is a structured arrow $X \xrightarrow{u} |A|$ with domain X that has the following universal property: for each structured arrow $X \xrightarrow{f} |B|$ with domain X there exists a unique **A**-morphism $\hat{f} : A \to B$ such that the triangle

$$X \xrightarrow{u} |A|$$
$$f \searrow \quad \downarrow \hat{f}$$
$$|B|$$

commutes,

(2) a **free object** over an **X**-object X is an **A**-object A such that there exists a universal arrow (u, A) over X.

8.23 EXAMPLES
(1) In a construct, an object A is a free object
   (a) over the empty set if and only if A is an initial object.

(b) over a singleton set if and only if A represents the forgetful functor (6.9).

(2) In the construct **Vec** each object is a free object over any basis for it.

(3) In the constructs **Top** and **Pos** the free objects are precisely the discrete ones.

(4) In the construct **Ab** free objects over X are the free abelian groups generated by X. They can be constructed as the group of all functions $p: X \to \mathbb{Z}$ with finite carrier, where addition is carried out componentwise, and, if A is such a group of functions, a universal arrow $u: X \to |A|$ is given by

$$(u(x))(z) = \begin{cases} 1, & \text{if } z = x \\ 0, & \text{if } z \neq x. \end{cases}$$

Similarly, the familiar free group generated by a set X is a free object over X in the construct **Grp**.

A non-free group and a free object

(5) In the construct **Mon** a free monoid X^* over a set X is a free object over X; elements are words (= finite sequences, including the empty one) formed from members of X, with operation $\bullet$ of concatenation: $(x_1, x_2, \ldots, x_n) \bullet (y_1, y_2, \ldots, y_m) = (x_1, x_2, \ldots, x_n, y_1, y_2, \ldots, y_m)$, and $u: X \to |A|$ is given by $u(x) = (x)$. Analogously, in the construct **Sgr** a free semigroup generated by X consists of all nonempty words formed from members of X.

(6) In each construct $\mathbf{Alg}(\Omega)$ the free objects over a set X can be described in a way similar to the description of initial objects [cf. 7.2(5)] except that in step (a) each element of $\Omega_0 \uplus X$ should be required to be a "term". Notice that in any $\mathbf{Alg}(\Sigma)$ a

free unary algebra over a singleton set is, by coincidence, the algebra Σ^* of all words with unary operations given by concatenations. A free algebra over a set X is the disjoint union of $\mathrm{card}(X)$ copies of Σ^* (cf. Exercise 8H).

(7) In the construct **CLat** an initial object is the two-point chain, a free object over a one-element set is the three-point chain, and a free object over a two-element set is the six-element lattice formed by replacing the middle member of a three-element chain by a four-element boolean algebra. There are no free objects over sets that have more than two elements (cf. Exercise 8G). In the construct **CBoo** an initial object is the two-point chain, the four-element boolean algebra is a free object over a one-element set, and a free object over n ($< \aleph_0$) generators has 2^{2^n} elements. There is no free object over any infinite set. (See Exercise 8F.)

(8) In **JCPos** the free objects are, up to isomorphism, precisely the power sets (regarded as **JCPos** objects). For each set X, the structured arrow $X \xrightarrow{u} |\mathcal{P}(X)|$, defined by $u(x) = \{x\}$, is a universal arrow.
However, in the construct Λ-**JCPos** (with objects triples $(X, \leq, \lambda)$ consisting of a complete lattice $(X, \leq)$ and a unary operation λ on X, and morphisms the join-preserving $\{\lambda\}$-homomorphisms) there do not exist free objects over any set. In particular, Λ-**JCPos** has no initial object [even though Λ-**JCPos** is extremally co-wellpowered, the empty set extremally generates arbitrarily large objects!], and the forgetful functor is not representable.

(9) In the construct **Cat** a free object over the set $\{0\}$ is a category of the form $\bullet \xrightarrow{g} \bullet$ (where the universal arrow u sends 0 to g).

(10) In the construct **Rng** the polynomial ring $\mathbb{Z}[M]$ over a set M of variables is a free object over M.

(11) Let Y be a nonempty set and consider **Set** as a construct via the forgetful functor $U = \hom(Y, -)$. Then the structured arrow $X \xrightarrow{u} U(X \times Y) = \hom(Y, X \times Y)$, defined by $(u(x))(y) = (x, y)$, is universal.

(12) To construct a universal arrow in (**Ban**, O) over a set X, let $\ell_1(X)$ be the subspace of the vector space K^X consisting of all $r = (r_x)_{x \in X}$ in K^X whose norm $\|r\| = \sum_{x \in X} |r_x|$ is finite. Then $\ell_1(X)$ is a Banach space. Define $X \xrightarrow{u} O(\ell_1(X))$ at y by the Dirac function $u(y) = (\delta_{yx})_{x \in X}$. Then $(u, \ell_1(X))$ is a universal arrow over X. Observe, for comparison, that for the construct (**Ban**, U) the only set having a universal arrow is the empty set, and that for the construct **Ban**$_b$ the only sets having universal arrows are the finite ones.

(13) In an abstract category, considered as a concrete category over itself via the identity functor, every object is free over itself.

(14) In an preordered set **A**, considered as a concrete category over **1**, an object a is free over the single object of **1** if and only if a is the smallest element of **A**.

(15) In the concrete category of **Rng** over **Mon** in which the forgetful functor "forgets addition" a universal arrow over a monoid M is given by the monoid ring $\mathbb{Z}[M]$ of M

over the additive group $(\mathbb{Z}, +)$ and the arrow $u : M \to |\mathbb{Z}[M]|$ defined by $u(x) = x$. Likewise in the concrete category of **Rng** over **Ab**, where multiplication is forgotten, a universal arrow over an abelian group G is given by the tensor ring over G.

(16) No acceptor is free in the construct Σ-**Seq**. (In fact, if A is an acceptor on n states and if B is the $(n+1)$-state acceptor in which the next-state maps $\delta(-, \sigma)$ are $(n+1)$-cycles, then there is no simulation from A to B.)

8.24 PROPOSITION
Every universal arrow is extremally generating.

Proof: Let $X \xrightarrow{u} |A|$ be a universal arrow. By the uniqueness requirement, (u, A) is generating. Let $A' \xrightarrow{m} A$ be an **A**-monomorphism, and let $X \xrightarrow{g} |A'|$ be an **X**-morphism with $u = m \circ g$. Since (u, A) is universal, there exists an **A**-morphism $A \xrightarrow{\hat{g}} A'$ with $g = \hat{g} \circ u$. Hence $A \xrightarrow{id_A} A$ and $A \xrightarrow{m \circ \hat{g}} A$ are **A**-morphisms with $id_A \circ u = (m \circ \hat{g}) \circ u$. By the uniqueness requirement in the definition of universal arrow, this implies that $id_A = m \circ \hat{g}$. Hence m is simultaneously an **A**-retraction and an **A**-monomorphism, and so is an **A**-isomorphism (7.36). □

8.25 PROPOSITION
*For any **X**-object X, universal arrows over X are essentially unique; i.e., any two universal arrows with domain X are isomorphic, and conversely, if $X \xrightarrow{u} |A|$ is a universal arrow and $A \xrightarrow{k} A'$ is an **A**-isomorphism, then $X \xrightarrow{k \circ u} |A'|$ is also universal.* $\boxed{A\ 4.19}$[49]

8.26 DEFINITION
A concrete category over **X** is said to **have free objects** provided that for each **X**-object X there exists a universal arrow over X.

8.27 EXAMPLES
By Examples 8.23 the constructs **Vec**, **Grp**, **Ab**, **Mon**, **Sgr**, **Alg**(Ω), **Top**, **Pos**, and (**Ban**, O) have free objects; but the constructs **CLat**, **CBoo**, and (**Ban**, U) don't. Neither do the constructs **Ban**$_b$ and **Met**; however, **PMet** does. A partially ordered set, considered as a concrete category over **1**, has free objects if and only if it has a smallest element.

8.28 PROPOSITION
*If a concrete category **A** over **X** has free objects, then an **A**-morphism is an **A**-monomorphism if and only if it is an **X**-monomorphism.*

---

[49] The symbol $\boxed{A\ 4.19}$ indicates that a proof of the preceding result can be obtained as a straightforward analogue of the proof of Proposition 4.19.

Proof: The sufficiency holds for all concrete categories [cf. Proposition 7.37(2)]. To see the necessity, let $A \xrightarrow{f} B$ be an **A**-monomorphism, and let $X \underset{s}{\overset{r}{\rightrightarrows}} |A|$ be a pair of **X**-morphisms with $f \circ r = f \circ s$. If $X \xrightarrow{u} |C|$ is a universal arrow over X, then there exist **A**-morphisms $C \underset{\hat{s}}{\overset{\hat{r}}{\rightrightarrows}} A$ with $\hat{r} \circ u = r$ and $\hat{s} \circ u = s$. Thus $(f \circ \hat{r}) \circ u = (f \circ \hat{s}) \circ u$. By the uniqueness requirement in the definition of universal arrow, this implies that $f \circ \hat{r} = f \circ \hat{s}$. Since f is an **A**-monomorphism, it follows that $\hat{r} = \hat{s}$, so that $r = s$. Hence f is an **X**-monomorphism. □

8.29 PROPOSITION

*If a construct **A** has a free object over a singleton set, then the monomorphisms in **A** are precisely those morphisms that are injective functions.*

Proof: By Example 8.23(1) the forgetful functor for **A** is representable. Hence the result follows from Corollary 7.38. □

OBJECTS AND MORPHISMS WITH RESPECT TO A FUNCTOR

8.30 DEFINITION

Let $G : \mathbf{A} \to \mathbf{B}$ be a functor, and let B be a **B**-object.

(1) A *G-structured arrow with domain B* is a pair (f, A) consisting of an **A**-object A and a **B**-morphism $f : B \to GA$.

(2) A *G*-structured arrow (f, A) with domain B is called

  (a) **generating** provided that for any pair of **A**-morphisms $A \underset{s}{\overset{r}{\rightrightarrows}} \hat{A}$, the equality $Gr \circ f = Gs \circ f$ implies that $r = s$,

  (b) **extremally generating** provided that it is generating and whenever $A' \xrightarrow{m} A$ is an **A**-monomorphism and (g, A') is a *G*-structured arrow with $f = G(m) \circ g$, then m is an **A**-isomorphism,

  (c) ***G*-universal for B** provided that for each *G*-structured arrow (f', A') with domain B there exists a unique **A**-morphism $A \xrightarrow{\hat{f}} A'$ with $f' = G(\hat{f}) \circ f$, i.e., such that the triangle

commutes.

8.31 EXAMPLES

(1) For concrete categories $(\mathbf{A}, U)$ over $\mathbf{X}$, we have the following:

$$U\text{-structured arrow} = \text{structured arrow},$$
$$U\text{-generating arrow} = \text{generating arrow},$$
$$\text{extremally } U\text{-generating arrow} = \text{extremally generating arrow},$$
$$U\text{-universal arrow} = \text{universal arrow}.$$

(2) If $\mathbf{A}$ is a subcategory of $\mathbf{B}$ and $E : \mathbf{A} \to \mathbf{B}$ is the associated inclusion functor, then an E-structured arrow (u, A) with domain B is E-universal if and only if $B \xrightarrow{u} A$ is an $\mathbf{A}$-reflection arrow for B.

(3) Let $\mathbf{A}$ be a category and let G be the unique functor from $\mathbf{A}$ to $\mathbf{1}$. Then a G-structured arrow (u, A) is G-universal if and only if A is an initial object in $\mathbf{A}$.

(4) If $G = \hom(A, -) : \mathbf{A} \to \mathbf{Set}$, then a G-structured arrow $X \xrightarrow{u} G(B) = \hom(A, B)$ is G-universal if and only if for every $\mathbf{A}$-object C and every family $(A \xrightarrow{f_x} C)_{x \in X}$ of $\mathbf{A}$-morphisms there exists a unique $\mathbf{A}$-morphism $B \xrightarrow{f} C$ with $f_x = f \circ u(x)$ for each $x \in X$. [By the definition of *coproducts* in §10, this is equivalent to $((u(x))_{x \in X}, B)$ being a coproduct of the family consisting of X copies of A; i.e., to B being an X-copower of A with injections $A \xrightarrow{u(x)} B$.]

(5) Minimal realization is universal. This means that for the category $\mathbf{Beh}$ of behaviors (i.e., triples (Σ, Y, b), where $\Sigma^* \xrightarrow{b} Y$ is a function) and behavior morphisms [i.e., $(f, g) : (\Sigma, Y, b) \to (\Sigma', Y', b')$, where $\Sigma \xrightarrow{f} \Sigma'$ and $Y \xrightarrow{g} Y'$ are functions with $(g \circ b)(\sigma_1 \sigma_2 \cdots \sigma_n) = b'(f(\sigma_1) f(\sigma_2) \cdots f(\sigma_n))$], the minimal realization functor $M : \mathbf{Beh} \to \mathbf{Aut}_r$ has universal arrows. Here $\mathbf{Aut}_r$ denotes the full subcategory of $\mathbf{Aut}$ [3.3(4)(b)] formed by those automata for which each state can be reached from the initial one, and M assigns to each behavior its minimal realization. An M-universal arrow $A \xrightarrow{\eta} M(\Sigma, Y, b_A)$ (where $b_A : \Sigma^* \to Y$ is the external behavior of A) is the unique simulation onto the minimal realization of b_A.

8.32 PROPOSITION

If $G : \mathbf{A} \to \mathbf{B}$ is a functor, then the following are equivalent:

(1) G is faithful,

(2) each $\mathbf{A}$-epimorphism, considered as a G-structured arrow, is generating,

(3) each $\mathbf{A}$-identity, considered as a G-structured arrow, is generating. □

8.33 PROPOSITION

Every G-universal arrow is extremally generating. $\boxed{A\ 8.24}$

8.34 DEFINITION
G-structured arrows (f, A) and (f', A') with the same domain are said to be **isomorphic** provided that there exists an **A**-isomorphism $k : A \to A'$ with $G(k) \circ f = f'$.

8.35 PROPOSITION
*For any functor $G : \mathbf{A} \to \mathbf{B}$ and any **B**-object B, G-universal arrows for B are essentially unique; i.e., any two G-universal arrows with domain B are isomorphic, and, conversely, if $B \xrightarrow{u} GA$ is a G-universal arrow and $A \xrightarrow{k} A'$ is an isomorphism, then $B \xrightarrow{Gk \circ u} GA'$ is also G-universal.* $\boxed{A\ 4.19}$

8.36 PROPOSITION
Let $G : \mathbf{A} \to \mathbf{B}$ be a functor. If the triangle

*commutes, where (u, A) is a G-universal arrow and $A \xrightarrow{\hat{f}} B$ is an **A**-morphism, then the following hold:*

(1) (f, B) is generating if and only if $\hat{f}$ is an epimorphism.

(2) (f, B) is extremally generating if and only if $\hat{f}$ is an extremal epimorphism.

Proof: It is clearly sufficient to show that whenever $\hat{f}$ is an extremal epimorphism, then (f, B) is extremally generating. Let $X \xrightarrow{g} GC$ be a G-structured arrow and let $C \xrightarrow{m} B$ be an **A**-monomorphism with $f = Gm \circ g$. Then there exists an **A**-morphism $A \xrightarrow{\hat{g}} C$ with $g = G\hat{g} \circ u$. Thus the equality $G(m \circ \hat{g}) \circ u = f = G\hat{f} \circ u$ implies that $m \circ \hat{g} = \hat{f}$. Hence m is an isomorphism. □

8.37 DEFINITION
(1) A functor $G : \mathbf{A} \to \mathbf{B}$ is called (**extremally**) **co-wellpowered** provided that for any **B**-object B, any class of pairwise non-isomorphic (extremally) generating G-structured arrows with domain B is a set.

(2) A faithful functor $G : \mathbf{A} \to \mathbf{B}$ is called **concretely co-wellpowered** provided that the concrete category $(\mathbf{A}, G)$ is concretely co-wellpowered.

8.38 PROPOSITION
If a faithful functor $G : \mathbf{A} \to \mathbf{B}$ is co-wellpowered, then so is $\mathbf{A}$. $\boxed{A\ 8.21}$

8.39 REMARK

If a faithful functor $G: \mathbf{A} \to \mathbf{B}$ is extremally co-wellpowered, then $\mathbf{A}$ need not be extremally co-wellpowered. Conversely, if $\mathbf{A}$ is co-wellpowered, then $G: \mathbf{A} \to \mathbf{B}$ need not even be extremally co-wellpowered. Consider, e.g., the unique functor from a large discrete category to **1**.
More can be said if, e.g., for every **B**-object B there exists a universal G-structured arrow for B. See 18.11 and 18B.

COSTRUCTURED ARROWS

All of the concepts relating to G-structured arrows have duals. In particular:

8.40 DEFINITION

Let $G: \mathbf{A} \to \mathbf{B}$ be a functor and let B be a **B**-object.

(1) A G-**costructured arrow with codomain** B is a pair (A, f) consisting of an **A**-object A and a **B**-morphism $GA \xrightarrow{f} B$.

(2) A G-costructured arrow (A, f) with codomain B is called G-**co-universal for** B provided that for each G-costructured arrow (A', f') with codomain B there exists a unique **A**-morphism $A' \xrightarrow{\hat{f}} A$ with $f' = f \circ G(\hat{f})$.

8.41 EXAMPLES

(1) If $\mathbf{A}$ is a subcategory of $\mathbf{B}$ and $E: \mathbf{A} \to \mathbf{B}$ is the associated inclusion functor, then an E-costructured arrow (A, u) with codomain B is E-co-universal if and only if $A \xrightarrow{u} B$ is an **A**-coreflection arrow for B.

(2) For forgetful functors U of familiar concrete categories, U-co-universal arrows are relatively rare. For example, **Grp** has U-co-universal arrows only for one-point sets, and **Pos** has U-co-universal arrows only for sets with at most one point. However,

(a) in the construct **Top** for every set X there exists a co-universal arrow $(X, \tau) \xrightarrow{id} X$, where τ is the indiscrete topology,

(b) if $T: \mathbf{X} \to \mathbf{Set}$ is a functor and $\mathbf{Spa}(T)$ is the associated concrete category over $\mathbf{X}$, then for every **X**-object X there exists a co-universal arrow $(X, T(X)) \xrightarrow{id_X} X$,

(c) in the constructs $\mathbf{Alg}(\Sigma)$ of unary algebras there exist co-universal arrows for each set X: consider the Σ-algebra X^{Σ^*} of all functions from Σ^* (the set of all words over Σ) into X, with the operation $\sigma \in \Sigma$ sending a function $g: \Sigma^* \to X$ to the function $g(\sigma_-)$, defined by: $g(\sigma_-)((\sigma_1 \cdots \sigma_n)) = g(\sigma\sigma_1 \cdots \sigma_n)$. Then $\varepsilon: X^{\Sigma^*} \to X$ given by $\varepsilon(g) = g(\emptyset)$ is co-universal. [In fact, given a Σ-algebra A and a function $f: A \to X$, the unique homomorphism $\hat{f}: A \to X^{\Sigma^*}$ with $f = \varepsilon \circ \hat{f}$ is given by $\hat{f}(a)(\sigma_1 \cdots \sigma_n) = f(\sigma_1 \cdots \sigma_n a)$.]

Suggestions for Further Reading

Samuel, P. On universal mappings and free topological groups. *Bull. Amer. Math. Soc.* **54** (1948): 591–598.

Sonner, J. Universal and special problems. *Math. Z.* **82** (1963): 200–211.

Hales, A. W. On the non-existence of free complete Boolean algebras. *Fund. Math.* **54** (1964): 45–66.

Gaifman, H. Infinite Boolean polynomials. *Fund. Math.* **54** (1964): 230–250.

Solovay, R. New proof of a theorem of Gaifman and Hales. *Bull. Amer. Math. Soc.* **72** (1966): 282–284.

Ehresmann, C. Construction de structures libres. *Springer Lect. Notes Math.* **92** (1969): 74–104.

Pumplün, D. Universelle und spezielle Probleme. *Math. Ann.* **198** (1972): 131–146.

EXERCISES

8A. Regular Monomorphisms vs. Embeddings

(a) Show that in a construct a regular monomorphism need not be an embedding. [Cf. Proposition 8.7.]

(b) Prove that each regular monomorphism in a construct that has a free object over a nonempty set must be an embedding.

(c) Prove that each embedding in a construct that has a two-element indiscrete object must be a regular monomorphism.

8B. Initial Monomorphisms vs. Embeddings

Show that in constructs the embeddings are precisely the initial monomorphisms, but that in concrete categories initial monomorphismss may fail to be embeddings.

8C. An Initial, Non-injective Morphism in Sgr

Show that if A resp. B are the semigroups with underlying sets $\{0,1,2\}$ resp. $\{0,1\}$ and multiplication defined by $x \cdot y = 0$ for all x and y, then the non-injective map $A \xrightarrow{f} B$, defined by $f(x) = \text{Min}\{x,1\}$, is an initial morphism in **Sgr**.

8D. A Characterization of Concretely Co-wellpowered Constructs

Show that a uniquely transportable construct is concretely co-wellpowered if and only if it is fibre-small and for every cardinal number k there exists a cardinal number $\overline{k}$ such that every object that is concretely generated by a set of cardinality not exceeding k has an underlying set with cardinality not exceeding $\overline{k}$.

* 8E. CBoo is Extremally Co-wellpowered, but Not Concretely Co-wellpowered

Let **CBoo** be the construct of complete boolean algebras and boolean homomorphisms that preserve arbitrary meets and joins. Show that:

(a) **CBoo** is wellpowered and extremally co-wellpowered,

(b) **CBoo** is not concretely co-wellpowered,

(c) the forgetful functor **CBoo** → **Set** is not extremally co-wellpowered.

[Hint for (b) and (c): Let X be a topological space. A subset A of X is called **regular open** provided that $int(cl A) = A$ (where "int" designates "interior" and "cl" designates "closure"). The set $R(X)$ of all regular open subsets of X is a complete boolean algebra with respect to the following operations:

$$\bigvee M = int(cl(\bigcup M)), \quad \text{for } M \subseteq R(X)$$
$$\bigwedge M = int(\bigcap M), \quad \text{for } \emptyset \neq M \subseteq R(X)$$
$$A' = int(X - A), \quad \text{for } A \in R(X).$$

Let K be an infinite cardinal number. Let X be the set of all ordinal numbers with cardinality less than K, considered as a discrete topological space. Let $P = X^{\mathbb{N}}$ be the topological product of countably many copies of X, with projections $\pi_n : X^{\mathbb{N}} \to X$. The complete boolean algebra $R(X^{\mathbb{N}})$ is extremally generated by the family $\{\pi_n^{-1}(\xi) \mid n \in \mathbb{N}, \xi \in X\}$, hence by the countable set $\{A_{m,n} \mid m, n \in \mathbb{N}\}$, where $A_{m,n} = \{x \in X^{\mathbb{N}} \mid \pi_m(x) \leq \pi_n(x)\}$.]

8F. Free Objects in CBoo

Show that in the construct **CBoo**

(a) there exists a free object over each finite set.

(b) there does not exist a free object over any infinite set. [Hint: 8E.]

8G. Free Objects in JPos, JCPos, CLat, and Fram

Show that

* (a) the constructs **JCPos** and **Fram** have free objects.

(b) in the construct **JPos** there exists a free object over X if and only if card $X \neq 1$.

* (c) in the construct **CLat** there exists a free object over X if and only if card $X \leq 2$.

8H. Free Objects in Alg(Ω)

As outlined in Example 8.23(6) free Ω-algebras can be constructed as algebras of terms. Another, more graphic, description can be achieved via labeled trees: Consider the infinite regular tree ω^* of all words in ω (with root $\emptyset$, the empty word; the first level consists of all natural numbers $i \in \omega$; in the second level, the successors of i are all words ij with $j \in \omega$, etc.). An Ω-**labeled tree** in a set X is defined to be a "labeling" partial function $t : \omega^* \to \Omega \uplus X$ such that

(a) the domain of definition D_t of t is finite, and it has $\emptyset$ as a member;

(b) if $t(i_1 i_2 \cdots i_n)$ is an operation symbol of arity k, then $i_1 i_2 \cdots i_n i_{n+1}$ belongs to D_t if and only if $i_{n+1} \in \{0, 1, \ldots, k-1\}$;

(c) if $t(i_1 i_2 \cdots i_n)$ is an element of X, then $i_1 i_2 \cdots i_n i_{n+1}$ does not belong to D_t for any i_{n+1}.

Show that the set of all Ω-labeled trees is an Ω-algebra that is isomorphic to the term algebra, and hence is free.

[Given $\sigma \in \Omega$ of arity k, and given Ω-labeled trees $t_0, t_1, \ldots, t_{k-1}$, then the tree $\sigma(t_0, t_1, \ldots, t_{k-1}) = t$ is defined on $i_0 i_1 \cdots i_n$ by

$$t(i_0 i_1 \cdots i_n) = t_{i_0}(i_1 \cdots i_n) \quad \text{for} \quad i_0 < k, \quad i_1 \cdots i_n \in D_{t_{i_0}};$$

otherwise it is undefined.]

8I. Free Objects in $\mathbf{Set}^{\text{op}}$, $\mathbf{Top}_0^{\text{op}}$, and $\mathbf{HComp}^{\text{op}}$

Show that the constructs $\mathbf{Set}^{\text{op}}$ (see 5.2(4)), $\mathbf{Top}_0^{\text{op}}$ (see 5L), and $\mathbf{HComp}^{\text{op}}$ (see 5M) have free objects and, in each case, describe them explicitly.

8J. Free Objects in BooSp and HComp

Show that the constructs **BooSp** and **HComp** have free objects and, in each case, describe them explicitly.

8K. Free Objects in (Ban, O) and (Ban, U)

Show that (**Ban**, O) has free objects (cf. 8.23(12)), but that (**Ban**, U) has free objects only over the empty set.

* 8L. Isomorphic Free Objects

Show that it can happen that a construct **A** has free objects in such a way that any two free objects over finite, nonempty sets are isomorphic (as objects). [Hint: Consider the concrete full subcategory of $\mathbf{Alg}(\Omega)$, where $\Omega = (1, 1, 2)$, consisting of those Ω-algebras $(X, (\omega_1, \omega_2, \omega_3))$ that satisfy the equations $\omega_1(\omega_3(x, y)) = x$, $\omega_2(\omega_3(x, y)) = y$ and $\omega_3(\omega_1(x), \omega_2(x)) = x$. Show that whenever an object $(X, (\omega_1, \omega_2, \omega_3))$ of the construct **A** is free over a set $Y \uplus \{x, y\}$, then it is free over $Y \uplus \{\omega_3(x, y)\}$.]

8M. Discrete Objects

Show that for concrete categories the following hold:

(a) Retracts of discrete objects are discrete.

(b) An object A is discrete if and only if the structured arrow $|A| \xrightarrow{id} |A|$ is universal.

8N. A Characterization of Faithfulness

Show that a functor $\mathbf{A} \xrightarrow{G} \mathbf{B}$ is faithful if and only if every **A**-epimorphism $A \xrightarrow{f} \hat{A}$, considered as a G-structured arrow $(Gf, \hat{A})$, is generating.

8O. Regular Epimorphisms and Finality

Let **A** be a concrete category over **X** that has free objects, and let f be an **A**-morphism such that Uf is a regular epimorphism in **X**. Show that f is a regular epimorphism in **A** if and only if f is final.

8P. Free Automata

(a) Show that the construct Σ-**Seq** has free objects if and only if $\Sigma = \emptyset$.

(b) Show that **Aut** considered as a concrete category over **Set** $\times$ **Set** $\times$ **Set** has only the (trivial) free objects over (Σ, Q, Y) with $Q = Y = \emptyset$.

9 Injective objects and essential embeddings

Earlier it has been shown that many familiar constructions, particularly "completions" such as the completion of a metric space or the Čech-Stone compactification of a Tychonoff space, can be naturally regarded as *reflections*. However, there also exist familiar completions that cannot be (or only artificially can be) regarded as such. Examples are the Mac Neille completion of a poset and the algebraic closure of a field.[50] In both cases, and in several others, the construction in question can be regarded rather naturally as an *injective hull*, a concept that will be studied in this section. Roughly speaking, an object C is called *injective* provided that for any object A, any morphism from a subobject of A into C can be extended to a morphism from A into C. Since a satisfactory concept of subobjects is available for concrete categories, but not for arbitrary categories, we will first define injective objects for concrete categories only. Later, for arbitrary categories **A** and arbitrary classes M of **A**-morphisms, M-*injective* objects will be introduced in such a way that for concrete categories the injective objects are precisely the Emb(**A**)-injective objects.

INJECTIVITY IN CONCRETE CATEGORIES

For concrete categories we will use the notational conventions described in Remark 6.22.

9.1 DEFINITION
In a concrete category an object C is called **injective** provided that for any embedding $A \xrightarrow{m} B$ and any morphism $A \xrightarrow{f} C$ there exists a morphism $B \xrightarrow{g} C$ extending f, i.e., such that the triangle

commutes.

9.2 REMARK
The morphism g in the above definition is *not* required to be uniquely determined by m and f. This contrasts sharply with many categorical definitions in which existence and uniqueness requirements are coupled (see 4.16 for a typical example).

9.3 EXAMPLES
In Examples (1)–(5) below, we consider injective objects in various constructs:

---
[50]Observe, e.g., that there exist two different automorphisms of the field of complex numbers (the algebraic closure of the field of real numbers) that keep the reals pointwise fixed.

(1) In **Set** the injective objects are precisely the nonempty sets.

(2) In **Pos** the injective objects are precisely the complete lattices. [Injectivity of a complete lattice C follows from the fact that, for given m and f as in Definition 9.1, the map $g : B \to C$ defined by $g(b) = \sup\{f(a) \mid a \in A \text{ and } m(a) \leq b\}$ is order-preserving. Completeness of an injective object C follows from 9.5 and the fact that C is a retract of its Mac Neille completion.] Similarly the injective objects

  (a) in **Boo** (and in the construct **DLat** of distributive lattices) are precisely the complete Boolean algebras,

  (b) in the construct **SLat** (posets with finite meets and maps that preserve finite meets) are precisely the frames [see Exercise 5L(b)],

  (c) in **JCPos** are precisely the completely distributive complete lattices.

(3) (Algebra)

  (a) In **Vec** every object is injective.

  (b) In **Ab** the injective objects are precisely the divisible abelian groups.

  (c) In **Alg**(1) the injective objects are precisely those unary algebras whose single unary operation is surjective and has a fixed point. Analogously, a $\{\sigma\}$-acceptor is injective in $\{\sigma\}$-**Seq** if and only if every state is final and its next state function is a permutation with a fixed point.

  (d) In **Grp** only the terminal objects are injective [since each group can be properly embedded into a simple group]. Analogously, terminal objects are the only injective objects in **Lat**, **Mon**, **Sgr**, and **Rng**.

(4) (Topology)

  (a) In **HComp** the injective objects are precisely the retracts of powers $[0, 1]^I$ of the unit interval $[0, 1]$. In particular, $[0, 1]$ is injective (Tietze-Urysohn Theorem).

  (b) In **BooSp** the injective objects are precisely the retracts of Cantor spaces, i.e., of powers of the two-element Boolean space $B = (\{0, 1\}, \mathcal{P}\{0, 1\})$.

  (c) In **Top**$_0$ the injective objects are precisely the retracts of powers of the Sierpinski space $S = (\{0, 1\}, \{\emptyset, \{0\}, \{0, 1\}\})$.

  (d) In **Top** the injective objects are precisely the retracts of powers C^I of the space $C = (\{0, 1, 2\}, \{\emptyset, \{0, 1\}, \{0, 1, 2\}\})$.

  (e) In **Met** an object (X, d) is injective if and only if it is hyperconvex, i.e., if for any superadditive map[51] $f : X \to \mathbb{R}^+$ there exists $z \in X$ with $d(z, x) \leq f(x)$ for all $x \in X$.

  (f) In the construct **Unif** of uniform spaces and uniformly continuous functions, the unit interval $[0, 1]$ is injective, but the real line $\mathbb{R}$ is not.

---

[51] $f : X \to \mathbb{R}^+$ is **superadditive** provided that $d(x, y) \leq f(x) + f(y)$ for all $x, y \in X$.

(g) In some constructs (e.g., for **Top$_1$**, **Haus**, and **Tych**) only the terminal objects are injective [since a map f from $\mathbb{R} \setminus \{0\}$ into a T_1-space X, sending all $r < 0$ to x and all $r > 0$ to y, can be extended to a continuous map $f : \mathbb{R} \to X$ only if $x = y$].

(5) (Analysis)

In **Ban**(K) (Banach spaces over K) the injective objects are (up to isomorphism) precisely the function spaces $C(X, K)$ for extremally disconnected compact Hausdorff spaces X. [In case K is the field $\mathbb{R}$ of real numbers, these are precisely the hyperconvex spaces (see (4)(e) above). In case K is the field $\mathbb{C}$ of complex numbers, the zero-space is the only hyperconvex injective object.] In particular, K itself is injective (Hahn-Banach Theorem).

(6) In a partially ordered set, considered as a concrete category over **1**, every object is injective.

9.4 PROPOSITION

Every terminal object is injective. □

9.5 PROPOSITION

Every retract of an injective object is injective.

Proof: Let $C \xrightarrow{r} D$ be a retraction with C an injective object. Then there exists $D \xrightarrow{s} C$ with $r \circ s = id_D$. Let $A \xrightarrow{m} B$ be an embedding and $A \xrightarrow{f} D$ be a morphism. Since C is an injective object, there exists an extension $B \xrightarrow{g} C$ of $A \xrightarrow{s \circ f} C$; i.e., a morphism g such that

$$\begin{array}{ccc} A & \xrightarrow{m} & B \\ f \downarrow & & \downarrow g \\ D & \xrightarrow{s} & C \end{array}$$

commutes. Thus $r \circ g$ is an extension of f. □

9.6 DEFINITION

In a concrete category an object C is called an **absolute retract** provided that any embedding with domain C is a section.

9.7 PROPOSITION

Every injective object is an absolute retract. □

9.8 REMARK

One can easily provide constructs in which absolute retracts fail to be injective. However, as we shall see below (9.10), under reasonable assumptions, injective objects are precisely the absolute retracts. From Examples 9.17 we see that this is the case in the constructs **Set**, **Vec**, **Pos**, **Ab**, **Met**, (**Ban**, O), and **Field** (= fields and algebraic field extensions).

9.9 DEFINITION
A concrete category **has enough injectives** provided that each of its objects is an initial subobject of an injective object.

9.10 PROPOSITION
If a concrete category **A** *has enough injectives, then in* **A** *injective objects are precisely the absolute retracts.*

Proof: Let C be an absolute retract in **A**, and let $C \xrightarrow{m} D$ be an initial monomorphism with D an injective object. Since C is an absolute retract, m is a section, i.e., there exists a retraction $D \xrightarrow{r} C$ with $r \circ m = id_C$. By Proposition 9.5, C is injective. □

9.11 REMARK
If a concrete category has enough injectives, one may ask whether for every object A there exists a distinguished **injective extension** (i.e., an embedding $A \xrightarrow{m} B$ with B an injective object), e.g., one which is in some sense minimal or smallest. The crucial concept needed in order to describe and analyze such "injective hulls" is that of *essential extensions*, i.e., extensions that are in a certain sense "dense" (but not necessarily epimorphic).

9.12 DEFINITION
In a concrete category an embedding $A \xrightarrow{m} B$ is called **essential** provided that a morphism $B \xrightarrow{f} C$ is an embedding, whenever $A \xrightarrow{f \circ m} C$ is an embedding.

9.13 EXAMPLES
In Examples (1)–(8) below, we consider essential embeddings for various constructs:

(1) In **Vec** the essential embeddings are the isomorphisms.

(2) In **Set** the only essential embeddings are the bijective functions and the maps $\emptyset \to \{a\}$ with empty domain and one-element codomain.

(3) In **Pos** the essential embeddings are the embeddings that are meet-dense and join-dense (e.g., the embedding of $\mathbb{Q}$ into $\mathbb{R}$).

(4) In **Boo** the essential embeddings are the join-dense embeddings (which are automatically meet-dense).

(5) In **Ab** an embedding $A \xrightarrow{m} B$ is essential if and only if every nontrivial subgroup of B meets $m[A]$ nontrivially (e.g., the embedding of $(\mathbb{Z}, +)$ into the rationals).

(6) In **Tych** the essential embeddings are precisely the homeomorphisms, the one-point compactifications of locally compact, noncompact spaces, and the embeddings of the empty space into one-point spaces. In particular this example shows that epimorphic embeddings need not be essential.

(7) In **Met** an embedding $A \xrightarrow{m} B$ is essential if and only if it is **tight**, i.e., if and only if it satisfies the following two conditions:

(a) $d_B(b_1, b_2) = \sup\{\, d_B(m(a), b_1) - d_B(m(a), b_2) \mid a \in A\,\}$ for all $b_1, b_2 \in B$.

(b) For each $b \in B$, the map $f \colon A \to \mathbb{R}^+$, defined by $f(a) = d_B(m(a), b)$, is a minimal superadditive map [cf. 9.3(4)(e)].

In particular, if A is a subspace of $\mathbb{R}$ with $\bigwedge A = a$ and $\bigvee A = b$, then the embedding $A \to [a, b]$ is essential.

(8) In **Lat** every object has arbitrarily large essential extensions (see Exercise 9H).

9.14 PROPOSITION

(1) Every isomorphism is essential.

(2) The composite of essential embeddings is essential.

(3) If f and g are embeddings with $g \circ f$ essential, then g is essential.

(4) If f and $g \circ f$ are essential embeddings, then g is an essential embedding. □

9.15 PROPOSITION

Injective objects have no proper essential extensions.

Proof: If $C \xrightarrow{m} D$ is an extension of an injective object C, then there is a retraction g with $id_C = g \circ m$. If m is essential, then g is an embedding, and hence an isomorphism. Thus m is an isomorphism as well. □

9.16 DEFINITION

An **injective hull** of A is an extension $A \xrightarrow{m} B$ of A such that B is injective and m is essential.

9.17 EXAMPLES

In Examples (1)–(6) below, we describe injective hulls for various constructs:

(1) In **Vec** every object A has an injective hull, namely, $A \xrightarrow{id_A} A$.

(2) In **Set** every object A has an injective hull, namely,
$$\begin{cases} A \xrightarrow{id_A} A, & \text{in case } A \neq \emptyset; \\ A \longrightarrow \{a\}, & \text{in case } A = \emptyset. \end{cases}$$

(3) In **Pos** every object has an injective hull, namely, its Mac Neille completion (= completion by cuts). Likewise in **Boo** every object has an injective hull, its Mac Neille completion.

(4) In **Ab** every object has an injective hull. The embedding $\mathbb{Z} \hookrightarrow \mathbb{Q}$ is an example.

(5) In the construct **Field** every object has an injective hull, namely, its algebraic closure. The embedding $\mathbb{R} \hookrightarrow \mathbb{C}$ is an example.

(6) In **Met** every object has an injective hull, namely, its hyperconvex envelope. If A is a subspace of $\mathbb{R}$ with $\bigwedge A = a$ and $\bigvee A = b$, then the embedding $A \hookrightarrow [a,b]$ is an injective hull. Likewise in (**Ban**, O) every object has an injective hull.

9.18 REMARK

It is sometimes difficult to decide whether or not every object in a given concrete category has an injective hull. The existence of enough injectives is a necessary but not sufficient condition. For example, the construct **BooSp** has enough injectives, since every Boolean space is a subspace of some Cantor space [cf. 9.3(4)(b)], but no nonempty Boolean space has a proper essential extension. A similar situation occurs for **HComp** (cf. Exercise 9D). In Exercise 9C we will formulate conditions that guarantee the existence of injective hulls.

9.19 PROPOSITION

Injective hulls are essentially unique, i.e.,

(1) *if (m, B) and (m', B') are injective hulls of A, then there exists an isomorphism $B \xrightarrow{k} B'$ with $m' = k \circ m$,*

(2) *if (m, B) is an injective hull of A, and if $B \xrightarrow{k} B'$ is an isomorphism, then $(k \circ m, B')$ is an injective hull of A.*

Proof:

(1). Since m is an embedding and B' is injective, there exists a morphism $B \xrightarrow{k} B'$ with $m' = k \circ m$. By Proposition 9.14(4), k is an essential embedding. Since B has no proper essential extension, k is an isomorphism.

(2). Obvious. □

9.20 PROPOSITION

If an object A has an injective hull, then for any extension (m, B) of A the following conditions are equivalent:

(1) (m, B) *is an injective hull of A,*

(2) (m, B) *is a **maximal essential extension** of A (i.e., (m, B) is an essential extension of A, and B has no proper essential extension),*

(3) (m, B) *is a **largest essential extension** of A (i.e., (m, B) is an essential extension of A, and for every essential extension (m', B') of A there exists an essential embedding $B' \xrightarrow{\overline{m}} B$ with $m = \overline{m} \circ m'$),*

(4) (m, B) *is a **smallest injective extension** of A (i.e., (m, B) is an injective extension of A, and for every injective extension (m', B') of A there exists an embedding $B \xrightarrow{\overline{m}} B'$ with $m' = \overline{m} \circ m$),*

(5) (m, B) *is a **minimal injective extension** of A (i.e., (m, B) is an injective extension of A and whenever $A \xrightarrow{m} B = A \xrightarrow{m'} B' \xrightarrow{\overline{m}} B$ with m' and $\overline{m}$ embeddings, and B' an injective object, then $\overline{m}$ is an isomorphism).*

Proof: Let (m_0, B_0) be an injective hull of A.

(1) $\Rightarrow$ (2). Immediate from Proposition 9.15.

(2) $\Rightarrow$ (1). Let (m, B) be a maximal essential extension of A. Since B_0 is injective, there exists a morphism f with $m_0 = f \circ m$. By Proposition 9.14(4), f is an essential embedding. Since (m, B) is a maximal essential extension, f must be an isomorphism. Hence (m, B) is an injective hull of A.

(1) $\Rightarrow$ (3). Let (m, B) be an essential extension of A. By the injectivity of B_0, there exists a morphism f with $m_0 = f \circ m$. By Proposition 9.14(4), f is an essential embedding.

(3) $\Rightarrow$ (1). Let (m, B) be a largest essential extension of A. Then there exists a morphism g with $m = g \circ m_0$. By Proposition 9.14(4) g is an essential embedding, and so by Proposition 9.15 is an isomorphism. Hence (m, B) is an injective hull of A.

(1) $\Rightarrow$ (4). Immediate.

(4) $\Rightarrow$ (1). If (m, B) is a smallest injective extension of A, then there exists an embedding $\overline{m}$ with $m_0 = \overline{m} \circ m$. By Proposition 9.14(3), $\overline{m}$ is essential and hence, since B has no proper essential extension, $\overline{m}$ is an isomorphism. Thus (m, B) is an injective hull of A.

(1) $\Rightarrow$ (5). Let (m, B) be an injective extension of A and let $(\overline{m}, B_0)$ be an extension of B with $m_0 = \overline{m} \circ m$. Then by Proposition 9.14(3), $\overline{m}$ is an essential embedding; hence, since B has no proper essential extensions, $\overline{m}$ is an isomorphism.

(5) $\Rightarrow$ (1). Let (m, B) be a minimal injective extension of A. Since B is injective, there exists a morphism f with $m = f \circ m_0$. Since m_0 is essential, f is an embedding; hence, by (5) an isomorphism. Thus (m, B) is an injective hull of A. □

9.21 REMARK

If A has no injective hull, then the concepts mentioned in the above proposition may fall apart. For example, in **Tych** every object has a simultaneously largest and maximal essential extension, but no space with more than one point has an injective hull [cf. 9.13(6) and 9.3(4)]. For further "negative" examples see Exercises 9F and 9G.

M-INJECTIVES IN ABSTRACT CATEGORIES

9.22 DEFINITION

Let M be a class of morphisms in a category **A**.

(1) An object C is called M-**injective** provided that for every morphism $A \xrightarrow{m} B$ in M and every morphism $A \xrightarrow{f} C$ there exists a morphism $B \xrightarrow{g} C$ with $f = g \circ m$.

(2) A morphism $A \xrightarrow{m} B$ in M is called M-**essential** provided that a morphism $B \xrightarrow{f} C$ belongs to M whenever $f \circ m$ does.

(3) An M-**injective hull** of an object A is a pair (m, B) consisting of an M-injective object B and an M-essential morphism $A \xrightarrow{m} B$.

(4) **A has enough M-injectives** provided that for each object A there exists an M-injective object C and a morphism $A \xrightarrow{m} C$ in M.

9.23 REMARK
If M is the class of all embeddings in a concrete category **A**, then the concepts defined above specialize to those of injective objects, essential embeddings, and injective hulls. Also the previous results of this section carry over to the more general context of M-injectivity, provided that suitable assumptions are imposed on M, e.g.,

(a) $Iso(\mathbf{A}) \subseteq M \subseteq Mono(\mathbf{A})$; i.e., M is a class of **A**-monomorphisms that contains all **A**-isomorphisms,

(b) $M \circ M \subseteq M$; i.e., M is closed under composition.

We refrain from explicitly formulating the corresponding results.

9.24 EXAMPLES
(1) In any category, if M consists of all sections, then every object is M-injective.

(2) In **Met**, if M consists of all dense embeddings, then the M-injective objects are the complete metric spaces, and the M-injective hulls are the metric completions.

(3) In **Top**$_0$, if M consists of all front-dense embeddings, then the M-injective objects are the sober spaces, and the M-injective hulls are the sober reflections.

(4) In **Top**, if M consists of all embeddings $X \to X \cup \{p\}$ of infinite discrete spaces into ultrafilter spaces (i.e., p is a point of the Čech-Stone compactification of X), then the M-injective objects are the compact spaces.

(5) In **Top**, if M consists of the single embedding $\{0,1\} \to [0,1]$, then the M-injective objects are the pathwise connected spaces. Similarly, if M consists of the single embedding of the unit circle into the unit disc, then a complex domain is an M-injective object if and only if it is simply connected.

(6) In **Top** a metrizable space X satisfies $\dim X \leq n$ if and only if the n-sphere S^n is an M-injective object, where M consists of all embeddings of closed subspaces of X into X.

(7) In **Top**, if M consists of all embeddings of closed subspaces of normal spaces, then $[0,1]$ and $\mathbb{R}$ are M-injective objects [Tietze-Urysohn Theorem].

(8) In **Ab**, if M consists of all pure embeddings, then the M-injective objects are the algebraically compact abelian groups, or, equivalently, are all the direct summands of direct products of cocyclic abelian groups.

(9) In **CAT(X)**, if M consists of all concrete full embeddings, then the M-injective objects are the "topological" concrete categories over **X**. Such categories will be defined and studied in detail in §21. In particular, the constructs **Top**, **Unif**, **Rel**, **Prost**, and each **Spa**(T) are topological.

(10) In a (meet) semilattice C, considered as a category, an object is Mono-injective if and only if it is the largest element of C.

9.25 PROPOSITION

*If **B** is a reflective, isomorphism-closed, full subcategory of **A** and M is the class of all **B**-reflection arrows, then*

*(1) the M-injective objects of **A** are precisely the **B**-objects, and*

*(2) the M-injective hulls are precisely the **B**-reflections.*

Proof:

(1). Clearly each **B**-object is M-injective. Let C be an M-injective object, and let $C \xrightarrow{r} B$ be an **B**-reflection arrow for C. Then there exists some $B \xrightarrow{f} C$ for which $id_C = f \circ r$. The fact that $id_B \circ r = (r \circ f) \circ r$ and the uniqueness in the definition of reflection arrow show that $id_B = r \circ f$. Thus r is an isomorphism, so that since **B** is isomorphism-closed, C is a **B**-object.

(2). By (1) we need only show that each **B**-reflection arrow $A \xrightarrow{r} B$ is M-essential. Suppose that $A \xrightarrow{r} B \xrightarrow{f} B'$ belongs to M, i.e., is also a **B**-reflection arrow. Since **B**-reflection arrows for A are essentially unique (4.19), we can conclude that $B \xrightarrow{f} B'$ is an isomorphism, and hence is a member of M. □

9.26 REMARK

We may call a class B of **A**-objects (resp. the associated full subcategory **B** of **A**) an **injectivity class** in **A**, whenever there exists a class M of **A**-morphisms such that B is precisely the class of M-injective objects. The above proposition shows that every reflective, isomorphism-closed, full subcategory of **A** is an injectivity class. The converse is not true. By Example 9.24(4) compact spaces form an injectivity class in **Top**, but the associated subcategory is not reflective in **Top**. A characterization of injectivity classes by suitable stability properties is unknown even for "nice" categories (e.g., for **Top**). Connected spaces, which have stability properties similar to those of compact spaces (e.g., they are closed under the formation of products and of continuous images), do not form an injectivity class in **Top**. In fact the only injectivity class in **Top** that contains all connected spaces is **Top** itself.

Sec. 9]   Injective objects and essential embeddings   161

PROJECTIVITY

9.27 TABLE OF DUAL CONCEPTS

The following table provides the names for the concepts dual to those investigated in this section:

Concept	Dual Concept
embedding	quotient morphism
injective object	projective object
essential embedding	coessential quotient morphism
injective hull	projective cover
M-injective object	M-projective object
M-essential morphism	M-coessential morphism
M-injective hull	M-projective cover

9.28 EXAMPLES

In Examples (1)–(4) below, we consider projective objects in various constructs:

(1) In **Set** and in **Vec** every quotient morphism is a retraction. Hence in either construct every object is projective, and the coessential quotient morphisms are the isomorphisms. Thus, in these constructs, each object has a projective cover.

(2) In any of the constructs **Top**, **Pos**, **Ab**, and **Grp** the projective objects are precisely the free objects. Projective covers generally fail to exist.

(3) In **BooSp** and in **HComp** the projective objects are the extremally disconnected compact Hausdorff spaces (= the retracts of Čech-Stone compactifications of discrete spaces), and the coessential quotient morphisms are the irreducible quotient morphisms (= those continuous surjections $f : X \to Y$ that map no proper closed subset of X onto Y). Projective covers exist for each object (and are called projective resolutions).

(4) In **Alg**(Ω) the projective objects are precisely the retracts of the free objects. This follows from the next proposition:

9.29 PROPOSITION

If $(\mathbf{A}, U)$ is a concrete category over $\mathbf{X}$ that has free objects, and E is the class of all $\mathbf{A}$-morphisms f for which Uf is a retraction, then the following are equivalent:

(1) A is an E-projective object,

(2) A is a retract of a free object.

Proof: (1) ⇒ (2). Let (u, B) be a universal arrow over UA. Then there exists a unique **A**-morphism $e: B \to A$ with $id_{UA} = Ue \circ u$. So e belongs to E. Hence, by (1), there exists an **A**-morphism $f: A \to B$ with $id_A = e \circ f$. Consequently A is a retract of the free object B.

(2) ⇒ (1). Let A be a free object (with universal arrow (u, A)), let $e: B \to C$ be a morphism in E, and let $f: A \to C$ be a **A**-morphism. Since in **X** every object is UE-projective there exists an **X**-morphism g such that the diagram

$$\begin{array}{ccc} X & \xrightarrow{u} & UA \\ {\scriptstyle g}\downarrow & & \downarrow{\scriptstyle Uf} \\ UB & \xrightarrow[Ue]{} & UC \end{array}$$

commutes. Since (u, A) is universal, there exists a unique **A**-morphism $k: A \to B$ with $g = Uk \circ u$. Hence $Uf \circ u = Ue \circ g = Ue \circ Uk \circ u = U(e \circ k) \circ u$, which implies that $f = e \circ k$. Hence A is E-projective. By the dual of Proposition 9.5 every retract of A is also E-projective. □

Projectivity and retracts of a free object

9.30 COROLLARY

If in a construct with free objects every surjective morphism is a quotient morphism, then the projective objects are precisely the retracts of the free objects. □

9.31 EXAMPLES
The above corollary applies to such constructs as **Grp**, **Lat**, **HComp** (and many others).

Suggestions for Further Reading

Baer, R. Abelian groups that are direct summands of every containing abelian group. *Bull. Amer. Math. Soc.* **46** (1940): 800–806.

Eckmann, B., and A. Schopf. Über injektive Moduln. *Archiv Math.* **9** (1953): 75–78.

Gleason, A. M. Projective topological spaces. *Illinois J. Math.* **2** (1958): 482–489.

Isbell, J. R. Six theorems about injective metric spaces. *Comment. Math. Helv.* **39** (1964): 65–76.

Cohen, H. B. Injective envelopes of Banach spaces. *Bull. Amer. Math. Soc.* **70** (1964): 723–726.

Maranda, J. M. Injective structures. *Trans. Amer. Math. Soc.* **110** (1964): 98–135.

Banaschewski, B., and G. Bruns. Categorical characterization of the Mac Neille completion. *Archiv Math.* **18** (1967): 369–377.

Banaschewski, B. Projective covers in categories of topological spaces and topological algebras. *Proceedings of the Kanpur Topological Conference 1968*, Academia, Prague, 1971, 63–91.

Banaschewski, B. Injectivity and essential extensions in equational classes of algebras. *Queen's Papers Pure Appl. Math.* **25** (1970): 131–147.

Scott, D. Continuous lattices. *Springer Lect. Notes Math.* **274** (1972): 97–136.

Banaschewski, B. Essential extensions of T_0-spaces. *Gen. Topol. Appl.* **7** (1977): 233–246.

Blass, A. Injectivity and projectivity of abelian groups and the axiom of choice. *Trans. Amer. Math. Soc.* **255** (1979): 31–59.

Tholen, W. Injective objects and cogenerating sets. *J. Algebra* **73** (1981): 139–155.

Porst, H.-E. Characterization of injective envelopes. *Cahiers Topol. Geom. Diff.* **22** (1981): 399–406.

EXERCISES

* 9A. The Axiom of Choice

Many results of this book can be expressed (in localized form) in the realm of Zermelo-Fraenkel set theory (ZF). But if the axiom of choice for sets (AC) is not assumed, several results fail to be true. Consider the following:

(ET) In **Set** every epimorphism is a retraction.

(PT) In **Set** every product of injective objects is injective.

(BT) The injective objects in **Ab** are precisely the divisible abelian groups.

(ST) The injective objects in **Boo** are precisely the complete Boolean algebras.

(UT) In **Boo** the two-element Boolean algebra is injective.

(GT) The projective objects in **HComp** are the extremally disconnected compact Hausdorff spaces.

(a) that in ZF the following implications hold:
(AC) $\iff$ (ET) $\iff$ (PT) $\iff$ (BT) $\implies$ (ST) $\implies$ (UT).

(b) Show that in ZF the following holds: (ST) $\iff$ [(GT) and (UT)].

(c) Does (ST) imply (AC)? [Unsolved.]

9B. A Characterization of Injective Objects

Let **A** be a construct satisfying

(1) **A** has enough injectives, and

(2) for every embedding $A \xrightarrow{m} B$ there exists a morphism $B \xrightarrow{f} C$ such that $A \xrightarrow{f \circ m} C$ is an essential embedding.

Show that in **A** injective objects are precisely those objects that have no proper essential extension.

9C. Existence of M-Injective Hulls

Let M be a class of morphisms in a category **A** and let M^* be the class of M-essential morphisms. Assume that the following conditions are satisfied:

(1) $Iso(\mathbf{A}) \subseteq M \subseteq Mono(\mathbf{A})$,

(2) $M \circ M \subseteq M$,

(3) for any $m \in M$, there exists an **A**-morphism f such that $f \circ m \in M^*$,

(4) for every 2-source $B \xleftarrow{m} A \xrightarrow{f} C$ with $m \in M$, there exists a 2-sink $B \xrightarrow{\overline{f}} D \xleftarrow{\overline{m}} C$ with $\overline{m} \in M$ and $\overline{f} \circ m = \overline{m} \circ f$,

(5) every well-ordered system in M has an upper bound in M,

(6) **A** is M^*-co-wellpowered (in the obvious sense; M^* need not consist of epimorphisms only).

Show that

(a) For **A**-objects A the following conditions are equivalent:

  (a1) A is M-injective,

  (a2) every $A \xrightarrow{m} B$ in M is a section,

  (a3) every $A \xrightarrow{m} B$ in M^* is an isomorphism.

(b) Every **A**-object has an M-injective hull.

* 9D. Enough Injectives and Injective Hulls in HComp

Show that **HComp** has enough injectives, but that the only compact Hausdorff space that has a proper essential extension is the empty space.

9E. Regular Projective Objects

Show that the RegEpi-projective objects

* (a) in **Top** are the discrete topological spaces,

* (b) in **Haus** are the discrete topological spaces,

  (c) in **Set**^{2} are the injective functions,

  (d) in (**Set**^{2})^{op} are the surjective functions with nonempty domain.

9F. Injective Objects and Maximal Essential Extensions in Top$_1$

Show that

(a) In **Top**$_1$ the following conditions are equivalent:

  (1) X is injective,

  (2) X is an absolute retract,

  (3) X is a terminal object.

(b) If a T_1-space X has a point such that each of its neighborhoods is cofinite in X, then X has no proper essential extension. If a T_1-space X has no such point, then it has an essentially unique proper essential extension. [Add a point p to X with each neighborhood of p cofinite in $X \cup \{p\}$.]

(c) In **Top**$_1$ every object has an essentially unique maximal essential extension.

(d) In **Top**$_1$ the only objects that have injective hulls are the initial object and the terminal objects.

(e) If M is the class of embeddings in **Top**$_1$, then all the conditions of Exercise 9C are satisfied except condition (3).

* **9G. Minimal and Smallest Injective Extensions**

Show that

(a) Minimal injective extensions need not be smallest injective extensions. [In the construct that consists of all sets X with $\mathrm{card}(X) \neq 1$ together with all identities and constant maps as morphisms, every $\emptyset \to B$ with $B \neq \emptyset$ is a minimal injective extension. However, $\emptyset$ has no smallest injective extension.]

(b) Smallest injective extensions need not be minimal. [In the full subconstruct of **Set** that contains $\emptyset$ and all infinite sets, $\emptyset \to \mathbf{N}$ is a smallest injective extension. However, $\emptyset$ has no minimal injective extension.]

(c) Injective extensions that are simultaneously smallest and minimal need not be injective hulls. [In the full subconstruct of **Set** that consists of all sets X with $\mathrm{card}(X) \neq 1$, the inclusion $\emptyset \hookrightarrow \{1,2\}$ is simultaneously a smallest and a minimal injective extension. However, $\emptyset$ has no M-injective hull.]

9H. Essential Extensions in Lat

Show that every lattice that has at least two elements has arbitrarily large essential extensions. [Hint: First assume that L is a lattice with smallest element 0 and largest element 1. Show that for any set A that has at least three elements and is disjoint from L, if A is ordered by equality, then the embedding

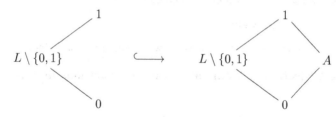

is an essential extension of L.]

9I. Injective Automata

(a) Prove that the following are equivalent for each object A of $\{\sigma\}$-**Seq**:

(1) A is injective,

(2) A is an absolute retract,

(3) all states in A are final, and $\delta(-, \sigma)$ is a permutation with a fixed point.

Whenever Σ has more than one element, show that there exists an object of Σ-**Seq** that is not an absolute retract, although every state is final and each $\delta(-, \sigma)$ is a permutation with a fixed point.

(b) Characterize those objects of Σ-**Seq** that have an injective hull.

Chapter III

SOURCES AND SINKS

A source

A sink

10 Sources and sinks

SOURCES

A basic categorical concept that simultaneously generalizes the concepts of objects and of morphisms is that of sources.

10.1 DEFINITION
A **source** is a pair $(A, (f_i)_{i \in I})$ consisting of an object A and a family of morphisms $f_i : A \to A_i$ with domain A, indexed by some class I. A is called the **domain of the source** and the family $(A_i)_{i \in I}$ is called the **codomain of the source**.

10.2 REMARK
(1) Whenever convenient we use more concise notations, such as $(A, f_i)_I$, (A, f_i) or $(A \xrightarrow{f_i} A_i)_I$.

(2) The indexing class I of a source $(A, f_i)_I$ may be a proper class, a nonempty set, or the empty set. In case $I = \emptyset$, the source is determined by A. In case $I \neq \emptyset$, the source is determined by the family $(f_i)_I$.

(3) Sources indexed by the empty set are called **empty sources** and are denoted by $(A, \emptyset)$. Whenever convenient, objects may be regarded as empty sources.

(4) Sources that are indexed by a set are called **set-indexed** or **small**.

(5) Sources that are indexed by the set $\{1, \ldots, n\}$ are called n-**sources** and are denoted by $(A, (f_1, \ldots, f_n))$. Whenever convenient, morphisms $f : A \to B$ may be regarded as 1-sources (A, f).

(6) There are properties of sources that depend heavily on the fact that $(f_i)_I$ is a *family*, i.e., an indexed collection (e.g., the property of being a product). There are other properties of sources (A, f_i), depending on the domain A and the associated class $\{ f_i \mid i \in I \}$ only (e.g., the property of being a mono-source). In order to avoid a clumsy distinction between indexed and non-indexed sources, we will sometimes regard classes as families (indexed by themselves via the corresponding identity function). Hence for any object A and any class S of morphisms with domain A, the pair (A, S) will be considered as a source. A particularly useful example is the **total source** (A, S_A), where S_A is the class of *all* morphisms with domain A.

10.3 DEFINITION
If $\mathcal{S} = (A \xrightarrow{f_i} A_i)_I$ is a source and, for each $i \in I$, $\mathcal{S}_i = (A_i \xrightarrow{g_{ij}} A_{ij})_{J_i}$ is a source, then the source

$$(\mathcal{S}_i) \circ \mathcal{S} = (A \xrightarrow{g_{ij} \circ f_i} A_{ij})_{i \in I, \, j \in J_i}$$

is called the **composite** of $\mathcal{S}$ and the family $(\mathcal{S}_i)_I$.

10.4 REMARKS

(1) For a source $\mathcal{S} = (A \xrightarrow{f_i} A_i)_I$ and a morphism $f : B \to A$ we use the notation $\mathcal{S} \circ f = (B \xrightarrow{f_i \circ f} A_i)_I$.

(2) The composition of morphisms can be regarded as a special case of the composition of sources.

MONO-SOURCES

10.5 DEFINITION

A source $\mathcal{S} = (A, f_i)_I$ is called a **mono-source** provided that it can be cancelled from the left, i.e., provided that for any pair $B \underset{s}{\overset{r}{\rightrightarrows}} A$ of morphisms the equation $\mathcal{S} \circ r = \mathcal{S} \circ s$ (i.e., $f_i \circ r = f_i \circ s$ for each $i \in I$) implies $r = s$.

10.6 EXAMPLES

(1) An empty source $(A, \emptyset)$ is a mono-source if and only if for each object B there is at most one morphism from B to A.

(2) A 1-source (A, f) is a mono-source if and only if f is a monomorphism.

(3) In **Set** mono-sources are precisely the **point-separating sources** $(A, f_i)_I$, i.e., sources $(A, f_i)_I$ such that for any two different elements a and b of A there exists some $i \in I$ with $f_i(a) \neq f_i(b)$.

(4) In many familiar constructs, e.g., in **Vec**, **Grp**, **Top**, and **Pos**, a source is a mono-source if and only if it is point-separating. [See Corollary 10.8.]

(5) In every preordered class, considered as a category, every source is a mono-source. This property characterizes thin categories. [Consider the empty sources.]

10.7 PROPOSITION

(1) Representable functors **preserve mono-sources** *(i.e., if $G : \mathbf{A} \to \mathbf{Set}$ is a representable functor and $\mathcal{S}$ is a mono-source in $\mathbf{A}$, then $G\mathcal{S}$ is a mono-source in $\mathbf{Set}$).*

(2) Faithful functors **reflect mono-sources** *(i.e., if $G : \mathbf{A} \to \mathbf{B}$ is a faithful functor, $\mathcal{S} = (A, f_i)$ is a source in $\mathbf{A}$, and $G\mathcal{S} = (GA, Gf_i)$ is a mono-source in $\mathbf{B}$, then $\mathcal{S}$ is a mono-source in $\mathbf{A}$).*

Proof: [52]

(1). If a functor preserves mono-sources, then, clearly, so does every functor that is naturally isomorphic to it. Thus it suffices to show that each mono-source $(B \xrightarrow{f_i} B_i)_I$ is sent by each hom-functor $\hom(A, -) : \mathbf{A} \to \mathbf{Set}$ into a point-separating source:

$$(\hom(A, B) \xrightarrow{\hom(A, f_i)} \hom(A, B_i))_I.$$

---

[52]Even though the proof is immediate by arguments analogous to those used in the proof of Proposition 7.37, we nevertheless sketch a proof so that the reader may gain some familiarity with notation concerning sources.

But this is immediate from the definition of mono-source.

(2). Let G and $\mathcal{S}$ be as described. If $B \underset{s}{\overset{r}{\rightrightarrows}} A$ is a pair of **A**-morphisms with $\mathcal{S} \circ r = \mathcal{S} \circ s$, then $G\mathcal{S} \circ Gr = G(\mathcal{S} \circ r) = G(\mathcal{S} \circ s) = G\mathcal{S} \circ Gs$. Since $G\mathcal{S}$ is a mono-source, this implies $Gr = Gs$. Since G is faithful, this gives $r = s$. □

10.8 COROLLARY
In a construct $(\mathbf{A}, U)$ every point-separating source is a mono-source. The converse holds whenever U is representable. □

10.9 PROPOSITION
Let $\mathcal{T} = (\mathcal{S}_i) \circ \mathcal{S}$ be a composite of sources.

(1) If $\mathcal{S}$ and all $\mathcal{S}_i$ are mono-sources, then so is $\mathcal{T}$.

(2) If $\mathcal{T}$ is a mono-source, then so is $\mathcal{S}$. □

10.10 PROPOSITION
Let $(A, f_i)_I$ be a source.

(1) If $(A, f_j)_J$ is a mono-source for some $J \subseteq I$, then so is $(A, f_i)_I$.

(2) If f_j is a monomorphism for some $j \in I$, then $(A, f_i)_I$ is a mono-source. □

10.11 DEFINITION
A mono-source $\mathcal{S}$ is called **extremal** provided that whenever $\mathcal{S} = \overline{\mathcal{S}} \circ e$ for some epimorphism e, then e must be an isomorphism.

10.12 EXAMPLES
(1) A 1-source (A, f) is an extremal mono-source if and only if f is an extremal monomorphism.

(2) In balanced categories (e.g. in **Set**, **Vec**, and **Grp**) every mono-source is extremal. [This follows immediately from Proposition 10.9(2).] Conversely, if every mono-source in **C** is extremal, then **C** is balanced (cf. Proposition 7.67).

(3) A source $(A \xrightarrow{f_i} A_i)_I$ in **Pos** is an extremal mono-source provided that the following equivalence holds:
$$a \leq b \Leftrightarrow \forall i \in I \quad f_i(a) \leq f_i(b).$$

(4) A mono-source $(A \xrightarrow{f_i} A_i)_I$ in **Top** is extremal if and only if A carries the initial (= weak) topology with respect to (f_i).

(5) A mono-source $(A \xrightarrow{f_i} A_i)_I$ in Σ-**Seq** is extremal if and only if a state q of A is final whenever each state $f_i(q)$ is final in A_i.

(6) In a poset, considered as a category, a source $(A \to A_i)_I$ is an extremal mono-source if and only if A is a maximal lower bound of $\{A_i \mid i \in I\}$.

10.13 PROPOSITION
(1) If a composite source $(\mathcal{S}_i) \circ \mathcal{S}$ is an extremal mono-source, then so is $\mathcal{S}$.

(2) If $\mathcal{S} \circ f$ is an extremal mono-source, then f is an extremal monomorphism. □

10.14 REMARK
If $\mathcal{S}$ and each $\mathcal{S}_i$ are extremal mono-sources, then $(\mathcal{S}_i) \circ \mathcal{S}$ need not be extremal. See Exercise 7N.

10.15 PROPOSITION
Let $(A, f_i)_I$ be a source.

(1) If $(A, f_j)_J$ is an extremal mono-source for some $J \subseteq I$, then so is $(A, f_i)_I$.

(2) If f_j is an extremal monomorphism for some $j \in I$, then $(A, f_i)_I$ is an extremal mono-source. □

10.16 REMARK
The concept of source allows a simple description of coseparators: namely, A is a coseparator if and only if, for any object B, the source $(B, \hom(B, A))$ is a mono-source. This suggests the following definition:

10.17 DEFINITION
An object A is called an **extremal coseparator** provided that for any object B the source $(B, \hom(B, A))$ is an extremal mono-source.

10.18 EXAMPLES
(1) In a balanced category every coseparator is extremal.

(2) In the nonbalanced category **Pos** every coseparator is extremal [cf. 7.18(3)].

(3) A topological space is an extremal coseparator in **Top** if and only if it contains an indiscrete subspace with two elements and a Sierpinski subspace (i.e., a nondiscrete T_0-space with two elements) [cf. 7.18(4)].

(4) The unit interval $[0, 1]$ is an extremal coseparator in **HComp**.

(5) The category **Tych** of Tychonoff spaces and continuous maps has no extremal coseparator [cf. 7.18(7)].

(6) Let $\Sigma\text{-}\mathbf{Seq}_0$ denote the category of sequential Σ-acceptors that have no initial state and that are observable (i.e., the observability equivalence of Example 4.17(7) is equality). Then $\Sigma\text{-}\mathbf{Seq}_0$ has an extremal coseparator: the acceptor (Rat, δ, F) of all rational languages in Σ, where $\delta(L, \sigma) = \{\sigma\sigma_1 \cdots \sigma_n \mid \sigma_1 \cdots \sigma_n \in L\}$ and $F = \{L \in Rat \mid \emptyset \in L\}$. [For each acceptor A consider the simulation $f : A \to (Rat, \delta, F)$ that assigns to each state q the language $f(q) \in Rat$ accepted by A in the initial state q. Then f is an extremal monomorphism.]

PRODUCTS

Cartesian products of pairs of sets or, more generally, of families of sets (resp. direct products of families of vector spaces, resp. topological products of families of topological spaces) can be regarded as objects together with families of (projection) morphisms emanating from them, i.e., as sources. As such — but not as objects alone — they can be characterized, up to isomorphism, by the following categorical property:

10.19 DEFINITION
A source $\mathcal{P} = (P \xrightarrow{p_i} A_i)_I$ is called a **product** provided that for every source $\mathcal{S} = (A \xrightarrow{f_i} A_i)_I$ with the same codomain as $\mathcal{P}$ there exists a unique morphism $A \xrightarrow{f} P$ with $\mathcal{S} = \mathcal{P} \circ f$. A product with codomain $(A_i)_I$ is called a **product of the family** $(A_i)_I$.

10.20 EXAMPLES
(1) In the category **Set**, given sets A_1 and A_2, the projections from the cartesian product $\pi_1 : A_1 \times A_2 \to A_1$ and $\pi_2 : A_1 \times A_2 \to A_2$ (given by: $\pi_i(x_1, x_2) = x_i$) form a product source $(A_1 \times A_2 \xrightarrow{\pi_i} A_i)_{i=1,2}$. Indeed, given a source $(A \xrightarrow{f_i} A_i)_{i=1,2}$, there is a unique $A \xrightarrow{f} A_1 \times A_2$ with $f_i = \pi_i \circ f$, namely, $f(a) = (f_1(a), f_2(a))$. More generally, let $(A_i)_I$ be a family of sets indexed by a set I, and let $\prod_{i \in I} A_i$ be its cartesian product, i.e., the set of all functions $g : I \to \bigcup_I A_i$ with the property that $g(i) \in A_i$. Then the family of projection functions $\pi_j : \prod_{i \in I} A_i \to A_j$, given by: $g \mapsto g(j)$, is a product in **Set**.

(2) Likewise in the categories **Vec**, **Ab**, and **Grp** the "direct products", in **Pos** the "ordinal products", and in **Top** the "topological products", considered as sources via the projections, are products.

(3) If $(A_i)_{i \in I}$ is a set-indexed family of objects in the category **AbTor** of abelian torsion groups, its direct product $\prod_{i \in I} A_i$ need not be a torsion group. However, a product of the family does exist in **AbTor**. Let P be the torsion-subgroup of $\prod_{i \in I} A_i$ (i.e., the subgroup consisting of all torsion-elements), and for each $j \in I$ let $p_j : P \to A_j$ be the restriction of the jth projection $\pi_j : \prod_{i \in I} A_i \to A_j$. Then $(P \xrightarrow{p_i} A_i)_{i \in I}$ is a product in **AbTor**. [Indeed, given a source $(A \xrightarrow{f_i} A_i)_I$ in **AbTor**, each $a \in A$ is a torsion-element of A, so that $(f_i(a))_I$ is a torsion-element of $\prod A_i$. Thus the function $A \xrightarrow{f} P$, given by $f(a) = (f_i(a))_I$, satisfies the above definition.]

(4) Similarly, in **Ban** products can be obtained as the subspaces of the direct products of the corresponding vector spaces, consisting of those elements $a = (a_i)_{i \in I}$ with $\|a\| = \sup_{i \in I} \|a_i\| < \infty$, supplied with the restrictions of the projection-maps.

(5) In Σ-**Seq** the product of two acceptors $A_1 \times A_2$ is their parallel connection (the state of which is determined by knowing the state of both A_1 and A_2). Thus finite products have a clear interpretation in Σ-**Seq**, although infinite products usually don't exist, since acceptors are by definition finite.

(6) In a partially ordered class, considered as a category [3.3(4)(d)], a source $(P \xrightarrow{p_i} A_i)_I$ is a product if and only if $P = \bigwedge_{i \in I} A_i$.

(7) An empty source $(P, \emptyset)$ is a product if and only if P is a terminal object. Thus, regarding empty sources as objects, one might say: terminal objects are empty products.

(8) A 1-source (P, p) is a product if and only if p is an isomorphism. Thus, regarding 1-sources as morphisms, one might say: isomorphisms are 1-products.

10.21 PROPOSITION
Every product is an extremal mono-source.

Proof: If $\mathcal{P} = (P, p_i)$ is a product and $A \underset{s}{\overset{r}{\rightrightarrows}} P$ is a pair of morphisms with $\mathcal{P} \circ r = \mathcal{P} \circ s$, then $\mathcal{S} = \mathcal{P} \circ r = \mathcal{P} \circ s$ is a source with the same codomain as $\mathcal{P}$. The uniqueness requirement in the definition of product implies that $r = s$. Hence $\mathcal{P}$ is a mono-source. To show that $\mathcal{P}$ is extremal, let $\mathcal{P} = \mathcal{Q} \circ e$ for some epimorphism e. Since $\mathcal{P}$ and $\mathcal{Q}$ have the same codomain, there exists a unique morphism f with $\mathcal{Q} = \mathcal{P} \circ f$. Since $\mathcal{P}$ is a mono-source, the equation $\mathcal{P} \circ id_p = \mathcal{P} = \mathcal{Q} \circ e = \mathcal{P} \circ (f \circ e)$ implies that $id_p = f \circ e$. Consequently, e is a section and an epimorphism, hence an isomorphism. □

10.22 PROPOSITION
For any family $(A_i)_I$ of objects, products of $(A_i)_I$ are essentially unique; i.e., if $\mathcal{P} = (P \xrightarrow{p_i} A_i)_I$ is a product of $(A_i)_I$, then the following hold:

(1) for each product $\mathcal{Q} = (Q \xrightarrow{q_i} A_i)_I$ there exists an isomorphism $Q \xrightarrow{h} P$ with $\mathcal{Q} = \mathcal{P} \circ h$,

(2) for each isomorphism $A \xrightarrow{h} P$ the source $\mathcal{P} \circ h$ is a product of $(A_i)_I$.

Proof:
(1). Since $\mathcal{P}$ and $\mathcal{Q}$ are products with the same codomain, there exist unique morphisms h and k with $\mathcal{Q} = \mathcal{P} \circ h$ and $\mathcal{P} = \mathcal{Q} \circ k$. Therefore $\mathcal{Q} \circ id_Q = \mathcal{Q} \circ (k \circ h)$ and $\mathcal{P} \circ id_P = \mathcal{P} \circ (h \circ k)$. Since $\mathcal{P}$ and $\mathcal{Q}$ are mono-sources, these equations imply that $id_Q = k \circ h$ and $id_P = h \circ k$. Hence h is an isomorphism.

(2). Obvious. □

10.23 REMARK
The above uniqueness result allows us to introduce special notations for products (provided that they exist):

(1) Products of $(A_i)_I$ will be denoted by $(\prod_{i \in I} A_i \xrightarrow{\pi_j} A_j)_{j \in I}$, or, more simply, by $(\prod A_i \xrightarrow{\pi_j} A_j)_I$, and the morphisms π_j will be called **projections**.

(2) If $(\prod A_i \xrightarrow{\pi_j} A_j)_I$ is a product and $(A \xrightarrow{f_j} A_j)_I$ is a source with the same codomain, then the unique morphism $f : A \to \prod A_i$ with $f_j = \pi_j \circ f$ for each $j \in I$ will be denoted by $\langle f_i \rangle$:

$$A \xrightarrow{\langle f_i \rangle} \prod A_i \qquad (*)$$
$$f_j \searrow \quad \downarrow \pi_j$$
$$A_j$$

(3) In case $I = \{1, 2, \ldots, n\}$ the following notation will often be used instead of that above:

$$A \xrightarrow{\langle f_1, f_2, \ldots, f_n \rangle} A_1 \times A_2 \times \cdots \times A_n \qquad (**)$$
$$f_j \searrow \quad \downarrow \pi_j$$
$$A_j$$

10.24 REMARK
The above diagram (*) makes visible why products are useful. The correspondence $(A, f_i) \mapsto \langle f_i \rangle$ provides a bijection from the collection of all sources with domain A and codomain $(A_i)_I$ to the set of all morphisms with domain A and codomain $\prod_{i \in I} A_i$. Hence *products allow one to treat sources as if they were morphisms*. Propositions 10.26 and 10.38 below show how this correspondence works.

10.25 PROPOSITION
Let $\mathcal{Q} = (\mathcal{P}_i) \circ \mathcal{P}$ be a composite of sources.

(1) If $\mathcal{P}$ and all $\mathcal{P}_i$ are products, then so is $\mathcal{Q}$.

(2) If $\mathcal{Q}$ is a product and all $\mathcal{P}_i$ are mono-sources, then $\mathcal{P}$ is a product.

Proof: This follows immediately from the definition of products and the fact that products are mono-sources (10.21). □

10.26 PROPOSITION
Consider

$$A \xrightarrow{\langle f_i \rangle} \prod A_i$$
$$f_j \searrow \quad \downarrow \pi_j$$
$$A_j$$

Then

(1) $(A, f_i)_I$ is a mono-source if and only if $\langle f_i \rangle$ is a monomorphism.

(2) $(A, f_i)_I$ is an extremal mono-source if and only if $\langle f_i \rangle$ is an extremal monomorphism.

(3) $(A, f_i)_I$ is a product if and only if $\langle f_i \rangle$ is a product; i.e., an isomorphism.

Proof: (1) follows from Proposition 10.9, (3) follows from Propositions 10.21 and 10.25, and the "only if" part of (2) follows from Proposition 10.13. To show the "if" part of (2), let $\langle f_i \rangle$ be an extremal monomorphism. By (1), (A, f_i) is a mono-source. Let $A \xrightarrow{e} B$ be an epimorphism and $(B, g_i)_I$ be a source with $f_i = g_i \circ e$ for each $i \in I$. Then, since the product is a mono-source,

$$\pi_i \circ \langle f_i \rangle = g_i \circ e = \pi_i \circ \langle g_i \rangle \circ e \quad \text{implies that} \quad \langle f_i \rangle = \langle g_i \rangle \circ e.$$

Hence e is an isomorphism. □

10.27 REMARK
Whereas each product is an extremal mono-source, the single projections are usually retractions, as the following result shows. That this need not always be the case is demonstrated in **Set** by the projection: $\emptyset \times \mathbb{N} \to \mathbb{N}$.

10.28 PROPOSITION
If $(P \xrightarrow{p_i} A_i)_I$ is a product and if $i_0 \in I$ is such that $\hom(A_{i_0}, A_i) \neq \emptyset$ for each $i \in I$, then p_{i_0} is a retraction.

Proof: For each $i \in I$ choose $f_i \in \hom(A_{i_0}, A_i)$ with $f_{i_0} = id_{A_{i_0}}$. Then $\langle f_i \rangle : A_{i_0} \to P$ is a morphism with $p_{i_0} \circ \langle f_i \rangle = f_{i_0} = id_{A_{i_0}}$. □

10.29 DEFINITION
(1) A category **has products** provided that for every set-indexed family $(A_i)_I$ of objects there exists a product $(\prod A_i \xrightarrow{\pi_j} A_j)_I$.

(2) A category **has finite products** provided that for every finite family $(A_i)_I$ of objects there exists a product $(\prod A_i \xrightarrow{\pi_j} A_j)_I$.

10.30 PROPOSITION
A category has finite products if and only if it has terminal objects and products of pairs of objects.

Proof: The result follows from the observations that

(1) empty products are terminal objects [10.20(7)],

(2) products of singleton families always exist [10.20(8)],

(3) products of n-indexed families for $n \geq 3$ can be constructed via induction by composing products of pairs [cf. 10.25(1)]:

$$A_1 \times A_2 \times \cdots \times A_n = (\cdots((A_1 \times A_2) \times A_3) \times \cdots \times A_n). \quad \square$$

10.31 EXAMPLES
(1) The categories **Set**, **Vec**, **Grp**, **Top**, **Rel**, **Pos**, **Alg**(Ω), and **Aut** have products.

(2) **Met** and **Σ-Seq** have finite products, **Met**$_u$ and **Met**$_c$ have countable products, and **PMet** has products. [In fact, a set-indexed family $(X_i)_{i \in I}$ of nonempty metric spaces each having finite diameter $diam(X_i)$ has a product in **Met** if and only if the set $\{\, diam(X_i) \,|\, i \in I \,\}$ is bounded.]

(3) **Ban**$_b$ has finite products, and **Ban** has products.

(4) A poset, considered as a category, has products if and only if it is a complete lattice.

10.32 THEOREM

(1) A category that has products for all class-indexed families must be thin.

(2) A small category has products if and only if it is equivalent to a complete lattice.

Proof:

(1). Assume that **A** has all class-indexed products, but that the set $\hom(A, B)$ has at least two elements. Consider the family $(B_i)_{i \in I}$ with $I = Mor(\mathbf{A})$ and $B_i = B$ for each $i \in I$. Since $\hom(A, B)$ has at least two members, there are at least as many distinct sources with domain A and codomain $(B_i)_I$ as there are subclasses of I. Hence $\hom(A, \prod B_i)$ contains at least as many members as this (cf. 10.24), contradicting the fact that $\hom(A, \prod B_i)$ is contained in I.

(2). Since **A** is small, $Mor(\mathbf{A})$ is a set, so that if **A** has products, the above proof shows that it is thin. Thus it is a preordered set with meets of all subsets, and, hence, is equivalent to a complete lattice [cf. 10.31(4)]. The converse is clear. □

10.33 REMARK

The above theorem shows why in Definition 10.29(1) we didn't require the existence of products for families of objects indexed by *arbitrary* (hence also proper) classes. Such a requirement would be far too strong. None of our familiar constructs [e.g., from Examples 10.31 (1)–(3)] satisfies this condition, yet many of them do satisfy the weaker condition of having products of all *set*-indexed families. This observation demonstrates strikingly that when working with categories one needs to distinguish carefully between sets and proper classes (resp. between "small" and "large" collections).

10.34 DEFINITION

If $(A_i \xrightarrow{f_i} B_i)_I$ is a family of morphisms and if $(\prod A_i \xrightarrow{\pi_j} A_j)_I$ and $(\prod B_i \xrightarrow{p_j} B_j)_I$ are products, then the unique morphism $\prod A_i \to \prod B_i$ that makes the following diagram commute for each $j \in I$

$$\begin{array}{ccc} \prod A_i & \xrightarrow{\Pi f_i} & \prod B_i \\ \pi_j \downarrow & & \downarrow p_j \\ A_j & \xrightarrow{f_j} & B_j \end{array}$$

is denoted by Πf_i and is called the **product** of the family $(f_i)_I$. If $I = \{1, \ldots, n\}$ then Πf_i is usually written as $f_1 \times f_2 \times \cdots \times f_n$.

10.35 PROPOSITION

Let $(f_i)_I$ be a set-indexed family of morphisms with product Πf_i. If each f_i has any of the following properties, then so does Πf_i:

(1) isomorphism,

(2) section,

(3) retraction,

(4) monomorphism,

(5) regular monomorphism (provided that the category in question has products).

Proof: (1), (2), and (3) follow immediately from the observation that $\Pi g_i \circ \Pi f_i = \Pi(g_i \circ f_i)$. (4) follows from Propositions 10.9 and 10.21.

(5) follows from the next proposition. (Observe that products of regular monomorphisms are always extremal monomorphisms. Cf. also 10D) □

10.36 PROPOSITION

In a category with products, if I is a set and if $E_i \xrightarrow{e_i} A_i$ is an equalizer of $A_i \underset{g_i}{\overset{f_i}{\rightrightarrows}} B_i$ for each $i \in I$, then $\prod E_i \xrightarrow{\Pi e_i} \prod A_i$ is an equalizer of $\prod A_i \underset{\Pi g_i}{\overset{\Pi f_i}{\rightrightarrows}} \prod B_i$.

Proof:

$$\prod E_i \xrightarrow{\Pi e_i} \prod A_i \underset{\Pi g_i}{\overset{\Pi f_i}{\rightrightarrows}} \prod B_i$$
$$\pi_j \downarrow \quad\quad p_j \downarrow \quad\quad q_j \downarrow$$
$$E_j \xrightarrow{e_j} A_j \underset{g_j}{\overset{f_j}{\rightrightarrows}} B_j$$

That $\Pi f_i \circ \Pi e_i = \Pi g_i \circ \Pi e_i$ follows from the fact that the product $(\prod B_i \xrightarrow{q_j} B_j)$ is a mono-source. If $C \xrightarrow{h} \prod A_i$ is a morphism such that $\Pi f_i \circ h = \Pi g_i \circ h$, then $f_j \circ p_j \circ h = g_j \circ p_j \circ h$ for each $j \in I$. Hence for each $j \in I$ there exists a morphism $h_j : C \to E_j$ with $p_j \circ h = e_j \circ h_j$. Consequently, $\langle h_i \rangle : C \to \prod E_i$ is a morphism with $p_j \circ \Pi e_i \circ \langle h_i \rangle = e_j \circ \pi_j \circ \langle h_i \rangle = e_j \circ h_j = p_j \circ h$; hence with $\Pi e_i \circ \langle h_i \rangle = h$. Therefore h factors through Πe_i. Since, by Proposition 10.35(4), Πe_i is a monomorphism, h factors uniquely. □

10.37 DEFINITION

If I is a set and $(\prod A_i \xrightarrow{\pi_j} A_j)_I$ is a product with $A_i = A$ for each $i \in I$, then $\prod A_i$ is denoted by A^I and called an *I*th **power of** A.

10.38 PROPOSITION

In a category that has products, an object A is an (extremal) coseparator if and only if every object is an (extremal) subobject of some power A^I of A.

Proof: Let A and B be objects. Consider the source $\mathcal{S} = (B, \hom(B, A))$ and the morphism $B \xrightarrow{m} A^{\hom(B,A)}$ defined by $f = \pi_f \circ m$ for each $f \in \hom(B, A)$. By Proposition 10.26, $\mathcal{S}$ is an (extremal) mono-source if and only if m is an (extremal) monomorphism. Hence A is an (extremal) coseparator if and only if $B \xrightarrow{m} A^{\hom(B,A)}$ is an (extremal) monomorphism for each object B. Finally, $B \xrightarrow{g} A^I$ is an (extremal) monomorphism for some I [i.e., the source $(B, \pi_i \circ g)_{i \in I}$ is an (extremal) mono-source] if and only if $\mathcal{S} = (B, \hom(B, A))$ is an (extremal) mono-source. Cf. Propositions 10.10(1) and 10.15(1). □

10.39 REMARKS
(1) In a category **A** that has products and an (extremal) coseparator the above result provides a useful description of the **A**-objects. For example, since the two-element chain **2** is a coseparator for **Pos**, posets are precisely the subobjects of powers of **2** in **Pos**, and since the unit interval $[0,1]$ is an extremal coseparator for **HComp**, compact Hausdorff spaces are precisely the extremal subobjects (= closed subspaces) of powers of $[0,1]$ in **HComp**.

(2) In a category that has products, M-**coseparators** may be defined (for any class M of monomorphisms) as objects A such that each object is an M-subobject of some power of A.

10.40 PROPOSITION
For any class M of morphisms, every product of M-injective objects is M-injective.[53] □

SOURCES IN CONCRETE CATEGORIES

Next we turn our attention to sources in concrete categories. As before, in the context of concrete categories, we use the notational conventions of Remark 6.22.

INITIAL SOURCES

10.41 DEFINITION
Let **A** be a concrete category over **X**. A source $(A \xrightarrow{f_i} A_i)$ in **A** is called **initial** provided that an **X**-morphism $f : |B| \to |A|$ is an **A**-morphism whenever each composite $f_i \circ f : |B| \to |A_i|$ is an **A**-morphism.

10.42 EXAMPLES
(1) An empty source $(A, \emptyset)$ is initial if and only if A is indiscrete.

(2) A 1-source (A, f) is initial if and only if f is an initial morphism (cf. 8.6).

---
[53] The special case for products that are empty sources yields Proposition 9.4.

(3) A source $(A, f_i)_I$ in **Top** is initial if and only if A carries the initial (= weak) topology with respect to the family $(f_i)_I$. In particular, a topological space X is completely regular if and only if the source $\mathcal{S}(X, \mathbb{R})$, consisting of all continuous maps from X to the real line, is initial (in the construct **Top**); and X is a Tychonoff space if and only if $\mathcal{S}(X, \mathbb{R})$ is an initial mono-source.

(4) In **Spa**(T) a source $((X, \alpha) \xrightarrow{f_i} (X_i, \alpha_i))_I$ is initial if and only if $\alpha = \bigcap_{i \in I}(Tf)^{-1}[\alpha_i]$. In particular, a source $((X, \rho) \xrightarrow{f_i} (X_i, \rho_i))_I$ in **Rel** is initial if and only if the equivalence: $x \rho y \Leftrightarrow \forall i \in I\ f_i(x) \rho_i f_i(y)$ holds.

(5) In **PMet** a source $((X, d) \xrightarrow{f_i} (X_i, d_i))_I$ is initial if and only if
$$d(x, y) = \sup_{i \in I} d_i(f_i(x), f_i(y)).$$

(6) In **Pos** and in Σ-**Seq** the initial sources are precisely the extremal mono-sources (cf. 10.12).

(7) In any of the constructs **Vec**, **Ab**, **Grp**, **Mon**, **Rng**, **Boo**, and **Alg**(Ω) the initial sources are precisely the mono-sources.

(8) In a preordered class, considered as a concrete category over **1**, the initial sources are precisely the products [cf. 10.20(6)]. This fact is not as accidental as it may seem. As we will see later in this section (cf. 10.58), a source in an arbitrary category **A** is a product if and only if it is T-initial, where $T : \mathbf{A} \to \mathbf{1}$ is the unique functor from **A** to **1**.

10.43 PROPOSITION
If $(A \xrightarrow{f_i} A_i)_I$ is an initial source in **A**, then
$A = \max\{ B \in Ob(\mathbf{A}) \mid |B| = |A| \text{ and all } |B| \xrightarrow{f_i} |A_i| \text{ are } \mathbf{A}\text{-morphisms}\}$.[54] □

10.44 REMARK
The above property often characterizes initial sources, e.g., in such constructs as **Top** or **Spa**(T). However, in the construct **Top**$_1$, there are non-initial sources with the above property. In fact, as shown in Example 8.5(2), for each set X the fibre of X has a largest element A_X, hence the empty source $(A_X, \emptyset)$ satisfies the above property. But $(A_X, \emptyset)$ is initial (i.e., A_X is indiscrete) only for $\mathrm{card}(X) \leq 1$.

10.45 PROPOSITION
Let $\mathcal{T} = (\mathcal{S}_i) \circ \mathcal{S}$ be a composite of sources in a concrete category.

(1) If $\mathcal{S}$ and all $\mathcal{S}_i$ are initial, then so is $\mathcal{T}$.

(2) If $\mathcal{T}$ is initial, then so is $\mathcal{S}$. □

---
[54] Recall the order on the fibre of $|A|$ [5.4(1)].

10.46 PROPOSITION
Let $(A, f_i)_I$ be a source in a concrete category. If $(A, f_i)_J$ is initial for some $J \subseteq I$, then so is $(A, f_i)_I$. □

10.47 DEFINITION
A concrete functor $F : \mathbf{A} \to \mathbf{B}$ over $\mathbf{X}$ is said to **preserve initial sources** provided that for every initial source $\mathcal{S}$ in $\mathbf{A}$, the source $F\mathcal{S}$ is initial in $\mathbf{B}$.

10.48 EXAMPLES
(1) The forgetful concrete functor $\mathbf{Rng} \to \mathbf{Ab}$ over $\mathbf{Set}$, which forgets multiplication, preserves initial sources [cf. 10.42(7)].

(2) The concrete functor $\mathbf{Top} \to \mathbf{Rel}$ over $\mathbf{Set}$, which assigns to each topological space (X, τ) the object (X, ρ) with $x\rho y$ if and only if $x \in cl\{y\}$, preserves initial sources.

(3) An order-preserving map between preordered classes, considered as a concrete functor over $\mathbf{1}$, preserves initial sources if and only if it preserves all existing meets.

10.49 PROPOSITION
If (F, G) is a Galois correspondence, then G preserves initial sources.

Proof: Let $G : \mathbf{A} \to \mathbf{B}$ and $F : \mathbf{B} \to \mathbf{A}$ be concrete functors over $\mathbf{X}$ such that (F, G) is a Galois correspondence over $\mathbf{X}$. Let $(A \xrightarrow{f_i} A_i)_I$ be an initial source in $\mathbf{A}$ and let $B \xrightarrow{h} GA$ be an $\mathbf{X}$-morphism such that all $B \xrightarrow{h} GA \xrightarrow{f_i} GA_i$ are $\mathbf{B}$-morphisms. Then by Proposition 6.28, all $FB \xrightarrow{h} A \xrightarrow{f_i} A_i$ are $\mathbf{A}$-morphisms. Hence $FB \xrightarrow{h} A$ is an $\mathbf{A}$-morphism. Again by Proposition 6.28, $B \xrightarrow{h} GA$ is a $\mathbf{B}$-morphism. □

10.50 COROLLARY
Embeddings of concretely reflective subcategories preserve initial sources. □

10.51 REMARK
Let $G : A \to B$ be a monotone map between posets, considered as a concrete functor over $\mathbf{1}$. If A is a complete lattice, then G preserves initial sources (= meets) if and only if there exists a (necessarily unique) monotone map $F : B \to A$ such that (F, G) is a Galois connection. [Namely, $F(b) = \bigwedge \{ a \in A \mid b \leq G(a) \}$]. Theorem 21.24 is a corresponding result for concrete functors over arbitrary base categories $\mathbf{X}$.

CONCRETE PRODUCTS

10.52 DEFINITION
Let $\mathbf{A}$ be a concrete category over $\mathbf{X}$. A source $\mathcal{S}$ in $\mathbf{A}$ is called a **concrete product** in $\mathbf{A}$ if and only if $\mathcal{S}$ is a product in $\mathbf{A}$ and $|\mathcal{S}|$ is a product in $\mathbf{X}$.

10.53 PROPOSITION
A source S in a concrete category $\mathbf{A}$ over $\mathbf{X}$ is a concrete product if and only if it is initial in $\mathbf{A}$ and $|S|$ is a product in $\mathbf{X}$. □

10.54 DEFINITION
A concrete category $\mathbf{A}$ **has concrete products** if and only if for every set-indexed family $(A_i)_I$ of $\mathbf{A}$-objects there exists a concrete product $(P \xrightarrow{p_i} A_i)_I$ in $\mathbf{A}$, i.e., if and only if $\mathbf{A}$ has products and the forgetful functor preserves them.

10.55 EXAMPLES
(1) Many familiar constructs, e.g., **Vec**, **Grp**, **Ab**, **Rng**, **Top**, **Rel**, **Pos**, and **Alg**(Ω), have concrete products. According to Proposition 10.53 they can be constructed in two steps: Given a family of objects $(A_i)_I$ in $\mathbf{A}$, first form the cartesian product $(\prod |A_i| \xrightarrow{\pi_j} |A_j|)_I$ of the underlying sets, and then supply the set $\prod |A_i|$ with the initial structure with respect to $(\pi_j)_I$.

(2) The construct (**Ban**, O) has concrete products, but the products in (**Ban**, U) generally fail to be concrete [cf. 10.20(4)].

(3) Products in the construct **AbTor** generally fail to be concrete [cf. 10.20(3)].

(4) In a concrete category over **1** every product is concrete.

10.56 PROPOSITION
Let $\mathcal{Q} = (\mathcal{P}_i) \circ \mathcal{P}$ be a composite of sources in a concrete category $\mathbf{A}$.
(1) If $\mathcal{P}$ and all $\mathcal{P}_i$ are concrete products, then so is $\mathcal{Q}$.
(2) If $\mathcal{Q}$ is a concrete product and each $|\mathcal{P}_i|$ is a mono-source, then $\mathcal{P}$ is a concrete product. □

G-INITIAL SOURCES

Now we investigate sources not only in relation to forgetful functors of concrete categories, but also in relation to arbitrary functors. This will throw additional light on products.

10.57 DEFINITION
Let $G: \mathbf{A} \to \mathbf{B}$ be a functor. A source $S = (A \xrightarrow{f_i} A_i)_I$ in $\mathbf{A}$ is called G-**initial** provided that for each source $T = (B \xrightarrow{g_i} A_i)_I$ in $\mathbf{A}$ with the same codomain as S and each $\mathbf{B}$-morphism $GB \xrightarrow{h} GA$ with $GT = GS \circ h$ there exists a unique $\mathbf{A}$-morphism $B \xrightarrow{\overline{h}} A$ with $T = S \circ \overline{h}$ and $h = G\overline{h}$.

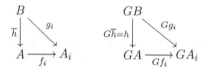

10.58 EXAMPLES

(1) If $(\mathbf{A}, U)$ is a concrete category, then U-initial sources are precisely the initial sources in $(\mathbf{A}, U)$.

(2) If $\mathbf{A}$ is a category and $G : \mathbf{A} \to \mathbf{1}$ is the unique functor from $\mathbf{A}$ to $\mathbf{1}$, then G-initial sources are precisely the products in $\mathbf{A}$.

10.59 PROPOSITION
For a functor $G : \mathbf{A} \to \mathbf{B}$ the following conditions are equivalent:

(1) G is faithful,

(2) for each $\mathbf{A}$-object A the 2-source $(A, (id_A, id_A))$ is G-initial,

(3) whenever $(A, f_i)_I$ is a source in $\mathbf{A}$ and $(A, f_j)_J$ is G-initial for some $J \subseteq I$, then $(A, f_i)_I$ is G-initial.

Proof: Obviously (1) $\Rightarrow$ (3) $\Rightarrow$ (2). To show that (2) implies (1) let $A \underset{s}{\overset{r}{\rightrightarrows}} B$ be a pair of $\mathbf{A}$-morphisms with $Gr = Gs$. Consider the 2-sources $\mathcal{S} = (B, (id_B, id_B))$ and $\mathcal{T} = (A, (r, s))$ in $\mathbf{A}$ and the $\mathbf{B}$-morphism $h = Gr : GA \to GB$. Then $G\mathcal{T} = G\mathcal{S} \circ h$. Hence, by G-initiality of $\mathcal{S}$, there exists an $\mathbf{A}$-morphism $\overline{h} : A \to B$ with $\mathcal{T} = \mathcal{S} \circ \overline{h}$, i.e., with $r = id_B \circ \overline{h} = s$. □

10.60 PROPOSITION
If $G : \mathbf{A} \to \mathbf{B}$ is a functor such that each mono-source in $\mathbf{A}$ is G-initial, then the following hold:

(1) G is faithful,

(2) G reflects products,

(3) G reflects isomorphisms.

Proof:

(1). follows from Proposition 10.59, since in $\mathbf{A}$ each 2-source $(A, (id_A, id_A))$ is a mono-source.

(2). If $\mathcal{P}$ is a source in $\mathbf{A}$ such that $G\mathcal{P}$ is a product in $\mathbf{B}$, then $G\mathcal{P}$ is a mono-source in $\mathbf{B}$. Hence, by (1), $\mathcal{P}$ is a mono-source in $\mathbf{A}$ and, consequently, is G-initial. This, together with faithfulness of G, immediately implies that $\mathcal{P}$ is a product in $\mathbf{A}$.

(3). follows from (2), since isomorphisms are 1-products. □

10.61 REMARK
The property that all mono-sources be initial, is not unfamiliar. As we will see in §23, it is typical for "algebraic" categories.

SINKS

The concept dual to that of source is called *sink*. Whereas the concepts of sources and sinks are dual to each other, frequently sources occur more naturally than sinks (cf. §15 and §17). However, there are cases where the opposite is true. Cf. in particular Definition 10.69 and Proposition 10.71 below.

10.62 DEFINITION
A **sink** is a pair $((f_i)_{i \in I}, A)$ [sometimes denoted by $(f_i, A)_I$ or $(A_i \xrightarrow{f_i} A)_I$] consisting of an object A (the **codomain** of the sink) and a family of morphisms $f_i : A_i \to A$ indexed by some class I. The family $(A_i)_{i \in I}$ is called the **domain** of the sink. Composition of sinks is defined in the (obvious) way dual to that of composition of sources.

10.63 TABLE OF DUAL CONCEPTS
The following table provides the names for the concepts dual to those investigated in this section:

Concept	Dual Concept
source	sink
mono-source	epi-sink
extremal mono-source	extremal epi-sink
extremal coseparator	extremal separator
initial source	final sink
G-initial source	G-final sink
product $(\prod A_i, \pi_j)_{j \in I}$	coproduct $(\mu_j, \coprod A_i)_{j \in I}$
projection π_j	injection μ_j
$A \xrightarrow{\langle f_i \rangle} \prod A_i$, $f_j \searrow \downarrow \pi_j$, A_j	$A \xleftarrow{[f_i]} \coprod A_i$, $f_j \searrow \uparrow \mu_j$, A_j
power A^I	copower $^I A$
$C \xrightarrow{\langle f,g \rangle} A \times B$	$C \xleftarrow{[f,g]} A + B$
$\prod f_i$; $f \times g$	$\coprod f_i$; $f + g$

10.64 EXAMPLES OF EPI-SINKS

(1) In **Set** a sink $(A_i \xrightarrow{f_i} A)_I$ is an epi-sink if and only if it is jointly surjective, i.e., if and only if $A = \bigcup_{i \in I} f_i[A_i]$.

In every construct, all jointly surjective sinks are epi-sinks. The converse implication holds, e.g., in **Vec**, **Pos**, **Top**, and **Σ-Seq**.

(2) In **Sgr** we have seen that there are epimorphisms that are not surjective [7.40(5)]. Thus, there are epi-sinks that are not jointly surjective.

(3) A category **A** is thin if and only if every sink in **A** is an epi-sink.

10.65 EXAMPLES OF EXTREMAL EPI-SINKS

(1) Every epi-sink (= jointly surjective sink) is an extremal epi-sink in **Set**, **Vec**, and **Ab**.

(2) In **Top** an epi-sink $(A_i \xrightarrow{f_i} A)$ is extremal if and only if A carries the final topology with respect to $(f_i)_{i \in I}$.

(3) In **Pos** an epi-sink $(A_i \xrightarrow{f_i} A)_I$ is extremal if and only if the ordering of A is the transitive closure of the relation consisting of all pairs $(f_i(x), f_i(y))$ with $i \in I$ and $x \leq y$ in A_i.

(4) In **Σ-Seq** an epi-sink $(A_i \xrightarrow{f_i} A)_I$ is extremal if and only if each final state of A has the form $f_i(q)$ for some $i \in I$ and some final state q of A_i.

10.66 EXAMPLES OF EXTREMAL SEPARATORS

(1) Every separator is extremal in **Set**, **Vec**, and **Ab**.

(2) In **Pos** the separators are precisely the nonempty posets, whereas the extremal separators are precisely the nondiscrete posets.

(3) **Top** has no extremal separator.

10.67 EXAMPLES OF COPRODUCTS

Many familiar categories have coproducts. However, as opposed to the situation for products, coproducts in familiar constructs often fail to be concrete.

(1) If $(A_i)_I$ is a pairwise-disjoint family of sets, indexed by a set I, then the sink of inclusion maps $(A_j \xrightarrow{\mu_j} \bigcup_{i \in I} A_i)_{j \in I}$ is a coproduct in **Set**. If $(A_i)_I$ is an arbitrary set-indexed family of sets, then it can be "made disjoint" by pairing each member of A_i with the index i, i.e., by working with $A_i \times \{i\}$, rather than A_i. The union $\bigcup_{i \in I}(A_i \times \{i\})$ is called the **disjoint union** of the family $(A_i)_I$ and is denoted by $\biguplus_{i \in I} A_i$. The sink of natural injections $(A_j \xrightarrow{\mu_j} \biguplus_{i \in I} A_i)_{j \in I}$ [where $\mu_j(a) = (a, j)$] is a coproduct[55] in **Set** of the the family $(A_i)_I$. [Indeed, given a sink $(A_j \xrightarrow{f_i} B)_{j \in I}$ in **Set**, the unique function $f : \biguplus_{i \in I} A_i \to B$ satisfying $f_i = f \circ \mu_i$ for all i is defined by: $f(a, i) = f_i(a)$.]

---

[55]For a disjoint family of sets, the union and the disjoint union (together with associated injections) each form a coproduct. This is not a contradiction since coproducts are determined only up to isomorphism. Compare this with the corresponding result for products: Proposition 10.22.

(2) In the construct **Top** coproducts are called "topological sums" and can be constructed as for sets by supplying the disjoint union with the final topology. Thus, **Top** has concrete coproducts.

(3) In the construct **Pos** coproducts are called "cardinal sums" and can be constructed as for sets by supplying the disjoint union of the underlying sets with the order that agrees on each $A_i \times \{i\}$ with the order on A_i, and where members of distinct summands are incomparable. Thus, **Pos** has concrete coproducts.

(4) **Vec** has nonconcrete coproducts called direct sums. The direct sum $\bigoplus_{i \in I} A_i$ of a family $(A_i)_I$ of vector spaces is the subspace of the direct product $\prod_{i \in I} A_i$ consisting of all elements $(a_i)_{i \in I}$ with finite carrier (i.e., $\{i \in I \,|\, a_i \neq 0\}$ is finite), together with the injections $\mu_j : A_j \to \bigoplus A_i$ given by:

$$\mu_j(a) = (a_i)_{i \in I} \quad \text{with} \quad a_i = \begin{cases} a, & \text{if } i = j \\ 0, & \text{if } i \neq j. \end{cases}$$

The description of coproducts in **Ab** is analogous.

(5) **HComp** has nonconcrete coproducts, namely the Čech-Stone compactifications of the topological sums.

(6) **Grp** has nonconcrete coproducts, called "free products".

(7) The constructs $\mathbf{Alg}(\Omega)$ all have coproducts. They are concrete if and only if all the operations are unary.

(8) In a poset **A**, considered as a category, coproducts are joins. Thus **A** has coproducts if and only if it is a complete lattice.

(9) In every category every 1-indexed family (A) has a coproduct $A \xrightarrow{id} A$; an empty sink with codomain A is a coproduct if and only if A is an initial object.

10.68 EXAMPLES OF FINAL SINKS

(1) An empty sink $(\emptyset, A)$ in a concrete category is final if and only if A is discrete.

(2) A singleton sink $A \xrightarrow{f} B$ in a concrete category is final if and only if f is a final morphism.

(3) A sink $((X_i, \tau_i) \xrightarrow{f_i} (X, \tau))_I$ in **Top** is final if and only if τ is the final topology with respect to the maps $(f_i)_I$, i.e., $\tau = \{U \subseteq X \,|\, \text{for all } i \in I, \; f_i^{-1}[U] \in \tau_i\}$.

(4) A sink $((X_i, \leq_i) \xrightarrow{f_i} (X, \leq))_I$ in **Pos** is final if and only if ρ is the transitive closure of the relation $\{(x,x) \,|\, x \in X\} \cup \bigcup_I \{(f_i(x), f_i(y)) \,|\, x \leq_i y\}$.

(5) A sink $((X_i, \alpha_i) \xrightarrow{f_i} (X, \alpha))_I$ in $\mathbf{Spa}(T)$ is final if and only if $\alpha = \bigcup_I T f_i[\alpha_i]$.

10.69 DEFINITION
A full concrete subcategory **A** of a concrete category **B** is said to be **finally dense** in **B** provided that for every **B**-object B there is a final sink $(A_i \xrightarrow{f_i} B)_I$ in **B** with A_i in **A** for all $i \in I$.

DUAL NOTION: **initially dense**.

10.70 EXAMPLES
A. Final density in constructs

(1) In **Met** the full subcategory consisting of all two-element metric spaces is finally dense.

(2) In **Pos** the full subcategory whose only object is a 2-chain is finally dense.

(3) In **Vec** the full subcategory whose only object is $\mathbb{R}^2$ is finally dense. In general, in a construct of algebras with operations of arity at most n, any free algebra on n generators, considered as a full subcategory with one object, is finally dense.

(4) **Top** does not have any small, finally dense subcategory.

(5) Σ-**Seq** does not have any finite, finally dense subcategory. [Consider an automaton whose transitions form a cycle larger than the number of objects of the given finite subcategory.]

B. Final density in concrete categories

(6) For a poset **B**, considered as a concrete category over **1**, a subset **A** is finally dense if and only if it is join-dense (i.e., each $b \in \mathbf{B}$ is a join of a subset of **A**).

(7) For a category **A** considered as concrete over itself, the empty subcategory is finally dense.

10.71 PROPOSITION
*If **A** is a finally dense full concrete subcategory of a concrete category **B**, then every initial source in **A** is initial in **B**.*

Proof: Let $\mathcal{A} = (A \xrightarrow{f_i} A_i)_I$ be an initial source in **A**, let B be a **B**-object, and $|B| \xrightarrow{f} |A|$ be an **X**-morphism, such that each $|B| \xrightarrow{f_i \circ f} |A_i|$ is a **B**-morphism. To show that $|B| \xrightarrow{f} |A|$ is a **B**-morphism, let $\mathcal{B} = (C_j \xrightarrow{g_j} B)_J$ be a final sink in **B** with each C_j belonging to **A**. Then each

$$|C_j| \xrightarrow{g_j} |B| \xrightarrow{f} |A| \xrightarrow{f_i} |A_i|$$

is an **A**-morphism. Since $\mathcal{A}$ is initial in **A** and each C_j belongs to **A**, each $|C_j| \xrightarrow{g_j} |B| \xrightarrow{f} |A|$ is an **A**-morphism. Since $\mathcal{B}$ is final, $|B| \xrightarrow{f} |A|$ is a **B**-morphism. □

10.72 REMARK
A full concrete embedding $E : \mathbf{A} \to \mathbf{B}$ is called **finally dense** if its image is a finally dense subcategory of $\mathbf{B}$. The above proposition states, more succinctly, that finally dense embeddings preserve initiality.

Suggestions for Further Reading

Mac Lane, S. Duality for groups. *Bull. Amer. Math. Soc.* **56** (1950): 485–516.

Taylor, J. C. Weak families of maps. *Canad. Math. Bull.* **8** (1965): 771–781.

Pumplün, D. Initial morphisms and monomorphisms. *Manuscr. Math.* **32** (1980): 309–333.

EXERCISES

10A. A Characterization of (Extremal) Mono-Sources

Let $\mathbf{A}$ be a category such that for any pair $A \underset{g}{\overset{f}{\rightrightarrows}} B$ of morphisms there exists a coequalizer in $\mathbf{A}$. Show that

(a) A source $\mathcal{S}$ in $\mathbf{A}$ is a mono-source provided that whenever $\mathcal{S} = \overline{\mathcal{S}} \circ e$ for some regular epimorphism e, it follows that e is an isomorphism.

(b) A source $\mathcal{S}$ in $\mathbf{A}$ is an extremal mono-source provided that whenever $\mathcal{S} = \overline{\mathcal{S}} \circ e$ for some epimorphism e, it follows that e is an isomorphism.

10B. A Characterization of Extremal (Co)Separators

Show that an object S of a category $\mathbf{A}$ is

(a) an extremal separator if and only if the functor $\mathbf{A} \xrightarrow{\hom(S,-)} \mathbf{Set}$ is faithful and reflects isomorphisms, and

(b) an extremal coseparator if and only if the functor $\mathbf{A}^{\mathrm{op}} \xrightarrow{\hom(-,S)} \mathbf{Set}$ is faithful and reflects isomorphisms.

10C. Extremal (Co)Separators in HComp, BooSpa, and Tych

Show that

(a) In **HComp** every nonempty space is an extremal separator, and $[0,1]$ is an extremal coseparator.

(b) In **BooSpa** every nonempty space is an extremal separator, and every space with at least two points is an extremal coseparator.

* (c) **Tych** has neither an extremal separator nor an extremal coseparator.

10D. Products of Special Morphisms

Show that

(a) Products of regular monomorphisms are extremal monomorphisms.

(b) In a category with products, products of regular monomorphisms are regular monomorphisms.

(c) Products of regular monomorphisms may fail to be regular.

(d) Products of extremal monomorphisms may fail to be extremal.

(e) Products of epimorphisms may fail to be epimorphisms.

10E. Hom-Functors Preserve and "Collectively Detect" Products

Show that a source S is a product in **A** if and only if for each **A**-object A the source $\hom(A, -)(S)$ is a product in **Set**.

10F. Concrete Products and Hom-Functors

Let A be a free object over a nonempty set in a construct **A**. Prove that a source S in **A** is a concrete product if and only if S is initial and $\hom(A, -)(S)$ is a product in **Set**.

10G. Products in Ab

Let $(G^2, (\pi_1, \pi_2))$ be the cartesian product of an abelian group G with itself (where the group operation is given by $G^2 \xrightarrow{+} G$). Show that each of $(G^2, (\pi_1, \pi_2))$, $(G^2, (\pi_1, +))$ and $(G^2, (\pi_2, +))$ is a product of the pair (G, G) in **Ab**.

10H. Products with Terminal Objects

In **A** let T be a terminal object, A be an arbitrary object, and $A \xrightarrow{t_A} T$ be the unique morphism from A to T. Show that $(A, (id_A, t_A))$ is a product of the pair (A, T) in **A**.

10I. No (Co)Products in Fields or in a Group

Show that neither **Field** nor a nontrivial group (nor a nontrivial finite monoid), considered as a category, has finite products or finite coproducts.

10J. Products for Banach Spaces

Show that

(a) $\mathbf{Ban_b}$ has finite products, but $\mathbf{Ban_b}$ does not have products.

(b) **Ban** has products, but $(\mathbf{Ban}, U)$ does not have concrete products.

(c) $(\mathbf{Ban}, O)$ has concrete products.

10K. Comparison of Powers

Let I be a subset of K and let A^I and A^K be powers of A in **A**. Show that:

(a) If $I \neq \emptyset$, then A^I is a retract of A^K.

(b) If $I = \emptyset$, then A^I need not even be a subobject of A^K. [Hint: Consider $A = \emptyset$ in **Set**.]

* 10L. A Characterization of Concretizable Categories Over Set

Show that a category that has equalizers and finite products is concretizable over **Set** if and only if it is regular wellpowered.

10M. Composites of Regular Monomorphisms With Sections

Show that in a category that has finite products the composite $s \circ m$ of a regular monomorphism m and a section s is a regular monomorphism. [Hint: If $r \circ s = id$ and m is an equalizer of f and g, then $s \circ m$ is an equalizer of $\langle f \circ r, id \rangle$ and $\langle g \circ r, s \circ r \rangle$.] Contrast this with Exercise 7S.

10N. Dualities and Representability

Let $(\mathbf{A}, U)$ and $(\mathbf{B}, V)$ be concrete categories that are dually equivalent; i.e., there are contravariant functors $G : \mathbf{A} \to \mathbf{B}$ and $F : \mathbf{B} \to \mathbf{A}$ such that $F \circ G \cong 1_\mathbf{A}$ and $G \circ F \cong 1_\mathbf{B}$. Suppose that the **A**-object A represents U and the **B**-object B represents V, and let $\tilde{A} = F(B)$ and $\tilde{B} = G(A)$. Prove that

(a) $U(\tilde{A}) \cong V(\tilde{B})$.

(b) $V \circ G \cong \hom_\mathbf{A}(-, \tilde{A})$ and $U \circ F \cong \hom_\mathbf{B}(-, \tilde{B})$.

If, moreover, **B** has products, then show that

(c) for each **A**-object X there exists a monomorphism $m_X : G(X) \to \tilde{B}^{U(X)}$ such that $V(m_X)$ is the embedding of $V(G(X)) \cong \hom_\mathbf{A}(X, \tilde{A})$ into $\hom_{\mathbf{Set}}(U(X), U(\tilde{A})) \cong \hom_{\mathbf{Set}}(U(X), V(\tilde{B})) = (V(\tilde{B}))^{U(X)} \cong V(\tilde{B}^{U(X)})$.

10O. Composites of F-Initial Sources

Let $F : \mathbf{A} \to \mathbf{B}$ be a functor and let $\mathcal{T} = (\mathcal{S}_i) \circ \mathcal{S}$ be a composite of sources in **A**. Show that

(a) If $\mathcal{S}$ and $\mathcal{S}_i$ are F-initial, then so is $\mathcal{T}$.

(b) If $\mathcal{T}$ is F-initial and all $\mathcal{S}_i$ are mono-sources, then $\mathcal{S}$ is F-initial.

(c) If $\mathcal{T}$ is F-initial, then $\mathcal{S}$ need not be F-initial. [Hint: Consider the unique functor F from

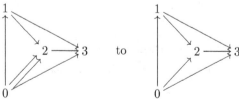

which leaves the four objects fixed. Then $1 \to 2$ is not F-initial, but $1 \to 2 \to 3$ is.]

10P. Universally Initial Morphisms

Let f be a morphism in a category **A**. Show that

* (a) f is an isomorphism if and only if f is F-initial with respect to every functor F that has domain **A**.

*(b) If f is a section, then f is F-initial with respect to every faithful functor $\mathbf{A} \xrightarrow{F} \mathbf{B}$. Does the converse hold?

(c) f is initial with respect to every full and faithful functor with domain $\mathbf{A}$.

10Q. A Characterization of Products

Show that a source in $\mathbf{A}$ is a product if and only if it is initial with respect to every functor $\mathbf{A} \xrightarrow{F} \mathbf{B}$ that preserves mono-sources.

10R. Copowers Versus Free Objects in Constructs

Let $(\mathbf{A}, U)$ be a construct such that U is representable by an object A. Show that for any set I and any $\mathbf{A}$-object B the following conditions are equivalent:

(1) B is a free object over I,

(2) B is an Ith copower $^I A$ of A.

10S. Copowers in CBoo and CLat

Show that neither **CBoo** nor **CLat** has small copowers. [Hint: Use 10R, 8.23(1)(b), 8.23(7), and 8F resp. 8G.]

10T. Mono-Sources in Σ-Seq

Prove that mono-sources in $\{\sigma\}$-**Seq** are precisely the point-separating sources — although the forgetful functor is not representable (cf. Corollary 10.8).

10U. Coproducts of Functors

Given functors $F_i : \mathbf{A} \to \mathbf{B}$, where $\mathbf{B}$ is a category with coproducts, then $F = \coprod F_i$ denotes the functor of "pointwise" coproducts, i.e.,

$$F(A \xrightarrow{f} A') = \coprod F_i A \xrightarrow{\coprod F_i f} \coprod F_i A'.$$

(a) Verify that F is a coproduct of the family (F_i) in the functor quasicategory $[\mathbf{A}, \mathbf{B}]$ (6.15).

(b) Verify that for each type Ω there exists a coproduct T of functors S^n [3.20(10)] such that $\mathbf{Alg}(\Omega)$ is concretely isomorphic to $\mathbf{Alg}(T)$.

10V. The Diagonal-Morphism Δ_A

$\Delta_A = \langle id_A, id_A \rangle : A \to A \times A$ is called the **diagonal morphism** of A. Consider the product $A \xleftarrow{\pi_1} A \times A \xrightarrow{\pi_2} A$ and show that (A, Δ_A) is an equalizer of (π_1, π_2).

10W. Products and Constant Morphisms

Consider the diagram

$$\begin{array}{ccc} A & \xrightarrow{f} & B \\ \uparrow_{\pi_1} & & \uparrow_{p_1} \\ A \times A & \xrightarrow{f \times f} & B \times B \\ \downarrow_{\pi_2} & & \downarrow_{p_2} \\ A & \xrightarrow{f} & B \end{array}$$

Show that the following are equivalent:

(1) f is constant,

(2) $f \circ \pi_1 = f \circ \pi_2$,

(3) $p_1 \circ (f \times f) = p_2 \circ (f \times f)$,

(4) $f \times f$ factors through Δ_B, i.e., for some g:

$$A \times A \xrightarrow{f \times f} B \times B = A \times A \xrightarrow{g} B \xrightarrow{\Delta_B} B \times B.$$

11 Limits and colimits

Many basic constructions in mathematics can be described as limits (or dually as colimits). Such constructions associate with a given *diagram* a distinguished object together with morphisms connecting this object with the objects of the diagram; i.e., a certain source called the *limit* of the diagram. This limit can be characterized, up to isomorphism, by a purely categorical property. We have encountered two special cases of limits already: equalizers and products. The main new type of limit introduced below is that of *pullback*. Unlike products, the usefulness of pullbacks was generally recognized only after the emergence of category theory.

LIMITS

11.1 DEFINITION
(1) A **diagram** in a category **A** is a functor $D : \mathbf{I} \to \mathbf{A}$ with codomain **A**. The domain, **I**, is called the **scheme** of the diagram.[56]

(2) A diagram with a small (or finite) scheme is said to be **small** (or **finite**).

11.2 EXAMPLES
(1) A diagram in **A** with discrete scheme is essentially just a family of **A**-objects.

(2) A diagram in **A** with scheme $\bullet \rightrightarrows \bullet$ is essentially just a pair of **A**-morphisms with common domain and common codomain.

11.3 DEFINITION
Let $D : \mathbf{I} \to \mathbf{A}$ be a diagram.

(1) An **A**-source $(A \xrightarrow{f_i} D_i)_{i \in Ob(\mathbf{I})}$ is said to be **natural** for D provided that for each **I**-morphism $i \xrightarrow{d} j$, the triangle

commutes.

---

[56] Although there is technically no difference between a *diagram* and a *functor*, or between a *scheme* and a *category*, we use the alternate terminology when treating limits and colimits, for reasons of both its historical development and to indicate a slight change of perspective. For example, we often denote the image of an object i under a diagram D by D_i rather than $D(i)$. This produces notation that is more consistent with that introduced earlier for sources and sinks.

(2) A **limit** of D is a natural source $(L \xrightarrow{\ell_i} D_i)_{i \in Ob(\mathbf{I})}$ for D with the (*universal*) property that each natural source $(A \xrightarrow{f_i} D_i)_{i \in Ob(\mathbf{I})}$ for D uniquely factors through it; i.e., for every such source there exists a unique morphism $f : A \to L$ with $f_i = \ell_i \circ f$ for each $i \in Ob(\mathbf{I})$.

11.4 EXAMPLES

(1) For a diagram $D : \mathbf{I} \to \mathbf{A}$ with a discrete scheme, every source with codomain $(D_i)_{i \in Ob(\mathbf{I})}$ is natural. A source is a limit of D if and only if it is a product of the family $(D_i)_{i \in Ob(\mathbf{I})}$. Expressed briefly: products are limits of diagrams with discrete schemes. In particular, an object, considered as an empty source, is a limit of the empty diagram (i.e., the one with empty scheme) if and only if it is a terminal object.

(2) For a pair of **A**-morphisms $A \xrightarrow[g]{f} B$, considered as a diagram D with scheme $\bullet \rightrightarrows \bullet$, a source $(A \xleftarrow{e} C \xrightarrow{h} B)$ is natural provided that $g \circ e = h = f \circ e$. Observe that in this case h is determined by e. Hence, $C \xrightarrow{e} A$ is an equalizer of $A \xrightarrow[g]{f} B$ if and only if the source $(A \xleftarrow{e} C \xrightarrow{f \circ e} B)$ is a limit of D. Thus we may say (imprecisely) that equalizers are limits of diagrams with scheme $\bullet \rightrightarrows \bullet$. If, in the above scheme, the two arrows are replaced by an arbitrary set of arrows, then limits of diagrams with such schemes are called **multiple equalizers**.

(3) If $\mathbf{I}$ is a down-directed[57] poset (considered as a category), then limits of diagrams with scheme $\mathbf{I}$ are called **projective** (or **inverse**) **limits**. If, e.g., $\mathbf{I} = \mathbb{N}^{op}$ is the poset of all non-negative integers with the opposite of the usual ordering, a diagram $D : \mathbf{I} \to \mathbf{A}$ with this scheme is essentially a sequence

$$\cdots \xrightarrow{d_2} D_2 \xrightarrow{d_1} D_1 \xrightarrow{d_0} D_0$$

of **A**-morphisms (where $D(n+1 \to n) = D_{n+1} \xrightarrow{d_n} D_n$, $D(n+2 \to n) = d_n \circ d_{n+1}$, etc.). A natural source for D is a source $(A \xrightarrow{f_n} D_n)_{n \in \mathbb{N}}$ with $f_n = d_n \circ f_{n+1}$ for each n. In **Set** a projective limit of a diagram D with scheme $\mathbb{N}^{op}$ is a source $(L \xrightarrow{\ell_n} D_n)_{n \in \mathbb{N}}$, where L is the set of all sequences $(x_n)_{n \in \mathbb{N}}$ with $x_n \in D_n$ and $d_n(x_{n+1}) = x_n$ for each $n \in \mathbb{N}$; and where each ℓ_m is a restriction of the mth projection $\pi_m : \prod_{n \in \mathbb{N}} D_n \to D_m$.

(4) If $D : \mathbf{A} \to \mathbf{A}$ is the identity functor, then a source $\mathcal{L} = (L \xrightarrow{\ell_A} A)_{A \in Ob\mathbf{A}}$ is a limit of D if and only if L is an initial object of $\mathbf{A}$. The sufficiency is obvious. For the necessity let $\mathcal{L}$ be a limit of D and let $L \xrightarrow{f} A$ be a morphism. By the naturality of $\mathcal{L}$ for D we obtain $f \circ \ell_L = \ell_A$. Application of this to $f = \ell_A$ yields $\ell_A \circ \ell_L = \ell_A = \ell_A \circ id_L$ for each object A. Hence, by the uniqueness requirement

---

[57] **Down-directed** means that every pair of elements has a lower bound. The dual notion is **up-directed**.

in the definition of limit, $\ell_L = id_L$. Consequently, $f = f \circ id_L = f \circ \ell_L = \ell_A$. Thus $\hom(L, A) = \{\ell_A\}$ for each object A.

(5) If the category $\mathbf{I}$ has an initial object i_0 with $\hom(i_0, i) = \{m_i\}$ for each $i \in Ob(\mathbf{I})$, then every diagram $D : \mathbf{I} \to \mathbf{A}$ with scheme $\mathbf{I}$ has a limit: $(D_{i_0} \xrightarrow{Dm_i} D_i)_{i \in Ob(\mathbf{I})}$.

11.5 REMARK
If $D : \mathbf{I} \to \mathbf{A}$ is a diagram, then natural sources for D can be regarded as natural transformations from constant functors $C : \mathbf{1} \to \mathbf{A}$ to the functor D.

11.6 PROPOSITION
Every limit is an extremal mono-source. [A 10.21]

11.7 PROPOSITION
Limits are essentially unique; i.e., if $\mathcal{L} = (L \xrightarrow{\ell_i} D_i)_{i \in Ob(\mathbf{I})}$ is a limit of $D : \mathbf{I} \to \mathbf{A}$, then the following hold:

(1) *for each limit $\mathcal{K} = (K \xrightarrow{k_i} D_i)_{i \in Ob(\mathbf{I})}$ of D, there exist an isomorphism $K \xrightarrow{h} L$ with $\mathcal{K} = \mathcal{L} \circ h$,*

(2) *for each isomorphism $A \xrightarrow{h} L$, the source $\mathcal{L} \circ h$ is a limit of D.* [A 10.22]

PULLBACKS

11.8 DEFINITION
(1) A square

$$\begin{array}{ccc} P & \xrightarrow{\bar{f}} & B \\ \bar{g} \downarrow & & \downarrow g \\ A & \xrightarrow{f} & C \end{array} \qquad (*)$$

is called a **pullback square** provided that it commutes and that for any commuting square of the form

$$\begin{array}{ccc} \hat{P} & \xrightarrow{\hat{f}} & B \\ \hat{g} \downarrow & & \downarrow g \\ A & \xrightarrow{f} & C \end{array}$$

there exists a unique morphism $\hat{P} \xrightarrow{k} P$ for which the following diagram commutes

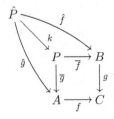

(2) If (∗) is a pullback square, then the 2-source $A \xleftarrow{\overline{g}} P \xrightarrow{\overline{f}} B$ is called a **pullback** of the 2-sink $A \xrightarrow{f} C \xleftarrow{g} B$ and $\overline{f}$ is called a **pullback of f along g**.

11.9 REMARKS

(1) A square (∗) is a pullback square if and only if the 3-source $(P, (\overline{g}, f \circ \overline{g}, \overline{f}))$ is a limit of the 2-sink $A \xrightarrow{f} C \xleftarrow{g} B$, considered as a diagram in **A** with scheme • → • ← •. Shortly (and imprecisely) pullbacks are limits of diagrams with scheme • → • ← •.

(2) If the square (∗) is a pullback, then $(P, (\overline{f}, \overline{g}))$ is a extremal mono-source. This follows from (1) and Proposition 11.6.

11.10 PROPOSITION
Let

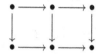

commute in **A**. Then

(1) if the squares are pullback squares, then so is the outer rectangle; i.e., pullbacks can be composed by "pasting" them together,

(2) if the outer rectangle and right-hand square are pullback squares, then so is the left-hand square. □

RELATIONSHIP OF PULLBACKS TO OTHER LIMITS

11.11 PROPOSITION (Canonical Construction of Pullbacks)

Let $A \xrightarrow{f} C \xleftarrow{g} B$ be a pair of morphisms with common codomain.

If $A \xleftarrow{\pi_A} A \times B \xrightarrow{\pi_B} B$ is a product of A and B, and $E \xrightarrow{e} A \times B$ is an equalizer of $A \times B \xrightarrow[g \circ \pi_B]{f \circ \pi_A} C$, then

$$\begin{array}{ccc} E & \xrightarrow{\pi_A \circ e} & A \\ {\scriptstyle \pi_B \circ e} \downarrow & & \downarrow {\scriptstyle f} \\ B & \xrightarrow{g} & C \end{array}$$

is a pullback square.

Proof: Let

$$\begin{array}{ccc} P & \xrightarrow{s} & A \\ r \downarrow & & \downarrow f \\ B & \xrightarrow{g} & C \end{array}$$

be a commuting square. Then $P \xrightarrow{\langle s,r \rangle} A \times B$ is a morphism for which $f \circ \pi_A \circ \langle s,r \rangle = g \circ \pi_B \circ \langle s,r \rangle$. Since e is an equalizer of $(f \circ \pi_A, g \circ \pi_B)$, there is a morphism $P \xrightarrow{k} E$ for which $\langle s,r \rangle = e \circ k$. Thus $\pi_A \circ e \circ k = s$ and $\pi_B \circ e \circ k = r$. That k is the only morphism with this property follows from the fact that $A \xleftarrow{\pi_A \circ e} E \xrightarrow{\pi_B \circ e} B$ is, as a composition of mono-sources, a mono-source. □

11.12 EXAMPLES

Application of the above construction provides concrete descriptions of pullbacks in many cases:

(1) Let $A \xrightarrow{f} C \xleftarrow{g} B$ be a pair of maps in **Set**, let $P = \{(a,b) \in A \times B \mid f(a) = g(b)\}$, and let $P \xrightarrow{\bar{g}} A$ and $P \xrightarrow{\bar{f}} B$ be the domain restrictions of the projections from $A \times B$. Then

$$\begin{array}{ccc} P & \xrightarrow{\bar{g}} & A \\ \bar{f} \downarrow & & \downarrow f \\ B & \xrightarrow{g} & C \end{array}$$

is a pullback square.

(2) In any of the constructs **Pos**, **Top**, or **Vec** pullbacks can be constructed as for sets by supplying P in each case with the initial structure with respect to its inclusion into the product $A \times B$; i.e.,

   (a) for **Pos**, by the pointwise order,

   (b) for **Top**, by the subspace topology,

   (c) for **Vec**, by defining operations coordinatewise.

11.13 PROPOSITION

If T is a terminal object, then the following are equivalent:

(1)

$$\begin{array}{ccc} P & \xrightarrow{p_A} & A \\ p_B \downarrow & & \downarrow \\ B & \longrightarrow & T \end{array}$$

is a pullback square,

(2) $(P, (p_A, p_B))$ is a product of A and B. □

11.14 PROPOSITION (Construction of Equalizers via Products and Pullbacks)
If $A \underset{g}{\overset{f}{\rightrightarrows}} B$ are morphisms, $(A \times B, \pi_A, \pi_B)$ is a product of (A, B), and

$$\begin{array}{ccc} P & \xrightarrow{p_1} & A \\ p_2 \downarrow & & \downarrow \langle id_A, f \rangle \\ A & \xrightarrow[\langle id_A, g \rangle]{} & A \times B \end{array}$$

is a pullback square, then $p_1 = p_2$ is an equalizer of f and g.

Proof: Since $\langle id_A, f \rangle \circ p_1 = \langle id_A, g \rangle \circ p_2$, we have

$$p_1 = \pi_A \circ \langle id_A, f \rangle \circ p_1 = \pi_A \circ \langle id_A, g \rangle \circ p_2 = p_2.$$

Similarly, $f \circ p_1 = \pi_B \circ \langle id_A, f \rangle \circ p_1 = \pi_B \circ \langle id_A, g \rangle \circ p_2 = g \circ p_2 = g \circ p_1$.
Suppose that $K \xrightarrow{k} A$ is a morphism such that $f \circ k = g \circ k$. Then

$$\pi_B \circ \langle id_A, f \rangle \circ k = \pi_B \circ \langle id_A, g \rangle \circ k, \quad \text{and}$$

$$\pi_A \circ \langle id_A, f \rangle \circ k = \pi_A \circ \langle id_A, g \rangle \circ k.$$

Since the product is a mono-source, we obtain $\langle id_A, f \rangle \circ k = \langle id_A, g \rangle \circ k$, so that since the square is a pullback square, there is a unique $h : K \to P$ such that $k = p_1 \circ h$. □

PULLBACKS RELATED TO SPECIAL MORPHISMS

11.15 LEMMA
Suppose that the diagram

commutes.

(1) *If the outer square is a pullback square, then so is* $\boxed{1}$.

(2) *If* $\boxed{1}$ *is a pullback square and h is a monomorphism, then the outer square is a pullback square.* □

11.16 PROPOSITION
$A \xrightarrow{f} B$ is a monomorphism if and only if

$$\begin{array}{ccc} A & \xrightarrow{id_A} & A \\ {\scriptstyle id_A}\downarrow & & \downarrow{\scriptstyle f} \\ A & \xrightarrow{f} & B \end{array}$$

is a pullback square. □

11.17 DEFINITION
A class M of morphisms in a category is called **pullback stable** (or **closed under the formation of pullbacks**) provided that for each pullback square

$$\begin{array}{ccc} P & \xrightarrow{\overline{f}} & B \\ {\scriptstyle \overline{g}}\downarrow & & \downarrow{\scriptstyle g} \\ A & \xrightarrow{f} & C \end{array} \qquad (*)$$

with $f \in M$, it follows that $\overline{f} \in M$.

11.18 PROPOSITION
Monomorphisms, regular monomorphisms, and retractions are pullback stable.

Proof: Let f be a monomorphism in the above pullback square $(*)$.

(1). If $h, k : Q \to P$ are morphisms such that $\overline{f} \circ h = \overline{f} \circ k$, then
$$f \circ (\overline{g} \circ h) = g \circ (\overline{f} \circ h) = g \circ (\overline{f} \circ k) = f \circ (\overline{g} \circ k),$$
so that $\overline{g} \circ h = \overline{g} \circ k$. Since pullbacks are mono-sources, this implies that $h = k$.

(2). Suppose that f is an equalizer of p and q. Then $(p \circ g) \circ \overline{f} = (q \circ g) \circ \overline{f}$. To show that $\overline{f}$ is an equalizer of $p \circ g$ and $q \circ g$, let $t : Q \to B$ be a morphism such that $(p \circ g) \circ t = (q \circ g) \circ t$. Then, by the definition of equalizer, there is some $u : Q \to A$ with $f \circ u = g \circ t$. Thus, by the definition of pullback, there is some $h : Q \to P$ such that $t = \overline{f} \circ h$. Uniqueness of h follows from the fact that $\overline{f}$ is a monomorphism.

(3). If f is a retraction, then there is some $C \xrightarrow{s} A$ with $f \circ s = id_C$. Hence $f \circ (s \circ g) = g \circ id_B$, so that by the definition of pullback there is some $h : B \to P$ with $id_B = \overline{f} \circ h$. □

11.19 REMARK
If (A, m) is a subobject of B, $C \xrightarrow{g} B$ is an arbitrary morphism and

$$\begin{array}{ccc} P & \xrightarrow{\overline{m}} & C \\ {\scriptstyle \overline{g}}\downarrow & & \downarrow{\scriptstyle g} \\ A & \xrightarrow{m} & B \end{array}$$

is a pullback square, then by the above proposition $(P, \overline{m})$ is a subobject of C. It is called an **inverse image** of (A, m) **under** g, since it corresponds to the concept of "inverse image" in familiar constructs. In particular, in **Set**, if $C \xrightarrow{g} B$ is a function and A is a subset of B, considered as a subobject of B via its inclusion function, then $g^{-1}[A]$, considered as a subobject of C via its inclusion function, is an inverse image of A under g.

CONGRUENCES

If $f : A \to B$ is a group homomorphism, then the group-theoretic congruence relation determined by f is the subset C of $A \times A$ consisting of all pairs (a, b) with $f(a) = f(b)$. Obviously, C can be regarded as a subgroup of $A \times A$, and if $m : C \hookrightarrow A \times A$ is the inclusion and $A \times A \underset{\pi_2}{\overset{\pi_1}{\rightrightarrows}} A$ are the projections, then (according to Proposition 11.11) the square

$$\begin{array}{ccc} C & \xrightarrow{\pi_1 \circ m} & A \\ {\scriptstyle \pi_2 \circ m} \downarrow & & \downarrow {\scriptstyle f} \\ A & \xrightarrow{f} & B \end{array}$$

is a pullback square.

This motivates our next definition.

11.20 DEFINITION
(1) If

$$\begin{array}{ccc} \bullet & \xrightarrow{p} & \bullet \\ {\scriptstyle q} \downarrow & & \downarrow {\scriptstyle f} \\ \bullet & \xrightarrow{f} & \bullet \end{array}$$

is a pullback square, then the pair (p, q) is called a **congruence relation of** f.

(2) A pair (p, q) of morphisms is called a **congruence relation** provided that there exists some morphism f such that (p, q) is a congruence relation of f.

11.21 LEMMA
Let (p, q) be a congruence relation of $A \xrightarrow{f} B$. Then

(1) (p, q) is a congruence relation of $A \xrightarrow{m \circ f} C$, for each monomorphism $B \xrightarrow{m} C$,

(2) if $f = g \circ h$ and $h \circ p = h \circ q$, then (p, q) is a congruence relation of h.

Proof: Apply Lemma 11.15. □

11.22 PROPOSITION

(1) If (p,q) is a congruence relation and c is a coequalizer of p and q, then (p,q) is a congruence relation of c.

(2) If c is a regular epimorphism and (p,q) is a congruence relation of c, then c is a coequalizer of p and q.

Proof:

(1). Immediate from Lemma 11.21(2).

(2). Let c be a coequalizer of r and s. By the pullback property there exists a morphism k with $p \circ k = r$ and $q \circ k = s$. To show that c is a coequalizer of p and q, let f be a morphism with $f \circ p = f \circ q$. Then $f \circ r = f \circ s$. Hence, since c is a coequalizer of r and s, there exists a unique morphism $\overline{f}$ with $f = \overline{f} \circ c$. □

INTERSECTIONS

11.23 DEFINITION

Let $\mathcal{A}$ be a family of subobjects (A_i, m_i) of an object B, indexed by a class I. A subobject (A, m) of B is called an **intersection of $\mathcal{A}$** provided that the following two conditions are satisfied:

(1) m factors through each m_i; i.e., for each i there exists an f_i with $m = m_i \circ f_i$,

(2) if a morphism $C \xrightarrow{f} B$ factors through each m_i, then it factors through m.[58]

11.24 EXAMPLES

(1) Let B be a set and let $(A_i)_I$ be a family of subsets of B, considered (via inclusions) as subobjects of B in **Set**. Then $\bigcap_{i \in I} A_i$, considered as a subobject of B, is an intersection in the sense of Definition 11.23. Similarly for **Vec**, **Top**, **Σ-Seq**, etc.

(2) In a poset, considered as a category, intersections are meets.

(3) For two subobjects (A_1, m_1) and (A_2, m_2) of B, an intersection is the diagonal morphism of a pullback of $A_1 \xrightarrow{m_1} B \xleftarrow{m_2} A_2$.

11.25 REMARKS

(1) Intersections can be regarded as limits (cf. Exercises 11F and 11L).

(2) Any two intersections of a family of subobjects of B are isomorphic subobjects of B.

(3) Let $A \xrightarrow{h} B$ be a morphism. Then (A, h) is an intersection of the empty family of subobjects of B if and only if h is an isomorphism.

---

[58]Observe that any intersection (A, m) of a family of subobjects (A_i, m_i) of an object B, indexed by a class I, depends only on the class $\{(A_i, m_i) \mid i \in I\}$ of subobjects of B and not on the indexing function. Hence we sometimes also speak of an intersection of a class of subobjects of B.

11.26 DEFINITION

A class M of monomorphisms is said to be **closed under the formation of intersections** provided that whenever (A, m) is an intersection of a family of subobjects (A_i, m_i) of B and each m_i belongs to M, then m belongs to M.

COLIMITS

The notion of colimit is dual to that of limit; namely,

11.27 DEFINITION

Let $D : \mathbf{I} \to \mathbf{A}$ be a diagram.

(1) An **A**-sink $(D_i \xrightarrow{f_i} A)_{i \in Ob(\mathbf{I})}$ is said to be **natural** for D provided that for each **I**-morphism $i \xrightarrow{d} j$, the triangle

commutes.

(2) A **colimit** of D is a natural sink $(D_i \xrightarrow{c_i} K)_{i \in Ob(\mathbf{I})}$ for D with the (*universal*) property that each natural sink for D uniquely factors through it.

11.28 EXAMPLES

(1) **Coproducts** are precisely the colimits of diagrams with discrete schemes. In particular, initial objects are the colimits of empty diagrams.

(2) For a pair of morphisms $A \xrightarrow[g]{f} B$, considered as a diagram D with scheme $\bullet \rightrightarrows \bullet$, $B \xrightarrow{c} C$ is a **coequalizer** of f and g if and only if $(B \xrightarrow{c} C \xleftarrow{cof} A)$ is a colimit of D.

(3) The dual of the concept of an intersection of a family of subobjects of B is the concept of a **cointersection** of a family of quotient objects of B. In particular,

(a) If B is a set and if $(\rho_i)_I$ is a family of equivalence relations on B considered (as in 7.86) as quotient objects of B in **Set**, then the smallest equivalence relation ρ on B generated by $\bigcup_{i \in I} \rho_i$, considered as a quotient object of B, is a cointersection of the family in **Set**.

(b) In a poset, considered as a category, cointersections are joins.

(4) The dual of the concept of projective limit is that of **directed colimit** (also called **inductive limit**). That is, directed colimits are colimits of diagrams whose schemes are up-directed posets.

Every diagram $D : \mathbf{I} \to \mathbf{Set}$, whose scheme is an up-directed poset, has a canonical colimit. Denote $D(i \to j)$ by $D_i \xrightarrow{d_{ij}} D_j$, let $C = \bigcup_{i \in I}(D_i \times \{i\})$ be the disjoint union of the family $(D_i)_{i \in I}$, and let $(D_j \xrightarrow{\mu_j} C)_{j \in Ob(\mathbf{I})}$ be the canonical coproduct of the family $(D_i)_{i \in Ob(\mathbf{I})}$ in $\mathbf{Set}$, as constructed in 10.67(1). Define an equivalence relation $\sim$ on C by: $(x, i) \sim (y, j)$ if and only if there exists $k \geq i, j$ with $d_{ik}(x) = d_{jk}(y)$. If $C \xrightarrow{q} Q$ is the natural map from C onto the set $Q = C/\sim$ of equivalence classes under $\sim$, then $(D_i \xrightarrow{q \circ \mu_i} Q)_{i \in Ob(I)}$ is a colimit of D.

Observe that directed colimits in any construct of the form $\mathbf{Alg}(\Omega)$ can be constructed as in $\mathbf{Set}$: Let $\Omega = (n_k)_{k \in K}$ and let $D : \mathbf{I} \to \mathbf{Alg}(\Omega)$ be a diagram for an up-directed poset $\mathbf{I}$. Denote $D(i \to j)$ by $(D_i, (\omega_k^i)_{k \in K}) \xrightarrow{d_{ij}} (D_j, (\omega_k^j)_{k \in K})$ and form (as above) the colimit $(D_i \xrightarrow{q \circ \mu_i} Q)_{i \in Ob(\mathbf{I})}$ of the diagram $\mathbf{I} \xrightarrow{U \circ D} \mathbf{Set}$, where $\mathbf{Alg}(\Omega) \xrightarrow{U} \mathbf{Set}$ is the forgetful functor. Since $\mathbf{I}$ is up-directed, for each $k \in K$ and each $(x_1, \ldots, x_{n_k}) \in Q^{n_k}$ there exists some $i \in Ob(\mathbf{I})$ and $(y_1, \ldots, y_{n_k}) \in D_i^{n_k}$ with $(q \circ \mu_i)^{n_k}(y_1, \ldots, y_{n_k}) = (x_1, \ldots, x_{n_k})$. Define $\omega_k(x_1, \ldots, x_{n_k}) = q \circ \mu_i \circ \omega_k^i(y_1, \ldots, y_{n_k})$. Then up-directedness of $\mathbf{I}$ implies that the so-defined map $Q^{n_k} \xrightarrow{\omega_k} Q$ is the unique function that makes the following diagram commute for each $i \in Ob(\mathbf{I})$:

$$\begin{array}{ccc} D_i^{n_k} & \xrightarrow{(q \circ \mu_i)^{n_k}} & Q^k \\ \omega_k^i \downarrow & & \downarrow \omega_k \\ D_i & \xrightarrow{q \circ \mu_i} & Q \end{array}$$

It follows that the sink $((D_i, (\omega_k^i)_{k \in K}) \xrightarrow{q \circ \mu_i} (Q, (\omega_k)_{k \in K}))_{i \in Ob(\mathbf{I})}$ is a colimit of D.

(5) The dual of the concept of pullback is that of **pushout**, explicitly described below.

11.29 PROPOSITION
Colimits are essentially unique and each colimit is an extremal epi-sink. □

11.30 DEFINITION
(1) The square

$$\begin{array}{ccc} C & \xrightarrow{f} & A \\ g \downarrow & & \downarrow \bar{g} \\ B & \xrightarrow{\bar{f}} & P \end{array} \qquad (*)$$

is called a **pushout square** provided that it commutes and that for any commuting square of the form

$$\begin{array}{ccc} C & \xrightarrow{f} & A \\ g \downarrow & & \downarrow \hat{g} \\ B & \xrightarrow{\hat{f}} & \hat{P} \end{array}$$

there exists a unique morphism $P \xrightarrow{k} \hat{P}$ for which the diagram

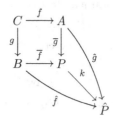

commutes.

(2) If $(*)$ is a pushout square, the 2-sink $((\overline{g}, \overline{f}), P)$ is called a **pushout** of the 2-source $(C, (f, g))$, and $\overline{f}$ is called a **pushout of f along g**.

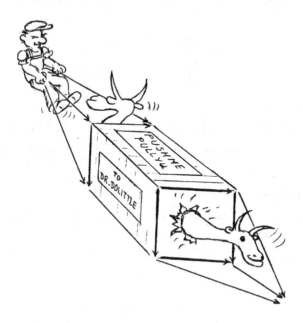

Pullback and Pushout

11.31 REMARK
Since pushouts are dual to pullbacks, there is a canonical construction of pushouts via coequalizers and coproducts (cf. 11.11).

11.32 DEFINITION
A square that is simultaneously a pullback square and pushout square is called a **pulation square**.

11.33 PROPOSITION
Consider a commuting square

$$(*)$$

(1) If $(*)$ is a pushout square, then c is a coequalizer of p and q.

(2) If (p, q) is a congruence relation of c, and c is a regular epimorphism, then $(*)$ is a pulation square.

Proof:
(1). Immediate.

(2). Let

$$\begin{array}{ccc} \bullet & \xrightarrow{p} & \bullet \\ q \downarrow & & \downarrow \hat{q} \\ \bullet & \xrightarrow{\hat{p}} & \bullet \end{array}$$

be a commutative square. Since $(*)$ is a pullback square and $c \circ id = c \circ id$, there exists a morphism k such that $id = p \circ k = q \circ k$. Hence $\hat{p} \circ q = \hat{q} \circ p$ implies that

$$\hat{p} = \hat{p} \circ q \circ k = \hat{q} \circ p \circ k = \hat{q}.$$

Hence $\hat{p} \circ p = \hat{p} \circ q$. Since by Proposition 11.22, c is a coequalizer of p and q, there exists a unique morphism h with $\hat{p} = h \circ c$; i.e., with $\hat{p} = h \circ c$ and $\hat{q} = h \circ c$. Hence $(*)$ is a pushout square. □

Suggestions for Further Reading

Maranda, J. M. Some remarks on limits in categories. *Canad. Math. Bull.* **5** (1962): 133–146.

Eckmann, B., and P. J. Hilton. Group-like structures in general categories II: Equalizers, limits, length. *Math. Ann.* **151** (1962): 150–186.

Kelly, G. M. Monomorphisms, epimorphisms, and pullbacks. *J. Austral. Math. Soc.* **9** (1969): 124–142.

Tholen, W. Amalgamations in categories. *Algebra Univ.* **14** (1982): 391–397.

EXERCISES

11A. Schemes With Initial Objects

Show that for categories **A** the following conditions are equivalent:

(1) **A** has an initial object,

(2) the diagram $\mathbf{A} \xrightarrow{id_{\mathbf{A}}} \mathbf{A}$ has a limit,

(3) each diagram with scheme **A** has a limit.

11B. Products as Projective Limits of Finite Products

Let $(A_i)_{i \in J}$ be a family of objects in **A**, indexed by an infinite set J. Let **I** be the set of all finite subsets of J, ordered by: $M \leq N \Leftrightarrow N \subseteq M$. For each $M \in \mathbf{I}$ let $(\prod_M A_i, (p_i^M))$ be a product of $(A_i)_{i \in M}$ in **A** (where $\prod_{\{i\}} A_i \xrightarrow{p_i^{\{i\}}} A_i = A_i \xrightarrow{id} A_i$ for each $i \in J$). Show that

(a) There exists a diagram $D : \mathbf{I} \to \mathbf{A}$ that associates with $M \leq N$ in **I** the unique morphism $\prod_M A_i \xrightarrow{f_{MN}} \prod_N A_i$ that satisfies $p_i^N \circ f_{MN} = p_i^M$ for each $i \in N$.

(b) If $(P, (p_N)_{N \in \mathbf{I}})$ is a (projective) limit of D, then $(P, (p_{\{i\}})_{i \in J})$ is a product of $(A_i)_{i \in I}$.

11C. Pullbacks as Products in Comma Categories

Show that a commuting square

$$\begin{array}{ccc} P & \xrightarrow{p_A} & A \\ {\scriptstyle p_B}\downarrow & & \downarrow{\scriptstyle f} \\ B & \xrightarrow{g} & C \end{array} \quad \text{with } f \circ p_A = g \circ p_B = p$$

is a pullback square in **A** if and only if $((P, p), (p_A, p_B))$ is a product of f and g in the comma category $(\mathbf{A} \downarrow C)$.

11D. Products of Morphisms as Pullbacks

Show that each diagram of the form

$$\begin{array}{ccc} A \times B & \xrightarrow{\pi_B} & B \\ {\scriptstyle id_A \times f}\downarrow & & \downarrow{\scriptstyle f} \\ A \times C & \xrightarrow{\pi_C} & C \end{array}$$

is a pullback square.

11E. Kernels as Pullbacks

Show that in a category **A** with a zero-object 0, k is a kernel of f if and only if

is a pullback square.

11F. Intersections as Pullbacks

Let m_1 and m_2 be monomorphisms and let

$$\begin{array}{ccc} A & \xrightarrow{\overline{m}_2} & A_1 \\ {\scriptstyle \overline{m}_1}\downarrow & & \downarrow{\scriptstyle m_1} \\ A_2 & \xrightarrow{m_2} & B \end{array}$$

be a commuting square with $m = m_1 \circ \overline{m}_2 = m_2 \circ \overline{m}_1$. Show that the square is a pullback square if and only if (A, m) is an intersection of the family $((A_i, m_i))_{i \in \{1,2\}}$ of subobjects of B.

11G. Pushouts as Pullbacks

Show that in a group, considered as a category, the pushout squares are precisely the pullback squares.

11H. Pullback Stability of Special Monomorphisms

Show that

(a) in any category the class of strict monomorphisms is pullback stable,

(b) in **Set** the class of sections is not pullback stable,

(c) in **DRail** the class of extremal monomorphisms is not pullback stable.

11I. Pullback Stability of Epi-Sinks in Set

For each $i \in I$ let

$$\begin{array}{ccc} B_i & \xrightarrow{k_i} & A_i \\ {\scriptstyle g_i}\downarrow & & \downarrow{\scriptstyle f_i} \\ B & \xrightarrow{f} & A \end{array}$$

be a pullback square in **Set**. Show that whenever $(A_i \xrightarrow{f_i} A)_I$ is an epi-sink, then so is $(B_i \xrightarrow{g_i} B)_I$.

11J. Stable Epimorphisms

A morphism f is called a **stable epimorphism** provided every pullback of f is an epimorphism. Show that

(a) Every retraction is a stable epimorphism.

(b) Every stable epimorphism is an epimorphism.

* (c) In **Cat** not all regular epimorphisms are stable.

* (d) In **Haus** and in **Met**: ExtrEpi $\subsetneq$ StableEpi = SurjMorph $\subsetneq$ Epi.

(e) Stable epimorphisms are pullback stable.

(f) If **A** has pullbacks of all 2-sinks, then in **A** stable epimorphisms are closed under composition.

11K. Intersections
Show that

(a) in **Set** an intersection of two sections need not be a section,

(b) if **A** has products, then in **A** an intersection of a set-indexed family of regular subobjects is a regular subobject.

[Hint: If (A, m) is an intersection of the family $(A_i, m_i)_I$, where each m_i is an equalizer of $B \underset{s_i}{\overset{r_i}{\rightrightarrows}} C_i$, then m is an equalizer of $B \underset{\langle s_i \rangle}{\overset{\langle r_i \rangle}{\rightrightarrows}} \prod C_i$.]

11L. Multiple Pullbacks
A pair $(f, \mathcal{S})$, consisting of a morphism $A \xrightarrow{f} B$ and a source $\mathcal{S} = (A \xrightarrow{f_i} A_i)_I$, is called a **multiple pullback** of a sink $(A_i \xrightarrow{g_i} B)_I$ provided that

(1) $f = g_i \circ f_i$ for each $i \in I$, and

(2) for each pair $(f', \mathcal{S}')$, with $A' \xrightarrow{f'} B$ a morphism and $\mathcal{S}' = (A' \xrightarrow{f'_i} A_i)_I$ a source for which $f' = g_i \circ f'_i$ for each $i \in I$, there exists a unique morphism $A' \xrightarrow{g} A$ with $f' = f \circ g$ and $f'_i = f_i \circ g$ for each $i \in I$.

(a) Interpret multiple pullbacks as limits.

(b) Interpret pullbacks as multiple pullbacks of 2-sinks.

(c) Interpret intersections as multiple pullbacks. In particular, show that whenever $\mathcal{R} = ((A_i, m_i))_I$ is a family of subobjects of B and $A \xrightarrow{f} B$ is a morphism, then the following are equivalent:

(1) (A, f) is an intersection of $\mathcal{R}$,

(2) there exists a (unique) source $\mathcal{S}$ such that $(f, \mathcal{S})$ is a multiple pullback of the sink $(A_i \xrightarrow{m_i} B)_{i \in I}$.

(d) Show that each sink that consists of isomorphisms alone has a multiple pullback.

11M. Products as Multiple Pullbacks

Let **A** have a terminal object T, and let $(A_i)_I$ be a family of **A**-objects. Show that a source $\mathcal{P} = (P \xrightarrow{p_i} A_i)_I$ is a product of $(A_i)_I$ if and only if $(P \to T, \mathcal{P})$ is a multiple pullback of the sink $(A_i \to T)_I$.

11N. A Characterization of Monomorphisms by Multiple Pullbacks

Show that a morphism $A \xrightarrow{m} B$ is a monomorphism if and only if, for each class I, the sink $(A_i \xrightarrow{m_i} B)_I$, defined by $m_i = m$ for each $i \in I$, has a multiple pullback.

11O. Directed Colimits in Ab

Show that an abelian group is torsion free if and only if it is a directed colimit in **Ab** of free abelian groups.

11P. Pushout Stability of Monomorphisms and Mono-Sources in Set

(a) Show that in **Set** the class of monomorphisms is pushout stable.

(b) For each $i \in I$ let

$$\begin{array}{ccc} A & \xrightarrow{f} & B \\ m_i \downarrow & & \downarrow n_i \\ A_i & \xrightarrow{g_i} & B_i \end{array}$$

be a pushout square in **Set**. Show that the fact that $(A \xrightarrow{m_i} A_i)_I$ is a mono-source does not imply that the source $(B \xrightarrow{n_i} B_i)_I$ is a mono-source. [Contrast this with 11I.]

11Q. Epimorphisms and Pulation Squares

(a) Let (p, q) be a congruence relation of e. Show that e is a regular epimorphism if and only if the square

$$\begin{array}{ccc} \bullet & \xrightarrow{p} & \bullet \\ q \downarrow & & \downarrow e \\ \bullet & \xrightarrow{e} & \bullet \end{array}$$

is a pulation square (cf. 11.32).

(b) Dualize part (a).

11R. A Characterization of Monomorphisms

Show that a morphism is a monomorphism if and only if it has a congruence relation of the form (p, p).

11S. A Construction of Equalizers
Show that if

$$\begin{array}{ccc} \bullet & \xrightarrow{p_2} & B \\ {\scriptstyle p_1}\downarrow & & \downarrow {\scriptstyle \langle id,id \rangle} \\ A & \xrightarrow[\langle f,g \rangle]{} & B \times B \end{array}$$

is a pullback square, then p_1 is an equalizer of f and g.

12 Completeness and cocompleteness

In this section we consider the existence of limits in a given category **A**. Only rarely does every diagram in **A** have a limit. [Recall that any such **A** must be thin (cf. 10.32 and 10.33).] However, in many familiar categories **A**, every *small* diagram has a limit, and for each object A in **A** every (possibly large) family of subobjects of A has an intersection. This leads to the following definitions:

12.1 DEFINITION
A category is said to

(1) **have (finite) products** provided that for each (finite) set-indexed family of objects there exists a product (cf. Definition 10.29),

(2) **have equalizers** provided that for each parallel pair of morphisms there exists an equalizer,

(3) **have pullbacks** provided that for each 2-sink there exists an pullback,

(4) **have (finite) intersections** provided that for each object A, and every (finite) family of subobjects of A, there exists an intersection.

DUAL NOTIONS: **have (finite) coproducts, have coequalizers**, and **have (finite) cointersections**.

12.2 DEFINITION
A category **A** is said to be

(1) **finitely complete** if for each finite diagram in **A** there exists a limit,

(2) **complete** if for each small diagram in **A** there exists a limit,

(3) **strongly complete** if it is complete and has intersections.

DUAL NOTIONS: **finitely cocomplete, cocomplete**, and **strongly cocomplete** categories.

12.3 THEOREM
*For each category **A** the following conditions are equivalent:*

*(1) **A** is complete,*

*(2) **A** has products and equalizers,*

*(3) **A** has products and finite intersections.*

Proof:

(1) ⇒ (3) is immediate.

(3) ⇒ (2) follows from Proposition 11.14 and Example 11.24(3).

(2) ⇒ (1). Suppose that **A** has products and equalizers, and let $D : \mathbf{I} \to \mathbf{A}$ be a small diagram. For each morphism $t : i \to j$ in **I** let $d(t) = i$ and $c(t) = j$. Form the products $(\prod_{i \in Ob(\mathbf{I})} D_i \xrightarrow{\pi_j} D_j)_{j \in Ob(\mathbf{I})}$ and $(\prod_{t \in Mor(\mathbf{I})} D_{c(t)} \xrightarrow{\hat{\pi}_s} D_{c(s)})_{s \in Mor(\mathbf{I})}$. For each $t \in Mor(\mathbf{I})$ there is a pair of morphisms $\pi_{c(t)}, Dt \circ \pi_{d(t)} : \prod_{i \in Ob(\mathbf{I})} D_i \to D_{c(t)}$, and these define a pair of morphisms

$$\langle \pi_{c(t)} \rangle, \langle Dt \circ \pi_{d(t)} \rangle : \prod_{i \in Ob(\mathbf{I})} D_i \longrightarrow \prod_{t \in Mor(\mathbf{I})} D_{c(t)}.$$

Let $e : E \to \prod D_i$ be an equalizer of this pair. We claim that $(E \xrightarrow{\pi_i \circ e} D_i)_{i \in Ob(\mathbf{I})}$ is a limit of D in **A**. Indeed,

(a) It is a natural source for D because for each morphism $i \xrightarrow{s} j$ in **I**, we have

$$\pi_j \circ e = \pi_{c(s)} \circ e = \hat{\pi}_s \circ \langle \pi_{c(t)} \rangle \circ e = \hat{\pi}_s \circ \langle Dt \circ \pi_{d(t)} \rangle \circ e = Ds \circ \pi_{d(s)} \circ e = Ds \circ \pi_i \circ e.$$

(b) Suppose that $(A \xrightarrow{f_i} D_i)$ is a natural source for D and let $f = \langle f_i \rangle : A \to \prod D_i$. For each morphism s of **I**, naturality of the source implies that $f_{c(s)} = Ds \circ f_{d(s)}$. Consequently,

$$\hat{\pi}_s \circ \langle \pi_{c(t)} \rangle \circ f = \pi_{c(s)} \circ f = f_{c(s)} = Ds \circ f_{d(s)} = Ds \circ \pi_{d(s)} \circ f = \hat{\pi}_s \circ \langle Dt \circ \pi_{d(t)} \rangle \circ f.$$

Since the product is a mono-source, $\langle \pi_{c(t)} \rangle \circ f = \langle Dt \circ \pi_{d(t)} \rangle \circ f$. Thus there is a unique $A \xrightarrow{f'} E$ such that $f = e \circ f'$. Thus $f_i = \pi_i \circ e \circ f'$ for each i, and f' is clearly seen to be unique with respect to the latter property. □

12.4 THEOREM

*For each category **A** the following conditions are equivalent:*

(1) **A** *is finitely complete,*

(2) **A** *has finite products and equalizers,*

(3) **A** *has finite products and finite intersections,*

(4) **A** *has pullbacks and a terminal object.*

Proof: (1) ⇔ (2) ⇔ (3) can be proved as in Theorem 12.3.

(1) ⇒ (4) is immediate.

(4) ⇒ (3). Recall that products of pairs of objects can be formed via pullbacks of morphisms to a terminal object (11.13) and that intersections of pairs of subobjects are formed via pullbacks [11.24(3)]. □

12.5 PROPOSITION
Each complete and wellpowered category is strongly complete.

Proof: Let **A** be complete and wellpowered. For each family $(A_i \xrightarrow{m_i} A)_I$ of subobjects of some object A, there exists a subset $J \subseteq I$ such that each subobject $A_i \xrightarrow{m_i} A$, $i \in I$, is isomorphic to some $A_j \xrightarrow{m_j} A$, $j \in J$. Obviously, any intersection (B, m) of the small sink $(A_j \xrightarrow{m_j} A)_J$ is an intersection of the original sink $(A_i \xrightarrow{m_i} A)_I$. □

12.6 EXAMPLES
(1) Each of the categories **Set**, **Vec**, **Top**, **HComp**, **Pos**, **Grp**, **Cat**, and Σ-**Seq** is strongly complete and strongly cocomplete.

Observe however that the quasicategory **CAT** fails to have coequalizers.

(2) **Met** has equalizers, but not products. **JPos** has products, but not equalizers.

(3) Every non-trivial group, considered as a category, has pullbacks and pushouts, but not equalizers or coequalizers, nor products or coproducts of pairs, nor terminal or initial objects.

(4) A poset, considered as a category, is (co)complete if and only if it is a complete lattice. Thus, for posets, completeness and cocompleteness coincide.

However, the partially ordered class **Ord** of all ordinal numbers is not complete (it has no terminal object) even though it is cocomplete (but not strongly cocomplete).

(5) The categories **CLat** and **CBoo** (= complete boolean algebras and boolean homomorphisms) are strongly complete, but not cocomplete (cf. Exercise 12I).

(6) The category of finite sets is finitely complete and finitely cocomplete, but is neither complete nor cocomplete.

(7) The category **Field** is neither finitely complete nor finitely cocomplete.

COCOMPLETENESS ALMOST IMPLIES COMPLETENESS

Although completeness and cocompleteness are not equivalent, the constructions of limits and colimits are intimately related. In fact, as is shown below, under suitable "smallness conditions", completeness and cocompleteness are equivalent.

12.7 THEOREM
A small category is complete if and only if it is cocomplete.

Proof: Immediate from Theorem 10.32(2). □

12.8 PROPOSITION (Canonical Construction of Limits via Large Colimits)
For a small diagram $D : \mathbf{I} \to \mathbf{A}$, let $\mathbf{S}^D$ be the category whose objects are all natural sources (A, f_i) for D, whose morphisms $(A, f_i) \xrightarrow{g} (A', f'_i)$ are all those $\mathbf{A}$-morphisms

$A \xrightarrow{g} A'$ with $(A, f_i) = (A', f'_i) \circ g$, and whose identity morphisms and composition law are as in **A**. If $D^* : \mathbf{S}^D \to \mathbf{A}$ is the diagram given by:

$$D^*((A, f_i) \xrightarrow{g} (A', f'_i)) = A \xrightarrow{g} A',$$

then for each **A**-object L, the following conditions are equivalent:

(1) D has a limit $\mathcal{L} = (L \xrightarrow{\ell_i} D_i)_{i \in Ob(\mathbf{I})}$,

(2) D^* has a colimit $\mathcal{K} = (D^*(\mathcal{S}) \xrightarrow{k_\mathcal{S}} L)_{\mathcal{S} \in Ob(\mathbf{S}^D)}$,

(3) $\mathbf{S}^D$ has a terminal object $\mathcal{L} = (L \xrightarrow{\ell_i} D_i)_{i \in Ob(\mathbf{I})}$.

Proof: (1) $\Leftrightarrow$ (3) is immediate.

(3) $\Rightarrow$ (2) follows from the dual of Example 11.4(5).

(2) $\Rightarrow$ (3). Given an object j of **I** and an object $\mathcal{S} = (A, f_i)$ in $\mathbf{S}^D$ let $D^*(\mathcal{S}) \xrightarrow{f_\mathcal{S}} D_j$ be the morphism $A \xrightarrow{f_j} D_j$ in the source $\mathcal{S}$ that is indexed by j. Then for each $j \in Ob(\mathbf{I})$

$$(D^*(\mathcal{S}) \xrightarrow{f_\mathcal{S}} D_j)_{\mathcal{S} \in Ob(\mathbf{S}^D)}$$

is a natural sink for D^*, so that by the definition of colimit there is a unique morphism $L \xrightarrow{\ell_j} D_j$ with the property that $f_j = f_\mathcal{S} = \ell_j \circ k_\mathcal{S}$ for each $\mathcal{S}$ in $\mathbf{S}^D$.

To see that the source $\mathcal{L} = (L \xrightarrow{\ell_i} D_i)_{i \in Ob(\mathbf{I})}$ is natural for D, let $i \xrightarrow{d} j$ be any morphism in **I**. If $\mathcal{S}$ is any natural source for D, then $f_j = D(d) \circ f_i$, so that $\ell_j \circ k_\mathcal{S} = D(d) \circ \ell_i \circ k_\mathcal{S}$. Since $\mathcal{K}$, being a colimit, is an epi-sink, it follows that $\ell_j = D(d) \circ \ell_i$. Consequently, $\mathcal{L}$ is an object of $\mathbf{S}^D$.

To see that $\mathcal{L}$ is a terminal object, first notice that if $\mathcal{S} = (A \xrightarrow{f_i} D_i)_{i \in Ob(\mathbf{I})}$ is a natural source for D, then the **A**-morphism $A \xrightarrow{k_\mathcal{S}} L$ is also a morphism $\mathcal{S} \xrightarrow{k_\mathcal{S}} \mathcal{L}$ in $\mathbf{S}^D$. Moreover if $\mathcal{S} \xrightarrow{g} \mathcal{L}$ is any $\mathbf{S}^D$-morphism, then since $\mathcal{K}$ is a natural sink for D^*, this implies that $k_\mathcal{L} \circ g = k_\mathcal{S}$. In particular, if for each object $\mathcal{S}$ in $\mathbf{S}^D$ one lets g take on the role of $k_\mathcal{S}$, this yields:

$$k_\mathcal{L} \circ k_\mathcal{S} = k_\mathcal{S} = id_L \circ k_\mathcal{S} \quad \text{for each } \mathcal{S} \text{ in } \mathbf{S}^D.$$

Since $\mathcal{K}$ is an epi-sink, this gives $k_\mathcal{L} = id_L$. Consequently, $g = id_L \circ g = k_\mathcal{L} \circ g = k_\mathcal{S}$, so that $k_\mathcal{S}$ is the unique morphism from $\mathcal{S}$ to $\mathcal{L}$. □

12.9 PROPOSITION

*A cocomplete category **A** has a terminal object if and only if it has a **weak terminal object** K; i.e., for each **A**-object A, there exists at least one morphism from A to K.*

Proof: Let K be a weak terminal object in a cocomplete category **A**. Then let **K** be the full subcategory of **A** that consists of the single object K, let $D : \mathbf{K} \to \mathbf{A}$ be the inclusion functor, and let $(K \xrightarrow{k} T)$ be a colimit of D. We shall prove that T is a terminal object of **A**.

For each **A**-object A there is some morphism $f : A \to K$, and hence a morphism $g = k \circ f : A \to T$. Suppose that $g' : A \to T$. Then to show that $g = g'$, form a coequalizer $T \xrightarrow{c} C$ of $A \underset{g'}{\overset{g}{\rightrightarrows}} T$. One need only show that c is an monomorphism. Indeed, there exists a morphism $f' : C \to K$, and since $f' \circ c \circ k$ is an element of $\hom(K, K)$, the naturality of the colimit of D implies that $k \circ f' \circ c \circ k = k$. Furthermore, k is an epimorphism (since it is a singleton colimit sink) and, hence, $k \circ f' \circ c = id_T$; so c is a section, and thus a monomorphism. □

12.10 DEFINITION
A full subcategory **B** of a category **A** with embedding $E : \mathbf{B} \to \mathbf{A}$ is called **colimit-dense** in **A** provided that for every **A**-object A there exists a diagram $D : \mathbf{I} \to \mathbf{B}$ such that the diagram $E \circ D : \mathbf{I} \to \mathbf{A}$ has a colimit with codomain A.

12.11 EXAMPLES
(1) A full subcategory of **Set** is colimit-dense in **Set** if and only if it contains at least one nonempty set as an object. [Observe that every set X is a coproduct of $\mathrm{card}(X)$ copies of $\{0\}$.]

(2) A full subcategory of **Pos** is colimit-dense in **Pos** if and only if it contains at least one nondiscrete poset as an object.

(3) A full subcategory of **Vec** is colimit-dense in **Vec** if and only if it contains at least one nonzero vector space as an object.

(4) Any full subcategory of **Ab** that contains $\mathbb{Z} \times \mathbb{Z} \times \mathbb{Z}$ is colimit-dense in **Ab**.

(5) The full subcategory of **Top** that consists of all zero-dimensional Hausdorff spaces is colimit-dense in **Top**. [Observe that every topological space is a (regular) quotient of a zero-dimensional Hausdorff space.]

12.12 THEOREM
Every cocomplete category with a small colimit-dense subcategory is complete.

Proof: Let **B** be a small, colimit-dense subcategory of a cocomplete category **A**. For each small diagram $D : \mathbf{I} \to \mathbf{A}$, consider the category $\mathbf{S}^D$ of all natural sources for D (cf. 12.8).

First we establish that $\mathbf{S}^D$ is cocomplete. Let $D_0 : \mathbf{J} \to \mathbf{S}^D$ be a small diagram (such that for each object j of **J**, $D_0(j) = (A_j, f_{ji})_I$) and let $D^* : \mathbf{S}^D \to \mathbf{A}$ be the diagram of Proposition 12.8. Form the colimit $(A_j \xrightarrow{c_j} K)_{j \in Ob(\mathbf{J})}$ of the composite diagram $\mathbf{J} \xrightarrow{D_0} \mathbf{S}^D \xrightarrow{D^*} \mathbf{A}$. Since for each object i of **I** $(A_j \xrightarrow{f_{ji}} D_i)_{Ob(\mathbf{J})}$ is a natural sink for $D^* \circ D_0$, we have the existence of a unique morphism $g_i : K \to D_i$ such that $g_i \circ c_j = f_{ji}$. Since $(c_j, K)_{Ob(\mathbf{J})}$ is an epi-sink, $(K, g_i)_{Ob(\mathbf{I})}$ is a natural source for D and $((A_j, f_{ji}) \xrightarrow{c_j} (K, g_i))_{Ob(\mathbf{J})}$ is a colimit of D_0. Thus $\mathbf{S}^D$ is cocomplete.

Now to show that the small diagram $D : \mathbf{I} \to \mathbf{A}$ has a limit, by Propositions 12.8 and 12.9 it is sufficient to prove that the category $\mathbf{S}^D$ has a weak terminal object. Let $\mathbf{I}^*$ be the full subcategory of $\mathbf{S}^D$ given by all those natural sources $\mathcal{S} = (A, f_i)_{Ob(\mathbf{I})}$ whose domain A is a member of $\mathbf{B}$. Since $\mathbf{B}$ is small, so is $\mathbf{I}^*$, and since $\mathbf{S}^D$ is cocomplete, the inclusion $\mathbf{I}^* \hookrightarrow \mathbf{S}^D$ has a colimit $(\mathcal{S} \xrightarrow{c_\mathcal{S}} \mathcal{K})_{\mathcal{S} \in Ob(\mathbf{I}^*)}$ in $\mathbf{S}^D$. We claim that $\mathcal{K} = (K, p_i)$ is a weak terminal object in $\mathbf{S}^D$.

To see this, let $\hat{\mathcal{S}} = (A, f_i)_{Ob(\mathbf{I})}$ be any object of $\mathbf{S}^D$. By the colimit density, there exists a small diagram $\hat{D} : \mathbf{K} \to \mathbf{B}$ with a colimit $(\hat{D}_k \xrightarrow{a_k} A)_{Ob(\mathbf{K})}$ in $\mathbf{A}$. Now for each object k of $\mathbf{K}$, $\mathcal{S}_k = (\hat{D}_k, f_i \circ a_k)_{Ob(\mathbf{I})}$ is an object of $\mathbf{I}^*$. Thus $c_{\mathcal{S}_k} : \mathcal{S}_k \to \mathcal{K}$ is an $\mathbf{S}^D$-morphism, and since $(c_\mathcal{S}, \mathcal{K})$ is a colimit, for each $\mathbf{K}$-morphism $k \xrightarrow{g} k'$ we have $c_{\mathcal{S}_k} = c_{\mathcal{S}_{k'}} \circ \hat{D}(g)$. Thus $(c_{\mathcal{S}_k}, K)_{k \in Ob(\mathbf{K})}$ is natural for $\hat{D}$, so that since $(a_k, A)_{Ob(\mathbf{K})}$ is a colimit of $\hat{D}$, there is a unique $A \xrightarrow{t_A} K$ such that $t_A \circ a_k = c_{\mathcal{S}_k}$. Since (a_k, A) is an epi-sink, it follows that t_A is a morphism in $\mathbf{S}^D$. Thus $\mathcal{K}$ is a weak terminal object. □

12.13 THEOREM

Every co-wellpowered cocomplete category with a separator is wellpowered and complete.

Proof: Let $\mathbf{A}$ be a co-wellpowered cocomplete category with a separator S. First we prove the following two facts about $\mathbf{A}$:

(1) For every source $\mathcal{S}$ there exist an epimorphism e and a mono-source $\mathcal{M}$ with $\mathcal{S} = \mathcal{M} \circ e$.

(2) If $\mathcal{M} = (A \xrightarrow{m_i} A_i)_I$ is a small mono-source, then A is a quotient object of $\prod_I \text{hom}(S,A_i) S$ or a quotient object of $^\emptyset S$.

(1). Let $(e_j)_J$ be the collection of those epimorphisms e_j, for which there exists a source $\mathcal{S}_j$ with $\mathcal{S} = \mathcal{S}_j \circ e_j$. If e is a cointersection of $(e_j)_J$, then there exists a source $\mathcal{M}$ with $\mathcal{S} = \mathcal{M} \circ e$. To show that $\mathcal{M}$ is a mono-source, let r and s be morphisms with $\mathcal{M} \circ r = \mathcal{M} \circ s$. If c is a coequalizer of r and s, then there exists a source $\mathcal{T}$ with $\mathcal{M} = \mathcal{T} \circ c$. This implies that $\mathcal{S} = \mathcal{T} \circ (c \circ e)$. Thus there exists some $\hat{\jmath} \in J$ with $c \circ e = e_{\hat{\jmath}}$. By the definition of e, there exists a morphism $f_{\hat{\jmath}}$ with $e = f_{\hat{\jmath}} \circ e_{\hat{\jmath}}$. Since e is an epimorphism, the equations $(f_{\hat{\jmath}} \circ c) \circ e = f_{\hat{\jmath}} \circ e_{\hat{\jmath}} = e = id \circ e$ imply that $f_{\hat{\jmath}} \circ c = id$. Thus c is a section and, consequently, an isomorphism. Hence $r = s$.

(2). Since $\mathcal{M}$ is a small mono-source, the map $\varphi : \text{hom}(S, A) \to \prod_{i \in I} \text{hom}(S, A_i)$, given by $(\varphi(f))(i) = m_i \circ f$, is injective. Thus either $\text{hom}(S, A) = \emptyset$ or there exists an epimorphism $\prod_I \text{hom}(S,A_i) S \to \text{hom}(S,A) S$. (Cf. the dual of Exercise 10K.) Since S is a separator, there exists an epimorphism $\text{hom}(S,A) S \to A$ (cf. the proof of Proposition 10.38). Thus A is a quotient object of $^\emptyset S$ or a quotient object of $\prod_I \text{hom}(S,A_i) S$.

Since **A** is co-wellpowered, (2) immediately implies that **A** is wellpowered. It remains to be shown that every small diagram $D : \mathbf{I} \to \mathbf{A}$ has a limit. Let D be such a diagram and let Z be a set that represents all quotient objects of $^\emptyset S$ and of $\Pi_{i \in Ob(\mathbf{I})} \hom(S, D(i)) S$. Then the collection $\mathcal{G} = \{\, \mathcal{S}_j \mid j \in J \,\}$ of all natural sources $\mathcal{S}_j = (Q_j \xrightarrow{f_{ji}} D(i))_{i \in Ob(\mathbf{I})}$ for D, whose domain Q_j belongs to Z, is a set. Let $(\mu_j, \coprod Q_j)_{j \in J}$ be a coproduct of the small family $(Q_j)_{j \in J}$. Then there exists a unique source $\mathcal{T} = (\coprod Q_j \xrightarrow{f_i} D(i))_{i \in Ob(\mathbf{I})}$ such that $\mathcal{S}_j = \mathcal{T} \circ \mu_j$ for each $j \in J$. By (1) there exist an epimorphism e and a mono-source $\mathcal{L}$ with $\mathcal{T} = \mathcal{L} \circ e$. Since each $\mathcal{S}_j$ is natural for D, so are $\mathcal{T}$ and $\mathcal{L}$. To verify that $\mathcal{L}$ is a limit of D it remains to be shown that every natural source $\mathcal{S}$ for D factors uniquely through $\mathcal{L}$. By (1) there exist an epimorphism $\bar{e}$ and a mono-source $\mathcal{M}$ with $\mathcal{S} = \mathcal{M} \circ \bar{e}$. Since $\mathcal{S}$ is natural for D, so is $\mathcal{M}$. By (2) there exist $\hat{\jmath} \in J$ and an isomorphism h with $\mathcal{M} = \mathcal{S}_{\hat{\jmath}} \circ h$. Thus $f = e \circ \mu_{\hat{\jmath}} \circ h \circ \bar{e}$ is a morphism with $\mathcal{S} = \mathcal{L} \circ f$. That f is uniquely determined by $\mathcal{S} = \mathcal{L} \circ f$ follows from the fact that $\mathcal{L}$ is a mono-source. □

12.14 REMARK
For slight modifications of the last two theorems see Exercise 12N.

Suggestions for Further Reading

Isbell, J. R. Structure of categories. *Bull. Amer. Math. Soc.* **72** (1966): 619–655.

Trnková, V. Limits in categories and limit preserving functors. *Comment. Math. Univ. Carolinae* **7** (1966): 1–73.

Trnková, V. Completions of small categories. *Comment. Math. Univ. Carolinae* **8** (1967): 581–633.

Isbell, J. R. Small subcategories and completeness. *Math. Systems Theory* **2** (1968): 27–50.

Kennison, J. Normal completions of small categories. *Canad. J. Math.* **21** (1969): 196–201.

Pumplün, D., and W. Tholen. Covollständigkeit vollständiger Kategorien. *Manuscr. Math.* **11** (1974): 127–140.

Börger, R., W. Tholen, M. B. Wischnewsky, and H. Wolff. Compact and hypercomplete categories. *J. Pure Appl. Algebra* **21** (1981): 129–144.

Adámek, J., and V. Koubek. Completion of concrete categories. *Cahiers Topol. Geom. Diff.* **22** (1981): 209–228.

Herrlich, H. Universal completions of concrete categories. *Springer Lect. Notes Math.* **915** (1982): 127–135.

Kelly, G. M. A survey of totality for enriched and ordinary categories. *Cahiers Topol. Geom. Diff. Cat.* **27** (1986): 109–132.

EXERCISES

12A. Regular Monomorphisms vs. Strict Monomorphisms
Show that

(a) The class of strict monomorphisms is closed under the formation of intersections. Neither the class of regular monomorphisms nor the class of extremal monomorphisms need be closed under the formation of intersections.

(b) Intersections of regular monomorphisms are strict monomorphisms.

(c) If **A** has equalizers, then in **A** strict subobjects are precisely the intersections of families of regular subobjects.

(d) If **A** is strongly complete, then in **A** StrictMono = RegMono.

(e) If **A** has pushouts, then in **A** StrictMono = RegMono.

12B. Regular Monomorphisms via Extremal Monomorphisms
Show that

(a) If **A** is strongly complete and in **A** regular monomorphisms are closed under composition, then in **A** RegMono = ExtrMono [cf. Corollary 14.20].

(b) If **A** has equalizers and pushouts and in **A** regular monomorphisms are closed under composition, then in **A** RegMono = ExtrMono [cf. Proposition 14.22].

12C. Multiple Pullbacks and Completeness
Show that:

(a) A category is complete if and only if it has a terminal object and multiple pullbacks of small sinks.

(b) A group, considered as a category, has multiple pullbacks for all (even large) sinks, but — if nontrivial — it fails to have equalizers or finite products.

12D. Dense Subcategories
A full subcategory **B** of a category **A** is called **dense** provided that each **A**-object A is a colimit of its "canonical diagram" with scheme $\mathbf{B} \downarrow A$ (see 3K) given on objects by $(X \to A) \mapsto X$ and on morphisms by $h \mapsto h$ (with the canonical colimit sink).

(a) Verify that the colimit-dense subcategories of Example 12.11(1), (2), (4), and (5) are in fact dense.

(b) Verify that $\mathbb{R}$ is colimit-dense in **Vec**, but is not dense in **Vec**.

(c) Let **B** be a small subcategory of **A**. Define $E : \mathbf{A} \to [\mathbf{B}^{\mathrm{op}}, \mathbf{Set}]$ by assigning to each **A**-object A the restriction of $\hom(-, A) : \mathbf{A}^{\mathrm{op}} \to \mathbf{Set}$ to $\mathbf{B}^{\mathrm{op}}$. Prove that E is full and faithful if and only if **B** is dense.

12E. Limit-Dense and Codense Subcategories

(a) Formulate the concepts of limit-density (dual to 12.10) and codensity (dual to 12D).

(b) Let **A** be the full subcategory of **Set** that consists of all finite sets, and let **B** be the full subcategory of **A** that consists of $\emptyset$ and of all finite sets that have cardinality a power of 2. Show that:

(1) **A** has no proper limit-closed subcategory (cf. 13.26) that contains $\{0, 1\}$. However, $\{0, 1\}$ is not limit-dense in **A**.

(2) $\{0, 1\}$ is limit-dense in **B**, but is not codense in **B**.

* (c) Prove the equivalence of the following statements:

(1) the monoid of all functions from $\mathbb{N}$ to $\mathbb{N}$ is a codense subcategory of **Set**,

(2) no cardinal is measurable; i.e., every ultrafilter that is closed under the formation of countable meets is closed under the formation of all meets.

[Hints: For (1) $\Rightarrow$ (2), let $\mathcal{F}$ be an ultrafilter on X that is closed under the formation of countable meets. For each function $f : X \to \mathbb{N}$ define $\hat{f} : \{0\} \to \mathbb{N}$ by: $\hat{f}(0) = n$, where n is the unique member of $\mathbb{N}$ with $f^{-1}(n) \in \mathcal{F}$. For each natural source $(X \xrightarrow{f_i} \mathbb{N})_I$ the source $(\hat{f_i})_I$ is also natural. Thus, it factors through $(f_i)_I$ by a unique $g : \{0\} \to X$. Then $g(0) \in \bigcap \mathcal{F}$.
For (2) $\Rightarrow$ (1), notice that given a set X, a natural source of the canonical diagram $X \downarrow \{\mathbb{N}\} \to$ **Set** is a set A and a function $\varphi : \hom(X, \mathbb{N}) \to \hom(A, \mathbb{N})$ with $\varphi(p \circ f) = p \circ \varphi(f)$ for all $p : \mathbb{N} \to \mathbb{N}$. Given $a \in A$, prove that the collection of all subsets $Y \subseteq X$ with $\varphi(f_Y)(a) = 1$ (where $f_Y(x) = 1$ if $x \in Y$, else 0) is an ultrafilter closed under countable meets. Define $h : A \to X$ by $\{h(a)\} = \bigcap\{Y \mid \varphi(f_Y)(t) = 1\}$, then $\varphi(f) = f \circ h$ for each f.]

* (d) Prove that the following statements are equivalent:

(1) **Set** has a small codense subcategory,

(2) there do not exist arbitrarily large measurable cardinals; i.e., for some n, every ultrafilter closed under n-meets is closed under all meets.

12F. Weakly Terminal Sets of Objects

Define the concept of **weakly terminal set of objects**. Prove that a cocomplete category has a terminal object if and only if it has a weakly terminal set of objects.

12G. Existence of Coproducts

Let **A** be complete and wellpowered. Show that:

(a) If an **A**-object A has an Ith copower $^I A$, then it has a Kth copower $^K A$ for each nonempty subset K of I.

(b) If **A** is connected and $\coprod_I X_i$ exists for some family $(X_i)_I$, then $\coprod_K X_i$ exists for each subset K of I (including the case $K = \emptyset$).

* 12H. Copowers and Cocompleteness

Let **A** be complete, wellpowered, co-wellpowered, and have a separator S. Prove that the following are equivalent:

(1) **A** is cocomplete,

(2) for each set I there exists an Ith copower $^I S$ of S in **A**.

*12I. Strongly Complete but not Cocomplete

(a) Show that the categories **CBoo** and **CLat** are each complete, wellpowered, and extremally co-wellpowered, but not cocomplete. (Cf. 8E and 10S.)

(b) Show that the construct **A**, whose objects are triples (X, λ, μ), where $X \xrightarrow{\lambda} X$ and $\mathcal{P}X \setminus \{\emptyset\} \xrightarrow{\mu} X$ are functions, and whose morphisms $(X, \lambda, \mu) \xrightarrow{f} (X', \lambda', \mu')$ are those functions $X \xrightarrow{f} X'$ that make the diagram

$$\begin{array}{ccccc} \mathcal{P}X \setminus \{\emptyset\} & \xrightarrow{\mu} & X & \xrightarrow{\lambda} & X \\ {\scriptstyle f[\,]}\downarrow & & \downarrow {\scriptstyle f} & & \downarrow {\scriptstyle f} \\ \mathcal{P}X' \setminus \{\emptyset\} & \xrightarrow{\mu'} & X' & \xrightarrow{\lambda'} & X' \end{array}$$

commute, is complete, wellpowered, and co-wellpowered, but not cocomplete.

12J. Completeness and Coequalizers

Show that a category **A** has coequalizers whenever **A** is complete, wellpowered, and extremally co-wellpowered. (Cf. 15.25 and 15.7.)

*12K. Cocomplete Subcategories of HComp

Show that a full subcategory of **HComp** is reflective in **HComp** if and only if it is cocomplete.

*12L. Completions of Abstract Categories

(a) Show that the following category **A** cannot be fully embedded into a category **B** such that the pair (A, B) has a product in **B** and the embedding preserves products of pairs:

$$\mathbf{A} \quad \begin{array}{c} A \xrightarrow{r_\beta} A_\alpha \\ \uparrow \qquad \uparrow \\ C_\beta \xrightarrow{f_{\alpha\beta}} A_\alpha \times B_\alpha \xrightarrow{g_\alpha} D \\ \downarrow \qquad \downarrow \\ B \xrightarrow[s_\beta]{} B_\alpha \end{array}$$

where α and β are ordinals and $g_\alpha \circ f_{\alpha\beta} = \begin{cases} r_\beta, & \text{if } \alpha = \beta \\ s_\beta, & \text{if } \alpha \neq \beta. \end{cases}$

[Hint: If $(P, (\pi_A, \pi_B))$ were a product of (A, B) in **B**, then $\hom_\mathbf{B}(P, D)$ would be a proper class.]

(b) Show that the following category **A** cannot be fully embedded into a category **B** such that the pair (r, s) has an equalizer in **B** and the embedding preserves equalizers:

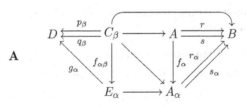

where α and β are ordinals, $r = r_\alpha \circ f_\alpha$, $s = s_\alpha \circ f_\alpha$, and

$$g_\alpha \circ f_{\alpha\beta} = \begin{cases} p_\beta, & \text{if } \alpha = \beta \\ q_\beta, & \text{if } \alpha \neq \beta. \end{cases}$$

[Hint: If $E \xrightarrow{e} A$ were an equalizer of r and s in **B**, then $\hom_\mathbf{B}(E, D)$ would be a proper class.]

(c) Show that for every category **A** that has a separating set there exists a full limit-dense embedding, $\mathbf{A} \xrightarrow{E} \mathbf{B}$, where **B** is complete and E preserves small limits.

(d) Show that every category can be fully embedded into a complete quasicategory.

* **12M. Universal Completions of Concrete Categories**

A full concrete embedding $(\mathbf{A}, U) \xrightarrow{E} (\mathbf{B}, V)$ is called a **universal completion** of $(\mathbf{A}, U)$ provided that the following conditions are satisfied:

(1) $(\mathbf{B}, V)$ is **concretely complete** [i.e., every small diagram has a concrete limit in $(\mathbf{B}, V)$] and is uniquely transportable,

(2) E preserves concrete limits of small diagrams,

(3) E is **concrete limit-dense**, i.e., for every **B**-object B there exists a small diagram $D : \mathbf{I} \to \mathbf{A}$, such that B is (the object part of) a concrete limit of $E \circ D$,

(4) for every concretely complete and uniquely transportable category $(\mathbf{C}, W)$ and every concrete functor $(\mathbf{A}, U) \xrightarrow{G} (\mathbf{C}, W)$ that preserves concrete limits of small diagrams, there exists a unique concrete functor $(\mathbf{B}, V) \xrightarrow{\overline{G}} (\mathbf{C}, W)$ that preserves concrete limits of small diagrams, with $G = \overline{G} \circ E$.

Show that

(a) Every amnestic concrete category over a complete category has a universal completion that is uniquely determined up to concrete isomorphism.

(b) If **A** is the construct of finite Boolean spaces (equivalently: the construct of finite sets), then the inclusion $\mathbf{A} \hookrightarrow \mathbf{BooSp}$ is a universal completion.

* **12N. Re: Theorems 12.12 and 12.13**
  (a) Show that every cocomplete category with a small colimit-dense subcategory is strongly complete.
  (b) Show that a strongly cocomplete category with a separating set is strongly complete.
  (c) Construct a strongly cocomplete category with a separator that is neither well-powered nor co-wellpowered.

12O. Completeness of Functor-Categories
  (a) Show that limits and colimits in the category [**A**, **B**] are formed componentwise.
  (b) Show that for each small category **A** and each strongly (co)complete category **B**, [**A**, **B**] is strongly (co)complete.

13 Functors and limits

In this section the behavior of limits with respect to functors $\mathbf{A} \xrightarrow{F} \mathbf{B}$ is investigated. The following problems receive special attention:

(A) Does F "preserve" limits?

The information that a functor $\mathbf{A} \xrightarrow{F} \mathbf{B}$ preserves limits is particularly useful if the domain category $\mathbf{A}$ is known to have enough limits. See, e.g., the adjoint functor theorems in §18.

In many cases, particularly for embedding functors of full subcategories and for forgetful functors of concrete categories, it is not immediately clear whether or not the domain category $\mathbf{A}$ of the functor $\mathbf{A} \xrightarrow{F} \mathbf{B}$ in question has enough limits, but it is known that the codomain category $\mathbf{B}$ does. In these cases the following questions arise naturally:

(B) Can limits in $\mathbf{B}$ be lifted along F to limits of $\mathbf{A}$?

If so, to which extent is the lifting unique?
If not, are there suitable properties of F that at least guarantee the existence of limits in $\mathbf{A}$ provided that limits exist in $\mathbf{B}$?

PRESERVATION OF LIMITS

13.1 DEFINITION

(1) A functor $F: \mathbf{A} \to \mathbf{B}$ is said to **preserve a limit** $\mathcal{L} = (L \xrightarrow{\ell_i} D_i)$ of a diagram $D: \mathbf{I} \to \mathbf{A}$ provided that $F\mathcal{L} = (FL \xrightarrow{F\ell_i} FD_i)$ is a limit of the diagram $F \circ D: \mathbf{I} \to \mathbf{B}$.

(2) F is said to **preserve limits over a scheme I** provided that F preserves all limits of diagrams $D: \mathbf{I} \to \mathbf{A}$ with scheme $\mathbf{I}$.

(3) F **preserves equalizers** if and only if F preserves all limits over the scheme $\bullet \rightrightarrows \bullet$; F **preserves products** if and only if F preserves all limits over small discrete schemes; F **preserves small limits** if and only if F preserves limits over all small schemes; F **preserves strong limits** if and only if F preserves small limits and arbitrary intersections; etc.

DUAL NOTIONS: F **preserves colimits** (over a scheme I), **coequalizers, coproducts, small colimits, strong colimits**, etc.

13.2 EXAMPLES

(1) For the constructs **Top**, **Rel**, and **Alg**(Σ), the forgetful functors preserve limits and colimits.

(2) For constructs of the form **Alg**(Ω) the forgetful functors preserve limits and directed colimits [see Example 11.28(4)], but generally fail to preserve coproducts or coequalizers. Also for the following constructs, the forgetful functors preserve limits and directed colimits, but neither coproducts nor coequalizers: **Vec, Grp, Sgr, Mon, Rng**, and Σ-**Seq**.

(3) For the constructs **AbTor** and (**Ban**,U) the forgetful functors preserve equalizers, but not products (cf. 10.55).

(4) The full embeddings **Haus** $\to$ **Top** and **Pos** $\to$ **Rel** preserve limits and coproducts, but not coequalizers.

(5) The covariant power-set functor $\mathcal{P}$: **Set** $\to$ **Set** preserves neither products (of pairs), nor coproducts (of pairs), nor equalizers, nor coequalizers.

13.3 PROPOSITION
If $F : \mathbf{A} \to \mathbf{B}$ is a functor and $\mathbf{A}$ is finitely complete, then the following conditions are equivalent:

(1) F preserves finite limits,

(2) F preserves finite products and equalizers,

(3) F preserves pullbacks and terminal objects.

Proof: Immediate from the characterizations of finite limits (12.4). □

13.4 PROPOSITION
For a complete category $\mathbf{A}$, a functor $F : \mathbf{A} \to \mathbf{B}$ preserves small limits if and only if it preserves products and equalizers.

Proof: This follows immediately from the construction of small limits via products and equalizers, presented in the proof of Theorem 12.3. □

13.5 PROPOSITION
(1) If a functor preserves finite limits, then it preserves monomorphisms and regular monomorphisms.

(2) If a functor preserves (small) limits, then it preserves (small) mono-sources.

Proof: (1) follows from Proposition 11.16.

(2) follows from the observation that a source $(A \xrightarrow{f_i} A_i)_I$ is a mono-source if and only if the source

$$A \xrightarrow{id_A} A$$
$$\downarrow{id_A} \quad \searrow{f_i} \quad (i \in I)$$
$$A \quad A_i$$

is a limit of the diagram

$$\begin{array}{c} A \\ \downarrow f_i \\ A \xrightarrow{f_i} A_i \end{array} \quad (i \in I)$$

13.6 REMARK
If a functor F preserves equalizers, then F (obviously) preserves regular monomorphisms. The converse is not true: the forgetful functors from the categories $\mathbf{Vec}^{op}$, $\mathbf{Grp}^{op}$, $\mathbf{Pos}^{op}$, and $\mathbf{Haus}^{op}$ to $\mathbf{Set}^{op}$ preserve regular monomorphisms, but don't preserve equalizers.

13.7 PROPOSITION
Hom-functors preserve limits.

Proof: Let $F = \hom(A, -) : \mathbf{A} \to \mathbf{Set}$ be a hom-functor, let $D : \mathbf{I} \to \mathbf{A}$ be a diagram, and let $\mathcal{L} = (L \xrightarrow{\ell_i} D_i)$ be a limit of D. Then $F\mathcal{L}$ is obviously a natural source for $F \circ D$. Let $\mathcal{S} = (X \xrightarrow{f_i} \hom(A, D_i))$ be an arbitrary natural source for $F \circ D$. Then, for each element x of X, $(A \xrightarrow{f_i(x)} D_i)$ is a natural source for D. Hence for each $x \in X$ there exists a unique morphism $f(x) : A \to L$ with $f_i(x) = \ell_i \circ f(x)$ for each $i \in Ob(\mathbf{I})$. This defines a function $f : X \to \hom(A, L)$ which is the unique function $f : X \to FL$ that satisfies $\mathcal{S} = F\mathcal{L} \circ f$. □

13.8 PROPOSITION
If F and G are naturally isomorphic functors, then F preserves limits over a scheme $\mathbf{I}$ if and only if G does.

Proof: Straightforward computation. □

13.9 COROLLARY
Representable functors preserve limits. □

13.10 REMARK
Since the forgetful functors for constructs are quite often representable, the above corollary explains why limits in constructs are much more often concrete (Definition 13.12) than are colimits. In §18 we will show that adjoint functors (in particular embeddings of reflective subcategories) preserve limits. Next we will show that embeddings of "sufficiently big" subcategories preserve limits as well.

13.11 PROPOSITION
Embeddings of colimit-dense subcategories preserve limits.

Proof: Let **A** be a colimit-dense subcategory of **B** with embedding $E : \mathbf{A} \to \mathbf{B}$, and let $D : \mathbf{I} \to \mathbf{A}$ be a diagram with a limit $\mathcal{L} = (L \xrightarrow{\ell_i} D_i)$. Then $\mathcal{L}$ is a natural source for $E \circ D$. Let $\mathcal{S} = (B \xrightarrow{f_i} D_i)$ be an arbitrary natural source for $E \circ D$. By colimit-density there exists a diagram $G : \mathbf{J} \to \mathbf{A}$ and a colimit $(G_j \xrightarrow{c_j} B)$ of $E \circ G$. For each object j of $\mathbf{J}$, $(G_j \xrightarrow{f_i \circ c_j} D_i)_{i \in Ob\mathbf{I}}$ is a natural source for D. Hence for each $j \in Ob(\mathbf{J})$ there exists a unique morphism $g_j : G_j \to L$ with $f_i \circ c_j = \ell_i \circ g_j$ for each $i \in I$.

$$\begin{array}{ccc} G_j & \xrightarrow{c_j} & B \\ {\scriptstyle g_j}\downarrow & & \downarrow{\scriptstyle f_i} \\ L & \xrightarrow{\ell_i} & D_i \end{array}$$

Since $\mathcal{L}$ is a mono-source in **A**, it follows that $(G_j \xrightarrow{g_j} L)$ is a natural sink for G and hence for $E \circ G$. Consequently, there exists a unique morphism $f : B \to L$ with $g_j = f \circ c_j$ for each j. Since $(G_j \xrightarrow{c_j} B)$ is an epi-sink, this implies that $f_i = \ell_i \circ f$ for each i. To show that f is the unique morphism with this property, let $f' : B \to L$ be a morphism with $f_i = \ell_i \circ f'$ for each i. Since $(L \xrightarrow{\ell_i} D_i)$ is a mono-source in **A**, this implies that $g_j = f' \circ c_j$ for each j, and hence that $f' = f$. □

CONCRETE LIMITS

13.12 DEFINITION

(1) Let $(\mathbf{A}, U)$ be a concrete category. A limit $\mathcal{L}$ of a diagram $D : \mathbf{I} \to \mathbf{A}$ is called a **concrete limit** of D in $(\mathbf{A}, U)$ provided that it is preserved by U.

(2) A concrete category $(\mathbf{A}, U)$ **has (small) concrete limits**, resp. **concrete products**, etc., if and only if **A** has (small) limits, resp. products, etc., and U preserves them.

DUAL NOTIONS: **(has) (small) concrete colimits, concrete coproducts**, etc.

13.13 EXAMPLES

(1) The constructs **Top**, **Rel**, **Prost**, and **Alg**(Σ) have small concrete limits and small concrete colimits. So have all constructs of the form **Spa**(T).

(2) The constructs **Vec**, **Grp**, **Pos**, and **Haus** have small concrete limits and non-concrete small colimits. **Alg**(Ω) has small concrete limits; it has small concrete colimits only if the defining operations are all unary.

(3) The construct Σ-**Seq** has finite concrete limits and finite colimits that are not concrete.

(4) The concrete category **Aut** over **Set** $\times$ **Set** $\times$ **Set** has small concrete limits and small colimits that are not concrete.

13.14 PROPOSITION
A concrete category has small concrete limits if and only if it has concrete products and concrete equalizers.
$\boxed{A\ 12.3}$

13.15 PROPOSITION
If $(\mathbf{A}, U)$ is a concrete category and $D : \mathbf{I} \to \mathbf{A}$ is a diagram, then $\mathcal{L} = (L \xrightarrow{\ell_i} D_i)_{i \in Ob(\mathbf{I})}$ is a concrete limit in $(\mathbf{A}, U)$ if and only if $U(\mathcal{L})$ is a limit of $U \circ D$ and $\mathcal{L}$ is an initial source in $(\mathbf{A}, U)$.
$\boxed{A\ 10.53}$

13.16 REMARK
By the above, a concrete limit can be constructed in two steps: first form the limit of the underlying diagram in the base category, and then provide an initial lift of this (underlying) limit.

LIFTING OF LIMITS

13.17 DEFINITION
A functor $F : \mathbf{A} \to \mathbf{B}$ is said to

(1) **lift limits (uniquely)** provided that for every diagram $D : \mathbf{I} \to \mathbf{A}$ and every limit $\mathcal{L}$ of $F \circ D$ there exists a (unique) limit $\mathcal{L}'$ of D with $F(\mathcal{L}') = \mathcal{L}$,

(2) **create limits** provided that for every diagram $D : \mathbf{I} \to \mathbf{A}$ and every limit $\mathcal{L}$ of $F \circ D$ there exists a unique source $\mathcal{S} = (L \xrightarrow{f_i} D_i)$ in $\mathbf{A}$ with $F(\mathcal{S}) = \mathcal{L}$, and that, moreover, $\mathcal{S}$ is a limit of D.

Similarly, one has **lifts small limits**, **lifts products**, **creates equalizers**, **creates finite limits**, etc.

DUAL NOTIONS: **lift colimits (uniquely)**, **create colimits**, etc.

13.18 EXAMPLES
The forgetful functors for the constructs

(1) of the form $\mathbf{Alg}(\Sigma)$ create limits and colimits,

(2) of the form $\mathbf{Alg}(\Omega)$ create limits and directed colimits [see Example 11.28(4)],

(3) **V̌ec**, **Grp**, **Sgr**, **Mon**, **Lat**, **CAT**$_{of}$, and **HComp** create limits,

(4) **Top**, **Rel**, and **Prost** lift limits and colimits uniquely, but create neither,

(5) **Met**$_c$ and **Ban**$_b$ lift finite limits, but do not lift them uniquely,

(6) Σ-**Seq** lifts finite limits uniquely.

13.19 THEOREM

If a functor $\mathbf{A} \xrightarrow{F} \mathbf{B}$ lifts limits and $\mathbf{B}$ is (strongly) complete, then $\mathbf{A}$ is (strongly) complete and F preserves small limits (and arbitrary intersections).

Proof: If F lifts limits and $\mathbf{B}$ is complete, then obviously $\mathbf{A}$ is complete and F preserves small limits. Hence, by Proposition 13.5, F preserves monomorphisms. Thus, if $\mathbf{B}$ has arbitrary intersections, so does $\mathbf{A}$, and F preserves them. □

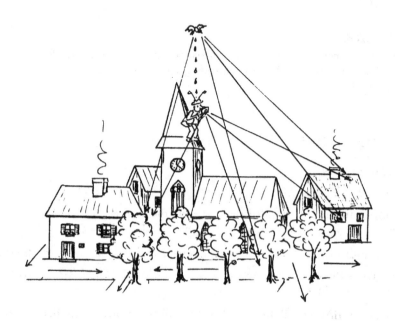

Lifting of limits

13.20 REMARKS

(1) The obvious implications

F creates limits $\Rightarrow$ F lifts limits uniquely $\Rightarrow$ F lifts limits

cannot be reversed.

(2) If $\mathbf{B}$ has certain limits and $F : \mathbf{A} \to \mathbf{B}$ lifts them, then $\mathbf{A}$ has these limits and F preserves them. Hence the concepts of lifting and creating limits are useful for functors $\mathbf{A} \xrightarrow{F} \mathbf{B}$ with range categories $\mathbf{B}$ that have enough limits. In general, functors that create limits need not even preserve them. An example of this is the embedding of the terminal category $\mathbf{1} = \{\bullet\} = \mathbf{A}$ into $\mathbf{B} = \bullet \rightrightarrows \circ$.

13.21 PROPOSITION

For functors $F : \mathbf{A} \to \mathbf{B}$ the following conditions are equivalent:

(1) F lifts limits uniquely,

(2) F lifts limits and is amnestic.

Proof: (1) $\Rightarrow$ (2). To show amnesticity, let $A' \xrightarrow{f} A$ be an isomorphism in **A** with $Ff = id_B$. The 1-indexed sources $(A' \xrightarrow{f} A)$ and $(A \xrightarrow{id_A} A)$ are products of the 1-indexed family (A). Since they are mapped via F to the product $(B \xrightarrow{id_B} B)$ of the 1-indexed family $(B) = (FA)$, uniqueness implies that $f = id_A$.

(2) $\Rightarrow$ (1). If $\mathcal{L} = (L \xrightarrow{\ell_i} D_i)$ and $\mathcal{L}' = (L' \xrightarrow{\ell_i'} D_i)$ are limits of a diagram $D : \mathbf{I} \to \mathbf{A}$, then there exists an isomorphism $h : L \to L'$ with $\mathcal{L} = \mathcal{L}' \circ h$. If $F(\mathcal{L}) = F(\mathcal{L}')$ is a limit of $F \circ D$, then $Fh = id_{FL}$. Hence, by amnesticity, $h = id_L$. Consequently, $\mathcal{L} = \mathcal{L}'$. $\square$

13.22 DEFINITION

A functor $F : \mathbf{A} \to \mathbf{B}$ is said to

(1) **reflect limits** provided that for each diagram $D : \mathbf{I} \to \mathbf{A}$ an **A**-source $\mathcal{S} = (A \xrightarrow{f_i} D_i)_{i \in Ob(\mathbf{I})}$ is a limit of D whenever $F(\mathcal{S})$ is a limit of $F \circ D$,

(2) **detect limits**, provided that a diagram $D : \mathbf{I} \to \mathbf{A}$ has a limit whenever $F \circ D$ has one.

Similarly, one has **reflect equalizers**, **detect products**, etc.

DUAL NOTIONS: **reflect colimits**, **detect colimits**, etc.

13.23 EXAMPLES

(1) Every functor that lifts limits, detects them.

(2) The forgetful functor $U : \mathbf{Top} \to \mathbf{Set}$ lifts limits uniquely (and detects and preserves them), but does not reflect them.

(3) Full embeddings reflect limits, but they need not lift, preserve, nor detect them.

13.24 PROPOSITION

A functor that reflects equalizers is faithful.

Proof: Let $F : \mathbf{A} \to \mathbf{B}$ be a functor and let $A \underset{g}{\overset{f}{\rightrightarrows}} B$ be a pair of **A**-morphisms with $Ff = Fg$. Then $\mathcal{S} = (A \xrightarrow{id_A} A)$ is a 1-source and $F\mathcal{S}$ is an equalizer of Ff and Fg. If F reflects equalizers, then $\mathcal{S}$ is an equalizer of f and g. Hence $f = id_A \circ f = id_A \circ g = g$. $\square$

13.25 PROPOSITION

For any functor $F : \mathbf{A} \to \mathbf{B}$ the following conditions are equivalent:

(1) F creates limits,

(2) F lifts limits uniquely and reflects limits,

(3) F lifts limits, is faithful and amnestic, and reflects isomorphisms in the sense that whenever Ff is a **B**-isomorphism, then f is an **A**-isomorphism.

Proof: (1) $\Leftrightarrow$ (2) is obvious.

(2) $\Rightarrow$ (3) follows from Propositions 13.21, 13.24, and Example 10.20(8).

(3) $\Rightarrow$ (2). To show that F reflects limits, let $D : \mathbf{I} \to \mathbf{A}$ be a diagram and let $\mathcal{S} = (A \xrightarrow{f_i} D_i)_{i \in Ob(\mathbf{I})}$ be a source in **A** such that $F(\mathcal{S})$ is a limit of $F \circ D$. Since F lifts limits, there exists a limit $\mathcal{L} = (L \xrightarrow{\ell_i} D_i)$ of D with $F(\mathcal{L}) = F(\mathcal{S})$. Since F is faithful, S is a natural source for D. Hence there exists a morphism $A \xrightarrow{f} L$ with $\mathcal{S} = \mathcal{L} \circ f$. Consequently, $F(\mathcal{L}) = F(\mathcal{S}) = F(\mathcal{L}) \circ Ff$, which implies that $Ff = id_{FL}$. Since F reflects isomorphisms, f is an isomorphism. Hence, by amnesticity, $f = id_L$. Consequently, $\mathcal{S} = \mathcal{L}$. Thus $\mathcal{S}$ is a limit of D. □

13.26 REMARK
Embeddings $E : \mathbf{A} \to \mathbf{B}$ of full subcategories obviously reflect limits. Hence they lift limits if and only if they create them. A more suggestive term for such full subcategories is that they are **closed under the formation of limits** (or just **limit-closed**) in **B**.

13.27 PROPOSITION
A full reflective subcategory **A** of **B** is limit-closed in **B** if and only if **A** is isomorphism-closed in **B**.

Proof: If **A** is limit-closed in **B**, then obviously **A** is isomorphism-closed in **B**. For the converse, consider a diagram $D : \mathbf{I} \to \mathbf{A}$ such that $E \circ D : \mathbf{I} \to \mathbf{B}$ has a limit $\mathcal{L} = (L \xrightarrow{\ell_i} D_i)$, where $E : \mathbf{A} \to \mathbf{B}$ denotes the embedding. Let $L \xrightarrow{r} A$ be an **A**-reflection arrow for L. Then for each i there exists a morphism $f_i : A \to D_i$ with $\ell_i = f_i \circ r$. By the uniqueness property of reflection arrows and the fact that all D_i belong to **A**, $\mathcal{S} = (A \xrightarrow{f_i} D_i)$ must be a natural source for D resp. for $E \circ D$. Consequently, there exists a morphism $f : A \to L$ with $\mathcal{S} = \mathcal{L} \circ f$. Hence $\mathcal{L} = \mathcal{S} \circ r = \mathcal{L} \circ f \circ r$, which implies that $f \circ r = id_L$. Therefore $(r \circ f) \circ r = r \circ id_L = id_A \circ r$, which implies that $r \circ f = id_A$. Hence r is an isomorphism, and, consequently, L belongs to **A**. □

13.28 COROLLARY
If a category has certain limits, then so does each of its isomorphism-closed full reflective subcategories. □

13.29 REMARK
Isomorphism-closed full reflective subcategories **A** of **B** usually fail to be colimit-closed. However, the following proposition shows that the associated inclusion functors detect colimits.

13.30 PROPOSITION
Let $\mathbf{A}$ be a full subcategory of $\mathbf{B}$ with embedding $E: \mathbf{A} \to \mathbf{B}$, and let $D: \mathbf{I} \to \mathbf{A}$ be a diagram. If $\mathcal{C} = (D_i \xrightarrow{c_i} C)$ is a colimit of $E \circ D$, and if $C \xrightarrow{r} A$ is an $\mathbf{A}$-reflection arrow for C, then $\mathcal{C}' = r \circ \mathcal{C}$ is a colimit of D.

Proof: Obviously, $\mathcal{C}'$ is a natural sink for D. Let $\mathcal{S} = (D_i \xrightarrow{f_i} A')$ be an arbitrary natural sink for D. Then $\mathcal{S}$ is a natural sink for $E \circ D$. Hence there exists a unique morphism $C \xrightarrow{f} A'$ with $\mathcal{S} = f \circ \mathcal{C}$. Since r is an $\mathbf{A}$-reflection arrow, there exists a unique morphism $A \xrightarrow{g} A'$ with $f = g \circ r$. Hence $\mathcal{S} = f \circ \mathcal{C} = g \circ r \circ \mathcal{C} = g \circ \mathcal{C}'$. Uniqueness of g is obtained by retracing the steps of the above construction. □

13.31 EXAMPLES
(1) **HComp** is a full reflective subcategory of **Top**. The construction of coproducts in **HComp** given in Example 10.67(5) is a special case of the above result.

(2) **AbTor** is a full coreflective subcategory of **Ab**. The construction of products in **AbTor** given in Example 10.20(3) is a special case of the dual of the above result.

13.32 COROLLARY
Embeddings of full reflective subcategories detect colimits. □

13.33 REMARKS
(1) Embeddings of nonfull reflective subcategories may be quite awkward. See Exercise 13K.

(2) As will be seen in Chapter VI, all "reasonable" forgetful functors of concrete categories lift limits uniquely and detect colimits.

13.34 PROPOSITION
If a functor $\mathbf{A} \xrightarrow{F} \mathbf{B}$ preserves limits, then the following conditions are equivalent:

(1) F lifts limits (uniquely),

(2) F detects limits and is (uniquely) transportable.

Proof: (1) $\Rightarrow$ (2) follows immediately from the fact that isomorphisms are products of 1-indexed families.

(2) $\Rightarrow$ (1). Let $D: \mathbf{I} \to \mathbf{A}$ be a diagram and let $\mathcal{S} = (L \xrightarrow{\ell_i} FD_i)_I$ be a limit of $F \circ D$. Since F detects limits, D has a limit $\mathcal{L} = (A \xrightarrow{f_i} D_i)_I$. Since F preserves limits, $F\mathcal{L} = (FA \xrightarrow{Ff_i} FD_i)_I$ is a limit of $F \circ D$. Hence there exists a **B**-isomorphism $FA \xrightarrow{h} L$ with $\mathcal{S} = h \circ F\mathcal{L}$. By transportability, h can be lifted to an **A**-isomorphism $A \xrightarrow{k} B$. Hence $(B \xrightarrow{f_i \circ k^{-1}} D_i)_I$ is a limit of D that lifts $\mathcal{S}$. The uniqueness part follows from Proposition 13.21. □

CREATION AND REFLECTION OF ISOMORPHISMS

Since isomorphisms can be regarded as the limits of diagrams with scheme **1**, the concepts of creation and reflection of limits specialize to the following concepts:

13.35 DEFINITION
A functor $\mathbf{A} \xrightarrow{G} \mathbf{B}$ is said to

(1) **create isomorphisms** provided that whenever $h : X \to GA$ is a G-structured **B**-isomorphism, there exists precisely one **A**-morphism $\hat{h} : B \to A$ with $G(\hat{h}) = h$, and, moreover, $\hat{h}$ is an isomorphism,

(2) **reflect isomorphisms** provided that an **A**-morphism f is an **A**-isomorphism whenever Gf is a **B**-isomorphism.

13.36 PROPOSITION
(1) If G creates (resp. reflects) limits, then G creates (resp. reflects) isomorphisms.

(2) G creates isomorphisms if and only if G reflects isomorphisms and is uniquely transportable.

(3) If G creates isomorphisms, then G reflects identities. □

13.37 EXAMPLES
(1) Let $E : \mathbf{A} \to \mathbf{B}$ be the embedding of a full subcategory. Then E reflects isomorphisms. E creates isomorphisms if and only if **A** is isomorphism-closed in **B**.

(2) Let **A** be a monoid, considered as a category. Then the unique functor $\mathbf{A} \xrightarrow{G} \mathbf{1}$ reflects isomorphisms if and only if **A** is a group. G creates isomorphisms if and only if **A** is a trivial (one-element) group.

(3) Each of the forgetful functors from **Vec**, **Grp**, and **Mon** to **Set** creates isomorphisms. None of the forgetful functors from **Top**, **Rel**, and **Pos** to **Set** reflects isomorphisms.

13.38 REMARK
The following diagram summarizes some of the relationships between functors and limits.

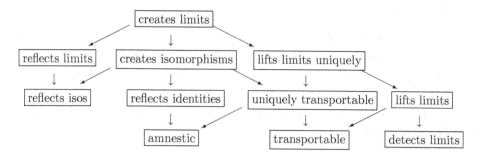

Suggestions for Further Reading

Trnková, V. Some properties of set functors. *Comment. Math. Univ. Carolinae* **10** (1969): 323–352.

Trnková, V. When the product-preserving functors preserve limits. *Comment. Math. Univ. Carolinae* **11** (1970): 365–378.

Adámek, J., J. Rosický, and V. Trnková. Are all limit-closed subcategories of locally presentable categories reflective? *Springer Lect. Notes Math.* **1348** (1988): 1–18.

EXERCISES

13A. Preservation of Multiple Pullbacks

Let **A** be a complete category. Show that a functor $\mathbf{A} \xrightarrow{F} \mathbf{B}$ preserves small limits if and only if F preserves small multiple pullbacks and terminal objects.

13B. Preservation of Limits by Set-Valued Functors

Show that

(a) If a functor $\mathbf{A} \xrightarrow{G} \mathbf{Set}$ has a G-universal arrow with a nonempty domain, then G preserves limits. Deduce Corollary 13.9.

(b) If **A** is a full subcategory of **Set** that contains at least one nonempty set, then the embedding $\mathbf{A} \to \mathbf{Set}$ preserves limits.

* (c) Every functor $\mathbf{Set} \to \mathbf{Set}$ preserves nonempty finite intersections.

(d) A functor $\mathbf{Set} \to \mathbf{Set}$ preserves products if and only if it is representable.

(e) A functor $\mathbf{Set} \to \mathbf{Set}$ preserves coproducts if and only if it is naturally equivalent, for some set A, to the functor $A \times - : \mathbf{Set} \to \mathbf{Set}$, defined by

$$(A \times -)(X \xrightarrow{f} Y) = A \times X \xrightarrow{id_A \times f} A \times Y.$$

13C. Reflection of Concrete Limits

Let $(\mathbf{A}, U)$ have small concrete limits. Show that U reflects small limits if and only if U reflects isomorphisms.

13D. Lifting of Limits and Faithfulness

Show that

(a) A functor that lifts equalizers is faithful if and only if it reflects epimorphisms.

(b) A functor that lifts limits need not be faithful.

13E. Lifting of Limits and Transportability

Show that a functor that lifts limits (uniquely) is (uniquely) transportable.

13F. Reflection of Isomorphisms
Show that a functor that reflects equalizers (or finite products) reflects isomorphisms.

13G. Initial Mono-Sources and Reflection of Limits
Show that if mono-sources are G-initial, then G reflects limits. [Cf. Proposition 10.60.]

13H. Hom-Functors and Limits
Show that

(a) If $D : \mathbf{I} \to \mathbf{A}$ is a diagram, then a source $\mathcal{S} = (L \xrightarrow{\ell_i} D_i)_{i \in Ob(\mathbf{I})}$ is a limit of D if and only if, for each $\mathbf{A}$-object A, $\hom(A, -)(\mathcal{S})$ is a limit of $\hom(A, -) \circ D$.

(b) If A is a nonempty set, then $\hom(A, -) : \mathbf{Set} \to \mathbf{Set}$ reflects limits.

13I. Limit- and Colimit-Closed Full Subcategories

(a) Show that

    (1) **Set** has no proper full subcategory that is both limit-closed and colimit-closed.

  * (2) **Top** has no proper full subcategory that is both limit-closed and colimit-closed.

    (3) **Vec** has precisely two full subcategories that are both limit-closed and colimit-closed.

(b) Determine all full subcategories of **Rel** that are both limit-closed and colimit-closed.

13J. Limit-Closed vs. Reflective Full Subcategories
Show that

(a) Every full subcategory of the partially ordered set $\mathbb{Z}$, considered as a category, is limit- and colimit-closed in $\mathbb{Z}$, but generally fails to be either reflective or coreflective in $\mathbb{Z}$.

(b) A full subcategory of a complete lattice $\mathbf{A}$, considered as a category, is reflective in $\mathbf{A}$ if and only if it is limit-closed in $\mathbf{A}$.

* (c) There exist limit-closed full subcategories of **Top** that are not reflective in **Top**. [Cf. 16D.]

13K. A Misbehaved Nonfull Reflective Embedding
Consider the following nonfull embedding $\mathbf{A} \xrightarrow{E} \mathbf{B}$ of preordered sets, considered as categories:

Show that

(a) $\mathbf{A}$ is reflective in $\mathbf{B}$,

(b) $\mathbf{B}$ is complete and cocomplete,

(c) **A** is neither complete nor cocomplete,

(d) E detects neither limits nor colimits.

13L. Colimit-Dense Full Embeddings
Show that colimit-dense full embeddings preserve limits.

13M. Creation of Isomorphisms
Show that

(a) G creates limits if and only if G is faithful, lifts limits, and creates isomorphisms.

(b) G creates isomorphisms if and only if G creates 1-indexed products.

(c) If G creates isomorphisms, then G reflects identities.

(d) If G reflects identities, then G is amnestic.

13N. Creation of Limits of T-Algebras
Prove that for each functor $T : \mathbf{X} \to \mathbf{X}$ the forgetful functor of $\mathbf{Alg}(T)$ creates limits.

Chapter IV

FACTORIZATION STRUCTURES

Every function $A \xrightarrow{f} B$ can be factored through its image, i.e., written as a composite $A \xrightarrow{f} B = A \xrightarrow{e} f[A] \xrightarrow{m} B$, where $A \xrightarrow{e} f[A]$ is the codomain-restriction of f and $f[A] \xrightarrow{m} B$ is the inclusion. This fact, though simple, is often useful. Similarly, in constructs such as **Vec**, **Grp**, and **Top** every morphism can be factored through its "image". Since

(a) for categories in general, no satisfactory concept of "embedding of subobjects" and hence of "image of a morphism" is available,

and

(b) for certain constructs, factorizations of morphisms different from the one through the image are of interest (e.g., in **Top** the one through the closure of the image: $A \xrightarrow{f} B = A \to cl_B f[A] \hookrightarrow B$),

categorists have created an axiomatic theory of factorization structures (E, M) for morphisms of a category **A**. Here E and M are classes of **A**-morphisms[59] such that each **A**-morphism has an (E, M)-factorization $A \xrightarrow{f} B = A \xrightarrow{e \in E} C \xrightarrow{m \in M} B$. Naturally, without further assumptions on E and M such factorizations might be quite useless. A careful analysis has revealed that the crucial requirement that causes (E, M)-factorizations to have appropriate characteristics is the so-called "unique (E, M)-diagonalization" condition, described in Definition 14.1. Such factorization structures for morphisms have turned out to be useful, especially for "well-behaved" categories (e.g., those having products and satisfying suitable smallness conditions). They have been transformed into powerful categorical tools by two successive generalizations

(a) factorization structures for sources in a category, and

(b) factorization structures for G-structured sources with respect to a functor G.

Instead of describing the most general concept first and then specializing to the others, our presentation will follow the historical development described above.

---

[59]The requirements $E \subseteq \text{Epi}(\mathbf{A})$ and $M \subseteq \text{Mono}(\mathbf{A})$ were originally included, but later dropped.

14 Factorization structures for morphisms

In this section factorization structures for morphisms are defined and investigated. In particular, it is shown that every strongly complete category is simultaneously (ExtrEpi, Mono)-structured and (Epi, ExtrMono)-structured (see 14.21).

14.1 DEFINITION
Let E and M be classes of morphisms in a category **A**.

(E, M) is called a **factorization structure for morphisms** in **A** and **A** is called (E, M)-**structured** provided that

(1) each of E and M is closed under composition with isomorphisms,[60]

(2) **A** has (E, M)-**factorizations (of morphisms)**; i.e., each morphism f in **A** has a factorization $f = m \circ e$, with $e \in E$ and $m \in M$, and

(3) **A** has the **unique** (E, M)-**diagonalization property**; i.e., for each commutative square

$$\begin{array}{ccc} A & \xrightarrow{e} & B \\ f \downarrow & & \downarrow g \\ C & \xrightarrow{m} & D \end{array} \qquad (*)$$

with $e \in E$ and $m \in M$ there exists a unique **diagonal**, i.e., a morphism d such that the diagram

$$\begin{array}{ccc} A & \xrightarrow{e} & B \\ f \downarrow & \swarrow d & \downarrow g \\ C & \xrightarrow{m} & D \end{array}$$

commutes (i.e., such that $d \circ e = f$ and $m \circ d = g$).

14.2 EXAMPLES
(1) For any category, (Iso, Mor) and (Mor, Iso)[61] are (trivial) factorization structures for morphisms.

---

[60] Condition (1) can be replaced by the following conditions (1a) and (1b):
(1a) if $e \in E$ and $h \in \text{Iso}(\mathbf{A})$, and $h \circ e$ exists, then $h \circ e \in E$,

(1b) if $m \in M$ and $h \in \text{Iso}(\mathbf{A})$, and $m \circ h$ exists, then $m \circ h \in M$.
This follows from Proposition 14.6 below since for its proof not the full strength of (1) but only that of (1a) and (1b) is used. When generalizing factorization structures to sources or to functors, the formulation in terms of (1a) and (1b) turns out to be more appropriate.

[61] Recall that Iso resp. Iso(**A**) is the class of all isomorphisms (in the category **A**); similarly for Mor, Mono, etc. See the Table of Symbols.

239

(2) For **Set**, **Vec**, **Grp**, **Mon**, Σ-**Seq**, each $\mathbf{Alg}(\Omega)$ and many other categories of algebras, (RegEpi, Mono) is a factorization structure for morphisms.

(3) For **Cat**, (RegEpi, Mono) is *not* a factorization structure for morphisms. [Regular epimorphisms are not closed under composition (cf. 14.6 and 7.76).] However, (ExtrEpi, Mono) is a factorization structure for morphisms in **Cat**.

(4) **Set** has precisely four factorization structures for morphisms: besides the three mentioned above, namely, (Epi, Mono) = (RegEpi, RegMono) and the two trivial ones, the following (pathological) (E, M), where

$$E = \{X \xrightarrow{e} Y \mid X = \emptyset \Rightarrow Y = \emptyset\}; \quad \text{and}$$
$$M = \{X \xrightarrow{m} Y \mid m \text{ is a bijection or } X = \emptyset\}.$$

(5) **Top** has a proper class (even an illegitimate conglomerate) of factorization structures for morphisms. Each of

(Epi, RegMono) = (surjection, embedding),
(RegEpi, Mono) = (quotient, injection),
(dense, closed embedding), and
(front-dense, front-closed embedding),

is a factorization structure for morphisms in **Top**, but (Epi, Mono) is not.

(6) (dense C^*-embedding, perfect map) is a factorization structure for morphisms in **Tych**.

14.3 PROPOSITION
A is (E, M)-*structured if and only if* $\mathbf{A}^{\mathrm{op}}$ *is* (M, E)-*structured*. $\square$

14.4 PROPOSITION
If **A** *is* (E, M)-*structured, then* (E, M)-*factorizations are essentially unique, i.e.*,

(1) *if* $A \xrightarrow{e_i} C_i \xrightarrow{m_i} B$ *are* (E, M)-*factorizations of* $A \xrightarrow{f} B$ *for* $i = 1, 2$, *then there exists a (unique) isomorphism* h, *such that the diagram*

$$\begin{array}{ccc} A & \xrightarrow{e_1} & C_1 \\ {\scriptstyle e_2}\downarrow & \swarrow h & \downarrow {\scriptstyle m_1} \\ C_2 & \xrightarrow{m_2} & B \end{array}$$

commutes,

(2) *if* $A \xrightarrow{f} B = A \xrightarrow{e} C \xrightarrow{m} B$ *is an* (E, M)-*factorization and* $C \xrightarrow{h} D$ *is an isomorphism, then* $A \xrightarrow{f} B = A \xrightarrow{h \circ e} D \xrightarrow{m \circ h^{-1}} B$ *is also an* (E, M)-*factorization*.

Sec. 14] Factorization structures for morphisms

Proof:

(1). By the diagonalization-property there exist morphisms h and k such that the diagrams

commute. Hence, also the diagrams

$$
\begin{array}{ccc}
A & \xrightarrow{e_1} & C_1 \\
e_1 \downarrow & {\scriptstyle koh} \nearrow & \downarrow m_1 \\
C_1 & \xrightarrow{m_1} & B
\end{array}
\quad \text{and} \quad
\begin{array}{ccc}
A & \xrightarrow{e_2} & C_2 \\
e_2 \downarrow & {\scriptstyle hok} \nearrow & \downarrow m_2 \\
C_2 & \xrightarrow{m_2} & B
\end{array}
$$

commute. By uniqueness we conclude $k \circ h = id_{C_1}$ and $h \circ k = id_{C_2}$. Therefore h is an isomorphism.

(2). This follows directly from the assumption that each of E and M is closed under composition with isomorphisms. □

14.5 LEMMA

Let $\mathbf{A}$ be (E, M)-structured and let $e \in E$ and $m \in M$. If the diagram

commutes, then e is an isomorphism and $f \in M$.

Proof: The diagram

commutes for $x = id$ and for $x = e \circ d$. Hence, by uniqueness, $e \circ d = id$. Consequently, e is an isomorphism. Thus $f = m \circ e$ belongs to M by our assumptions on M. □

14.6 PROPOSITION

If $\mathbf{A}$ is (E, M)-structured, then the following hold:

(1) $E \cap M = Iso(\mathbf{A})$,

(2) each of E and M is closed under composition,

(3) E and M determine each other via the diagonalization-property;[62] in particular, a morphism m belongs to M if and only if for each commutative square of the form $(*)$ (see Definition 14.1) with $e \in E$ there exists a diagonal.

Proof:
(1). For any $f \in E \cap M$ there exists a diagonal d that makes the diagram

commute. Hence $f \in \text{Iso}(\mathbf{A})$. Conversely, if $f \in \text{Iso}(\mathbf{A})$ and $f = m \circ e$ is an (E, M)-factorization of f, then for $d = f^{-1} \circ m$ the diagram

commutes. Hence, by Lemma 14.5, $f \in M$. By duality $f \in E$.

(2). Let $A \xrightarrow{m_1} B$ and $B \xrightarrow{m_2} C$ be morphisms in M. If $m_2 \circ m_1 = m \circ e$ is an (E, M)-factorization, then diagonals d_1 and d_2 can be constructed successively such that the diagrams

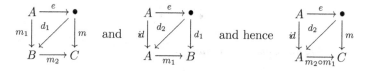

commute. By Lemma 14.5 this implies $m_2 \circ m_1 \in M$. By duality, E is also closed under composition.

(3). This follows for M immediately from Lemma 14.5, and for E via duality. □

14.7 PROPOSITION

If E and M are classes of morphisms in $\mathbf{A}$, then $\mathbf{A}$ is (E, M)-structured if and only if the following conditions are satisfied

(1) $\text{Iso}(\mathbf{A}) \subseteq E \cap M$,

(2) each of E and M is closed under composition,

---
[62] Here the diagonal needn't be required to be unique.

Sec. 14] Factorization structures for morphisms

(3) **A** *has the (E, M)-factorization property, unique in the sense that for any pair of (E, M)-factorizations $m_1 \circ e_1 = f = m_2 \circ e_2$ of a morphism f there exists a unique isomorphism h, such that the diagram*

commutes.

Proof: By Propositions 14.4 and 14.6, conditions (1)–(3) are necessary. To show that they suffice, the unique (E, M)-diagonalization property must be established. Let

$$\begin{array}{ccc} \bullet & \xrightarrow{e} & \bullet \\ f \downarrow & & \downarrow g \\ \bullet & \xrightarrow{m} & \bullet \end{array} \qquad (*)$$

be a commutative square with $e \in E$ and $m \in M$. Let $f = m' \circ e'$ and $g = m'' \circ e''$ be (E, M)-factorizations. Then there exists a unique isomorphism h that makes the diagram

$$\begin{array}{ccc} \bullet & \xrightarrow{e} & \bullet \\ e' \downarrow & & \downarrow e'' \\ \bullet & \xleftarrow{h} & \bullet \\ m' \downarrow & & \downarrow m'' \\ \bullet & \xrightarrow{m} & \bullet \end{array} \qquad (**)$$

commute. Hence $d = m' \circ h \circ e''$ is a diagonal for $(*)$. To show uniqueness, let $\tilde{d}$ be a diagonal for $(*)$. If $\tilde{d} = \tilde{m} \circ \tilde{e}$ is an (E, M)-factorization, then there exist isomorphisms h' and h'' such that the diagrams

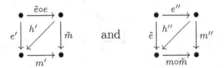

commute. Hence in $(**)$ h can be replaced by $h' \circ h''$, which implies that $h = h' \circ h''$. Consequently, $\tilde{d} = \tilde{m} \circ \tilde{e} = m' \circ h' \circ h'' \circ e'' = m' \circ h \circ e'' = d$. □

14.8 REMARK

If condition (3) in the above proposition would be weakened by just requiring the existence but not the uniqueness of h, then (1)–(3) would not imply that (E, M)-factorizations are unique. To see this, let **A** be a category with three objects A, B, and C, and four non-identity morphisms, three of which are depicted in the commutative diagram

$$\begin{array}{ccc} A & \xrightarrow{e} & B \\ {\scriptstyle e}\downarrow & {\scriptstyle f}\searrow & \downarrow{\scriptstyle m} \\ B & \xrightarrow{m} & C \end{array}$$

and for which the fourth is $B \xrightarrow{h} B$ with $h \circ h = h$. If $E = \mathrm{Iso}(\mathbf{A}) \cup \{e\}$ and $M = \mathrm{Iso}(\mathbf{A}) \cup \{m\}$, then both h and id_B serve as diagonals of the above square. See also Exercise 14B.

14.9 PROPOSITION

Let $\mathbf{A}$ be (E, M)-structured and let $f \circ g \in M$.

(1) If $f \in M$, then $g \in M$.

(2) If f is a monomorphism, then $g \in M$.

(3) If g is a retraction, then $f \in M$.

Proof:

(1). Let $g = m \circ e$ be an (E, M)-factorization. Then $(f \circ m) \circ e$ and $(f \circ g) \circ id$ are (E, M)-factorizations of $f \circ g$. By Proposition 14.4 there is an isomorphism h with $h \circ e = id$. Consequently, e is an isomorphism, and thus $g = m \circ e$ belongs to M.

(2). Let $g = m \circ e$ be an (E, M)-factorization. Then there exists a diagonal d that makes the diagram

commute. Since f is a monomorphism, the diagram

commutes as well. Hence Lemma 14.5 implies that $g \in M$.

(3). Let $f = m \circ e$ be an (E, M)-factorization, and let s be a morphism with $id = g \circ s$. Then there exists a diagonal d that makes the diagram

commute. Hence

commutes, so that, by Lemma 14.5, $f \in M$. □

RELATIONSHIP TO SPECIAL MORPHISM CLASSES

14.10 PROPOSITION

If **A** is (E, M)-structured, then the following hold:

(1) $E \subseteq Epi(\mathbf{A})$ implies that $ExtrMono(\mathbf{A}) \subseteq M$.

If, moreover, **A** has (Epi,Mono)-factorizations, then

(2) $Epi(\mathbf{A}) \subseteq E$ implies that $M \subseteq ExtrMono(\mathbf{A})$.

(3) $Epi(\mathbf{A}) = E$ implies that $M = ExtrMono(\mathbf{A})$.

Proof:
(1). If $f = m \circ e$ is an (E, M)-factorization of an extremal monomorphism, then $e \in Epi(\mathbf{A})$ implies that $e \in Iso(\mathbf{A})$. Hence $f \in M$.

(2). We will show that whenever $m \in M$ is factored as $m = f \circ e$, where e is an epimorphism, then e is an isomorphism. This will imply that: (a) m is a monomorphism (consider an (Epi, Mono)-factorization of m), and (b) m is extremal. It is clearly sufficient to show that e is a section, and this follows from the existence of a diagonal d for the diagram

(3) follows from (1) and (2). □

14.11 PROPOSITION

If **A** is (E, M)-structured and has products of pairs, then the following conditions are equivalent:

(1) $E \subseteq Epi(\mathbf{A})$,

(2) $ExtrMono(\mathbf{A}) \subseteq M$,

(3) $Sect(\mathbf{A}) \subseteq M$,

(4) for each object **A**, the diagonal morphism $\Delta_A = \langle id_A, id_A \rangle : A \to A \times A$ belongs to M,

(5) $f \circ g \in M$ implies that $g \in M$,

(6) $f \circ e \in M$ and $e \in E$ imply that $e \in \text{Iso}(\mathbf{A})$,

(7) $M = \{f \in \text{Mor}(\mathbf{A}) \mid f = g \circ e$ and $e \in E$ imply that $e \in \text{Iso}(\mathbf{A})\}$.

Proof: The implication (1) $\Rightarrow$ (2) follows from Proposition 14.10(1). The implications (2) $\Rightarrow$ (3) $\Rightarrow$ (4), are obvious.

(4) $\Rightarrow$ (1). Consider $A \xrightarrow{e} B$ in E and $B \underset{s}{\overset{r}{\rightrightarrows}} C$ with $r \circ e = s \circ e \ (= h)$. Then there exists a diagonal morphism d that makes the diagram

commute. Hence $r = d = s$, so that e is an epimorphism.

Therefore conditions (1)–(4) are equivalent.

(1) $\Rightarrow$ (6). If $e \in E$ and $f \circ e \in M$, then there exists a diagonal morphism d that makes the diagram

$$\begin{array}{ccc} \bullet & \xrightarrow{e} & \bullet \\ {\scriptstyle id}\downarrow & \overset{d}{\nearrow} & \downarrow{\scriptstyle f} \\ \bullet & \xrightarrow[f \circ e]{} & \bullet \end{array}$$

commute. Hence e is a section and an epimorphism, i.e., an isomorphism.

(6) $\Rightarrow$ (7) is obvious.

(7) $\Rightarrow$ (5). If $f \circ g \in M$ and $g = m \circ e$ is an (E, M)-factorization, then e is an isomorphism. Hence $g \in M$.

(5) $\Rightarrow$ (3) follows from $\text{Iso}(\mathbf{A}) \subseteq M$. □

14.12 COROLLARY

If $\mathbf{A}$ is (E, Mono)-structured and has products of pairs, then $E = \text{ExtrEpi}(\mathbf{A})$.

Proof: Immediate from the Proposition 14.11 and the dual of Proposition 14.10(3). □

14.13 REMARKS

(1) If $\mathbf{A}$ does not have products of pairs, the conditions of the above proposition need no longer be equivalent. Consider, e.g., a category $\mathbf{A}$ with $Ob(\mathbf{A}) = \mathbb{Z}$, and

$$\hom_{\mathbf{A}}(n,m) \quad \text{having exactly} \quad \begin{Bmatrix} 0 \\ 2 \\ 1 \end{Bmatrix} \quad \text{elements if} \quad \begin{cases} m < n \\ m = n+1 \\ \text{otherwise} \end{cases}$$

If $E = \text{Mor}(\mathbf{A})$ and $M = \text{Iso}(\mathbf{A}) = \text{Mono}(\mathbf{A}) = \text{Epi}(\mathbf{A})$, then $\mathbf{A}$ is (E, M)-structured and conditions (2), (3), (5), (6), and (7) of Proposition 14.11 are satisfied, but (1) is not.

The fact that $\mathbf{A}$ is (M, E)-structured also shows that the results (2) and (3) of Proposition 14.10 need not hold for factorization structures for morphisms in arbitrary categories.

(2) As will be seen in Theorem 15.4, factorization structures $(E, \mathbf{M})$ for sources in a category $\mathbf{A}$ always satisfy $E \subseteq \text{Epi}(\mathbf{A})$.

14.14 PROPOSITION
If $\mathbf{A}$ has (RegEpi, Mono)-factorizations, then the following hold:

(1) $\mathbf{A}$ is (RegEpi, Mono)-structured,

(2) $RegEpi(\mathbf{A}) = ExtrEpi(\mathbf{A})$,

(3) the class of regular epimorphisms in $\mathbf{A}$ is closed under composition,

(4) if $f \circ g$ is a regular epimorphism in $\mathbf{A}$, then so is f.

Proof:
(1). Obviously, every category has the unique (RegEpi, Mono)-diagonalization property.

(2). This follows from the dual of Proposition 14.10(3).

(3). This follows from (1) and Proposition 14.6(2).

(4). This follows from (2). □

RELATIONSHIP TO LIMITS

14.15 PROPOSITION
If $\mathbf{A}$ is (E, M)-structured, then M is closed under the formation of products and pullbacks, and $M \cap \text{Mono}(\mathbf{A})$ is closed under the formation of intersections.[63]

Proof:
(1). Let $\prod A_i \xrightarrow{\Pi m_i} \prod B_i$ be a product of morphisms m_i in M, and let
$\prod A_i \xrightarrow{\Pi m_i} \prod B_i = \prod A_i \xrightarrow{e} C \xrightarrow{m} \prod B_i$ be an (E, M)-factorization. Then for each index j there exists a diagonal morphism d_j that makes the following diagram commute (where π_j and ρ_j are the projection morphisms):

$$\begin{array}{ccc} \prod A_i & \xrightarrow{e} & C \\ {\scriptstyle \pi_j}\downarrow & {\scriptstyle d_j}\nearrow & \downarrow{\scriptstyle \rho_j \circ m} \\ A_j & \xrightarrow{m_j} & B_j \end{array}$$

---

[63] In fact, M is closed under the formation of multiple pullbacks (cf. Exercise 11L).

This implies that $\langle d_i \rangle \circ e = id_{\Pi A_i}$ and $\Pi m_i \circ \langle d_i \rangle = m$. Therefore Lemma 14.5 implies that $\Pi m_i \in M$.

(2). Let the diagram

$$\begin{array}{ccc} A & \xrightarrow{\overline{m}} & B \\ \overline{f} \downarrow & & \downarrow f \\ D & \xrightarrow{m} & C \end{array}$$

be a pullback with $m \in M$, and let $A \xrightarrow{\overline{m}} B = A \xrightarrow{e} A' \xrightarrow{\tilde{m}} B$ be an (E, M)-factorization. Then there exists a diagonal morphism d that makes the diagram

$$\begin{array}{ccc} A & \xrightarrow{e} & A' \\ \overline{f} \downarrow & \swarrow d & \downarrow f \circ \tilde{m} \\ D & \xrightarrow{m} & C \end{array}$$

commute. By the pullback-property there exists a morphism $A' \xrightarrow{g} A$ with $\tilde{m} = \overline{m} \circ g$ and $d = \overline{f} \circ g$. This implies that $\overline{m} \circ (g \circ e) = \overline{m}$ and $\overline{f} \circ (g \circ e) = \overline{f}$ and hence that $g \circ e = id_A$. By Lemma 14.5 this implies that $\overline{m} \in M$.

(3). Let $A \xrightarrow{f} B = A \xrightarrow{f_i} A_i \xrightarrow{m_i} B$ be an intersection with each m_i being a monomorphism in M, and let $A \xrightarrow{f} B = A \xrightarrow{e} C \xrightarrow{m} B$ be an (E, M)-factorization. Then for each index i there exists a diagonal morphism d_i that makes the diagram

$$\begin{array}{ccc} A & \xrightarrow{e} & C \\ f_i \downarrow & \swarrow d_i & \downarrow m \\ A_i & \xrightarrow{m_i} & B \end{array}$$

commute. Thus m factors through each m_i. Hence there exists a morphism $C \xrightarrow{d} A$ with $m = f \circ d$. Consequently, $f \circ d \circ e = f$, which implies that $d \circ e = id_A$. So, by Lemma 14.5, $f \in M$. □

We conclude this section with several results showing that the existence of suitable limits or colimits in a category guarantees the existence of distinguished factorization structures for morphisms.

14.16 FACTORIZATION LEMMA

Let **A** have intersections and equalizers, let $C \xrightarrow{f} D$ be an **A**-morphism, and let $M \subseteq$ Mono(**A**) satisfy the following conditions:

(a) intersections of families of M-subobjects of D belong to M,

(b) if $f = \hat{m} \circ g \circ h$ with $\hat{m} \in M$ and $g \in$ RegMono(**A**), then $\hat{m} \circ g \in M$.

Then there exist $m \in M$ and $e \in$ Epi(**A**), such that

Sec. 14] Factorization structures for morphisms 249

(1) $f = m \circ e$,

(2) if $f = \overline{m} \circ g$ with $\overline{m} \in M$, then there exists a diagonal d that makes the diagram

commute,

(3) if $e = \overline{m} \circ g$, where $m \circ \overline{m} \in M$, then $\overline{m} \in \mathrm{Iso}(\mathbf{A})$.

Proof: Let $\mathcal{S}$ be the sink consisting of all M-subobjects $D_i \xrightarrow{m_i} D$ of D through which f factors (i.e., $f = m_i \circ g_i$ for some g_i), and let (B, m) be the intersection of $\mathcal{S}$. Then $m \in M$ and there exists a morphism e with $f = m \circ e$. By construction, (2) is immediate. To establish (3) consider a factorization $e = \overline{m} \circ g$ of e with $m \circ \overline{m} \in M$. Then, by (2), there exists a diagonal d that makes the diagram

commute. Since m is a monomorphism, this implies that $\overline{m} \circ d = \mathrm{id}$. Hence $\overline{m}$ is a retraction and, being the first factor of the monomorphism $m \circ \overline{m}$, a monomorphism; thus an isomorphism. Hence (3) holds. It remains to be shown that e is an epimorphism. Consider a pair (r, s) of morphisms with $r \circ e = s \circ e$, and let g be an equalizer of (r, s). Then there exists a morphism h with $e = g \circ h$. Hence $f = m \circ g \circ h$. By (b) $m \circ g \in M$, so that by (3) g is an isomorphism. Thus $r = s$. □

14.17 THEOREM
If $\mathbf{A}$ has finite limits and intersections, then $\mathbf{A}$ is (ExtrEpi, Mono)-structured.

Proof: Application of Lemma 14.16 to $M = \mathrm{Mono}(\mathbf{A})$ yields that $\mathbf{A}$ is (ExtrEpi, Mono)-factorizable. To show that $\mathbf{A}$ has the (automatically unique) (ExtrEpi, Mono)-diagonalization property, consider a commutative diagram

with e an extremal epimorphism and m a monomorphism. Let be a pullback. Then there exists a unique morphism h such that the diagram

commutes. Since e is an extremal epimorphism and $\hat{m}$ (being a pullback of a monomorphism) is a monomorphism, $\hat{m}$ must be an isomorphism. Hence $d = \hat{g} \circ \hat{m}^{-1}$ is the desired diagonal. □

14.18 PROPOSITION

*If **A** has the (Epi, ExtrMono)-diagonalization property, then the class of extremal monomorphisms in **A** is closed under composition and intersections.*

Proof: Since monomorphisms are closed under composition and intersections, only the stability of the extremal-property needs to be verified. If m and $\overline{m}$ are extremal monomorphisms and $m \circ \overline{m} = f \circ e$ for some epimorphism e, then by assumption there exists a diagonal d that makes the diagram

commute. Hence $\overline{m} = d \circ e$ implies that e is an isomorphism.

Now let $\mathcal{A} = (A_i \xrightarrow{m_i} A)_I$ be a family of extremal subobjects of A. If $D \xrightarrow{d} A = D \xrightarrow{d_i} A_i \xrightarrow{m_i} A$ is an intersection of $\mathcal{A}$, and if $d = f \circ e$ for some epimorphism e, then for each $i \in I$ there exists a diagonal k_i that makes the diagram

commute. Since the intersection (being a limit) is an extremal mono-source, the epimorphism e must be an isomorphism. □

14.19 THEOREM

*If **A** has equalizers and intersections, then **A** is (Epi, ExtrMono)-structured.*

Proof: First we show that **A** has the (Epi, ExtrMono)-diagonalization property. Let be a commutative diagram, where c is an epimorphism and f is an extremal monomorphism. Let M be the class of all subobjects (D_m, m) of D with the property that there exist (necessarily unique) morphisms f_m and h_m that make the diagram

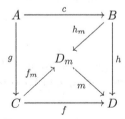

commute. Then M satisfies the conditions (a) and (b) of Lemma 14.16, so that there exists an (Epi, M)-factorization $f = m \circ e$. Since f is an extremal monomorphism, this implies that e ($= f_m$) is an isomorphism. Thus $d = e^{-1} \circ h_m$ is the desired diagonal. Consequently, **A** is (Epi, ExtrMono)-diagonalizable. This fact, together with Proposition 14.18, implies that the class $M = \mathrm{ExtrMono}(\mathbf{A})$ satisfies conditions (a) and (b) of Lemma 14.16 for every **A**-morphism f. Hence **A** is (Epi, M)-factorizable. □

14.20 COROLLARY

In a category with equalizers and intersections the class of extremal monomorphisms is the smallest class of morphisms that contains all regular monomorphisms and is closed under composition and intersections.

Proof: By the above results (14.18 and 14.19) and Corollary 7.63, ExtrMono(**A**) contains all regular monomorphisms and is closed under composition and intersections. If $\hat{M}$ is a class of morphisms with these properties, then $M = \hat{M} \cap Mono(\mathbf{A})$ satisfies conditions (a) and (b) of Lemma 14.16 for every **A**-morphism f, which implies that **A** is (Epi, M)-factorizable. If $f = m \circ e$ is an (Epi, M)-factorization of an extremal monomorphism f, then e is an isomorphism, and so belongs to M. Thus $f = m \circ e \in M$.□

14.21 COROLLARY

Every strongly complete category is (ExtrEpi, Mono)-structured and (Epi, ExtrMono)-structured. □

14.22 PROPOSITION

A category with pullbacks and coequalizers is (RegEpi, Mono)-structured if and only if regular epimorphisms are closed under composition.

Proof: The composition closure is a necessary condition, by Proposition 14.6(2). To show that it is sufficient, observe that by Proposition 14.14 it need only be shown that every morphism f has a (RegEpi, Mono)-factorization. Let (p, q) be a congruence-relation of f, and let c be a coequalizer of (p, q). Then there exists a unique morphism m with $f = m \circ c$. To show that m is a monomorphism, let (r, s) be a pair of morphisms with $m \circ r = m \circ s$ and let $\bar{c}$ be a coequalizer of (r, s). Then there exists a unique morphism g with $m = g \circ \bar{c}$. Since (p,q) is a congruence relation of $f = g \circ \bar{c} \circ c$ and since $c \circ p = c \circ q$, (p,q) is a congruence relation of c and of $\bar{c} \circ c$. Since c and $\bar{c} \circ c$ are both regular epimorphisms, each (by Proposition 11.22) is a coequalizer of (p,q). Hence, by essential uniqueness of coequalizers, $\bar{c}$ must be an isomorphism. So $r = s$. Thus m is a monomorphism. □

14.23 EXAMPLES

(1) Many familiar categories are (RegEpi, Mono)- and (Epi, RegMono)-structured, e.g., **Set**, **Vec**, **Grp**, **Pos**, **Top**, **Aut**, Σ-**Seq**, and **HComp**. So are any of the constructs **Alg**(Ω) and **Spa** (T).

(2) **Cat** is (ExtrEpi, Mono)-structured, but it has extremal epimorphisms that are not regular (cf. 7.76).

(3) **Sgr** is (Epi, ExtrMono)-structured, but it has extremal monomorphisms that are not regular (cf. 14I).

Suggestions for Further Reading

Kennison, J. F. Full reflective subcategories and generalized covering spaces. *Illinois J. Math.* **12** (1968): 353–365.

Ringel, C. M. Diagonalisierungspaare I, II. *Math. Z.* **117** (1970): 249-266, and **122** (1971): 10–32.

Freyd, P. J., and G. M. Kelly. Categories of continuous functors I. *J. Pure Appl. Algebra* **2** (1972): 169–191.

Bousfield, A. K. Constructions of factorization systems in categories. *J. Pure Appl. Algebra* **9** (1977): 207–220.

Manes, E. G. Compact Hausdorff objects. *Topol. Appl.* **4** (1979): 341–360.

Cassidy, C., M. Hébert, and G. M. Kelly. Reflective subcategories, localizations and factorization systems. *J. Austral. Math. Soc.* **38** (1985): 287–329, Corrigenda **41** (1986): 286.

EXERCISES

14A. Factorization Structures for Morphisms in Special Categories
Let $A \xrightarrow{f} B$ be a morphism in one of the categories **Set**, **Vec**, **Top**, **Pos**, **Sgr**, or **Rng**. Show that the familiar "image"-factorization $A \xrightarrow{f} B = A \xrightarrow{e} f[A] \xrightarrow{m} B$ is

(a) an (Epi, ExtrMono)-factorization in **Set**, **Vec**, **Top**, and **Pos**,

(b) an (ExtrEpi, Mono)-factorization in **Set**, **Vec**, **Sgr**, and **Rng**.

14B. Diagonals

1. Show that **Set** has

   (1) (Mono, Epi)-factorizations and -diagonalizations, but not unique ones,

   (2) (Section, Projection)-factorizations and -diagonalizations, but not unique ones.

2. Show that every category that has pullbacks has the unique (ExtrEpi, Mono)-diagonalization property.

14C. Strong Monomorphisms

A monomorphism m in **A** is called **strong** provided **A** has the unique (Epi, $\{m\}$)-diagonalization property. Show that

(a) Every strict monomorphism is strong, but not vice versa.

(b) Every strong monomorphism is extremal, but not vice versa.

(c) The class of strong monomorphisms is closed under composition, intersections, pullbacks, products, and left-cancellation.

(d) If **A** has pushouts, then in **A** StrongMono = ExtrMono.

(e) If **A** has equalizers and intersections, then in **A** StrongMono = ExtrMono [cf. Theorem 14.19].

(f) If **A** is (Epi, M)-structured for some class M of monomorphisms, then M = StrongMono = ExtrMono.

(g) If **A** is (Epi, M)-structured, then M need not consist of monomorphisms alone [cf. Remark 14.13(1)].

14D. (RegEpi, Mono)-Structured Categories

Let **A** have pullbacks and coequalizers. Show that

(a) **A** is (RegEpi, Mono)-structured if and only if RegEpi(**A**) = ExtrEpi(**A**). [Cf. Propositions 14.22 and 7.62(1).]

(b) If regular epimorphisms are stable in **A** (cf. Exercise 11J), then **A** is (RegEpi, Mono)-structured.

(c) If there exists a faithful functor from **A** into a (RegEpi, Mono)-structured category that preserves and reflects regular epimorphisms, then **A** is (RegEpi, Mono)-structured.

14E. Regular Categories

A category **A** is called **regular** provided that it satisfies the following conditions:

(1) **A** has finite limits,

(2) **A** is (RegEpi, Mono)-structured,

(3) In **A** regular epimorphisms are stable (cf. 11J).

Show that

(a) If **A** has coequalizers, then **A** is regular provided that it satisfies (1) and (3) above.

(b) If **A** is regular and **X** is small, then [**X**, **A**] is regular.

14F. Exact Categories

Pointed categories that are (NormalEpi, NormalMono)-structured are called **exact**. Show that

(a) A pointed category is exact if and only if it has (NormalEpi, NormalMono)-factorizations.

(b) If **A** is exact, then so is **A**^{op}.

(c) In an exact category every monomorphism is normal.

(d) An exact category has kernels and finite intersections.

(e) A nonempty exact category has a zero-object.

(f) An exact category is wellpowered if and only if it is co-wellpowered.

(g) Exact constructs are wellpowered and co-wellpowered.

(h) **Vec** and **Ab** are exact, but none of **Grp**, **Mon**, or **pSet** is exact.

14G. Closure Properties

Let **A** be (E, M)-structured. Show that

(a) M is closed under the formation of multiple pullbacks (cf. Proposition 14.15).

(b) M is closed under formation of retracts in the arrow category **A**^{2} (cf. Exercise 3K(b)).

* 14H. $(E, -)$-structured Categories

A is called $(E, -)$-**structured** provided that there exists some M such that **A** is (E, M)-structured. Show that a cocomplete category **A** is $(E, -)$-structured if and only if $E \subseteq Mor(\mathbf{A})$ satisfies the following conditions:

(1) $Iso(\mathbf{A}) \subseteq E$,

(2) E is closed under composition,

(3) E is closed under the formation of pushouts,

(4) E is closed under the formation of colimits,

(5) if $e = f \circ \bar{e}$ with e and $\bar{e}$ in E, then $f \in E$,

(6) (solution set condition) for each **A**-morphism f there exists a *set* of factorizations $f = g_i \circ e_i$, $i \in I$, with $e_i \in E$ such that for each factorization $f = g \circ e$ with $e \in E$ there is some $i \in I$ and some morphism h such that the following diagram commutes:

14I. Regular Monomorphisms in the Category of Semigroups

Let $C = \{0, a, b, c, d, e\}$ and consider the following multiplication table:

·	0	a	b	c	d	e
0	0	0	0	0	0	0
a	0	0	0	0	b	c
b	0	0	0	0	c	0
c	0	0	0	0	0	0
d	0	b	c	0	e	0
e	0	c	0	0	0	0

(a) Prove that C with the above multiplication $\cdot$ is a semigroup.

(b) $A = \{0, a, b\}$ and $B = \{0, a, b, c\}$ are subsemigroups of C. Let $f : A \hookrightarrow B$ and $g : B \hookrightarrow C$ be the inclusion homomorphisms. Prove that f and g are regular monomorphisms and that $g \circ f$ is an extremal monomorphism in **Sgr**.

(c) Prove that if $C \underset{k}{\overset{h}{\rightrightarrows}} D$ are morphisms in **Sgr** that coincide on A, then h and k coincide on B. [Hint: Use the equalities $b = da$ and $c = bd$.] Conclude that $g \circ f$ is *not* a regular monomorphism in **Sgr**.

(d) Let $\hat{B}$ be the free semigroup on three generators $\{\hat{a}, \hat{b}, \hat{c}\}$; let $\hat{A}$ be the subsemigroup of $\hat{B}$ generated by $\{\hat{a}, \hat{b}\hat{a}, \hat{a}\hat{c}\}$; let $\hat{C}$ be the quotient semigroup obtained by identifying the words $\hat{a}$ and $\hat{b}\hat{a}\hat{c}$; and let $i : \hat{A} \hookrightarrow \hat{B}$ and $p : \hat{B} \to \hat{C}$ be the inclusion map and natural map, respectively.

(1) Construct a semigroup $\hat{D}$ and homomorphisms $\hat{C} \underset{s}{\overset{r}{\rightrightarrows}} \hat{D}$ such that $p \circ i$ is an equalizer of r and s.

(2) Show that i is not a regular monomorphism in **Sgr**.

(e) Conclude that for the complete, wellpowered category **Sgr**, the following hold:

(1) The class of regular monomorphisms coincides with the class of strict monomorphisms.

(2) There exist extremal monomorphisms that are not regular.

(3) The class of regular monomorphisms is not closed under composition.

(4) **Sgr** is not (Epi, RegMono)-factorizable.

(5) The first factor of a regular monomorphism is not necessarily regular.

14J. Dominions

Let **A** be a wellpowered complete category, and let $X \xrightarrow{f} Y$ be an **A**-morphism. Prove that

(a) $X \xrightarrow{f} Y$ has a factorization $X \xrightarrow{g} D \xrightarrow{d} Y$, where d is a regular monomorphism that is characterized uniquely by any of the following equivalent conditions:

   (1) for all morphisms r and s, $r \circ f = s \circ f$ implies that $r \circ d = s \circ d$,

   (2) (D, d) is the smallest regular subobject of Y through which f can be factored,

   (3) d is the intersection of all regular subobjects of Y through which f can be factored.

   (D, d) is called a **dominion of** f.

(b) Any two dominions of f are isomorphic subobjects of Y.

(c) In **A** the extremal monomorphisms are precisely the regular monomorphisms if and only if the (Epi, ExtrMono)-factorization of any morphism is the same as its dominion factorization.

(d) f is an epimorphism if and only if (Y, id_Y) is a dominion of f.

(e) f is a regular monomorphism if and only if (X, f) is a dominion of f.

(f) If $f = r \circ s$, where s is an epimorphism, then the dominions of f and r coincide.

(g) Consider the pushout square

Then (D, d) is a dominion of f if and only if d is an equalizer of p and q.

15 Factorization structures for sources

Because of their generality, factorization structures for sources are frequently a more powerful categorical tool than factorizations for morphisms. However, in two respects they are more restrictive. Namely, if $(E, \mathbf{M})$ is a factorization structure for sources in **A**, then

(1) E must be contained in the class of epimorphisms of **A** (15.4), and

(2) certain colimits of diagrams in **A** involving morphisms in E must exist (15.14).

Such source factorization structures occur quite frequently (15.10 and 15.25), and for co-wellpowered categories with products there is no essential difference between these two approaches to factorization since for such categories **A** every factorization structure for morphisms, with $E \subseteq \text{Epi}(\mathbf{A})$, has a unique extension to a factorization structure for sources (15.21). In particular, every strongly complete, co-wellpowered category has both (Epi, ExtrMono-Source) and (ExtrEpi, Mono-Source) as factorization structures for sources (15.25).

15.1 DEFINITION
Let E be a class of morphisms and let **M** be a conglomerate of sources in a category **A**. $(E, \mathbf{M})$ is called a **factorization structure** on **A**, and **A** is called an $(E, \mathbf{M})$-**category** provided that

(1) each of E and **M** is closed under compositions with isomorphisms in the following sense:

 (1a) if $e \in E$ and $h \in \text{Iso}(\mathbf{A})$ and $h \circ e$ exists, then $h \circ e \in E$,

 (1b) if $\mathcal{S} \in \mathbf{M}$ and $h \in \text{Iso}(\mathbf{A})$ and $\mathcal{S} \circ h$ exists, then $\mathcal{S} \circ h \in \mathbf{M}$,

(2) **A** has $(E, \mathbf{M})$-**factorizations (of sources)**; i.e., each source $\mathcal{S}$ in **A** has a factorization $\mathcal{S} = \mathcal{M} \circ e$ with $e \in E$ and $\mathcal{M} \in \mathbf{M}$, and

(3) **A** has the **unique $(E, \mathbf{M})$-diagonalization property**; i.e., whenever $A \xrightarrow{e} B$ and $A \xrightarrow{f} C$ are **A**-morphisms with $e \in E$, and $\mathcal{S} = (B \xrightarrow{g_i} D_i)_I$ and $\mathcal{M} = (C \xrightarrow{m_i} D_i)_I$ are sources in **A** with $\mathcal{M} \in \mathbf{M}$, such that $\mathcal{M} \circ f = \mathcal{S} \circ e$, then there exists a unique **diagonal**, i.e., a morphism $B \xrightarrow{d} C$ such that for each $i \in I$ the diagram

$$\begin{array}{ccc} A & \xrightarrow{e} & B \\ {\scriptstyle f}\downarrow & \swarrow{\scriptstyle d} & \downarrow{\scriptstyle g_i} \\ C & \xrightarrow{m_i} & D_i \end{array}$$

commutes.

Factorization of sources

15.2 REMARKS

(1) As opposed to the concept of factorization structures for morphisms the concept of factorization structures (for sources) is not self-dual. The dual concept, that of **factorization structures for sinks**, will not be explicitly formulated here.

(2) Another distinction from factorization structures for morphisms is the fact that the uniqueness requirement for diagonals in the definition of factorization structures (for sources) is redundant. Cf. Exercise 14B on the one hand, and the proof of Theorem 15.4 on the other.

(3) If $(E, \mathbf{M})$ is a factorization structure on $\mathbf{A}$ and M is the class of those $\mathbf{A}$-morphisms that (considered as 1-sources) belong to $\mathbf{M}$, then (E, M) is a factorization structure for morphisms on $\mathbf{A}$. (Cf. footnote to Definition 14.1) Hence all the results of §14 apply.

(4) Factorization of sources applies to empty sources too. If $(E, \mathbf{M})$ is a factorization structure on $\mathbf{A}$, and $\mathbf{A_M}$ is the class of all $\mathbf{A}$-objects A such that the empty source $(A, \emptyset)$ with domain A belongs to $\mathbf{M}$, then conditions (2) and (3) in Definition 15.1 translate into:

(2^0) for every object A there exists a morphism $A \xrightarrow{e} B$ in E with $B \in \mathbf{A_M}$,

(3^0) for every morphism $A \xrightarrow{e} B$ in E and every morphism $A \xrightarrow{f} C$ with $C \in \mathbf{A_M}$ there exists a unique morphism $B \xrightarrow{d} C$ with $f = d \circ e$.

15.3 EXAMPLES

(1) Every category is an (Iso, Source)-category.[64] This factorization structure is called **trivial**. Also, every category has the (unique) (RegEpi, Mono-Source)-diagonalization property, although not all categories are (RegEpi, Mono-Source)-categories (cf. Proposition 15.13).

(2) **Set** is an (Epi, Mono-Source)-category. For **Set** this is the only nontrivial factorization structure. (Epi, Mono-Source)-factorizations of a source $(A \xrightarrow{f_i} A_i)_I$ in **Set** can be obtained via either of the following two constructions:

  (a) Define an equivalence relation on A by: "$a \sim b$ if and only if $f_i(a) = f_i(b)$ for each $i \in I$", and let $A \xrightarrow{e} A/\sim$ be the naturally associated surjection. Then for each $i \in I$ there exists a unique map $A/\sim \xrightarrow{m_i} A_i$ with $f_i = m_i \circ e$. Then

  $$A \xrightarrow{f_i} A_i = A \xrightarrow{e} A/\sim \xrightarrow{m_i} A_i$$

  is an (Epi, Mono-Source)-factorization of (f_i) in **Set**.

  (b) Consider the cartesian product $(\prod A_i, \pi_i)_I$ of the codomain $(A_i)_I$ — that is, the conglomerate[65] of all functions x with domain I such that $x(i) \in A_i$ for each $i \in I$. The function $A \xrightarrow{f} \prod A_i$ defined by $a \mapsto f(a)$, where $(f(a))(i) = f_i(a)$ for each $i \in I$, has a factorization $A \xrightarrow{f} \prod A_i = A \xrightarrow{e} B \xrightarrow{m} \prod A_i$, with e a surjective function and m an injective one. Since A is a set and $A \xrightarrow{e} B$ is surjective, B is codable by a set; i.e., there exists a set B' and a bijection $B' \xrightarrow{b} B$. Then for $B \xrightarrow{m_i} A_i$ defined by $m_i(b) = (m(b))(i)$, it follows that

  $$A \xrightarrow{f_i} A_i = A \xrightarrow{b^{-1} \circ e} B' \xrightarrow{m_i \circ b} A_i$$

  is an (Epi, Mono-Source)-factorization of (f_i) in **Set**.

  For various constructs, one of the above constructions can be used to obtain (Epi, Mono-Source)-factorizations.

(3) Many algebraic constructs such as **Vec**, **Grp**, **Mon**, **Rng**, and **Alg**(Ω) are (RegEpi, Mono-Source)-categories.

(4) **Cat** is an (ExtrEpi, Mono-Source)-category.

---

[64] Source (resp. Source(**A**)) is the conglomerate of all sources (in the category **A**). Similarly, Mono-Source is the conglomerate of all mono-sources in the given category, etc. See the Table of Symbols.

[65] One might be tempted to describe this construction more categorically by forming the quasicategory **Class** of all classes and all functions, and by letting $(\prod A_i, \pi_i)_I$ be the product of $(A_i)_I$ in that quasicategory. This, however, is not possible, since the conglomerate $(\prod A_i)_I$ frequently fails to be a class and, hence, to be an object of the quasicategory **Class**. In fact, the quasicategory **Class** behaves badly with respect to many familiar constructions (recall, e.g., that there is a *largest* class, $\mathcal{U}$). This problem could be resolved, e.g., by introducing in addition to "sets", "classes", and "conglomerates", one higher level of entities, say, "collections". Then the collection of all conglomerates and all functions between them would form a rather well-behaved "quasi-quasicategory". Since there are only a few occasions in this text where the use of something like this would be advantageous, we have refrained from complicating the foundations by introducing it.

(5) **Aut** is an (RegEpi, Mono-Source)-category.

(6) **Top** has a proper class (even an illegitimate conglomerate) of factorization structures (cf. Exercise 15L). In particular, **Top** is an (Epi, ExtrMono-Source)-category, an (ExtrEpi, Mono-Source)-category, and a (Bimorphism, Initial-Source)-category.

15.4 THEOREM
*If **A** is an $(E, \mathbf{M})$-category, then $E \subseteq Epi(\mathbf{A})$.*

Proof: Consider $e \in E$ and a pair (r, s) of morphisms with $r \circ e = s \circ e$. The source $(h_f)_{f \in \mathrm{Mor}(\mathbf{A})}$, defined by $h_f = r \circ e$, has an $(E, \mathbf{M})$-factorization $(h_f) = (m_f) \circ \overline{e}$. The source $(g_f)_{f \in \mathrm{Mor}(\mathbf{A})}$, defined by

$$g_f = \begin{cases} r, & \text{if } m_f \circ f = s \\ s, & \text{otherwise,} \end{cases}$$

satisfies the equation $(m_f) \circ \overline{e} = (g_f) \circ e$. Hence there exists a diagonal d that makes the diagram

commute for each $f \in \mathrm{Mor}(\mathbf{A})$. In particular,

$$m_d \circ d = g_d = \begin{cases} r, & \text{if } m_d \circ d = s \\ s, & \text{otherwise.} \end{cases}$$

This is possible only for $r = s$. Thus e is a epimorphism. □

15.5 PROPOSITION
*If **A** is an $(E, \mathbf{M})$-category, then the following hold:*

(1) $(E, \mathbf{M})$-factorizations are essentially unique,

(2) $E \subseteq Epi(\mathbf{A})$ and $ExtrMono\text{-}Source(\mathbf{A}) \subseteq \mathbf{M}$,

(3) $E \cap \mathbf{M} = Iso(\mathbf{A})$,

(4) each of E and $\mathbf{M}$ is closed under composition,

(5) if $f \circ g \in E$ and $g \in Epi(\mathbf{A})$, then $f \in E$,

(6) if $f \circ g \in E$ and $f \in Sect(\mathbf{A})$, then $g \in E$,

(7) if $(S_i) \circ S \in \mathbf{M}$, then $S \in \mathbf{M}$,

(8) if a subsource of S belongs to $\mathbf{M}$, then S belongs to $\mathbf{M}$,

(9) E and $\mathbf{M}$ determine each other via the diagonalization-property; moreover,

(a) a source belongs to **M** *if and only if every E-morphism through which it factors is an isomorphism,*

(b) if **M** *consists of mono-sources only, then a morphism f belongs to E if and only if $f = m \circ g$ with $m \in$ **M** *implies that $m \in Iso(\mathbf{A})$.*

Proof: (8) follows from (9). All other conditions are proved as in §14 by use of Theorem 15.4. □

RELATIONSHIP TO SPECIAL MORPHISMS AND SPECIAL SOURCES

15.6 PROPOSITION
If **A** *is a (RegEpi,* **M***)-category, then* **M** *contains all mono-sources of* **A**.

Proof: Let $\mathcal{S}$ be a mono-source and $\mathcal{S} = \mathcal{M} \circ e$ be a (RegEpi, **M**)-factorization of $\mathcal{S}$. Let e be a coequalizer of r and s. Then $\mathcal{S} \circ r = \mathcal{M} \circ e \circ r = \mathcal{M} \circ e \circ s = \mathcal{S} \circ s$. Since $\mathcal{S}$ is a mono-source, this implies that $r = s$. Hence e is an isomorphism, so that $\mathcal{S} \in \mathbf{M}$. □

15.7 PROPOSITION
For $(E, \mathbf{M})$-categories **A**, *the following are equivalent:*

(a) **M** $\subseteq$ *Mono-Source(***A***),*

(b) **A** *has coequalizers and RegEpi(***A***) $\subseteq E$.*

Proof: (a) ⇒ (b). Let $A \underset{s}{\overset{r}{\rightrightarrows}} B$ be a pair of morphisms. Consider the source $\mathcal{S} = (B \xrightarrow{f_i} B_i)_I$ that consists of all morphisms f_i with $f_i \circ r = f_i \circ s$. If $\mathcal{S} = \mathcal{M} \circ e$ is an $(E, \mathbf{M})$-factorization, then, since $\mathcal{M}$ is a mono-source, e is a coequalizer of (r, s). Hence **A** has coequalizers. If c is a coequalizer of some pair (r, s), and $c = m \circ e$ is an $(E, \mathbf{M})$-factorization, then e is a coequalizer of (r, s). Hence $m \in Iso(\mathbf{A})$, so that $c \in E$.

(b) ⇒ (a). Let $\mathcal{M}$ belong to **M** and let (r, s) be a pair of morphisms with $\mathcal{M} \circ r = \mathcal{M} \circ s$. Then $\mathcal{M}$ factors through the coequalizer c of (r, s). Hence $c \in E$ implies, by Proposition 15.5(9)(a), that c is an isomorphism. So $r = s$. □

15.8 PROPOSITION
For $(E, \mathbf{M})$-categories **A** *the following hold:*

(1) if **M** = *Mono-Source(***A***), then $E = ExtrEpi(\mathbf{A})$,*

(2) if **M** = *ExtrMono-Source(***A***), then $E = Epi(\mathbf{A})$,*

(3) if $E = Epi(\mathbf{A})$, then the following conditions are equivalent:

    (a) **M** = *ExtrMono-Source(***A***),*

(b) **A** has coequalizers,

(4) if $E = ExtrEpi(\mathbf{A})$ or $E = RegEpi(\mathbf{A})$, then the following conditions are equivalent:

(a) $\mathbf{M} = Mono\text{-}Source(\mathbf{A})$,

(b) **A** has coequalizers.

Proof:
(1). Apply Theorem 15.4 and Proposition 15.5(9)(b).

(2). $E \subseteq Epi(\mathbf{A})$ by Theorem 15.4. If f is an epimorphism and $f = m \circ e$ is an (E, M)-factorization, then m is an extremal monomorphism and (as the second factor of an epimorphism) an epimorphism; hence an isomorphism. Thus $f \in E$.

(3a) $\Rightarrow$ (3b) and (4a) $\Rightarrow$ (4b) follow from Proposition 15.7.

(3b) $\Rightarrow$ (3a) follows from Propositions 15.5(9)(a) and 15.7.

(4b) $\Rightarrow$ (4a). By Proposition 15.7, $\mathbf{M} \subseteq Mono\text{-}Source(\mathbf{A})$. If $\mathcal{S}$ is a mono-source and $\mathcal{S} = \mathcal{M} \circ e$ is an $(E, \mathbf{M})$-factorization, then e is an extremal epimorphism and a monomorphism; hence an isomorphism. Thus $\mathcal{S} \in \mathbf{M}$. □

15.9 EXAMPLE
The category **A** described in Example 14.13(1) is simultaneously an (Epi, Source)-category, an (ExtrEpi, Source)-category and a (RegEpi, Source)-category. But in **A** not every source is a mono-source.

Also, every group, considered as a category, is an (Epi, Source)-category; but for a nonzero group the empty source is not a mono-source.

EXISTENCE OF FACTORIZATION STRUCTURES

15.10 THEOREM
Every category that has (Epi, Mono-Source)-factorizations is an (ExtrEpi, Mono-Source)-category.

Proof: Let **A** have (Epi, Mono-Source)-factorizations and let $\mathcal{S} = (A \xrightarrow{f_i} A_i)_I$ be a source in **A**. Consider the source $\mathcal{T} = (A \xrightarrow{e_j} B_j)_J$ that consists of all epimorphisms e_j for which there exists a (necessarily unique) mono-source $\mathcal{M}_j = (B_j \xrightarrow{m_{ij}} A_i)_I$ with $\mathcal{S} = \mathcal{M}_j \circ e_j$. Let $\mathcal{T} = \mathcal{N} \circ e = (A \xrightarrow{e} B \xrightarrow{n_j} B_j)_J$ be an (Epi, Mono-Source)-factorization. Then the sources $(\mathcal{M}_j) \circ n_j$ do not depend upon j and thus each can be denoted by $\mathcal{M} = (B \xrightarrow{m_i} A_i)_I$. Hence $\mathcal{S} = \mathcal{M} \circ e$ is an (Epi, Mono-Source)-factorization. To show that e is an extremal epimorphism, consider a factorization $e = m \circ g$ with monomorphic m. Let $g = n \circ \bar{e}$ be an (Epi, Mono)-factorization of g. Then $\mathcal{S} = (\mathcal{M} \circ m \circ n) \circ \bar{e}$ is an (Epi, Mono-Source)-factorization of $\mathcal{S}$. Hence there exists some j in J with $\mathcal{M} \circ m \circ n = \mathcal{M}_j$ and $\bar{e} = e_j$. For this j, the equations $m \circ n \circ n_j \circ e = m \circ n \circ e_j = m \circ n \circ \bar{e} = m \circ g = e$ hold. Since e is an epimorphism, we conclude that $m \circ n \circ n_j = id$. Hence m is a monomorphic retraction, i.e., an isomorphism. Thus **A** is (ExtrEpi, Mono-Source)-factorizable. To show that **A** has the (ExtrEpi, Mono-Source)-diagonalization property, consider morphisms e_j and sources $\mathcal{S}_j$, for $j = 1, 2$, such that e_1 is an extremal epimorphism, $\mathcal{S}_2$ is a mono-source, and $\mathcal{S}_1 \circ e_1 = \mathcal{S}_2 \circ e_2$. If $(A \xrightarrow{e_j} B_j)_{j \in \{1,2\}} = (A \xrightarrow{e} B \xrightarrow{m_j} B_j)_{j \in \{1,2\}}$ is an (Epi, Mono-Source)-factorization, then $\mathcal{S}_1 \circ m_1 \circ e = \mathcal{S}_1 \circ e_1 = \mathcal{S}_2 \circ e_2 = \mathcal{S}_2 \circ m_2 \circ e$. Hence $\mathcal{S}_1 \circ m_1 = \mathcal{S}_2 \circ m_2$.

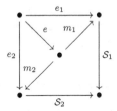

Let $C \underset{s}{\overset{r}{\rightrightarrows}} B$ be a pair of morphisms with $m_1 \circ r = m_1 \circ s$. Then

$$\mathcal{S}_2 \circ m_2 \circ r = \mathcal{S}_1 \circ m_1 \circ r = \mathcal{S}_1 \circ m_1 \circ s = \mathcal{S}_2 \circ m_2 \circ s.$$

Since $\mathcal{S}_2$ is a mono-source, this implies that $m_2 \circ r = m_2 \circ s$. Since $(B \xrightarrow{m_j} B_j)_{j \in \{1,2\}}$ is a mono-source, this implies that $r = s$. Thus m_1 is a monomorphism. Since $e_1 = m_1 \circ e$ is an extremal epimorphism, this implies that m_1 is an isomorphism. Consequently, $d = m_2 \circ m_1^{-1}$ is the desired diagonal. □

15.11 REMARK
As the above result shows, every (Epi, Mono-Source)-factorizable category is an $(E, \text{Mono-Source})$-category for a suitable class E. However, it need not be an (Epi, **M**)-category for any conglomerate **M**:

Every poset, considered as a category, is an (ExtrEpi, Mono-Source)-category. But a poset with a smallest element is an (Epi, **M**)-category for some **M** if and only if it is a complete lattice.

15.12 DEFINITION
A category is said to **have regular factorizations** provided that it is (RegEpi, Mono-Source)-factorizable.

15.13 PROPOSITION
If a category has regular factorizations, then it is a (RegEpi, Mono-Source)-category.

Proof: Every category has the (RegEpi, Mono-Source)-diagonalization property. □

15.14 THEOREM
*If E is a class of morphisms in **A**, then **A** is an $(E, \mathbf{M})$-category for some **M** if and only if the following conditions are satisfied:*

(1) $\mathrm{Iso}(\mathbf{A}) \subseteq E \subseteq \mathrm{Epi}(\mathbf{A})$,

(2) E is closed under composition,

(3) for every $A \xrightarrow{e} B$ in E and every morphism $A \xrightarrow{f} C$ there exists a pushout square

$$\begin{array}{ccc} A & \xrightarrow{e} & B \\ f\downarrow & & \downarrow \bar{f} \\ C & \xrightarrow{\bar{e}} & D \end{array}$$

for which $\bar{e} \in E$,

(4) for every source $(A \xrightarrow{e_i} A_i)_I$ that consists of E-morphisms, there exists a cointersection

$$A \xrightarrow{e} B = A \xrightarrow{e_i} A_i \xrightarrow{p_i} B$$

for which $e \in E$.

Proof: Necessity: (1) and (2) follow from Theorem 15.4 and Proposition 15.5. To see (3), let $A \xrightarrow{e} B$ and $A \xrightarrow{f} C$ be morphisms with $e \in E$. Consider the source $\mathcal{S} = (C \xrightarrow{f_i} C_i)_I$ consisting of those morphisms f_i for which there exists a (necessarily unique) morphism g_i with $f_i \circ f = g_i \circ e$. Let $\mathcal{S} = \mathcal{M} \circ \bar{e}$ be an $(E, \mathbf{M})$-factorization. Then there exists a diagonal that makes the diagram

$$\begin{array}{ccc} A & \xrightarrow{e} & B \\ \bar{e}\circ f \downarrow & \swarrow d & \downarrow g_i \\ \bullet & \xrightarrow{\mathcal{M}} & C_i \end{array}$$

commute. As can be seen easily, the diagram

$$\begin{array}{ccc} A & \xrightarrow{e} & B \\ f\downarrow & & \downarrow d \\ C & \xrightarrow{\bar{e}} & \bullet \end{array}$$

is a pushout.

To establish (4), let $(A \xrightarrow{e_i} A_i)_I$ be a source consisting of E-morphisms. Consider the source $\mathcal{S} = (A \xrightarrow{f_j} B_j)_J$ that consists of those morphisms f_j that have the property that for each $i \in I$ there exists a (necessarily unique) morphism f_{ij} with $f_j = f_{ij} \circ e_i$, and proceed as in the proof of part (3).

Sufficiency: Define $\mathbf{M}$ to be the conglomerate of all sources that do not factor through a non-isomorphic morphism in E. In order to show that $\mathbf{A}$ has $(E, \mathbf{M})$-factorizations, let $(A \xrightarrow{f_i} A_i)_I$ be a source in $\mathbf{A}$. Consider the source $\mathcal{S} = (A \xrightarrow{e_j} B_j)_J$ that consists of those morphisms $e_j \in E$ that have the property that for every $i \in I$ there exists a (necessarily unique) morphism f_{ij} with $f_i = f_{ij} \circ e_j$. Let

$$A \xrightarrow{e} B = A \xrightarrow{e_j} B_j \xrightarrow{p_j} B$$

be a cointersection of $\mathcal{S}$ such that $e \in E$. Then for each $i \in I$ there exists a morphism m_i that makes the diagram

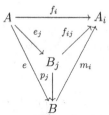

commute for each $j \in J$. To show that the source $\mathcal{M} = (B \xrightarrow{m_i} A_i)_I$ belongs to $\mathbf{M}$, let $\mathcal{M} = \mathcal{T} \circ \hat{e}$ be a factorization with $\hat{e} \in E$. Then there exists some $j \in J$ with $\hat{e} \circ e = e_j$. Hence, for this j the equality $e = p_j \circ e_j = p_j \circ \hat{e} \circ e$ implies that $p_j \circ \hat{e} = id$. Thus $\hat{e}$ is an epimorphic section; hence an isomorphism.

To show that $\mathbf{A}$ has the $(E, \mathbf{M})$-diagonalization property, let e and f be morphisms with $e \in E$, and let $\mathcal{M} = (m_i)_I$ and $\mathcal{S} = (f_i)_I$ be sources with $\mathcal{M} \in \mathbf{M}$ and $\mathcal{M} \circ f = \mathcal{S} \circ e$. Let

be a pushout square for which $\hat{e} \in E$. Then for each $i \in I$ there exists a morphism g_i that makes the diagram

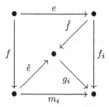

commute. Hence $\mathcal{M}$ factors through $\hat{e} \in E$, which implies that $\hat{e}$ is an isomorphism. Thus $\hat{e}^{-1} \circ \hat{f}$ is the desired diagonal. □

15.15 COROLLARY

Let $\mathbf{A}$ be a category with pushouts and cointersections. Then a class E of $\mathbf{A}$-morphisms is part of a factorization structure on $\mathbf{A}$ (i.e., $\mathbf{A}$ is an $(E, \mathbf{M})$-category for some $\mathbf{M}$) if and only if $Iso(\mathbf{A}) \subseteq E \subseteq Epi(\mathbf{A})$ and E is closed under composition, pushouts, and cointersections. □

15.16 COROLLARY

(1) **A** *is an (Epi,* **M***)-category for some* **M** *if and only if* **A** *has cointersections and has a pushout for every 2-source of the form* • $\xleftarrow{f}$ • $\xrightarrow{e}$ • *with epimorphic e.*

(2) **A** *is an (Epi, ExtrMono-Source)-category if and only if* **A** *has cointersections, pushouts for 2-sources of the form* • $\xleftarrow{f}$ • $\xrightarrow{e}$ • *with epimorphic e, and has coequalizers.* □

15.17 COROLLARY

Every strongly cocomplete category is an (Epi, ExtrMono-Source)-category. □

15.18 EXAMPLE

The partially ordered class of all ordinals, considered as a category, is a cocomplete (ExtrEpi, Mono-Source)-category, but is not (Epi, ExtrMono-Source)-factorizable. [Consider empty sources.] See also Exercise 15D.

Further results concerning the existence of factorization structures will follow after some necessary preparations, cf., e.g., Theorem 15.25.

EXTENSIONS OF FACTORIZATION STRUCTURES

Factorization structures may be considered for certain specified sources only; e.g., for

(1) empty sources (= objects), see Exercise 15G,

(2) 1-sources (= morphisms), see §14,

(3) 2-sources, see Exercise 15I,

(4) set-indexed sources (= small sources), see Exercise 15J.

Here we are concerned with the question: Under which conditions is it possible to extend factorization structures (E, M) for morphisms to factorization structures $(E, \mathbf{M})$ for sources, or at least for small sources?

15.19 PROPOSITION

(1) If **A** *has products, then every factorization structure (E, M) for morphisms on* **A** *can be uniquely extended to a factorization structure $(E, \mathbf{M})$ for small sources.*

(2) Conversely, if **A** *has an initial object and each factorization structure (E, M) for morphisms on* **A** *can be extended to a factorization structure $(E, \mathbf{M})$ for small sources, then* **A** *has products.*

Proof:

(1). Let (E, M) be a factorization structure for morphisms. If **M** is the class of all small sources of the form $\mathcal{P} \circ m$, where $\mathcal{P}$ is a product and $m \in M$, then $(E, \mathbf{M})$

is a factorization structure for small sources. Indeed, if $\mathcal{S} = (A \xrightarrow{f_i} A_i)_I$ is a set-indexed source, let $\langle f_i \rangle = A \xrightarrow{e} B \xrightarrow{m} \prod A_i$ be an (E, M) factorization. Then $\mathcal{S} = (B \xrightarrow{\pi_i \circ m} A_i)_I \circ e$ is an $(E, \mathbf{M})$-factorization.

Uniqueness of the extension follows from the observation that $\mathbf{M}$ is determined by E as in Proposition 15.5(9).

(2). Let (Mor($\mathbf{A}$), $\mathbf{M}$) be an extension of the factorization structure (Mor($\mathbf{A}$), Iso($\mathbf{A}$)) for morphisms to a factorization structure for small sources. Then $\mathbf{M}$ consists precisely of the set-indexed products in $\mathbf{A}$. In fact, if $(A_i)_I$ is a set-indexed family of objects, Q is an initial object, $\mathcal{S} = (Q \to A_i)_I$ is the associated source, and $\mathcal{S} = \mathcal{M} \circ e$ is a (Mor($\mathbf{A}$), $\mathbf{M}$)-factorization, then $\mathcal{M}$ is a product of $(A_i)_I$. □

15.20 PROPOSITION

If $(E, \mathbf{M})$ is a factorization structure for small sources on $\mathbf{A}$ and $\mathbf{A}$ is co-wellpowered, then the following conditions are equivalent:

(1) $(E, \mathbf{M})$ can be uniquely extended to a factorization structure $(E, \mathbf{N})$ on $\mathbf{A}$,

(2) $E \subseteq \text{Epi}(\mathbf{A})$,

(3) $\text{Sect}(\mathbf{A}) \subseteq \mathbf{M}$,

(4) for each object A the 2-source (id_A, id_A) belongs to $\mathbf{M}$,

(5) whenever a subsource of a small source $\mathcal{S}$ belongs to $\mathbf{M}$, then so does $\mathcal{S}$.

Proof: (1) $\Leftrightarrow$ (2). By Theorem 15.4, (1) implies (2). For the converse it suffices to verify that E satisfies the conditions (1)–(4) of Theorem 15.14. That 15.14(1) and 15.14(2) hold follows from Proposition 14.6. The validity of 15.14(3) can be established as in the corresponding part of the proof of Theorem 15.14 by replacing the source $\mathcal{S} = (C \xrightarrow{f_i} C_i)_I$ by a small subsource that contains a representative for each $f_i \in E$. Likewise the validity of 15.14(4) can be established as in the corresponding part of the proof of Theorem 15.14 by replacing the source $\mathcal{S} = (A \xrightarrow{f_j} B_j)_J$ by a small subsource that contains a representative for each $f_j \in E$.

The equivalence of the conditions (2)–(5) can be established as in Proposition 14.11 by using the fact that the assumption made there that products of pairs exist is not needed here, since we have $(E, \mathbf{M})$-factorizations for 2-sources. Details are left to the reader. □

15.21 COROLLARY

In a co-wellpowered category $\mathbf{A}$ with products, every factorization structure (E, M) for morphisms with $E \subseteq \text{Epi}(\mathbf{A})$ can be uniquely extended to a factorization structure $(E, \mathbf{M})$ for sources. □

15.22 EXAMPLE

The factorization structure (dense maps, closed embeddings) for morphisms

- cannot be extended to a factorization structure (dense maps, $\mathbf{M}$) on **Top**, but

• can be uniquely extended to a factorization structure (dense maps, **M**) on **Haus**.

The latter is one of the most important factorization structures on **Haus**.

15.23 REMARK
Proposition 15.19 shows that products are a suitable tool for extending factorization structures for morphisms to those for small sources. Proposition 15.20 shows that co-wellpoweredness is a suitable tool for extending factorization structures for small sources to those for sources. The next proposition shows that factorization structures themselves can be used to extend factorization structures of morphisms to those for sources.

15.24 PROPOSITION
In an $(E, \mathbf{M})$-category $\mathbf{A}$ every factorization structure (C, N) for morphisms with $C \subseteq E$ can be uniquely extended to a factorization structure $(C, \mathbf{N})$ for sources.

Proof: Let $\mathbf{N}$ be the conglomerate of all sources of the form $\mathcal{M} \circ n$ with $\mathcal{M} \in \mathbf{M}$ and $n \in N$. If $\mathcal{S}$ is a source with $(E, \mathbf{M})$-factorization $\mathcal{S} = \mathcal{M} \circ e$ and $e = n \circ c$ is a (C, N)-factorization, then $\mathcal{S} = (\mathcal{M} \circ n) \circ c$ is a $(C, \mathbf{N})$-factorization. If f and c are morphisms with $c \in C$, and $\mathcal{N} = \mathcal{M} \circ n$ and $\mathcal{S}$ are sources with $n \in N$ and $\mathcal{M} \in \mathbf{M}$ (i.e., $\mathcal{N} \in \mathbf{N}$) and if $\mathcal{N} \circ f = \mathcal{S} \circ c$, then diagonals d_1 and d_2 can be successively constructed such that the diagrams

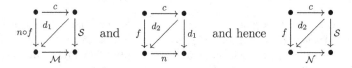

commute. Uniqueness follows from $c \in C \subseteq E \subseteq \mathrm{Epi}(\mathbf{A})$. □

FACTORIZATION STRUCTURES AND LIMITS

The fact that in an $(E, \mathbf{M})$-category every extremal mono-source belongs to $\mathbf{M}$ [cf. 15.5(2)] implies that all limit sources belong to $\mathbf{M}$ (cf. Proposition 11.6). Below, we further explore the relationship between factorization structures and limits.

15.25 THEOREM
Let $\mathbf{A}$ be a strongly complete, extremally co-wellpowered category. Then the following hold:

(1) $\mathbf{A}$ is an (ExtrEpi, Mono-Source)-category.

(2) If $\mathbf{A}$ is co-wellpowered, then $\mathbf{A}$ is an (Epi, ExtrMono-Source)-category.

(3) If in $\mathbf{A}$ regular epimorphisms are closed under composition, then $\mathbf{A}$ is a (RegEpi, Mono-Source)-category.

Proof:

(1). Let $\mathcal{S} = (A \xrightarrow{f_i} A_i)_I$ be a source. By Theorem 14.17, for each $i \in I$ there is an (ExtrEpi, Mono)-factorization $A \xrightarrow{f_i} A_i = A \xrightarrow{e_i} B_i \xrightarrow{m_i} A_i$. Since **A** is extremally co-wellpowered, there exists a subset J of I, and for each $i \in I$ there is a $j(i) \in J$ and an isomorphism h_i with $e_i = h_i \circ e_{j(i)}$ (where for $i \in J$ we can choose $j(i) = i$ and $h_i = id$). The family $(B_j)_{j \in J}$ has a product $(P \xrightarrow{p_j} B_j)_J$. Let

$$A \xrightarrow{\langle e_j \rangle} P = A \xrightarrow{e} B \xrightarrow{m} P$$

be an (Epi, Mono)-factorization. Then

$$A \xrightarrow{f_i} A_i = A \xrightarrow{e} B \xrightarrow{m_i \circ h_i \circ p_{j(i)} \circ m} A_i$$

is an (Epi, Mono-Source)-factorization of $\mathcal{S}$. Hence (1) follows from Theorem 15.10.

(2). Since **A** has equalizers and intersections, it is (Epi, ExtrMono)-structured (14.19). By Corollary 15.21, **A** is an (Epi, **M**)-category for a suitable conglomerate **M** of sources. Moreover, by (1), **A** is an (ExtrEpi, Mono-Source)-category. Hence by Proposition 15.8(4), **A** has coequalizers. Thus by Proposition 15.8(3), **M** is the conglomerate of all extremal mono-sources.

(3). By (1) and Proposition 15.7, **A** has coequalizers. Thus by Proposition 14.22, **A** is (RegEpi, Mono)-structured. By Proposition 15.24 this implies that **A** is a (RegEpi, **M**)-category for some conglomerate **M**. By 15.8(4) it follows that **M** is the conglomerate of all mono-sources in **A**. □

15.26 REMARK

If **A** is a strongly complete, extremally co-wellpowered category, there need not be any **M** for which **A** is an (Epi, **M**)-category. See Exercise 15D.

Suggestions for Further Reading

Strecker, G. E. Perfect sources. *Springer Lect. Notes Math.* **540** (1976): 605–624.

Hoffmann, R.-E. Factorization of cones. *Math. Nachr.* **87** (1979): 221–238.

Herrlich, H., G. Salicrup, and R. Vazquez. Dispersed factorization structures. *Canad. J. Math.* **31** (1979): 1059–1071.

Tholen, W. Factorizations, localizations, and the orthogonal subcategory problem. *Math. Nachr.* **114** (1983): 63–85.

EXERCISES

15A. Swell Epimorphisms

A morphism f in **A** is called a **swell epimorphism** provided that **A** has the unique $(\{f\}, \text{Mono-Source})$-diagonalization property. Show that

(a) Every strict epimorphism is swell, but not vice versa.

(b) Every swell epimorphism is an epimorphism.

(c) Every swell epimorphism is a strong epimorphism (cf. 14C), but not vice versa.

(d) The class of swell epimorphisms is closed under right-cancellation, composition, and the formation of cointersections, pushouts, and coproducts.

(e) Under each of the following assumptions the equality SwellEpi = ExtrEpi holds in **A**:

   (e1) In **A** 2-sources are (Epi, Mono-Source)-factorizable [but cf. Exercise 15B(b)].

   (e2) **A** is strongly complete and extremally co-wellpowered.

   (e3) **A** is complete and in **A** every mono-source contains a small mono-source.

   (e4) **A** is strongly cocomplete. [Cf. (d), Corollary 15.15, Proposition 15.7, and (f).]

(f) If **A** is an $(E, \text{Mono-Source})$-category, then $E = \text{SwellEpi} = \text{ExtrEpi}$.

15B. $(E, -)$-Categories

A is called an $(E, -)$-**category** provided it is an $(E, \mathbf{M})$-category for some **M**. Show that

(a) Every (Epi,−)-category is a (SwellEpi,−)-category [cf. Theorem 15.14].

(b) The category

is a (SwellEpi, −)-category, but neither an (Epi, −)-category nor an (ExtrEpi, −)-category. In particular, SwellEpi ≠ ExtrEpi.

(c) Every partially ordered set, considered as a category, is an (ExtrEpi, Mono-Source)-category [and SwellEpi = ExtrEpi]. But a partially ordered set $(X, \leq)$ is an (Epi, −)-category if and only if for each $x \in X$ the set $\{y \in X \mid x \leq y\}$ is a complete lattice. [E.g.,

is an (Epi, −)-category, but the addition of a smallest element destroys this property.]

Sec. 15]   Factorization structures for sources   271

(d) Let **A** be an $(E, -)$-category. Show that for any $D \subseteq E$, **A** is a $(D, -)$-category if and only if $\text{Iso}(\mathbf{A}) \subseteq D$ and D is closed under composition, pushouts, and cointersections.

15C. (ExtrEpi, Mono-Source)- vs. (Epi, ExtrMono-Source)-Categories

Show that

(a) Every (Epi, ExtrMono-Source)-category is an (ExtrEpi, Mono-Source)-category [cf. Ex. 15B(a) and 15A(e)].

(b) Not every (ExtrEpi, Mono-Source)-category is an (Epi, ExtrMono-Source)-category [cf. Ex. 15B(c), 15D, or 16A(b)].

* 15D. A Strongly Complete, Extremally Co-wellpowered, Non-(Epi, −)-Category

Let Λ-**CCPos** be the construct that has as objects all triples $(X, \leq, \lambda)$ with $(X, \leq)$ a partially ordered set in which each nonempty chain has a join and $\lambda : X \to X$ a unary (not necessarily order-preserving) operation, and as morphisms all λ-homomorphisms that preserve joins of nonempty chains. Show that

(a) Λ-**CCPos** is complete, cocomplete, and wellpowered, hence strongly complete.

(b) Λ-**CCPos** is extremally co-wellpowered but not concretely co-wellpowered, hence not co-wellpowered. [Hint: Consider for each limit ordinal α the natural injection $B_0 \xrightarrow{e_\alpha} A_\alpha$, where $B_0 = (\mathbb{N}, =, \lambda)$ with $\lambda(n) = n + 1$ and where $A_\alpha = (X_\alpha, \leq, \lambda_\alpha)$ with $(X_\alpha, \leq)$ the ordered set of all ordinals less than or equal to α and $\lambda_\alpha : X_\alpha \to X_\alpha$ defined by $\lambda_\alpha(\beta) = Min\{\beta + 1, \alpha\}$.]

(c) Λ-**CCPos** is an (ExtrEpi, Mono-Source)-category [cf. Theorem 15.25].

(d) Λ-**CCPos** does not have cointersections. [Hint: Consider for each ordinal α the natural injection $B_0 \xrightarrow{e_\alpha} B_\alpha$, where $B_\alpha = (Y_\alpha, \leq, \lambda_\alpha)$ with $(Y_\alpha, \leq)$ the set of all ordinals less than $\alpha + \omega$, ordered by $\beta < \gamma \Leftrightarrow (\beta < \gamma \leq \alpha$ w.r.t. the natural order of ordinals), and $\lambda_\alpha(\beta) = \beta + 1$.]

(e) Λ-**CCPos** is not an (Epi, −)-category [cf. 15.16(2)], but (Epi, Small ExtrMono-Source) is a factorization-structure for small sources on Λ-**CCPos** [cf. Proposition 15.19 and Corollary 14.21].

15E. Re: Theorem 15.14

Consider the three-element chain $\{0, 1, 2\}$ with its natural order as a category **A** and let E consist of all **A**-morphisms except $0 \to 2$. Show that **A** and E satisfy conditions (1), (3), and (4) of Theorem 15.14, but not condition (2).

15F. Existence of Coequalizers

(a) Show that each category **A** that satisfies one of the following conditions, has coequalizers:

   (1) **A** has (Epi, Mono-Source)-factorizations,

   (2) **A** is strongly complete and extremally co-wellpowered [cf. Theorem 15.25].

(b) Construct a strongly complete category that does not have coequalizers.

15G. Factorization Structures for Empty Sources

(a) Let $(E, \mathbf{M})$ be a factorization structure for empty sources on $\mathbf{A}$. Show that

   (1) An empty source $(A, \emptyset)$ belongs to $\mathbf{M}$ if and only if A is E-injective.

   (2) The full subcategory of $\mathbf{A}$ that consists of all E-injective objects is reflective in $\mathbf{A}$.

(b) Let E be a class of epimorphisms in $\mathbf{A}$ that is closed under composition with isomorphisms. Show that the following are equivalent:

   (1) There exists a class $\mathbf{M}$ of empty sources such that $(E, \mathbf{M})$ is a factorization structure for empty sources on $\mathbf{A}$.

   (2) $\mathbf{A}$ **has enough E-injectives**, i.e., for each $\mathbf{A}$-object A there exists an E-injective object B and a morphism $A \xrightarrow{e} B$ in E.

(c) Let (E, M) be a factorization structure for morphisms on a category $\mathbf{A}$ with a terminal object T. Let N be the class of all empty sources $(A, \emptyset)$ with $A \to T \in M$. Show that (E, N) is a factorization structure for empty sources on $\mathbf{A}$.

15H. E-Injectives

Show that every $(E, \mathbf{M})$-category has enough E-injectives.

15I. Factorization Structures for 2-Sources and 1-Sources

Let $(E, \mathbf{M})$ be a factorization structure for 2-sources and 1-sources on $\mathbf{A}$. Show that

(a) Conditions (1), (2), (3), (5), and (6) of Proposition 14.11 are equivalent to each other and to each of the following conditions:

   (7a) a morphism f belongs to $\mathbf{M}$ iff $f = g \circ e$ and $e \in E$ imply $e \in \mathrm{Iso}(\mathbf{A})$.

   (7b) a 1-source or 2-source F belongs to $\mathbf{M}$ iff $F = G \circ e$ and $e \in E$ imply $e \in \mathrm{Iso}(\mathbf{A})$.

   (8) for each object A the 2-source $A \xleftarrow{id} A \xrightarrow{id} A$ belongs to $\mathbf{M}$,

   (9) for each 1-source $A \xrightarrow{m} B$ in $\mathbf{M}$ the 2-source $B \xleftarrow{m} A \xrightarrow{m} B$ belongs to $\mathbf{M}$,

   (10) for each 1-source $A \xrightarrow{m} B$ in $\mathbf{M}$ and each morphism $A \xrightarrow{f} C$ the 2-source $B \xleftarrow{m} A \xrightarrow{f} C$ belongs to $\mathbf{M}$.

(b) The full subcategory $\mathbf{B}$ of $\mathbf{A}$ that consists of those objects A for which the 2-source $A \xleftarrow{id} A \xrightarrow{id} A$ belongs to $\mathbf{M}$ is closed under the formation of products and $\mathbf{M}$-subobjects.

15J. Factorization Structures for Small Sources

Show that

(a) (Mor, Small Product) is a factorization structure for small sources on **Set**, but **Set** is not a (Mor, $-$)-category.

(b) (Epi, Small ExtrMono-Source) is a factorization structure for small sources on Λ-**CCPos**, but Λ-**CCPos** is not an (Epi, −)-category.

15K. Source-Sink-Diagonals

Show that if $(X_i \xrightarrow{e_i} X)_I$ is an epi-sink, $(X_i \xrightarrow{f_i} Y)_I$ is an arbitrary sink, $(Y \xrightarrow{m_j} Y_j)_J$ is a mono-source, and $(X \xrightarrow{g_j} Y_j)_J$ is an arbitrary source in **Set**, then there exists a unique function $X \xrightarrow{d} Y$ such that the following diagram commutes:

$$\begin{array}{ccc} X_i & \xrightarrow{e_i} & X \\ {\scriptstyle f_i}\downarrow & {\scriptstyle d}\swarrow & \downarrow{\scriptstyle g_j} \\ Y & \xrightarrow{m_j} & Y_j \end{array}$$

* 15L. Dispersed Factorization Structures

Let $(E, \mathbf{M})$ be a fixed factorization structure on **A**. For a full subcategory **B** of **A** call an **A**-morphism f **B-concentrated** provided that $f \in E$ and each **B**-object is $\{f\}$-injective, and call an **A**-source $\mathcal{S}$ with domain A **B-dispersed** provided that the source obtained from $\mathcal{S}$ by adding all morphisms with domain A and codomain in **B** belongs to **M**. Show that

(a) **A** is a (**B**-Concentrated, **B**-Dispersed)-category.

(b) If **B** is E-reflective, then the following are equivalent for each **A**-object A:

 (1) A belongs to **B**,

 (2) the empty source with domain A is **B**-dispersed,

 (3) A is **B**-Concentrated-injective.

(c) There are at least as many factorization structures on **A** as there are E-reflective subcategories of **A**.

(d) Each of the categories **Top** and **Rere** has a proper class of factorization structures.

(e) If $(C, \mathbf{D})$ is a factorization structure on **A** with $C \subseteq E$, then the following are equivalent:

 (1) there exists a full subcategory **B** of **A** with $C = $ **B**-Concentrated morphisms and $\mathbf{D} = $ **B**-Dispersed sources,

 (2) if $g \circ f \in C$ and $f \in E$, then $f \in C$.

15M. Factorization Structures on Cat

Show that

(a) **Cat** is an (Epi, ExtrMono-Source)-category.

(b) **Cat** is an (ExtrEpi, Mono-Source)-category.

(c) **Cat** is not a (RegEpi, −)-category.

15N. Factorization Structures on $\mathbf{Set}^2$ and $(\mathbf{Set}^2)^{op}$

Show that $\mathbf{Set}^2$ and $(\mathbf{Set}^2)^{op}$ have regular factorizations.

15O. Re: Proposition 15.19

Show that one cannot omit the hypothesis in Proposition 15.19(2) that an initial object exists. [Consider discrete categories.]

16 E-reflective subcategories

In this section we will demonstrate that factorization structures provide a convenient tool to investigate reflective subcategories. The typical situation is this: **B** is an $(E, \mathbf{M})$-category and **A** is an isomorphism-closed full E-reflective subcategory of **B**.

16.1 DEFINITION
Let **B** be a category and let E be a class of **B**-morphisms. An isomorphism-closed, full subcategory **A** of **B** is called E**-reflective** in **B** provided that each **B**-object has an **A**-reflection arrow in E. In particular, we use the terms **epireflective** (resp. **monoreflective, bireflective**) in case E is the class of epimorphisms (resp. monomorphisms, bimorphisms) in **B**. Likewise, we use the terms **regular epireflective** (resp. **extremally epireflective**) in case E is the class of regular (resp. extremal) epimorphisms in **B**.

DUAL NOTION: E**-coreflective subcategory**.

16.2 EXAMPLES
(1) In **Met** the full subcategory of complete metric spaces is bireflective [cf. 4.17(8)].

(2) **HComp** is reflective but not epireflective in **Top**, even though **HComp** is epireflective in **Haus**, and **Haus** is epireflective in **Top**.

(3) **Ab** is regular epireflective in **Grp** [cf. 4.17(4)] and **Pos** is regular epireflective in **Prost** [cf. 4.17(3)].

(4) An isomorphism-closed full concrete subcategory of a concrete category **B** is (identity carried)-reflective in **B** if and only if it is concretely reflective in **B** (cf. 5.22).

(5) The construct of minimal acceptors is regular epireflective in the construct of reachable acceptors [cf. 4.17(7)].

Haus is epireflective in Top

16.3 PROPOSITION
Every monoreflective subcategory of **B** *is bireflective in* **B**.

Proof: Let **A** be a monoreflective (hence by Definition 16.1 a full) subcategory of **B**. Consider an **A**-reflection-arrow $B \xrightarrow{r} A$ for some **B**-object B and a pair $A \underset{q}{\overset{p}{\rightrightarrows}} B'$ of **B**-morphisms with $p \circ r = q \circ r$. Let $B' \xrightarrow{r'} A'$ be an **A**-reflection arrow for B'. Then $(r' \circ p) \circ r = (r' \circ q) \circ r$ and $A' \in Ob(\mathbf{A})$ imply that $r' \circ p = r' \circ q$. Since r' is assumed to be a monomorphism, this implies that $p = q$. Hence r is an epimorphism. □

16.4 PROPOSITION
Every coreflective isomorphism-closed full subcategory of **B** *that contains a* **B**-*separator is bicoreflective in* **B**.

Proof: Let S be a separator in **B**, let **A** be a coreflective full subcategory of **B** that contains S, and let $A \xrightarrow{c} B$ be an **A**-coreflection arrow. By the dual of Proposition 16.3 it suffices to show that c is an epimorphism. Let $B \underset{q}{\overset{p}{\rightrightarrows}} B'$ be distinct morphisms. Then there exists a morphism $S \xrightarrow{h} B$ with $p \circ h \neq q \circ h$. Since S belongs to **A**, there exists a unique morphism $S \xrightarrow{h'} A$ with $h = c \circ h'$. Consequently, $p \circ c \circ h' = p \circ h \neq q \circ h = q \circ c \circ h'$, which implies that $p \circ c \neq q \circ c$. □

16.5 EXAMPLES
The above proposition immediately implies that

(1) Every coreflective isomorphism-closed full subcategory of **Top** (resp. **Pos**) that contains a nonempty space (resp. a nonempty poset) is bicoreflective in **Top** (resp. **Pos**); hence it is a coreflective modification.

(2) Every reflective isomorphism-closed full subcategory of **Top$_0$** (resp. **Pos**) that contains the Sierpinski space (resp. a two-element chain) is bireflective in **Top$_0$** (resp. **Pos**); hence it is a reflective modification.

(3) **Set** (resp. **Vec**) contains precisely one proper coreflective, isomorphism-closed, full subcategory — namely, the one consisting of the empty set (resp. the zero vector spaces).

(4) **Set** contains precisely two proper reflective, isomorphism-closed, full subcategories — namely, the one consisting of all one-element sets and the one consisting of all sets with at most one element. **Vec** contains precisely one proper reflective, isomorphism-closed, full subcategory — namely, the one consisting of all zero vector spaces.

16.6 REMARK
It is not always easy to determine whether or not a given subcategory is reflective. (For example, which of the full subcategories consisting of regular spaces, or of connected spaces, or of compact spaces is reflective in **Top**?) Fortunately, for E-reflectivity we have powerful and easily applied criteria, as will be seen in the next results.

16.7 DEFINITION
Let **M** be a conglomerate of sources in a category **B**. A subcategory **A** of **B** is said to be **closed under the formation of M-sources** provided that whenever $(B \xrightarrow{f_i} A_i)_I$ is a source in **M** such that all A_i belong to **A**, then B belongs to **A**.

16.8 THEOREM
*If **A** is a full subcategory of an $(E, \mathbf{M})$-category **B**, then the following conditions are equivalent:*

*(1) **A** is E-reflective in **B**.*

*(2) **A** is closed under the formation of **M**-sources in **B**.*

*In the case that **B** has products and is E-co-wellpowered, the above conditions are equivalent to:*

*(3) **A** is closed under the formation of products and **M**-subobjects[66] in **B**.*

Proof: $(1) \Rightarrow (2)$. Let $\mathcal{S} = (B \xrightarrow{m_i} A_i)_I$ be a source in **M** such that all A_i belong to **A**. If $B \xrightarrow{e} A$ is an **A**-reflection arrow for B in E, then $\mathcal{S}$ factors through e. Hence, by Proposition 15.5(9)(a), e is an isomorphism. Thus B belongs to **A**.

$(2) \Rightarrow (1)$. For any **B**-object B, consider the source $\mathcal{S}$ with domain B, consisting of all morphisms with domain B and codomain in **A**. If $\mathcal{S} = \mathcal{M} \circ e$ is an $(E, \mathbf{M})$-factorization, then e is an **A**-reflection arrow for B.

$(2) \Rightarrow (3)$. Obvious.

$(3) \Rightarrow (2)$. Consider a source $\mathcal{M} = (B \xrightarrow{m_i} A_i)_I$ in **M** such that all A_i belong to **A**. For each $i \in I$ let $B \xrightarrow{m_i} A_i = B \xrightarrow{e_i} B_i \xrightarrow{n_i} A_i$ be an $(E, \mathbf{M})$-factorization. Then each B_i, being an **M**-subobject of A_i, belongs to **A**. Since **B** is E-co-wellpowered, we can select a subset J of I and for each $i \in I$ a $j(i) \in J$ and an isomorphism $h_i : B_{j(i)} \to B_i$ with $e_i = h_i \circ e_{j(i)}$. If $(P, \pi_j)_J$ is a product of the family $(B_j)_{j \in J}$, then P belongs to **A**. If
$$B \xrightarrow{\langle e_j \rangle} P = B \xrightarrow{e} C \xrightarrow{m} P$$
is an (E, M)-factorization, then C belongs to **A**, and we have commutativity of the diagram

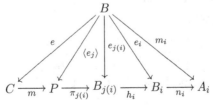

Since $\mathcal{M}$ factors through e, and $e \in E$, by Proposition 15.5(9)(a), e is an isomorphism. Thus B belongs to **A**. □

---
[66] An **M**-subobject is simply a singleton **M**-source. It need not be a monomorphism.

16.9 COROLLARY
A full subcategory of a co-wellpowered, strongly complete category **B** *is epireflective in* **B** *if and only if it is closed under the formation of products and extremal subobjects in* **B**.

Proof: Immediate from Theorems 16.8 and 15.25(2). □

16.10 EXAMPLE
The full subcategory **B** of **BiTop** that consists of all bitopological spaces with both topologies Hausdorff is strongly complete (but not co-wellpowered). The full subcategory **BiComp** of **B** [cf. 4F(b)] is closed under the formation of products and extremal subobjects in **B**, but is not reflective in **B**.

SUBCATEGORIES DEFINED BY EQUATIONS AND IMPLICATIONS

16.11 MOTIVATING REMARK
Many familiar mathematical objects (e.g., semigroups, monoids, groups, abelian groups, rings, lattices, boolean algebras, vector spaces, etc.) can be defined by means of *operations* and *equations*. Moreover, the corresponding categories (**Sgr**, **Mon**, **Grp**, etc.) can be obtained as full subcategories of categories of the form $\mathbf{Alg}(\Omega)$, consisting of those objects that satisfy suitable equations. This singling out of subcategories by equations (or, more generally, by implications) will be described here in some detail in order to motivate the much simpler and far more elegant categorical concepts introduced below. As an example, consider **Sgr**, which is the full subcategory of the category $\mathbf{Alg}(2)$ of algebras with one binary operation (usually written as multiplication) consisting of those algebras that satisfy the equation:

$$\forall x\ \forall y\ \forall z \quad x \cdot (y \cdot z) = (x \cdot y) \cdot z. \tag{e}$$

In order to find a categorical description of the above equation (e) and of those algebras that satisfy (e), observe that the expressions $x \cdot (y \cdot z)$ and $(x \cdot y) \cdot z$ can be interpreted as elements t and s of the free algebra F on the set $\{x, y, z\}$ of generators. If $\eta: \{x, y, z\} \to F$ is the corresponding universal arrow, then, for any algebra A in $\mathbf{Alg}(2)$, the following conditions are equivalent:

(1) A is a semigroup,

(2) A satisfies equation (e),

(3) for every map $f: \{x, y, z\} \to |A|$ (i.e., for every interpretation of the variables x, y, and z), the unique homomorphism $\overline{f}: F \to A$, determined by $f = \overline{f} \circ \eta$, satisfies $\overline{f}(t) = \overline{f}(s)$.

There is a further characterization: if ρ is the congruence on F generated by (t, s), (i.e., the smallest congruence relation on the algebra F such that $t\rho s$) and $F \xrightarrow{e} F/\rho$ is the natural map onto the quotient algebra F/ρ, then (3) is equivalent to

(4) A is $\{e\}$-injective (i.e., every homomorphism from F to A factors through e).

Hence an algebra satisfies equation (e) if and only if it is injective with respect to a suitable morphism e. In general, an equation in $\mathbf{Alg}(\Omega)$ is a pair (t, s) of elements in some free Ω-algebra F. An Ω-algebra A satisfies the equation (t, s) if and only if A is $\{e\}$-injective, where $F \xrightarrow{e} F/\rho$ is the quotient map corresponding to the congruence relation ρ on F generated by (t, s).

Slightly more complicated — but in its categorical formulation even simpler (!) — is the concept of *implications*. An implication $P \Rightarrow K$ consist of a set P of equations, called premises, and a set K of equations, called conclusions. For example, the implication

$$\forall x \ \forall y \ \forall z \quad x \cdot y = x \cdot z \quad \Rightarrow \quad y = z \tag{i}$$

describes, together with above equation (e), the left-cancellative semigroups. It is clear what it means to satisfy implication (i): whenever an interpretation of variables satisfies the premise, then it also satisfies the conclusion. (In this example, both the set of premises and the set of conclusions are singletons.) Formally, an implication $P \overset{X}{\Rightarrow} K$ in $\mathbf{Alg}(\Omega)$ consists of a set X (of "variables"), a subset P of $|F| \times |F|$ (called the set of "premises"), and a subset K of $|F|\times|F|$ (called the set of "conclusions"), where F denotes the canonical free Ω-algebra on X. (Cf. Example 8.23(6).) An Ω-algebra A *satisfies the implication* $P \overset{X}{\Rightarrow} K$ provided that under each interpretation of variables (i.e., for each map $f : X \to |A|$) such that each premise $(p, q) \in P$ becomes true (i.e., such that the homomorphic extension $\overline{f} : F \to A$ of f satisfies $\overline{f}(p) = \overline{f}(q)$ for each $(p, q) \in P$), each conclusion $(t, s) \in K$ must be true (i.e., $\overline{f}(t) = \overline{f}(s)$ for each $(t, s) \in K$). If π resp. γ are the congruence relations on F generated by P resp. by $P \cup K$, if $F \xrightarrow{e_P} F/\pi$ and $F \xrightarrow{e_K} F/\gamma$ are the associated quotient maps, and if $F/\pi \xrightarrow{e} F/\gamma$ is the unique homomorphism satisfying $e_K = e \circ e_P$, then the following conditions are easily seen to be equivalent for any Ω-algebra A:

(1) A satisfies the implication $P \overset{X}{\Rightarrow} K$,

(2) A is $\{e\}$-injective in $\mathbf{Alg}(\Omega)$.

Since in $\mathbf{Alg}(\Omega)$ morphisms of the form $F/\pi \xrightarrow{e} F/\gamma$ are, up to isomorphism, precisely the surjective homomorphisms (= epimorphisms = regular epimorphisms), we are led to the following definition:

16.12 DEFINITION

(1) (Regular) epimorphisms are called (**regular**) **implications**.

(2) An object Q **satisfies the implication** $A \xrightarrow{e} B$ provided that Q is $\{e\}$-injective (i.e., provided that for each morphism $A \xrightarrow{f} Q$ there exists a morphism $B \xrightarrow{\overline{f}} Q$ with $f = \overline{f} \circ e$).

(3) A full subcategory $\mathbf{A}$ of $\mathbf{B}$ is called **implicational** provided that there exists a class C of implications in $\mathbf{B}$ such that $\mathbf{A}$ consists precisely of those $\mathbf{B}$-objects that satisfy

each implication in C. Constructs that are concretely isomorphic to implicational subconstructs of $\mathbf{Alg}(\Omega)$ for some Ω are called **finitary quasivarieties**.[67]

In case C can be chosen to be a subclass of some class E of implications in $\mathbf{B}$, $\mathbf{A}$ is called **E-implicational**.

16.13 EXAMPLES

(1) **Pos** is an implicational subcategory of **Rel**. Consider the following implications:

REFLEXIVITY: $(\{x\}, \emptyset) \xrightarrow{id} (\{x\}, \{(x,x)\})$,

TRANSITIVITY: $(\{x,y,z\}, \{(x,y),(y,z)\}) \xrightarrow{id} (\{x,y,z\}, \{(x,y),(y,z),(x,z)\})$,

ANTISYMMETRY: $(\{x,y\}, \{(x,y),(y,x)\}) \xrightarrow{e} (\{x\}, \{(x,x)\})$,

where $e(x) = e(y) = x$.
A binary relation satisfies the above implications if and only if it is a poset.

(2) T_1-spaces form an implicational subcategory of **Top**. If P is the Sierpinski space and P' is a singleton space, then a topological space is T_1 if and only if it satisfies the implication $P \to P'$.

(3) In Σ-**Seq**, $\Sigma = \{\sigma\}$, consider all the automata with permutation transition (i.e., $\delta(\sigma, x) = \delta(\sigma, y) \Rightarrow x = y$). These form an implicational class given by $P_n \to P'_n$ ($n = 1, 2, \ldots$), where P_n has states $x, y, 1, \ldots, n$ such that $\delta(\sigma, x) = \delta(\sigma, y) = 1$ and $1, \ldots, n$ form a cycle, and P'_n is the quotient of P_n obtained by merging x with y.

(4) The constructs **Vec**, R-**Mod**, **Ab**, **Grp**, **Mon**, **Sgr**, **Rng**, **Lat**, **DLat**, **SLat**, **Boo**, **pSet**, and **AbTop** (= abelian topological groups) are finitary quasivarieties. However, **Field** is not a finitary quasivariety.

16.14 THEOREM
A full subcategory of an $(E, \mathbf{M})$-category $\mathbf{B}$ is E-implicational if and only if it is E-reflective in $\mathbf{B}$.

Proof: Let C be a subclass of E and let $\mathbf{A}$ be the full subcategory of $\mathbf{B}$ determined by C; i.e., given by those objects that satisfy each implication in C. Consider a source $(B \xrightarrow{m_i} A_i)_I$ in $\mathbf{M}$ such that all A_i belong to $\mathbf{A}$. By Theorem 16.8, it is sufficient to show that B belongs to $\mathbf{A}$. Let $Q \xrightarrow{e} R$ be an element of C and let $Q \xrightarrow{f} B$ be a morphism. Then for each $i \in I$ there exists a morphism $R \xrightarrow{g_i} A_i$ with $g_i \circ e = m_i \circ f$. Hence there exists a diagonal d that makes the diagram

$$\begin{array}{ccc} Q & \xrightarrow{e} & R \\ {\scriptstyle f}\downarrow & \swarrow{\scriptstyle d} & \downarrow{\scriptstyle g_i} \\ B & \xrightarrow{m_i} & A_i \end{array}$$

commute. In particular $f = d \circ e$. Hence B belongs to $\mathbf{A}$.

---
[67] Recall that in $\mathbf{Alg}(\Omega)$, Epi = RegEpi = surjective homomorphisms.

Conversely, let **A** be a E-reflective full subcategory of **B**. Consider the class C of all **A**-reflection arrows for **B**-objects, and let **C** be the implicational subcategory of **B** determined by C. Obviously, $\mathbf{A} \subseteq \mathbf{C}$. For the converse, let D be an object of **C** and let $D \xrightarrow{c} A$ be an **A**-reflection arrow for D. Since D satisfies the implication c, there exists a morphism f with $id_D = f \circ c$. Thus c is an epimorphic section (cf. 15.4); hence an isomorphism. Since **A** is isomorphism-closed, this implies that D belongs to **A**. Therefore $\mathbf{C} = \mathbf{A}$. □

16.15 REMARK

As indicated in Remark 16.11, equations in **Alg**(Ω) are those implications that have the empty set P of premises; equivalently, they are those implications $A \xrightarrow{e} B$ whose domain A is a free Ω-algebra. The concept of free Ω-algebras, unfortunately, refers to the construct **Alg**(Ω) and not to the abstract category **Alg**(Ω). Fortunately, however, as we will see below, a full subcategory **A** of **Alg**(Ω) is E-implicational, where E is the class of all (regular) implications in **Alg**(Ω) with free domain, if and only if **A** is $\overline{E}$-implicational, where $\overline{E}$ is the class of all (regular) implications in **Alg**(Ω) with (regular-) projective domain. This motivates the following definition:

16.16 DEFINITION

(1) Let E be a class of epimorphisms in a category **B**. An implication in E with E-projective domain is called an E-**equation**. Regular epimorphic equations are called **regular equations**. A full subcategory **A** of **B** is called E-**equational** provided that there exists a class C of E-equations in **B** such that **A** consists precisely of those **B**-objects that satisfy each E-equation in C.

(2) Let **B** be a construct. Regular implications with free domain are called **equations**. A full subcategory **A** of **B** is called **equational** provided that it can be defined as above by a class C of equations in **B**.

(3) Constructs that are concretely isomorphic to equational subconstructs of **Alg**(Ω) for some Ω are called **finitary varieties**.

16.17 THEOREM

*Let **B** be an (E, M)-category with enough E-projectives (9.22 dual). Then the following conditions are equivalent for any full subcategory **A** of **B**:*

*(1) **A** is E-equational in **B**.*

*(2) **A** is closed under the formation of **M**-sources and E-quotients in **B**.*

*In the case that **B** has products and is E-co-wellpowered, the above conditions are equivalent to:*

*(3) **A** is closed under the formation of products, **M**-subobjects, and E-quotients in **B**.*

Proof: (1) $\Rightarrow$ (2). By Theorems 16.8 and 16.14, **A** is closed under the formation of M-sources in **B**. Let A be in **A** and be $A \xrightarrow{e} B$ be in E. To show that B is in **A** we will verify that B satisfies any E-equation $C \xrightarrow{c} D$ that is satisfied by all **A**-objects. Since C is E-projective, for each morphism $C \xrightarrow{f} B$ there exists a morphism $\overline{f}$ with $f = e \circ \overline{f}$. Since A is in **A**, there exists a morphism $\tilde{f}$ with $\overline{f} = \tilde{f} \circ c$. Hence $f = (e \circ \tilde{f}) \circ c$. Thus B belongs to **A**.

$$C \xrightarrow{c} D$$
$$f \downarrow \; \overline{f} \searrow \; \downarrow \tilde{f}$$
$$B \xleftarrow{e} A$$

(2) $\Rightarrow$ (1). By Theorem 16.8, **A** is E-reflective in **B**. Let C be the class of all **A**-reflection arrows for E-projective objects P, and let **C** be the subcategory of **B** determined by C. Obviously, **A** $\subseteq$ **C**. For the converse, let D be an object of **C**. Then there exists an E-projective object P and a morphism $P \xrightarrow{e} D$ in E. Let $P \xrightarrow{c} A$ be an **A**-reflection arrow for P. Since D satisfies the E-equation c, there exists a morphism $A \xrightarrow{f} D$ with $e = f \circ c$. Since c is an epimorphism, this implies, by Proposition 15.5(5), that $A \xrightarrow{f} D$ belongs to E. Thus D is an E-quotient of an **A**-object, and so belongs to **A**.

(2) $\Leftrightarrow$ (3). Immediate from Theorem 16.8. □

16.18 THEOREM
*Let **B** be a fibre-small, transportable, complete construct that has free objects and for which the surjective morphisms, extremal epimorphisms and regular epimorphisms coincide. Then for full subconstructs **A** of **B** the following conditions are equivalent:*

*(1) **A** is equational in **B**,*

*(2) **A** is regular-equational in **B**,*

*(3) **A** is regular epireflective and closed under the formation of regular quotients (= homomorphic images) in **B**,*

*(4) **A** is closed under the formation of products, subobjects, and homomorphic images in **B**.*

Proof: In **B** the extremal epimorphisms are precisely the surjective morphisms, and, by Proposition 8.28 the monomorphisms are precisely the injective morphisms. Thus fibre-smallness and transportability imply wellpoweredness and extremal co-wellpoweredness. Hence completeness implies strong completeness. Since surjective morphisms are closed under composition, by Theorem 15.25(3), **B** is a (RegEpi, Mono-Source)-category. Consequently, Theorem 16.8 and Theorem 16.17 imply the equivalence of the conditions (2), (3), and (4). Since by Proposition 9.29 every free object is regular projective, (1) implies (2). To show that (3) implies (1), let C be the class of all **A**-reflection arrows $F \xrightarrow{c} A$ for free objects F in **B**, and let **C** be the subcategory of **B** determined by C. Since in **B** every object is a homomorphic image of a free object, the same argument as in the proof of Theorem 16.17 shows that **C** = **A**. □

16.19 COROLLARY

For full subcategories **A** *of* **Alg**(Ω), *the following hold:*

(1) **A** *is implicational in* **Alg**(Ω) *if and only if* **A** *is closed under the formation of products and subalgebras.*

(2) **A** *is equational in* **Alg**(Ω) *if and only if* **A** *is closed under the formation of products, subalgebras and homomorphic images.* □

E-REFLECTIVE HULLS

16.20 PROPOSITION

For $(E, \mathbf{M})$-*categories* **B**, *the following hold:*

(1) *The intersection of any conglomerate of* E-*reflective subcategories of* **B** *is* E-*reflective in* **B**.

(2) *For every full subcategory* **A** *of* **B** *there exists a smallest* E-*reflective subcategory of* **B** *that contains* **A**.

Proof: (1) follows from Theorem 16.8. (2) follows from (1). □

16.21 DEFINITION

If **A** is a full subcategory of an $(E, \mathbf{M})$-category **B**, then the smallest E-reflective subcategory of **B** that contains **A** is called the E-**reflective hull** of **A** in **B**.

16.22 PROPOSITION

If **A** *is a full subcategory of an* $(E, \mathbf{M})$-*category* **B**, *then a* **B**-*object* B *belongs to the* E-*reflective hull of* **A** *in* **B** *if and only if there exists a source* $(B \xrightarrow{f_i} A_i)_I$ *in* **M** *with all* A_i *in* **A**. □

16.23 Examples

B	**A**	Epireflective hull of **A** in **B**
Top	S (= Sierpinski-space)	**Top**$_0$
Top	$[0,1]$	**Tych**
Haus	$[0,1]$	**HComp**
Rel	$(\{0,1\}, \leq)$	**Pos**
Set	$\{0,1\}$	**Set**
Vec	$\mathbb{R}$	**Vec**

16.24 PROPOSITION (Reflectors as Composites of Epireflectors)

*If **A** is a full reflective subcategory of an (Epi, Mono-Source)-factorizable category **B**, and if **C** is the extremally epireflective hull of **A** in **B**, then **A** is epireflective in **C** and **C** is epireflective in **B**.*

Proof: By Theorem 15.10, **B** is an (ExtrEpi, Mono-Source)-category. Hence **A** has an ExtrEpi-reflective hull **C**. Obviously, **C** is epireflective in **B**, and **A** is reflective in **C**. For any C in **C** there exists, by Proposition 16.22, a mono-source $\mathcal{M} = (C \xrightarrow{m_i} A_i)_I$ with all A_i in **A**. Let $C \xrightarrow{r} A$ be an **A**-reflection arrow for C. Then $\mathcal{M}$ factors through r, which implies that r is a monomorphism in **B**, hence also in **C**. Thus **A** is monoreflective in **C**, so that by Proposition 16.3 it is epireflective in **C**.

16.25 REMARK

If we generalize the situation that we are working in just slightly, our results may break down completely. Here are two examples, the first very simple, the second very deep:

(1) Consider the ordered set of all natural numbers as a category **A**. Then **A** almost satisfies the assumptions of Theorem 15.25: **A** is well-powered, co-wellpowered, and "almost" strongly complete (every nonempty diagram has a limit). But **A** is not an (Epi, M)-category for any M. If **B** (resp. **C**) is the isomorphism-closed, epireflective subcategory consisting of all even (resp. all odd) numbers, then **B** ∩ **C** is empty — hence not even reflective.

(2) The category **Top** is strongly complete, wellpowered, and co-wellpowered. However, there exist two isomorphism-closed full reflective subcategories of **Top** whose intersection is not reflective in **Top**. [For a corresponding example in the category of bitopological spaces see Exercise 4F(b).]

If in a category **B** the intersection of two (epi)reflective isomorphism-closed full subcategories is not (epi)reflective, then this intersection obviously has no (epi)reflective hull in **B**.

If there is an isomorphism-closed full subcategory of **B** that does not have an (epi)reflective hull in **B**, then (epi)reflective subcategories of **B** cannot be characterized among the isomorphism-closed full subcategories of **B** by suitable closure properties.

Suggestions for Further Reading

Kennison, J. F. Reflective functors in general topology and elsewhere. *Trans. Amer. Math. Soc.* **118** (1965): 303–315.

Baron, S. Reflectors as compositions of epi-reflectors. *Trans. Amer. Math. Soc.* **136** (1969): 499–508.

Hatcher, W. S. Quasiprimitive subcategories. *Math. Ann.* **190** (1970): 93–96.

Ringel, C. M. Monofunctors as reflectors. *Trans. Amer. Math. Soc.* **161** (1971): 293–306.

Banaschewski, B., and H. Herrlich. Subcategories defined by implications. *Houston J. Math.* **2** (1976): 149–171.

Hušek, M. Lattices of reflections and coreflections in continuous structures. *Springer Lect. Notes Math.* **540** (1976): 404–424.

Nyikos, P. Epireflective categories of Hausdorff spaces. *Springer Lect. Notes Math.* **540** (1976): 452–481.

Marny, T. On epireflective subcategories of topological categories. *Gen. Topol. Appl.* **10** (1979): 175–181.

Hoffmann, R.-E. Reflective hulls of finite topological spaces. *Archiv Math.* **33** (1979): 258–262.

Németi, I., and I. Sain. Cone-implicational subcategories and some Birkhoff-type theorems. *Universal Algebra* (Esztergom, 1977), *Colloq. Math. Soc. János Bolyai* **29**, North Holland, Amsterdam, 1982, 535–578.

Hoffmann, R.-E. Cowell-powered reflective subcategories. *Proc. Amer. Math. Soc.* **90** (1984): 45–46.

Giuli, E., and M. Hušek. A diagonal theorem for epireflective subcategories of **Top** and cowell-poweredness. *Anal. Math. Pure Appl.* **145** (1986): 337–346.

EXERCISES

16A. Epireflective Subcategories with Bad Behavior

(a) Let I be a proper class, and let the category **B** have as objects A_0^0, A_1^1, A_2^i for $i \in I$, A_3^3, and A_4^4; and morphism-sets such that

$$\hom_\mathbf{B}(A_n^i, A_m^j) \text{ has } \begin{cases} \text{no element, if } m < n \text{ or } (m = n \text{ and } i \neq j) \\ \text{precisely two elements, if } (n, m) = (0, 1) \text{ or } (n, m) = (2, 3) \\ \text{precisely one element, otherwise.} \end{cases}$$

Let **A** be the full subcategory obtained from **B** by removing object A_3^3.
Show that

(1) **A** is regular epireflective in **B**.

(2) **B** is co-wellpowered, but **A** is not even extremally co-wellpowered.

(3) The inclusion functor $\mathbf{A} \hookrightarrow \mathbf{B}$ sends certain extremal epimorphisms to non-epimorphisms.

(b) Let the category **B** have as objects A_{2n}^0 for $n \in \mathbb{N}$, and A_n^1, for $n \in \mathbb{N}$; and morphism-sets such that

$$\hom_\mathbf{B}(A_n^i, A_m^j) \text{ has } \begin{cases} \text{precisely two elements, if } n = m \text{ and } i < j \\ \text{no elements, if } m < n \text{ or } j < i \\ \text{precisely one element, otherwise.} \end{cases}$$

Let **A** be the full subcategory of **B** whose objects are all A_{2n}^0 and all A_{2n}^1. Show that

(1) **A** is epireflective in **B**.

(2) **B** has regular factorizations, but **A** is not even (Epi, Mono-Source)-factorizable. [Consider empty sources.]

(3) The inclusion functor **A** $\hookrightarrow$ **B** sends certain regular epimorphisms to non-extremal epimorphisms.

16B. Inheritance of Factorization Structures

Let $E : \mathbf{A} \hookrightarrow \mathbf{B}$ be the embedding of a subcategory. Show that

(a) If **B** has regular factorizations and **A** is regular epireflective in **B**, then **A** has regular factorizations and E preserves and reflects regular epimorphisms.

(b) If **B** is an (ExtrEpi, Mono-Source)-category and **A** is extremally epireflective in **B**, then **A** is an (ExtrEpi, Mono-Source)-category and E preserves and reflects extremal epimorphisms.

(c) If **B** is an (Epi, ExtrMono-Source)-category and **A** is epireflective in **B**, then E need not preserve epimorphisms [cf. **Haus** and **Top**].

(d) If **B** has regular factorizations and **A** is epireflective in **B**, then **A** need not be an (ExtrEpi, Mono-Source)-category and E need neither preserve regular nor preserve extremal epimorphisms [cf. Exercise 16A(b)].

16C. Co-wellpoweredness and (Epi)Reflective Hulls

Let **A** be an isomorphism-closed full subcategory of a complete, well-powered, and co-wellpowered category **C**, and let **B** be the epireflective hull of **A** in **C**. Show that

(a) **A** is epireflective in **C** if and only if **A** is closed under the formation of products and regular subobjects in **C**.

(b) The embedding **A** $\hookrightarrow$ **B** preserves epimorphisms.

(c) If **D** is an isomorphism-closed full subcategory of **B** that contains **A**, then the following are equivalent:

  (1) **D** is reflective in **C**,

  (2) **D** is epireflective in **B**.

(d) **B** is strongly complete.

(e) If **B** is co-wellpowered, then the epireflective hull of **A** in **B** is a **reflective hull** of **A** in **C**, i.e., the smallest isomorphism-closed full reflective subcategory of **C** that contains **A**.

(f) The following are equivalent:

  (1) **A** is reflective in **C** and co-wellpowered,

(2) **A** is limit-closed in **C** and **B** is co-wellpowered.

(g) The following are equivalent:

(1) **A** has a co-wellpowered reflective hull in **C**,

(2) **A** has a co-wellpowered epireflective hull in **C**.

* **16D. Subcategories of Top**

(a) Show that a full reflective subcategory of **Top** is co-wellpowered if and only if its epireflective hull in **Top** is co-wellpowered.

(b) Construct an epireflective, non-co-wellpowered subcategory of **Top**. [Cf. Example 7.90(3).]

(c) Construct a full, limit-closed, non-reflective subcategory of **Top**.

16E. Smallest E-Reflective Subcategories

Let (E, M) be a factorization structure for empty sources (resp. for morphisms) on a category **A** (with a terminal object T). Show that

(a) A is E-injective if and only if $(A, \emptyset) \in M$ (resp. $A \to T \in M$).

(b) The full subcategory **B** of **A** that consists of all E-injective objects is the smallest E-reflective subcategory of **A**. Moreover, a morphism $A \xrightarrow{r} B$ is a **B**-reflection arrow if and only if $(A, \emptyset) = A \xrightarrow{r} (B, \emptyset)$ is an (E, M)-factorization [resp. if and only if $A \to T = A \xrightarrow{r} B \to T$ is an (E, M)-factorization].

16F. The Čech-Stone Compactification

Show that if E is the class of dense C^*-embeddings and M is the class of perfect maps, then (E, M) is a factorization structure for morphisms on **Tych**. Apply the results of Exercise 16E.

16G. A Generalized Birkhoff Theorem

Let **A** be a full subcategory of an $(E, \mathbf{M})$-category **B** and let Q be a class of epimorphisms in **B** such that

(α) **B** has enough Q-projectives,

(β) if $g \circ e \in Q$ and $e \in E$, then $g \in Q$.

(a) Show that the following conditions are equivalent:

(1) **A** is closed under the formation of **M**-sources and Q-quotients.

(2) **A** is definable by implications $B \xrightarrow{e} C$ in E with Q-projective domain B.

(b) Show that in the case that **A** is co-wellpowered and has products, the above conditions are equivalent to

(3) **A** is closed under the formation of products, **M**-subobjects, and Q-quotients.

16H. Epireflective Subcategories Are Closed Under Extremal Mono-Sources

Let **A** be an epireflective subcategory of **B** and let $(B \xrightarrow{m_i} B_i)_I$ be an extremal mono-source in **B**. Show that whenever all B_i belong to **A**, then so does B.

16I. The Lattice of Regular Epireflective Subcategories

Show that the lattice of regular epireflective subcategories of

(a) **Set** has precisely two elements,

(b) **Set**^{2} has the form

(c) (**Set**^{2})^{op} has the form

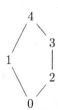

* 16J. Injective Hulls for Finitary Varieties

Show that in a finitary variety $(\mathbf{A}, U)$ every object has an injective hull if and only if the following conditions (cf. 9C) are satisfied:

(1) Monomorphisms are pushout stable,

(2) **A** is M-co-wellpowered, where M is the class of essential monomorphisms.

* 16K. Finitary Quasivarieties Isomorphic to Finitary Varieties

Show that a finitary quasivariety $(\mathbf{A}, U)$ is a finitary variety if and only if there exists a finitary variety $(\mathbf{B}, V)$ and an isomorphism $\mathbf{A} \xrightarrow{H} \mathbf{B}$.

* 16L. Strongly Limit-Closed vs. Epireflective Subcategories

A full subcategory **A** of **B** is called **strongly limit-closed in B** provided that whenever $(L \xrightarrow{f_i} D_i)_{Ob(\mathbf{I})}$ is a limit of a diagram $D : \mathbf{I} \to \mathbf{B}$ and for each $i \in Ob(\mathbf{I})$ there exists a $j \in Ob(\mathbf{I})$ with D_j in **A** and $\hom_{\mathbf{I}}(j, i) \neq \emptyset$, then L belongs to **A**. Show that:

(a) If **A** is epireflective in **B**, then **A** is strongly limit-closed in **B**.

(b) If **B** is complete, wellpowered, and co-wellpowered, then a full subcategory **A** of **B** is epireflective in **B** if and only if **A** is strongly limit-closed in **B**.

16M. A Characterization of Extremally Epireflective Subcategories

Show that a full subcategory of an extremally co-wellpowered, strongly complete category **B** is extremally epireflective in **B** if and only if it is closed under the formation of products and subobjects in **B**.

16N. Finitary Varieties and Finitary Quasivarieties

Show that

(a) All the finitary quasivarieties exhibited in Example 16.13(4) are finitary varieties.

(b) The full subconstruct **TfAb** of **Ab** is a finitary quasivariety, but is not a finitary variety. [Cf. 20.41(2) and 24.7.]

(c) Every regular epireflective subconstruct of a finitary quasivariety is a finitary quasivariety.

(d) The finitary variety **Grp** is a full, isomorphism-closed, reflective subconstruct of the finitary variety **Sgr**. However, **Grp** is not even implicational in **Sgr**.

17 Factorization structures for functors

In §8 we have generalized the concept of morphisms to that of G-structured arrows, where G is a functor. Here we will generalize, in an analogous way, the concept of sources to that of G-structured sources. Many (but by no means all) of the results on factorization structures of sources can be carried over to the context of G-structured sources. Sometimes even proofs of the more general results can be readily obtained by adapting the corresponding proofs from §15.

This section contains the basic results on factorization structures for functors. Special classes of functors, defined by means of particular factorization properties, will be studied in greater detail in the remaining chapters; especially: adjoint functors (Chapter V); and topological, algebraic, and topologically algebraic functors (Chapter VI).

Domain and codomain of a source

17.1 DEFINITION
Let $G\colon \mathbf{A} \to \mathbf{X}$ be a functor. A **G-structured source** S is a pair $(X, (f_i, A_i)_{i \in I})$ that consists of an $\mathbf{X}$-object X and a family of G-structured arrows $X \xrightarrow{f_i} GA_i$ with domain X, indexed by some class I.

X is called the **domain** of S and the family $(A_i)_{i \in I}$ is called the **codomain** of S.

17.2 REMARKS
(1) Whenever it is convenient, we use notations for G-structured sources such as
$(X \xrightarrow{f_i} GA_i)_{i \in I}$ or $(X \xrightarrow{f_i} GA_i)_I$ (which is reminiscent of notations used for sources in §15).

(2) As with sources, we use, whenever convenient, such expressions as
   (a) G-structured empty sources (i.e., **X**-objects),
   (b) G-structured 1-sources (i.e., G-structured arrows),
   (c) G-structured 2-sources (when the cardinality of the index class is 2), and
   (d) G-structured small sources (when the index class is a set).
(3) As with sources, every pair $(X, \mathcal{S})$ that consists of an **X**-object X and a class $\mathcal{S}$ of G-structured arrows with domain X can be considered to be a G-structured source via indexing $\mathcal{S}$ by itself.

17.3 DEFINITION
Let $G: \mathbf{A} \to \mathbf{X}$ be a functor, let E be a class of G-structured arrows, and let $\mathbf{M}$ be a conglomerate of **A**-sources. $(E, \mathbf{M})$ is called a **factorization structure** for G, and G is called an $(E, \mathbf{M})$-**functor** provided that

(1) E and $\mathbf{M}$ are closed under composition with isomorphisms in the following sense:
   (1a) if $X \xrightarrow{e} GA \in E$ and $A \xrightarrow{h} B \in \mathrm{Iso}(\mathbf{A})$, then $X \xrightarrow{e} GA \xrightarrow{Gh} GB \in E$,
   (1b) if $(A \xrightarrow{m_i} A_i)_I \in \mathbf{M}$ and $B \xrightarrow{h} A \in \mathrm{Iso}(\mathbf{A})$, then $(B \xrightarrow{h} A \xrightarrow{m_i} A_i)_I \in \mathbf{M}$.

(2) G has $(E, \mathbf{M})$-**factorizations**, i.e., for each G-structured source $(X \xrightarrow{f_i} GA_i)_I$ there exists $X \xrightarrow{e} GA \in E$ and $\mathcal{M} = (A \xrightarrow{m_i} A_i)_I \in \mathbf{M}$ such that
$$X \xrightarrow{f_i} GA_i = X \xrightarrow{e} GA \xrightarrow{Gm_i} GA_i \quad \text{for each } i \in I.$$

(3) G has the **unique** $(E, \mathbf{M})$-**diagonalization property**, i.e., whenever $X \xrightarrow{f} GA$ and $X \xrightarrow{e} GB$ are G-structured arrows with $(e, B) \in E$, and $\mathcal{M} = (A \xrightarrow{m_i} A_i)_I$ and $\mathcal{S} = (B \xrightarrow{f_i} A_i)_I$ are **A**-sources with $\mathcal{M} \in \mathbf{M}$, such that $(Gm_i) \circ f = (Gf_i) \circ e$ for each $i \in I$, then there exists a unique **diagonal**, i.e., an **A**-morphism $B \xrightarrow{d} A$ with $f = Gd \circ e$ and $\mathcal{S} = \mathcal{M} \circ d$, which will be expressed (imprecisely) by saying that the following diagram commutes:

$$\begin{array}{ccc} X & \xrightarrow{e} & GB \\ {\scriptstyle f}\downarrow & {\scriptstyle Gd}\swarrow & \downarrow{\scriptstyle Gf_i} \\ GA & \xrightarrow[Gm_i]{} & GA_i \end{array}$$

17.4 REMARKS
(1) A category **A** is an $(E, \mathbf{M})$-category if and only if the identity functor on **A** is an $(E, \mathbf{M})$-functor (provided that we identify each **A**-morphism $A \xrightarrow{e} B$ with the $id_\mathbf{A}$-structured arrow (e, B)). Hence $(E, \mathbf{M})$-categories can be considered as special cases of $(E, \mathbf{M})$-functors.

(2) The dual of a factorization structure on G is a factorization structure on G^{op}, hence it concerns factorizations of G-**costructured sinks**.

(3) A functor G is called an $(E,-)$-**functor** provided that there exists some **M** for which G is an $(E,\mathbf{M})$-functor. Analogously for $(-,\mathbf{M})$-functors.

(4) Whereas every category is an $(E,\mathbf{M})$-category for suitably chosen E and $\mathbf{M}$, a functor is an $(E,\mathbf{M})$-functor for suitably chosen E and $\mathbf{M}$ if and only if it is an adjoint functor (cf. 18.3).

17.5 EXAMPLES

(1) The forgetful functor $U : \mathbf{Top} \to \mathbf{Set}$ is a (Generating, Initial Mono-Source)-functor. The desired factorizations of U-structured sources of the form $(X \xrightarrow{f_i} (X_i, \tau_i))_I$ can be obtained in two steps: First, let $X \xrightarrow{f_i} X_i = X \xrightarrow{e} Y \xrightarrow{m_i} X_i$ be an (Epi, Mono-Source)-factorization in $\mathbf{Set}$. Second, let τ be the initial topology on Y with respect to the m_i and τ_i. Then

$$X \xrightarrow{f_i} (X_i, \tau_i) = X \xrightarrow{e} (Y, \tau) \xrightarrow{m_i} (X_i, \tau_i)$$

provides a factorization with the desired properties. By using similar arguments, U can be seen to be an (Extremally Generating, Mono-Source)-functor and a (Bijection, Initial Source)-functor.

(2) The forgetful functor $U : \mathbf{Grp} \to \mathbf{Set}$ is an (Extremally Generating, Mono-Source)-functor. (Recall that $X \xrightarrow{e} UG$ is extremally generating if and only if the set $e[X]$ generates the group G in the familiar algebraic sense.) The desired factorizations of U-structured sources $(X \xrightarrow{f_i} UG_i)_I$ can be obtained by either of the following two constructions, each being of independent interest:

(a) Let $X \xrightarrow{u} UF$ be a universal arrow in $\mathbf{Grp}$. Then each f_i can be uniquely extended to a homomorphism $F \xrightarrow{\overline{f}_i} G_i$. Let

$$F \xrightarrow{\overline{f}_i} G_i = F \xrightarrow{e} G \xrightarrow{m_i} G_i, \qquad i \in I,$$

be an (Epi, Mono-Source)-factorization of the source $(\overline{f}_i)_I$ in the category $\mathbf{Grp}$. Then the factorization

$$X \xrightarrow{f_i} UG_i = X \xrightarrow{Ue \circ u} UG \xrightarrow{Um_i} UG_i, \qquad i \in I,$$

has the desired properties.

(b) Form the cartesian product conglomerate $\prod_I G_i$ and consider this as a (possibly illegitimate) group by defining the operation coordinate wise. [That is, the members of $\prod_I G_i$ are functions $x : I \to \bigcup G_i$ with $x(i) \in G_i$, and $(x \cdot y)(i) = x(i) \cdot y(i)$ and $x^{-1}(i) = (x(i))^{-1}$.] Define a function $f : X \to \prod_I G_i$ by $(f(x))(i) = f_i(x)$. Then $f[X]$ is a sub*set* of $\prod_I G_i$, and it clearly generates a (small) subgroup H of the (possibly illegitimate) group $\prod_I G_i$. Thus the codomain restriction

$e : X \to UH$ of f together with the functions $m_i : H \to G_i$, defined by $m_i(x) = x(i)$, form the desired factorization.[68]

Observe that the second construction is more elementary than the first one, even though it uses the generalized group $\prod_I G_i$ (which often fails to be a group because it often fails to be a set).

(3) The forgetful functors U of many familiar constructs are (Extremally Generating, Mono-Source)-functors, e.g., those of **Top**, **Haus**, **Pos**, **Rel**, **Vec**, **Grp**, **Rng**, **Sgr**, **Mon**, **Lat**, **Boo**, **HComp**, **Cat**, and **Alg**(Ω). In each case a construction analogous to one of those given in (1) and (2) yields the desired factorizations. The diagonalization property holds automatically: cf. Theorem 17.10.

17.6 THEOREM
If G is an $(E, \mathbf{M})$-functor, then each member of E is generating. $\boxed{A\ 15.4}$

17.7 PROPOSITION
If $G : \mathbf{A} \to \mathbf{B}$ is an $(E, \mathbf{M})$-functor, then the following hold:

(1) $(E, \mathbf{M})$-factorizations are essentially unique,

(2) $\mathbf{M}$ determines E via the unique diagonalization property,

(3) if $\mathbf{A}$ is an $(\tilde{E}, \mathbf{M})$-category, $(e, A) \in E$ and $A \xrightarrow{\tilde{e}} B \in \tilde{E}$, then $((G\tilde{e}) \circ e, B) \in E$. $\boxed{A\ 15.5}$

17.8 REMARK
Even though $(E, \mathbf{M})$-functors are analogous to $(E, \mathbf{M})$-categories, not all of the properties of $(E, \mathbf{M})$-categories carry over to $(E, \mathbf{M})$-functors. In particular, for $(E, \mathbf{M})$-functors

(1) $\mathbf{M}$ need not be determined by E [cf. 15.5(9)]. [The canonical forgetful functor $U : \mathbf{Top} \to \mathbf{Set}$ is simultaneously an $(E, \mathbf{M}_i)$-functor for $i = 1, 2$, where E consists of all U-structured arrows (e, A) with e bijective and A discrete, $\mathbf{M}_1$ consists of all sources in **Top**, and $\mathbf{M}_2$ consists of all sources in **Top** with discrete domain.]

(2) Extremal mono-sources need not all belong to $\mathbf{M}$ [cf. 15.5(2)]. [In (1) above, there are limit sources — even isomorphisms (considered as 1-sources) — that do not belong to $\mathbf{M}_2$.]

(3) $\mathbf{M}$ need not be closed under composition [cf. 15.5(4)]. [If U and E are as in (1) above, and $\mathbf{M}_3 = \mathbf{M}_2 \cup \{\mathbb{Q} \hookrightarrow \mathbb{R}, \mathbb{R} \hookrightarrow \mathbb{C}\}$, then U is an $(E, \mathbf{M}_3)$-functor, but $\mathbb{Q} \hookrightarrow \mathbb{C}$ does not belong to $\mathbf{M}_3$.]

---

[68] If I is a set, then the entire construction above is just the image factorization of $f = \langle f_i \rangle$ (as $f = m \circ e$) composed with the product source (π_i), i.e., $m_i = \pi_i \circ m$. Even if I is a proper class, the construction could be performed in the quasicategory of "large groups", but in order to introduce this entity we would need to enrich our foundations.

17.9 PROPOSITION
If G is an $(E, \text{Mono-Source})$-functor, then E consists precisely of those structured arrows that are extremally generating. $\boxed{A\ 15.8(1)}$

17.10 THEOREM
If a functor G has (Generating, Mono-Source)-factorizations, then G is an (Extremally Generating, Mono-Source)-functor. $\boxed{A\ 15.10}$

FACTORIZATION STRUCTURES AND LIMITS

17.11 THEOREM
Let $\mathbf{A}$ be a strongly complete category and let $\mathbf{A} \xrightarrow{G} \mathbf{X}$ be a functor that preserves strong limits.

(1) If G is extremally co-wellpowered or if $\mathbf{A}$ has a coseparator, then G is an (ExtrGen, Mono-Source)-functor.

(2) If G is faithful and concretely co-wellpowered, then G is a (ConGen, Initial Mono-Source)-functor.

Proof:
(1). In case G is extremally co-wellpowered, the result is a straightforward generalization of Theorem 15.25(1). In case $\mathbf{A}$ has a coseparator C, we conclude as above that G has (ExtrGen, Mono-Source)-factorizations for set-indexed G-structured sources. As in Theorem 15.10 this implies that G has the (ExtrGen, Mono-Source)-diagonalization property (for arbitrary sources). Let X be an arbitrary $\mathbf{X}$-object. For each G-structured morphism $X \xrightarrow{h} G(C^{\hom(X, GC)})$ select an (ExtrGen, Mono)-factorization:

$$X \xrightarrow{h} G(C^{\hom(X,GC)}) = X \xrightarrow{g_h} GA_h \xrightarrow{Gm_h} G(C^{\hom(X,GC)}).$$

Likewise for each G-structured morphism $X \xrightarrow{k} G(C^\emptyset)$ select an (ExtrGen, Mono)-factorization:

$$X \xrightarrow{k} G(C^\emptyset) = X \xrightarrow{g_k} GA_k \xrightarrow{Gm_k} G(C^\emptyset).$$

Then $M = \{(g_h, A_h) \mid h \in \hom(X, G(C^{\hom(X,C)}))\} \cup \{(g_k, A_k) \mid k \in \hom(X, G(C^\emptyset))\}$ is a set. Next we show that every G-structured morphism $X \xrightarrow{f} GA$ has an (M, Mono)-factorization. Analogous to Theorem 15.25(1) this implies that G is (ExtrGen, Mono Source)-factorizable. Let $X \xrightarrow{f} GA = X \xrightarrow{e} GB \xrightarrow{Gm} GA$ be a (Gen, Mono)-factorization. Since C is a coseparator, there exists a monomorphism $B \xrightarrow{n} C^{\hom_\mathbf{A}(B,C)}$. In case $\hom_\mathbf{A}(B, C) = \emptyset$, choose $k = Gn \circ e$. Then there exists a diagonal $A_k \xrightarrow{d} B$ such that the diagram

$$\begin{array}{ccc} X & \xrightarrow{g_k} & GA_k \\ e \downarrow & \swarrow Gd & \downarrow Gm_k \\ GB & \xrightarrow{Gn} & G(C^{\emptyset}) \end{array}$$

commutes. Since m_k is a monomorphism, so is d. Thus

$$X \xrightarrow{f} GA = X \xrightarrow{g_k} GA_k \xrightarrow{G(m \circ d)} GA$$

is an (M, Mono)-factorization. In case $\hom_\mathbf{A}(B, C) \neq \emptyset$, there exists a monomorphism $C^{\hom_\mathbf{A}(B,C)} \xrightarrow{\tilde{n}} C^{\hom_\mathbf{X}(X,GC)}$, since the map

$$\varphi : \hom_\mathbf{A}(B, C) \to \hom_\mathbf{X}(X, GC),$$

defined by $\varphi(\ell) = G\ell \circ e$, is injective (cf. Exercise 10K). By choosing $h = G(\tilde{n} \circ n) \circ e$ and proceeding as before, one obtains an (M, Mono)-factorization of $X \xrightarrow{f} GA$.

(2). This follows as a ready adaptation of the proof of Theorem 15.25 since regular monomorphisms are G-initial [cf. Proposition 8.7(3)]. □

FACTORIZATIONS OF STRUCTURED 2-SOURCES

17.12 PROPOSITION
If G-structured 2-sources have (Generating, $-$)-factorizations, then G preserves mono-sources.

Proof: Let $\mathbf{A} \xrightarrow{G} \mathbf{X}$ be a functor. Let $\mathcal{M} = (A \xrightarrow{m_i} A_i)_I$ be a mono-source in $\mathbf{A}$, and let $X \underset{r_2}{\overset{r_1}{\rightrightarrows}} GA$ be a pair of $\mathbf{X}$-morphisms with $G\mathcal{M} \circ r_1 = G\mathcal{M} \circ r_2$. If

$$X \xrightarrow{r_i} GA = X \xrightarrow{e} GB \xrightarrow{Gs_i} GA, \quad i \in \{1, 2\}$$

is an (Generating, $-$)-factorization of the structured 2-source $(X \xrightarrow{r_i} GA)_{i \in \{1,2\}}$, then $G(m_i \circ s_1) \circ e = Gm_i \circ Gs_1 \circ e = Gm_i \circ r_1 = Gm_i \circ r_2 = Gm_i \circ Gs_2 \circ e = G(m_i \circ s_2) \circ e$ for each $i \in I$. Since (e, B) is generating, this implies that $m_i \circ s_1 = m_i \circ s_2$ for each $i \in I$. Since $(A \xrightarrow{m_i} A_i)_I$ is a mono-source, this implies that $s_1 = s_2$. Thus $r_1 = r_2$. □

17.13 PROPOSITION
If $G : \mathbf{A} \to \mathbf{B}$ is a functor such that G-structured 2-sources have (Generating, Mono-Source)-factorizations, then the following conditions are equivalent:

(1) G reflects isomorphisms,

(2) each mono-source is G-initial,

(3) G is faithful and reflects extremal epimorphisms,

(4) G reflects limits,

(5) G reflects equalizers.

Proof: (1) $\Rightarrow$ (2). Let $\mathcal{M} = (A \xrightarrow{m_i} A_i)_I$ be a mono-source in **A**, let $\mathcal{S} = (B \xrightarrow{f_i} A_i)_I$ be a source in **A**, and let $GB \xrightarrow{f} GA$ be an **B**-morphism with $G\mathcal{S} = G\mathcal{M} \circ f$. Consider a (Generating, Mono-Source)-factorization:

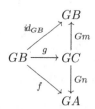

of the G-structured 2-source $GB \xleftarrow{id} GB \xrightarrow{f} GA$. Then $G(\mathcal{M} \circ n) \circ g = G(\mathcal{S} \circ m) \circ g$ implies that $\mathcal{M} \circ n = \mathcal{S} \circ m$. Thus, since (n, m) and $\mathcal{M}$ are mono-sources, this implies that m is a monomorphism. Hence, by Proposition 17.12, Gm is a monomorphism and a retraction, so an isomorphism. By (1), m is an isomorphism. Thus $\overline{f} = n \circ m^{-1} : B \to A$ is an **A**-morphism with $G\overline{f} = f$ and $\mathcal{S} = \mathcal{M} \circ \overline{f}$. It is the unique morphism with these properties, since $\mathcal{M}$ is a mono-source.

(2) $\Rightarrow$ (3). By Proposition 10.60 G is faithful and reflects isomorphisms. Let e be an **A**-morphism such that Ge is an extremal epimorphism. By faithfulness, e is an epimorphism. If $e = m \circ g$ is a $(-, \text{Mono})$-factorization of e, then, by Proposition 17.12, $Ge = Gm \circ Gg$ is a $(-, \text{Mono})$-factorization of Ge. Hence Gm is an isomorphism in **B**. Thus m is an isomorphism in **A**.

(3) $\Rightarrow$ (1). G reflects extremal epimorphisms and, by faithfulness, monomorphisms. Hence, by the dual of Proposition 7.66, G reflects isomorphisms.

(2) $\Rightarrow$ (4). By Proposition 10.60 G is faithful. Let $D : \mathbf{I} \to \mathbf{A}$ be a diagram and let $\mathcal{L} = (L \xrightarrow{\ell_i} D_i)_I$ be a source in **A** such that $G\mathcal{L}$ is a limit of $G \circ D$. By faithfulness $\mathcal{L}$ is a natural source for D. Let $\mathcal{S} = (A \xrightarrow{f_i} D_i)_I$ be an arbitrary natural source for D. Then $G\mathcal{S}$ is a natural source for $G \circ D$. Hence there exists a **B**-morphism $GA \xrightarrow{g} GL$ with $G\mathcal{S} = G\mathcal{L} \circ g$. Since $G\mathcal{L}$, being a limit, is a mono-source, the faithfulness of G implies that $\mathcal{L}$ is a mono-source; hence it is G-initial. Thus there exists an **A**-morphism $A \xrightarrow{f} L$ with $Gf = g$ and $\mathcal{S} = \mathcal{L} \circ f$. Since $\mathcal{L}$ is a mono-source, f is uniquely determined by the latter property. Hence $\mathcal{L}$ is a limit of D.

(4) $\Rightarrow$ (5). Obvious.

(5) $\Rightarrow$ (1). Immediate, since a morphism $A \xrightarrow{f} B$ is an equalizer of (id_B, id_B) if and only if f is an isomorphism. □

Sec. 17]   Factorization structures for functors   297

17.14 REMARK

Observe that in the diagram below, all arrows indicate implications that hold without any assumptions, whereas those labeled 1, 2, 3, and 4 are equivalences under the hypothesis of the preceding Proposition (cf. 17G).

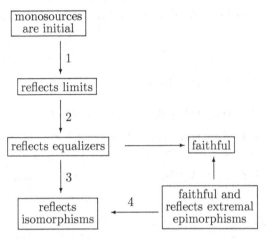

17.15 PROPOSITION

A functor $\mathbf{A} \xrightarrow{G} \mathbf{X}$ is faithful if and only if for each $\mathbf{A}$-object A the G-structured source $(GA \xleftarrow{id} GA \xrightarrow{id} GA)$ is (Generating, Initial Source)-factorizable.

Proof: If G is faithful, then

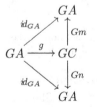

is, by Proposition 8.16(3) and Proposition 10.59, a (Generating, Initial Source)-factorization. Conversely, if $A \underset{s}{\overset{r}{\rightrightarrows}} B$ is a pair of $\mathbf{A}$-morphisms with $Gr = Gs$ and

$$
\begin{array}{c}
 & & GA \\
 & \overset{id_{GA}}{\nearrow} & \uparrow Gm \\
GA & \xrightarrow{g} & GC \\
 & \underset{id_{GA}}{\searrow} & \downarrow Gn \\
 & & GA
\end{array}
$$

is a (Generating, Initial Source)-factorization, then there exists an $\mathbf{A}$-morphism $A \xrightarrow{\bar{g}} C$ with $G\bar{g} = g$ and $m \circ \bar{g} = id_A = n \circ \bar{g}$. Since (g, C) is generating, the equalities $G(r \circ m) \circ g = Gr \circ Gm \circ g = Gr = Gs = Gs \circ Gn \circ g = G(s \circ n) \circ g$ imply that $r \circ m = s \circ n$. Thus $r = r \circ id_A = r \circ m \circ \bar{g} = s \circ n \circ \bar{g} = s \circ id_A = s$. □

17.16 COROLLARY
If G-structured 2-sources are (Generating, Initial Source)-factorizable, then G is faithful. □

Suggestions for Further Reading

Tholen, W. Factorizations of cones along a functor. *Quaest. Math.* **2** (1977): 335–353.

Herrlich, H., and G. E. Strecker. Semi-universal maps and universal initial completions. *Pacific J. Math.* **82** (1979): 407–428.

Nakagawa, R. A note on (E, M)-functors. *Springer Lect. Notes Math.* **719** (1979): 250–258.

EXERCISES

17A. A Characterization of Generating Arrows
Let $(\mathbf{A}, U)$ be a concrete category that has concrete equalizers. Show that a structured arrow is

(a) generating if and only if it does not factor through a non-isomorphic regular monomorphism,

(b) concretely generating if and only if it does not factor through a non-isomorphic embedding,

(c) extremally generating if and only if it does not factor through a non-isomorphic monomorphism.

17B. Composites of (Generating,−)-Factorizable Functors
Show that whenever $\mathbf{A} \xrightarrow{F} \mathbf{B}$ is (Generating, Mono-Source)-factorizable and $\mathbf{B} \xrightarrow{G} \mathbf{C}$ is (Generating, −)-factorizable, then $A \xrightarrow{G \circ F} \mathbf{C}$ is (Generating, Mono-Source)-factorizable.

Conclude that a functor with an (Epi, Mono-Source)-factorizable domain is (Generating, Mono-Source)-factorizable if and only if it is (Generating,−)-factorizable.

17C. Existence of Factorization Structures
Let $(\mathbf{A}, U)$ be a concrete category over $\mathbf{X}$. Show that

(a) If $(\mathbf{A}, U)$ has concrete equalizers and concrete intersections, then structured arrows have (Extremally Generating, Mono)-factorizations.

(b) If **X** is strongly complete and if U lifts limits and is concretely co-wellpowered, then U is a (Concretely Generating, Initial Mono-Source)-functor.

17D. Uniqueness of Diagonals

Show that a functor is an $(E, \mathbf{M})$-functor if it has $(E, \mathbf{M})$-factorizations and the (not necessary unique) (E, M)-diagonalization property.

17E. Properties of $(E, \mathbf{M})$-Functors

Consider the following properties of an $(E, \mathbf{M})$-functor $\mathbf{A} \xrightarrow{G} \mathbf{B}$:

(1) If $GA \xrightarrow{e} GA'$ is a **B**-morphism and (e, A') belongs to E, then there exists an **A**-morphism $A \xrightarrow{\bar{e}} A'$ with $G\bar{e} = e$.

(2) E is closed under composition.

(3) $(id_{GA}, A) \in E$ for each **A**-object A.

(4) Any source in **M** is G-initial.

(5) A source $(A \xrightarrow{m_i} A_i)_I$ belongs to **M** if and only if for every commutative diagram

$$\begin{array}{ccc} B & \xrightarrow{e} & G\overline{A} \\ f \downarrow & & \downarrow Gf_i \\ GA & \xrightarrow{Gm_i} & GA_i \end{array}$$

with $(e, \overline{A}) \in E$ there exists a unique diagonal.

Show that

(a) (1) implies (2), but not conversely.

(b) (2) implies (3) if each **A**-object is the codomain of a member of E.

(c) (3) and (4) are equivalent.

(d) (4) implies (5), but not conversely.

(e) (5) implies that every **A**-isomorphism, considered as source, belongs to **M**.

(f) (3) implies that G is faithful and that all extremal mono-sources belong to **M**.

17F. Factorizations of Structured 2-Sources

Show that

(a) If G has (Generating, Mono-Source)-factorizations for structured 2-sources, then G has the unique (Extremally Generating, Mono-Source)-diagonalization property (for all structured sources).

(b) If G has (Concretely Generating, Initial Mono-Source)-factorizations for structured 2-sources, then G has the unique (Concretely Generating, Initial Mono-Source)-diagonalization property.

(c) If G has the $(E,$ Extremal Mono-Source)-diagonalization property for structured 2-sources, then each (e, A) in E is generating.

(d) If G has the $(E,$ Mono-Source)-diagonalization property for structured 2-sources, then each (e, A) in E is extremally generating.

(e) If G is faithful and G has the $(E,$ Initial Mono-Source)-diagonalization property for 2-sources, then each (e, A) in E is concretely generating.

17G. Re: Remark 17.14

Show that

(a) A functor need not be faithful even though it reflects every type of monomorphism and every type of epimorphism introduced above (extremal, regular, strict, strong, section, retraction, isomorphism, etc.). [Hint: Consider the functor G from $0 \rightrightarrows 1$ to $0 \to 1$ with $G(0) = 0$ and $G(1) = 1$.]

(b) A faithful functor that reflects isomorphisms and extremal epimorphisms need not reflect equalizers. Hint: Consider the inclusion functor from

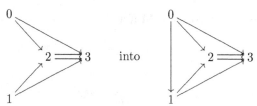

(c) A functor that reflects equalizers need reflect neither limits nor regular monomorphisms. Hint: For limits consider the inclusion functor from the discrete category with objects 0 and 1 into $0 \to 1$; for regular monomorphisms consider the inclusion functor from

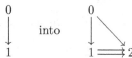

(d) If a functor $\mathbf{A} \xrightarrow{G} \mathbf{B}$ reflects limits, then mono-sources in $\mathbf{A}$ need not be G-initial. Hint: Consider the inclusion functor from

(e) If mono-sources are G-initial, then G need not reflect extremal epimorphisms. Hint: Consider the inclusion functor from

*17H. A Generalization of Theorem 17.11
Let **A** be a strongly complete category with a coseparating set. Show that every functor $\mathbf{A} \xrightarrow{G} \mathbf{B}$ that preserves strong limits is an (ExtrGen, Mono-Source)-functor.

17I.
Prove that if **A** has equalizers and intersections and $G : \mathbf{A} \to \mathbf{B}$ preserves these limits, then G-structured arrows have (Extremal Generating, Mono)-factorizations. Cf. Theorem 14.19.

17J.
Let $(\mathbf{A}, U)$ be a concrete category over a strongly complete category **X**. Show that if U is concretely co-wellpowered and preserves and detects limits, then U is (Concretely Generating, Initial Mono-Source)-factorizable, thus a $(-,-)$-functor.

17K.
Let $(\mathbf{A}, U)$ be a fibre-small, transportable concrete category over a wellpowered category **X**. Prove that if $(\mathbf{A}, U)$ has concrete limits and bounded generation, then U is an (Extremal Generating, Mono-Source)-functor.

17L.
Show that the nonfull embedding

$$\begin{array}{c} 3 \\ \diagup \diagdown \\ 1 \quad \quad 2 \\ \diagdown \diagup \\ 0 \end{array} \hookrightarrow \begin{array}{c} 3 \\ | \\ 2 \\ | \\ 1 \\ | \\ 0 \end{array}$$

satisfies all of the conditions of Proposition 17.13 except condition (2). Cf. also Proposition 10.60.

Chapter V

ADJOINTS AND MONADS

18 Adjoint functors

Perhaps the most successful concept of category theory is that of *adjoint functor*. Adjoint functors occur frequently in many branches of mathematics and the "adjoint functor theorems" have a surprising range of applications.

18.1 DEFINITION
A functor $G: \mathbf{A} \to \mathbf{B}$ is said to be **adjoint** provided that for every **B**-object B there exists a G-universal arrow with domain B (cf. 8.30).

Dually, a functor $G: \mathbf{A} \to \mathbf{B}$ is said to be **co-adjoint** provided that for every **B**-object B there exists a G-co-universal arrow with codomain B (cf. 8.40).

18.2 EXAMPLES
(1) A subcategory **A** of a category **B** is (co)reflective in **B** if and only if the associated inclusion $\mathbf{A} \hookrightarrow \mathbf{B}$ is a (co-)adjoint functor [cf. 8.31(2) and 8.41(1)].

(2) The forgetful functor U of a concrete category $(\mathbf{A}, U)$ over $\mathbf{X}$ is adjoint if and only if for each **X**-object X there exists a free object over X. In particular, the forgetful functors:

   (a) of the constructs **Rel**, **Top**, and **Alg**(Σ) are both adjoint and co-adjoint (cf. 8.23 and 8.41),

   (b) of the constructs **Vec**, **Grp**, **Pos**, and **Cat** are adjoint but not co-adjoint,

   (c) of the constructs **CLat** and **CBoo** are neither adjoint nor co-adjoint,

   (d) of the concrete categories of the form **Spa**(T) are both adjoint and co-adjoint,

   (e) **TopGrp** $\to$ **Grp**, **TopGrp** $\to$ **Top**, and **TopGrp** $\to$ **Set** are adjoint.

(3) $O: \mathbf{Ban} \to \mathbf{Set}$ is adjoint, but $U: \mathbf{Ban} \to \mathbf{Set}$ is not [cf. 8.23(12)].

(4) Each equivalence is adjoint and co-adjoint.

(5) If **A** is a category, then the unique functor $\mathbf{A} \to \mathbf{1}$ is adjoint if and only if **A** has an initial object [cf. 8.31(3)].

(6) A hom-functor $\hom(A, -): \mathbf{A} \to \mathbf{Set}$ is adjoint if and only if there exist arbitrary copowers of A in **A** [cf. 8.31(4)].

(7) For each set M, the endofunctor $(M \times -): \mathbf{Set} \to \mathbf{Set}$ that sends $A \xrightarrow{f} B$ to $M \times A \xrightarrow{id_M \times f} M \times B$ is co-adjoint. "Evaluation" $ev: M \times A^M \to A$ defined by $(m, f) \mapsto f(m)$ is an $(M \times -)$-co-universal arrow for A (cf. §27).

(8) The minimal realization functor $M: \mathbf{Beh} \to \mathbf{Aut}_r$ is adjoint [cf. 8.31(5)].

18.3 PROPOSITION

For a functor $G\colon \mathbf{A} \to \mathbf{B}$ the following conditions are equivalent:

(1) G is adjoint,

(2) G has (Generating[69],–)-factorizations,

(3) G is an $(E, \mathbf{M})$-functor for some E and $\mathbf{M}$,

(4) G is a (Universal[69], Source)-functor.

Proof: (1) $\Rightarrow$ (4) $\Rightarrow$ (3). Obvious.

(3) $\Rightarrow$ (2). Theorem 17.6.

(2) $\Rightarrow$ (1). For any **B**-object B let $B \xrightarrow{f_i} GA_i = B \xrightarrow{e} GA \xrightarrow{Gm_i} GA_i$ be a (Generating,–) factorization of the G-structured source that consists of all G-structured arrows (f_i, A_i) with domain B. Then the G-structured arrow (e, A) is universal for B. □

18.4 PROPOSITION

If $\mathbf{A}$ is an $(Epi, \mathbf{M})$-category, then for any functor $G: \mathbf{A} \to \mathbf{B}$ the following are equivalent:

(1) G is adjoint,

(2) G is a (Generating,$\mathbf{M}$)-functor.

Proof: (1) $\Rightarrow$ (2). Let $(B \xrightarrow{f_i} GA_i)_I$ be a G-structured source. If $B \xrightarrow{u} GA$ is a universal arrow, then for each $i \in I$ there exists an **A**-morphism $A \xrightarrow{\overline{f}_i} A_i$ with $f_i = G\overline{f}_i \circ u$. If $A \xrightarrow{\overline{f}_i} A_i = A \xrightarrow{e} \overline{A} \xrightarrow{m_i} A_i$ is an $(Epi,\mathbf{M})$-factorization, then $B \xrightarrow{f_i} GA_i = B \xrightarrow{(Ge)\circ u} G\overline{A} \xrightarrow{Gm_i} GA_i$ is a (Generating,$\mathbf{M}$)-factorization. To show the (Generating,$\mathbf{M}$)-diagonalization property, consider a commutative diagram

$$\begin{array}{ccc} B & \xrightarrow{g} & GA \\ {\scriptstyle f}\downarrow & & \downarrow{\scriptstyle Gf_i} \\ GA' & \xrightarrow{Gm_i} & GA_i \end{array} \qquad (*)$$

with (g, A) generating and $(A' \xrightarrow{m_i} A_i)_I$ in $\mathbf{M}$. If $B \xrightarrow{u} G\overline{A}$ is a universal arrow, then there exist **A**-morphisms $\overline{A} \xrightarrow{\overline{f}} A'$ and $\overline{A} \xrightarrow{\overline{g}} A$ with $f = G\overline{f} \circ u$ and $g = G\overline{g} \circ u$. This implies that $m_i \circ \overline{f} = f_i \circ \overline{g}$ for each $i \in I$. Since (g, A) is generating, $\overline{g}$ must be an epimorphism. Hence there exists an **A**-morphism $A \xrightarrow{d} A'$ that makes

$$\begin{array}{ccc} \overline{A} & \xrightarrow{\overline{g}} & A \\ {\scriptstyle \overline{f}}\downarrow & {\scriptstyle d}\swarrow & \downarrow{\scriptstyle f_i} \\ A' & \xrightarrow{m_i} & A_i \end{array}$$

---

[69] "Generating" in this context denotes the class of all generating G-structured arrows; likewise, "Universal" denotes the class of all universal G-structured arrows.

commute. Obviously, d is a diagonal for (*). Uniqueness follows from the fact that (g, A) is generating.

$(2) \Rightarrow (1)$. Proposition 18.3. □

PROPERTIES OF ADJOINT FUNCTORS

18.5 PROPOSITION
If $\mathbf{A} \xrightarrow{G_1} \mathbf{B}$ and $\mathbf{B} \xrightarrow{G_2} \mathbf{C}$ are adjoint, then so is $\mathbf{A} \xrightarrow{G_2 \circ G_1} \mathbf{C}$.

Proof: If $C \xrightarrow{u} G_2 B$ is universal for C, and $B \xrightarrow{v} G_1 A$ is universal for B, then $C \xrightarrow{(G_2 v) \circ u} G_2 G_1 A$ is universal for C. □

18.6 PROPOSITION
Adjoint functors preserve mono-sources.

Proof: Let $G \colon \mathbf{A} \to \mathbf{B}$ be an adjoint functor, let $\mathcal{S} = (A \xrightarrow{m_i} A_i)_I$ be a mono-source in $\mathbf{A}$, and let $B \underset{s}{\overset{r}{\rightrightarrows}} GA$ be a pair of $\mathbf{B}$-morphisms with $G\mathcal{S} \circ r = G\mathcal{S} \circ s$. If $B \xrightarrow{u} G\overline{A}$ is a universal arrow, then there exist $\mathbf{A}$-morphisms $\overline{A} \underset{\overline{s}}{\overset{\overline{r}}{\rightrightarrows}} A$ with $r = G\overline{r} \circ u$ and $s = G\overline{s} \circ u$. This implies that $G(\mathcal{S} \circ \overline{r}) \circ u = G(\mathcal{S} \circ \overline{s}) \circ u$; hence $\mathcal{S} \circ \overline{r} = \mathcal{S} \circ \overline{s}$, so that $\overline{r} = \overline{s}$, and thus $r = s$. □

18.7 COROLLARY
Embeddings of reflective subcategories preserve and reflect mono-sources. □

18.8 REMARK
(1) As shown in Example 7.33(5) embeddings of full subcategories need not preserve monomorphisms, hence they need not preserve mono-sources.

(2) Full, reflective embeddings preserve extremal monomorphisms. (Cf. 7F.)

18.9 PROPOSITION
Adjoint functors preserve limits.

Proof: Let $G \colon \mathbf{A} \to \mathbf{B}$ be an adjoint functor, let $D \colon \mathbf{I} \to \mathbf{A}$ be a diagram, and let $\mathcal{L} = (L \xrightarrow{\ell_i} D_i)_I$ be a limit of D. Then $G\mathcal{L}$ is a natural source for $G \circ D$. Let $\mathcal{S} = (B \xrightarrow{f_i} GD_i)_I$ be a natural source for $G \circ D$. If $B \xrightarrow{u} GA$ is a universal arrow for B, then for each $i \in I$ there exists an $\mathbf{A}$-morphism $\overline{f}_i \colon A \to D_i$ with $f_i = G\overline{f}_i \circ u$. Since (u, A) is generating, the source $\overline{\mathcal{S}} = (A \xrightarrow{\overline{f}_i} D_i)$ is natural for D. Hence there exists an $\mathbf{A}$-morphism $f \colon A \to L$ with $\overline{\mathcal{S}} = \mathcal{L} \circ f$. Consequently, $\overline{f} = Gf \circ u \colon B \to GL$ is a $\mathbf{B}$-morphism with $\mathcal{S} = G\mathcal{L} \circ \overline{f}$. Uniqueness follows from the fact that, by Proposition 18.6, $G\mathcal{L}$ is a mono-source. □

18.10 COROLLARY

If $(\mathbf{A}, U)$ is a concrete category over $\mathbf{X}$ that has free objects, then the following hold:

(1) all limits in $(\mathbf{A}, U)$ are concrete,

(2) U preserves and reflects mono-sources,

(3) if $(\mathbf{A}, U)$ is fibre-small and transportable, then wellpoweredness of $\mathbf{X}$ implies wellpoweredness of $\mathbf{A}$. □

18.11 PROPOSITION

If $G: \mathbf{A} \to \mathbf{B}$ is an adjoint functor and $\mathbf{A}$ is co-wellpowered, or extremally co-wellpowered, then so is G.

Proof: Immediate from Proposition 8.36. □

ADJOINT FUNCTOR THEOREMS

We have seen above that adjoint functors preserve limits. Next we will see that this property in conjunction with certain completeness and smallness conditions actually characterizes such functors. We have seen before that both general constructions and smallness conditions are essential ingredients of category theory. The adjoint functor theorems of this section are prime examples of this fact. Below we will see that each of them contains a completeness condition and a smallness condition, neither of which can be eliminated.

18.12 ADJOINT FUNCTOR THEOREM

A functor $G : \mathbf{A} \to \mathbf{B}$, whose domain $\mathbf{A}$ is complete, is adjoint if and only if G satisfies the following conditions:

(1) G preserves small limits,

*(2) for each $\mathbf{B}$-object B there exists a G-**solution set**, i.e., a set-indexed G-structured source $(B \xrightarrow{f_i} GA_i)_I$ through which each G-structured arrow factors (in the sense that given any $B \xrightarrow{f} GA$, there exists a $j \in I$ and a $A_j \xrightarrow{g} A$ in $\mathbf{A}$ such that*

commutes).

Proof: If G is an adjoint functor, then, by Proposition 18.9, G preserves limits. If $B \xrightarrow{u} GA$ is a G-universal arrow for B, then the G-structured source S_B consisting of (u, A) alone is a G-solution set for B. Hence conditions (1) and (2) are necessary.

To show the converse, suppose that (1) and (2) are satisfied. Let B be a **B**-object and let $S_B = ((u_i, A_i))_I$ be a G-solution set for B. If $(P, \pi_i)_I$ is a product of the family $(A_i)_I$ in **A**, then by (1) $(GP, G\pi_i)_I$ is a product of $(GA_i)_I$ in **B**. Thus $u = \langle u_i \rangle : B \to GP$ is a G-structured arrow such that the G-structured source, consisting of (u, P) alone, is a G-solution set for B. Consider the family $(g_j)_J$ consisting of all **A**-morphisms $P \xrightarrow{g_j} P$ with $(Gg_j) \circ u = u$. If $E \xrightarrow{e} P$ is a multiple equalizer of $(g_j)_J$, then by (1) $GE \xrightarrow{Ge} GP$ is a multiple equalizer of $(Gg_j)_J$. Thus there is a **B**-morphism $B \xrightarrow{v} GE$ with $u = Ge \circ v$. The G-structured source, consisting of the G-structured arrow (v, E) alone, is a G-solution set for B. Thus to prove that (v, E) is a G-universal arrow, it need only be shown that (v, E) is G-generating. Let $E \underset{s}{\overset{r}{\rightrightarrows}} A$ be a pair of **A**-morphisms with $Gr \circ v = Gs \circ v$. If $\overline{E} \xrightarrow{\overline{e}} E$ is an equalizer of r and s, then by (1) $G\overline{E} \xrightarrow{G\overline{e}} GE$ is an equalizer of Gr and Gs. Thus there exists a **B**-morphism $B \xrightarrow{f} G\overline{E}$ with $v = G\overline{e} \circ f$. Since $((u, P))$ is a G-solution set for B, there exists an **A**-morphism $P \xrightarrow{g} \overline{E}$ with $f = Gg \circ u$. Hence $P \xrightarrow{e \circ \overline{e} \circ g} P$ is an **A**-morphism with $G(e \circ \overline{e} \circ g) \circ u = G(e \circ \overline{e}) \circ f = Ge \circ v = u$. Therefore there exists some $j \in J$ with $g_j = e \circ \overline{e} \circ g$. There also exists some $k \in J$ with $g_k = \mathrm{id}_P$. Thus $g_j \circ e = g_k \circ e$ implies $e \circ \overline{e} \circ g \circ e = e$. Since e is a monomorphism, this implies that $\overline{e} \circ g \circ e = \mathrm{id}_E$. Thus $\overline{e}$ is a retraction, which by Proposition 7.54 implies that $r = s$. Consequently, (v, E) is a G-universal arrow for B. Thus G is an adjoint functor. □

18.13 EXAMPLES

The following functors have complete domains and preserve limits, but fail to be adjoint:

(1) the unique functor $\mathbf{Ord}^{\mathrm{op}} \to \mathbf{1}$, where **Ord** is the partially ordered class of all ordinal numbers, considered as a category,

(2) the forgetful functors $\mathbf{CLat} \to \mathbf{Set}$ and $\mathbf{CBoo} \to \mathbf{Set}$ [cf. 8.23(7)].

Whereas the "limit-preservation" condition in the above theorem is usually easy to check, the "solution set" condition is often rather cumbersome, particularly since the theorem gives no idea of how to find such a set. In the next results, attention is focused on "canonical" candidates for solution sets.

18.14 THEOREM

*If **A** is strongly complete and (extremally) co-wellpowered, then the following conditions are equivalent for any functor $\mathbf{A} \xrightarrow{G} \mathbf{B}$:*

(1) G is adjoint,

(2) G preserves small limits and is (extremally) co-wellpowered.

*Moreover the implication (2) $\Rightarrow$ (1) holds without the assumption that **A** be (extremally) co-wellpowered.*

Proof: (1) $\Rightarrow$ (2). Immediate from Propositions 18.9 and 18.11.

(2) $\Rightarrow$ (1). Immediate from Proposition 18.3 and Theorem 17.11(1). □

18.15 COROLLARY

Fibre-small, concretely co-wellpowered constructs that are concretely complete have free objects. □

An adjoint functor is co-wellpowered and preserves small limits

18.16 EXAMPLES

The conditions in the above theorem are carefully balanced as the following examples show:

1. There exist full embeddings $\mathbf{A} \xrightarrow{G} \mathbf{B}$ such that $\mathbf{A}$ is complete and co-wellpowered and G preserves limits and is extremally co-wellpowered, but G is neither adjoint nor co-wellpowered. [Let $\mathbf{C}$ be the partially ordered class of all ordinals, considered as a category, let $\mathbf{A} = \mathbf{C}^{op}$, let $\mathbf{B}$ be obtained from $\mathbf{A}$ by adding an initial object, and let $\mathbf{A} \xrightarrow{G} \mathbf{B}$ be the inclusion.]

2. There exist strongly complete categories $\mathbf{A}$ that are not co-wellpowered, even though the unique functor $\mathbf{A} \xrightarrow{G} \mathbf{1}$ is adjoint and co-wellpowered. [Consider Λ-**CCPos**.]

For suitable categories $\mathbf{A}$ the somewhat cumbersome solution set condition completely vanishes, as the following result shows:

18.17 SPECIAL ADJOINT FUNCTOR THEOREM

If $\mathbf{A}$ is a strongly complete category with a coseparator, then for any functor $G : \mathbf{A} \to \mathbf{B}$, the following conditions are equivalent:

(1) G is adjoint,

(2) G preserves strong limits.

Proof: Immediate from Theorem 17.11, Proposition 18.3, and Proposition 18.9. □

18.18 REMARK
The categories **Set**, **Vec**, **Pos**, **Top**, and **HComp** are complete, wellpowered, and have coseparators (cf. 7.18), so that the above theorem applies to them. Since many familiar categories have separators but fail to have coseparators, the dual of the Special Adjoint Functor Theorem is applicable even more often than the theorem itself.

18.19 CONCRETE ADJOINT FUNCTOR THEOREM
Let $G : (\mathbf{A}, U) \to (\mathbf{B}, V)$ be a concrete functor. If $(\mathbf{A}, U)$ is complete, wellpowered, co-wellpowered, and has free objects, then G is adjoint if and only if G preserves small limits.

Proof: As before, it suffices to show that the above assumptions imply that G is co-wellpowered. Let B be a **B**-object and let $VB \xrightarrow{u} UA_B$ be a (fixed) universal U-structured arrow with domain VB. It suffices to show that for every generating G-structured arrow $B \xrightarrow{g} GA$, A can be considered (via a suitable epimorphism) as a quotient-object of A_B. By universality there exists a unique **A**-morphism $\bar{g} : A_B \to A$ such that the triangle

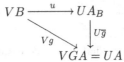

commutes. Since (g, A) is a generating G-structured arrow, $\bar{g}$ is an epimorphism; hence A is a quotient-object of A_B. □

18.20 REMARKS
(1) Faithfulness of U is not needed for the above result, but faithfulness of V is essential.

(2) If $(\mathbf{A}, U)$ and $(\mathbf{B}, V)$ satisfy the assumptions of the above result and if V reflects limits, then *every* concrete functor $(\mathbf{A}, U) \to (\mathbf{B}, V)$ preserves limits, and hence must be adjoint. Similar situations will be investigated in §23. (Observe that the reflection of limits is typical for forgetful functors of "algebraic" concrete categories.)

Suggestions for Further Reading

Kan, D. N. Adjoint functors. *Trans. Amer. Math. Soc.* **87** (1958): 294–329.

Freyd, P. *Functor theory.* Ph. D. dissertation, Princeton University, 1960.

Felscher, W. Adjungierte Funktoren und primitive Klassen. *Sitzungsberichte Heidelberger Akad. Wiss.* **4** (1965): 447–509.

Ehresmann, C. Sur l'existence de structures libres et de foncteurs adjoints. *Cahiers Topol. Geom. Diff.* **9** (1967): 33–180.

EXERCISES

18A. Functors that are Simultaneously Adjoint and Co-adjoint

Let $\mathbf{A} \xrightarrow{G} \mathbf{B}$ be a functor. Show that

(a) If G is an equivalence then G is adjoint and co-adjoint.

(b) If G is adjoint and co-adjoint and $\mathbf{A} = \mathbf{B} = \mathbf{Set}$, then G is an equivalence.

(c) If $\mathbf{A}$ and $\mathbf{B}$ are monoids, considered as categories, then G is adjoint iff G is co-adjoint.

(d) If $\mathbf{A}$ has an initial and a terminal object and $\mathbf{B} = \mathbf{1}$, then G is adjoint and co-adjoint.

(e) If $\mathbf{A} = \mathbf{1}$ and G maps the single object of $\mathbf{A}$ to a zero object in $\mathbf{B}$, then G is adjoint and co-adjoint.

(f) The forgetful functor $\mathbf{Top} \to \mathbf{Set}$ is adjoint and co-adjoint.

(g) The forgetful functor $\mathbf{Alg}(1) \to \mathbf{Set}$ is adjoint and co-adjoint, where the $\eta : X \to (X \times \mathbb{N}, \lambda)$, defined by $\eta(x) = (x, 0)$ and $\lambda(x, n) = (x, n+1)$, are universal arrows, and the $\varepsilon : (X^{\mathbb{N}}, \mu) \to X$, defined by $\varepsilon(x_n) = x_0$ and $\mu(x_n) = (x_{n+1})$ are couniversal arrows.

(h) If $\mathbf{A}$ is small and $\mathbf{C}$ is a category that is complete and cocomplete, then the functor $[\mathbf{B}, \mathbf{C}] \xrightarrow{[G, id]} [\mathbf{A}, \mathbf{C}]$ is adjoint and co-adjoint.

18B. Smallness Conditions For Adjoints

(a) Let $\mathbf{A} \xrightarrow{G} \mathbf{B}$ be a faithful adjoint functor. Show that $\mathbf{A}$ is co-wellpowered if and only if G is co-wellpowered.

(b) Let $\mathbf{A} \xrightarrow{G} \mathbf{B}$ be an adjoint functor such that mono-sources are G-initial. Show that $\mathbf{A}$ is extremally co-wellpowered if and only if G is extremally co-wellpowered.

(c) Show that the unique functor Λ-$\mathbf{CCPos} \to \mathbf{1}$ is adjoint and co-wellpowered, even though Λ-$\mathbf{CCPos}$ is not co-wellpowered. Cf. Exercise 15D.

18C. Adjoints via Representable Functors

Show that

(a) A functor $\mathbf{A} \xrightarrow{G} \mathbf{Set}$ is adjoint if and only if G is representable by an object for which arbitrary copowers exist (cf. Exercise 10R).

(b) A functor $\mathbf{A} \xrightarrow{G} \mathbf{B}$ is adjoint if and only if for each $\mathbf{B}$-object B the functor $\mathbf{A} \xrightarrow{G} \mathbf{B} \xrightarrow{\hom(B,-)} \mathbf{Set}$ is representable.

18D. (Co-)Adjoints and Colimits

(a) Show that a functor $\mathbf{Set} \to \mathbf{A}$ is co-adjoint if and only if it preserves coproducts.

(b) Let $\mathbf{A} \xrightarrow{G} \mathbf{Set}$ be a faithful and representable functor whose domain is a complete, wellpowered, and co-wellpowered category $\mathbf{A}$. Show that G is adjoint if and only if $\mathbf{A}$ is cocomplete.

18E. Power-Set Functors

Show that the covariant power-set functor $\mathcal{P} : \mathbf{Set} \to \mathbf{Set}$ is neither adjoint nor co-adjoint, but that the contravariant power-set functor $Q : \mathbf{Set}^{\mathrm{op}} \to \mathbf{Set}$ is adjoint.

18F. Upper Semicontinuity as Adjointness

Let $\mathbf{A}$ be a complete totally ordered set and let $\mathbf{R}$ be the ordered set of real numbers. Let $G : \mathbf{A} \to \mathbf{R}$ be an order-preserving map, considered as a functor. Show that G is adjoint if and only if G is upper semicontinuous.

18G. Complete Boolean Algebras

Show that (cf. Exercises 8E, 8F, and 10R):

(a) **CBoo** is complete, well-powered, and extremally co-wellpowered.

(b) **CBoo** is not cocomplete.

(c) The forgetful functor $\mathbf{CBoo} \to \mathbf{Set}$ is not adjoint.

18H. Adjoint Functors between Posets

Consider order-preserving functions $A \xrightarrow{f} B$ between posets as functors and show that

(a) f is adjoint if and only if for each $b \in B$, the set $\{\, a \in A \mid b \leq f(a) \,\}$ has a smallest element.

(b) If $A = B = \mathbb{N}$ (with the usual order), then

   (1) f is adjoint if and only if it is unbounded.

   (2) f is co-adjoint if and only if it is unbounded and $f(0) = 0$.

(c) If $A = B = \mathbb{Z}$ (with the usual order), then f is adjoint if and only if it is co-adjoint.

18I. Stable Epimorphisms and Adjoints

Show that faithful adjoints reflect stable epimorphisms.

* 18J. Extremal Monomorphisms and Adjoints

Show that adjoint functors need not preserve extremal monomorphisms.

18K. Compact Categories

A category $\mathbf{A}$ is called **compact** provided that each functor with domain $\mathbf{A}$ that preserves colimits is co-adjoint. Prove that

(a) Each cocomplete co-wellpowered category that has a separator is compact.

(b) Each compact category is complete. [Hint: For each small diagram D in **A** the functor $F: \mathbf{A} \to \mathbf{Set}^{op}$ that assigns to each **A**-object A the set of all natural sources for D with domain A preserves colimits. Thus F^{op} is representable.]

18L. Coseparating Sets

(a) Prove that a set S of objects of a category **A** that has products is an (extremally) coseparating set for **A** if and only if every **A**-object is an (extremal) subobject of some product of objects in S (cf. Proposition 10.38).

(b) Using (a) prove that the Special Adjoint Functor Theorem (18.17) holds also for all strongly complete categories **A** that have a coseparating set.

19 Adjoint situations

With every adjoint functor $\mathbf{A} \xrightarrow{G} \mathbf{B}$ there can be naturally associated (in an essentially unique way) a functor $\mathbf{B} \xrightarrow{F} \mathbf{A}$ and two natural transformations $id_{\mathbf{B}} \xrightarrow{\eta} G \circ F$ and $F \circ G \xrightarrow{\varepsilon} id_{\mathbf{A}}$. There is an intricate web of relationships between the functors G and F and the natural transformations η and ε, including an inherent duality, that is largely responsible for the importance of adjoint functors.

19.1 THEOREM

Let $G : \mathbf{A} \to \mathbf{B}$ be an adjoint functor, and for each $\mathbf{B}$-object B let $\eta_B : B \to G(A_B)$ be a G-universal arrow. Then there exists a unique functor $F : \mathbf{B} \to \mathbf{A}$ such that $F(B) = A_B$ for each $\mathbf{B}$-object B, and $id_{\mathbf{B}} \xrightarrow{\eta=(\eta_B)} G \circ F$ is a natural transformation.

Moreover, there exists a unique natural transformation $F \circ G \xrightarrow{\varepsilon} id_{\mathbf{A}}$ that satisfies the following conditions:

(1) $G \xrightarrow{\eta G} GFG \xrightarrow{G\varepsilon} G = G \xrightarrow{id_G} G$,

(2) $F \xrightarrow{F\eta} FGF \xrightarrow{\varepsilon F} F = F \xrightarrow{id_F} F$.

Proof:

(a). The existence of a unique F with the required properties follows analogously to the proof of Proposition 4.22.

(b). For each $\mathbf{A}$-object A, there exists a unique $\mathbf{A}$-morphism $FGA \xrightarrow{\varepsilon_A} A$ such that

$$GA \xrightarrow{\eta_{GA}} GFGA$$
$$id_{GA} \searrow \quad \downarrow G\varepsilon_A$$
$$GA$$

commutes. Thus if $\varepsilon = (\varepsilon_A)$ is a natural transformation from FG to $id_{\mathbf{A}}$, it is the unique one that satisfies (1). To show naturality, consider an $\mathbf{A}$-morphism $A \xrightarrow{f} A'$. Then it follows that $G(f \circ \varepsilon_A) \circ \eta_{GA} = Gf \circ G\varepsilon_A \circ \eta_{GA} = Gf = G\varepsilon_{A'} \circ \eta_{GA'} \circ Gf = G\varepsilon_{A'} \circ GFGf \circ \eta_{GA} = G(\varepsilon_{A'} \circ FGf) \circ \eta_{GA}$ by the naturality of η. Hence $f \circ \varepsilon_A = \varepsilon_{A'} \circ FGf$.

(c). Since $\eta : id_{\mathbf{B}} \to GF$ is natural, we get $\eta GF \circ \eta = GF\eta \circ \eta$. Hence (1) implies that $G(id_F) \circ \eta = id_G F \circ \eta = G\varepsilon F \circ \eta GF \circ \eta = G\varepsilon F \circ GF\eta \circ \eta = G(\varepsilon F \circ F\eta) \circ \eta$. Thus $id_F = \varepsilon F \circ F\eta$, i.e., (2) holds. □

19.2 REMARK

Below (cf. 19.7) it is shown that conditions (1) and (2) of Theorem 19.1 play a crucial role. In particular, given functors $\mathbf{A} \xrightarrow{G} \mathbf{B}$ and $\mathbf{B} \xrightarrow{F} \mathbf{A}$ and natural transformations $id_{\mathbf{B}} \xrightarrow{\eta} GF$ and $FG \xrightarrow{\varepsilon} id_{\mathbf{A}}$ that satisfy the above conditions, the following hold:

- G is adjoint.
- F is co-adjoint.
- Each $B \xrightarrow{\eta_B} G(FB)$ is a G-universal arrow.
- Each $F(GA) \xrightarrow{\varepsilon_A} A$ is an F-co-universal arrow.

19.3 DEFINITION
An **adjoint situation** $(\eta, \varepsilon): F \dashv G: \mathbf{A} \to \mathbf{B}$ consists of functors $\mathbf{A} \xrightarrow{G} \mathbf{B}$ and $\mathbf{B} \xrightarrow{F} \mathbf{A}$ and natural transformations $id_\mathbf{B} \xrightarrow{\eta} GF$ (called the **unit**) and $FG \xrightarrow{\varepsilon} id_\mathbf{A}$ (called the **co-unit**) that satisfy the following conditions:

(1) $G \xrightarrow{\eta G} GFG \xrightarrow{G\varepsilon} G = G \xrightarrow{id_G} G$,

(2) $F \xrightarrow{F\eta} FGF \xrightarrow{\varepsilon F} F = F \xrightarrow{id_F} F$.

19.4 EXAMPLES
By Theorem 19.1 every adjoint functor gives rise to an adjoint situation. (Cf. Examples 18.2 of adjoint functors.) In particular:

(1) every embedding $\mathbf{A} \xrightarrow{E} \mathbf{B}$ of a full reflective subcategory gives rise to an adjoint situation $(\eta, \varepsilon): R \dashv E: \mathbf{A} \to \mathbf{B}$, where R is a reflector, the η_B's are $\mathbf{A}$-reflection arrows, and the ε_A's are isomorphisms,

(2) every concrete category $(\mathbf{A}, U)$ over $\mathbf{X}$ with free objects gives rise to an adjoint situation $(\eta, \varepsilon): F \dashv U: \mathbf{A} \to \mathbf{X}$, where F is called a **free functor**, the η_X's are universal arrows (also called **insertions of generators**), and each ε_A expresses A as a retract of the free object generated by UA,

(3) for each set M, $(\eta, \varepsilon): (M \times -) \dashv \hom(M, -): \mathbf{Set} \to \mathbf{Set}$ is an adjoint situation, where $(M \times -)$ is the endofunctor of Example 18.2(7), each $B \xrightarrow{\eta_B} \hom(M, M \times B)$ is defined by $(\eta_B(b))(m) = (m, b)$, and each $M \times \hom(M, A) \xrightarrow{\varepsilon_A} A$ is defined by $\varepsilon_A(m, f) = f(m)$.

19.5 REMARK
We have seen that every adjoint functor gives rise to an adjoint situation. Below it will be shown that

(1) every adjoint situation arises in this way (cf. 19.7),

(2) every adjoint situation $(\eta, \varepsilon): F \dashv G: \mathbf{A} \to \mathbf{B}$ is, up to a natural isomorphism, uniquely determined by G (cf. 19.9).

First, however, we exhibit a duality that is inherent in the concept of adjoint situations.

19.6 DUALITY THEOREM FOR ADJOINT SITUATIONS
If $(\eta, \varepsilon): F \dashv G: \mathbf{A} \to \mathbf{B}$ *is an adjoint situation, then* $(\varepsilon^{op}, \eta^{op}): G^{op} \dashv F^{op}: \mathbf{B}^{op} \to \mathbf{A}^{op}$ *is an adjoint situation.*

Proof: If $\mathbf{A} \xrightarrow{G} \mathbf{B}$ and $\mathbf{B} \xrightarrow{F} \mathbf{A}$ are functors, then so are $\mathbf{A}^{\mathrm{op}} \xrightarrow{G^{\mathrm{op}}} \mathbf{B}^{\mathrm{op}}$ and $\mathbf{B}^{\mathrm{op}} \xrightarrow{F^{\mathrm{op}}} \mathbf{A}^{\mathrm{op}}$. If $id_\mathbf{B} \xrightarrow{\eta} GF$ and $FG \xrightarrow{\varepsilon} id_\mathbf{A}$ are natural transformations, then so are $G^{\mathrm{op}}F^{\mathrm{op}} \xrightarrow{\eta^{\mathrm{op}}} id_{\mathbf{B}^{\mathrm{op}}}$ and $id_{\mathbf{A}^{\mathrm{op}}} \xrightarrow{\varepsilon^{\mathrm{op}}} F^{\mathrm{op}}G^{\mathrm{op}}$. If $G\varepsilon \circ \eta G = id_G$, it follows that $\eta^{\mathrm{op}}G^{\mathrm{op}} \circ G^{\mathrm{op}}\varepsilon^{\mathrm{op}} = id_{G^{\mathrm{op}}}$, and if $\varepsilon F \circ F\eta = id_F$, it follows that $F^{\mathrm{op}}\eta^{\mathrm{op}} \circ \varepsilon^{\mathrm{op}}F^{\mathrm{op}} = id_{F^{\mathrm{op}}}$. □

19.7 PROPOSITION

If $(\eta, \varepsilon): F \dashv G: \mathbf{A} \to \mathbf{B}$ is an adjoint situation, then the following hold:

(1) G is an adjoint functor,

(2) for each $\mathbf{B}$-object B, $B \xrightarrow{\eta_B} GFB$ is a G-universal arrow,

(3) F is a co-adjoint functor,

(4) for each $\mathbf{A}$-object A, $FGA \xrightarrow{\varepsilon_A} A$ is a F-co-universal arrow.

Proof: It suffices to prove (2), since (2) implies (1), and (3) and (4) follow by the Duality Theorem 19.6 from (1) and (2). Let $B \xrightarrow{f} GA$ be a G-structured arrow. Then the commutative diagram

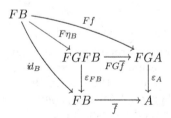

shows that $FB \xrightarrow{\varepsilon_A \circ Ff} A$ is an $\mathbf{A}$-morphism $\overline{f}$ with $f = G\overline{f} \circ \eta_B$. Conversely, if $FB \xrightarrow{\overline{f}} A$ is an $\mathbf{A}$-morphism with $f = G\overline{f} \circ \eta_B$, then the commutative diagram

shows that $\overline{f} = \varepsilon_A \circ Ff$. □

19.8 REMARKS

(1) Observe that Theorem 19.1 and Proposition 19.7 imply that $G: \mathbf{A} \to \mathbf{B}$ is an adjoint functor if and only if there exists an adjoint situation $(\eta, \varepsilon): F \dashv G: \mathbf{A} \to \mathbf{B}$.

(2) A functor $G: \mathbf{A} \to \mathbf{B}$ is an equivalence if and only if there exists an adjoint situation $(\eta, \varepsilon): F \dashv G: \mathbf{A} \to \mathbf{B}$ with natural isomorphisms η and ε. Cf. Proposition 6.8.

(3) If $G: \mathbf{A} \to \mathbf{B}$ and $F: \mathbf{B} \to \mathbf{A}$ are concrete functors over $\mathbf{X}$, then (F, G) is a Galois correspondence if and only if there exist concrete natural transformations η and ε for which $(\eta, \varepsilon): F \dashv G: \mathbf{A} \to \mathbf{B}$ is an adjoint situation. (Cf. 6.24 and 6.25.)

19.9 PROPOSITION
Adjoint situations associated with a given adjoint functor $G: \mathbf{A} \to \mathbf{B}$ are essentially unique, i.e., for each adjoint situation $(\eta, \varepsilon): F \dashv G: \mathbf{A} \to \mathbf{B}$, the following hold:

(1) if $(\overline{\eta}, \overline{\varepsilon}): \overline{F} \dashv G: \mathbf{A} \to \mathbf{B}$ is an adjoint situation, then there exists a natural isomorphism $F \xrightarrow{\tau} \overline{F}$ for which $\overline{\eta} = G\tau \circ \eta$ and $\overline{\varepsilon} = \varepsilon \circ \tau^{-1}G$,

(2) if $\overline{F}: \mathbf{B} \to \mathbf{A}$ is a functor and $F \xrightarrow{\tau} \overline{F}$ is a natural isomorphism, then $(G\tau \circ \eta, \varepsilon \circ \tau^{-1}G): \overline{F} \dashv G: \mathbf{A} \to \mathbf{B}$ is an adjoint situation.

Proof:
(1). By Proposition 19.7, for each **B**-object B, $B \xrightarrow{\eta_B} GFB$ and $B \xrightarrow{\overline{\eta}_B} G\overline{F}B$ are G-universal arrows for B. Hence by Proposition 8.25 there exists an isomorphism $FB \xrightarrow{\tau_B} \overline{F}B$ with $G\tau_B \circ \eta_B = \overline{\eta}_B$. It follows that $F \xrightarrow{\tau = (\tau_B)} \overline{F}$ is a natural isomorphism with $\overline{\eta} = G\tau \circ \eta$. For each **A**-object A the equalities $G\varepsilon_A \circ \eta_{GA} = id_{GA} = G\overline{\varepsilon}_A \circ \overline{\eta}_{GA} = G\overline{\varepsilon}_A \circ G\tau_{GA} \circ \eta_{GA} = G(\overline{\varepsilon}_A \circ \tau_{GA}) \circ \eta_{GA}$ imply that $\varepsilon_A = \overline{\varepsilon}_A \circ \tau_{GA}$. Hence $\varepsilon = \overline{\varepsilon} \circ \tau G$.

(2). Straightforward computations show that $G(\varepsilon \circ \tau^{-1}G) \circ (G\tau \circ \eta)G = id_G$ and that $(\varepsilon \circ \tau^{-1}G)\overline{F} \circ \overline{F}(G\tau \circ \eta) = id_{\overline{F}}$. □

19.10 DEFINITION
Let $\mathbf{A} \xrightarrow{G} \mathbf{B}$ and $\mathbf{B} \xrightarrow{F} \mathbf{A}$ be functors. Then F is called a **co-adjoint for G** and G is called an **adjoint for F** (in symbols: $F \dashv G$) provided that there exist natural transformations η and ε such that $(\eta, \varepsilon): F \dashv G: \mathbf{A} \to \mathbf{B}$ is an adjoint situation.

19.11 REMARKS
(1) Every adjoint functor G has, by Theorem 19.1, a co-adjoint F; and any two co-adjoints F and $\overline{F}$ of G are, by Proposition 19.9, naturally equivalent. Thus every adjoint functor has, up to natural equivalence, a unique co-adjoint. Conversely, if a functor G has a co-adjoint, then, by Proposition 19.7, G is an adjoint functor. Thus a functor *is adjoint* if and only if it *has a co-adjoint*. Moreover, by Proposition 19.7, the co-adjoint of a functor is a co-adjoint functor.

(2) By duality, a functor is co-adjoint if and only if it has an (essentially unique) adjoint. A functor F is a co-adjoint for G if and only if G is an adjoint for F. Thus adjoint functors and co-adjoint functors naturally come in pairs, and these pairs are parts of adjoint situations.

(3) The reader should be aware that the following alternative terminology is also used:
G is right adjoint = G has a left adjoint = G is adjoint.
F is left adjoint = F has a right adjoint = F is co-adjoint.

19.12 EXAMPLES

(1) Let $\mathbf{A}$ be a reflective subcategory of $\mathbf{B}$. Then a functor $\mathbf{B} \xrightarrow{R} \mathbf{A}$ is a co-adjoint for the inclusion functor $\mathbf{A} \hookrightarrow \mathbf{B}$ if and only if R is a reflector for $\mathbf{A}$.

(2) Let $\mathbf{A}$ be a coreflective subcategory of $\mathbf{B}$. Then a functor $\mathbf{B} \xrightarrow{C} \mathbf{A}$ is an adjoint for the inclusion functor $\mathbf{A} \hookrightarrow \mathbf{B}$ if and only if C is a coreflector for $\mathbf{A}$.

(3) If a construct $(\mathbf{A}, U)$ has free objects, i.e., if U is an adjoint functor, then co-adjoints for U are the free functors for $(\mathbf{A}, U)$.

(4) The discrete functor $D : \mathbf{Set} \to \mathbf{Top}$ is a co-adjoint for the forgetful functor $U : \mathbf{Top} \to \mathbf{Set}$, and the indiscrete functor $I : \mathbf{Set} \to \mathbf{Top}$ is an adjoint for U.

(5) If a functor G is an isomorphism, then G^{-1} is simultaneously an adjoint and a co-adjoint for G. More generally, for every equivalence $\mathbf{A} \xrightarrow{G} \mathbf{B}$ there exists a functor $\mathbf{B} \xrightarrow{F} \mathbf{A}$ that is simultaneously an adjoint and a co-adjoint for G. Cf. Proposition 6.8.

(6) Let $\mathbf{A}$ be a category with a zero-object. Then the zero-functor $\mathbf{1} \to \mathbf{A}$ is simultaneously an adjoint and a co-adjoint for the functor $\mathbf{A} \to \mathbf{1}$. Cf. Exercise 4H(b).

(7) A co-adjoint for the minimal realization functor [cf. 18.2(8)] is the behavior functor $B : \mathbf{Aut}_r \to \mathbf{Beh}$ that assigns to each reachable automaton its behavior function [cf. 8.31(5)].

19.13 PROPOSITION
Adjoint situations can be composed; specifically, if $(\eta, \varepsilon) : F \dashv G : \mathbf{A} \to \mathbf{B}$ *and* $(\overline{\eta}, \overline{\varepsilon}) : \overline{F} \dashv \overline{G} : \mathbf{B} \to \mathbf{C}$ *are adjoint situations, then so is*

$$(\overline{G}\eta\overline{F} \circ \overline{\eta}, \ \varepsilon \circ F\overline{\varepsilon}G) : F \circ \overline{F} \dashv \overline{G} \circ G : \mathbf{A} \to \mathbf{C}. \qquad \square$$

19.14 THEOREM
If $(\eta, \varepsilon) : F \dashv G : \mathbf{A} \to \mathbf{B}$ *is an adjoint situation, then the following hold:*

(1) The following are equivalent:

  (a) G is faithful,

  (b) G reflects epimorphisms,

  (c) ε is an Epi-transformation.

(2) The following are equivalent:

  (a) G is faithful and reflects isomorphisms,

  (b) G reflects extremal epimorphisms,

  (c) ε is an (Extremal Epi)-transformation.

(3) G is full if and only if ε is a Section-transformation.

(4) G is full and faithful if and only if ε is a natural isomorphism.

(5) If G reflects regular epimorphisms, then each mono-source is G-initial.[70]

Proof: (1a) $\Rightarrow$ (1b) $\Rightarrow$ (1c) is immediate, since $G\varepsilon$ is a Retraction-transformation.

(1c) $\Rightarrow$ (1a). If $A \underset{g}{\overset{f}{\rightrightarrows}} \hat{A}$ is a pair of **A**-morphisms with $Gf = Gg$, then the equations $f \circ \varepsilon_A = \varepsilon_{\hat{A}} \circ FGf = \varepsilon_{\hat{A}} \circ FGg = g \circ \varepsilon_A$ imply that $f = g$.

(2a) $\Rightarrow$ (2b) $\Rightarrow$ (2c) is immediate, since G, being adjoint, preserves monomorphisms.

(2c) $\Rightarrow$ (2a). By (1), G is faithful. Let $A \xrightarrow{f} \hat{A}$ be an **A**-morphism such that Gf is an isomorphism. Since $f \circ \varepsilon_A = \varepsilon_{\hat{A}} \circ FGf$ is a composite of an isomorphism with an extremal epimorphism, and so is an extremal epimorphism, the dual of Proposition 7.62(2) implies that f is an extremal epimorphism. By the faithfulness of G, f is a monomorphism as well; hence it is an isomorphism.

(3). Let G be full. Then for each **A**-object A there exists an **A**-morphism $A \xrightarrow{s_A} FGA$ with $Gs_A = \eta_{GA}$. Hence the equations $G(s_A \circ \varepsilon_A) \circ \eta_{GA} = Gs_A \circ G\varepsilon_A \circ \eta_{GA} = Gs_A = G(id_{FGA}) \circ \eta_{GA}$ imply that $s_A \circ \varepsilon_A = id_{FGA}$. Thus ε_A is a section. Conversely, let ε be a Section-transformation, let A and $\hat{A}$ be **A**-objects, and let $GA \xrightarrow{f} G\hat{A}$ be a **B**-morphism. Then there exists an **A**-morphism $A \xrightarrow{r} FGA$ with $r \circ \varepsilon_A = id_{FGA}$. Hence $A \xrightarrow{\hat{f}} \hat{A} = A \xrightarrow{r} FGA \xrightarrow{Ff} FG\hat{A} \xrightarrow{\varepsilon_{\hat{A}}} \hat{A}$ is an **A**-morphism. Since $\varepsilon_{\hat{A}}$ is an F-co-universal arrow, the equalities $\varepsilon_{\hat{A}} \circ FG\hat{f} = \hat{f} \circ \varepsilon_A = \varepsilon_{\hat{A}} \circ Ff \circ r \circ \varepsilon_A = \varepsilon_{\hat{A}} \circ Ff$ imply that $G\hat{f} = f$. Thus G is full.

(4). This follows from (1) and (3).

(5). If G reflects regular epimorphisms, then ε is a (Regular Epi)-transformation. Let $(A \xrightarrow{m_i} A_i)_I$ be a mono-source, $(\hat{A} \xrightarrow{f_i} A_i)_I$ be a source in **A**, and $G\hat{A} \xrightarrow{f} GA$ be a **B**-morphism such that $Gf_i = Gm_i \circ f$ for each $i \in I$. Then the equalities $f_i \circ \varepsilon_{\hat{A}} = \varepsilon_{A_i} \circ FGf_i = \varepsilon_{A_i} \circ FGm_i \circ Ff = m_i \circ \varepsilon_A \circ Ff$ imply that for each $i \in I$ the diagram

$$\begin{array}{ccc} FG\hat{A} & \xrightarrow{\varepsilon_{\hat{A}}} & \hat{A} \\ {\scriptstyle \varepsilon_A \circ Ff} \downarrow & & \downarrow {\scriptstyle f_i} \\ A & \xrightarrow{m_i} & A_i \end{array}$$

commutes. Since every category has the (Regular Epi, Mono-Source)-diagonalization property, there exists a diagonal $\hat{A} \xrightarrow{d} A$. In particular $f_i = m_i \circ d$. Since, by (1), G is faithful, this implies that the mono-source $(A \xrightarrow{m_i} A_i)_I$ is G-initial. □

---

[70]See Exercise 19B(c) for a characterization of adjoint functors G with the property that each mono-source is G-initial.

Suggestions for Further Reading

Booth, P. I. Sequences of adjoint functors. *Archiv Math.* **23** (1972): 489–493.

Hoffmann, R.-E. Sequences of adjoints for **Ens**-valued functors. *Manuscr. Math.* **32** (1980): 191–210.

EXERCISES

19A. Alternative Description of Adjoint Situations

Show that functors $\mathbf{A} \xrightarrow{G} \mathbf{B}$ and $\mathbf{B} \xrightarrow{F} \mathbf{A}$ yield an adjoint situation if and only if there exists an isomorphism $\hom(FA, B) \cong \hom(A, GB)$ that is natural in the variables A and B; more precisely,

Let $\mathbf{A} \xrightarrow{G} \mathbf{B}$ and $\mathbf{B} \xrightarrow{F} \mathbf{A}$ be functors. Define functors $\hom(F_-, \_): \mathbf{B}^{op} \times \mathbf{A} \to \mathbf{Set}$ and $\hom(\_, G_-): \mathbf{B}^{op} \times \mathbf{A} \to \mathbf{Set}$ by

$$\hom(F_-, \_)(B, A) = \hom_{\mathbf{A}}(FB, A)$$
$$\hom(F_-, \_)(f, g) = \hom(Ff, g), \text{ where } \hom(Ff, g)(k) = g \circ k \circ Ff$$
$$\hom(\_, G_-)(B, A) = \hom_{\mathbf{B}}(B, GA)$$
$$\hom(\_, G_-)(f, g) = \hom(f, Gg), \text{ where } \hom(f, Gg)(k) = Gg \circ k \circ f.$$

(a) Let $(\eta, \varepsilon): F \dashv G$ be an adjoint situation.
Define $\tau_{(B,A)}: \hom_{\mathbf{A}}(FB, A) \to \hom_{\mathbf{B}}(B, GA)$ by $\tau_{(B,A)}(f) = Gf \circ \eta_B$ and show that $\tau = (\tau_{B,A}): \hom(F_-, \_) \to \hom(\_, G_-)$ is a natural isomorphism.

(b) Let $\tau: \hom(F_-, \_) \to \hom(\_, G_-)$ be a natural isomorphism. Define $\eta = (\eta_B)$ by $\eta_B = \tau_{(B,FB)}(id_{FB})$ and $\varepsilon = (\varepsilon_A)$ by $\varepsilon_A = \tau^{-1}_{(GA,A)}(id_{GA})$ and show that $(\eta, \varepsilon): F \dashv G$ is an adjoint situation.

(c) The constructions described in (a) and (b) are essentially inverse to each other.

19B. Adjoints Reflecting Special Epimorphisms

Let $(\eta, \varepsilon): F \dashv G: \mathbf{A} \to \mathbf{B}$ be an adjoint situation. Show that

(a) If f is G-final and Gf is an extremal (resp. a regular) epimorphism, then f is an extremal (resp. a regular) epimorphism.

(b) G reflects regular epimorphisms if and only if every **A**-morphism f, for which Gf is a regular epimorphism, is G-final.

(c) The following are equivalent:

(1) mono-sources are G-initial,

(2) G reflects swell epimorphisms,

(3) ε is a (Swell Epi)-transformation.

(d) Each of the following conditions implies but is not equivalent to its immediate successor:

   (1) G reflects regular epimorphisms,
   (2) G reflects swell epimorphisms,
   (3) G reflects extremal epimorphisms,
   (4) G reflects epimorphisms.

(e) If G reflects swell epimorphisms, then G reflects limits.

(f) If G reflects isomorphisms and $\mathbf{A}$ has equalizers, then G is faithful.

(g) If G reflects isomorphisms, then G need not be faithful.

19C. Hom-functors Reflecting Epimorphisms

Let $\hom(A, -) : \mathbf{A} \to \mathbf{Set}$ be a covariant hom-functor that reflects epimorphisms. Show that

(a) If in $\mathbf{A}$ arbitrary copowers ^{I}A of A exist, then $\hom(A, -)$ is faithful.

(b) In general $\hom(A, -)$ need not be faithful.

19D. Galois Adjunctions

An adjoint situation $(\eta, \varepsilon) : F \dashv G$ is called a **Galois adjunction** provided that ηG is a natural isomorphism. Show that

(a) For adjoint situations the following conditions are equivalent:

   (1) $(\eta, \varepsilon) : F \dashv G$ is a Galois adjunction,
   (2) $F\eta$ is a natural isomorphism,
   (3) ηG is an Epi-transformation,
   (4) $F\eta$ is an Epi-transformation,
   (5) $GF\eta = \eta GF$.

(b) If $(\eta, \varepsilon) : F \dashv G$ is a Galois-adjunction, then so is $(\varepsilon^{\mathrm{op}}, \eta^{\mathrm{op}}) : G^{\mathrm{op}} \dashv F^{\mathrm{op}}$.

(c) If $(\eta, \varepsilon) : F \dashv G$ is an adjoint situation and if η is an Epi-transformation or if G is full, then $(\eta, \varepsilon) : F \dashv G$ is a Galois-adjunction.

(d) If $(\eta, \varepsilon) : F \dashv G$ is a Galois-adjunction, then G and GFG are naturally isomorphic.

19E. Galois Correspondences

(a) Let $(\mathbf{A}, U) \xrightarrow{G} (\mathbf{B}, V)$ and $(\mathbf{B}, V) \xrightarrow{F} (\mathbf{A}, U)$ be concrete functors over $\mathbf{X}$. Show that (F, G) is a Galois correspondence if and only if there exists an adjoint situation $(\eta, \varepsilon) : F \dashv G$ with concrete natural transformations η and ε.

* (b) Construct a concrete functor G that has a concrete co-adjoint, but is not a part of a Galois correspondence (F, G). Cf. Exercise 21E(b).

* 19F. Adjoint Sequences

Consider the poset $\mathbb{N}$ as a category and show that

(a) There is a sequence of functors $G_n : \mathbb{N} \to \mathbb{N}$, no two of which are naturally isomorphic, such that $G_0(x) = x + 1$ and $G_{n+1} \dashv G_n$ for each $n \in \mathbb{N}$.

(b) If $G_n : \mathbb{N} \to \mathbb{N}$ is a sequence of functors such that $G_n \dashv G_{n+1}$ for each $n \in \mathbb{N}$, then $G_n = id_\mathbb{N}$ for each $n \in \mathbb{N}$.

19G. Self-Adjoints

(a) Let $\mathbf{A} \xrightarrow{G} \mathbf{A}$ be an endofunctor of $\mathbf{A}$ with $G \dashv G$. Show that

(1) If $\mathbf{A}$ is a poset, considered as a category, then G is an isomorphism.

(2) If $\mathbf{A} = \mathbf{Set}$, then G is an equivalence.

(b) Let $\mathbf{A}$ be a category with a zero object A and let $G : \mathbf{A} \to \mathbf{A}$ be the constant functor with value A. Show that $G \dashv G$.

19H. Adjoint Situations and Equivalences

(a) Show that a functor $\mathbf{A} \xrightarrow{G} \mathbf{B}$ is an equivalence if and only if there exists an adjoint situation $(\eta, \varepsilon) : F \dashv G : \mathbf{A} \to \mathbf{B}$ with natural isomorphisms η and ε.

(b) Let $(\mathbf{A}, U)$ be a construct. Show that $F \dashv U \dashv F$ implies that F is an equivalence.

* 19I.

Show that if $F \dashv G \dashv H$, then F is full and faithful if and only if H is full and faithful.

19J. Units and Co-units

Let $(\eta, \varepsilon) : F \dashv G : \mathbf{A} \to \mathbf{B}$ be an adjoint situation. Show that

(a) $GF\eta \circ \eta = \eta GF \circ \eta$.

(b) $\varepsilon \circ \varepsilon FG = \varepsilon \circ FG\varepsilon$.

(c) If $B \xrightarrow{f} GA$ is a G-structured arrow, then $\overline{f} = \varepsilon_A \circ Ff$ is the unique $\mathbf{A}$-morphism $FB \xrightarrow{\overline{f}} A$ with $f = G\overline{f} \circ \eta_B$.

19K. Swell Separators

In $\mathbf{A}$ an object S is called a **swell separator** provided that mono-sources in $\mathbf{A}$ are $hom(S,-)$-initial. Show that

(a) Every swell separator is an extremal separator.

(b) If S is a swell separator, then $hom(S,-)$ reflects limits.

(c) If in $\mathbf{A}$ arbitrary copowers ^IS of S exist, then the following hold:

(1) S is a swell separator if and only if $hom(S,-)$ reflects swell epimorphisms.

(2) S is an extremal separator if and only if $\hom(S, -)$ reflects extremal epimorphisms.

(3) S is a separator if and only if $\hom(S, -)$ reflects epimorphisms. [But cf. 19C(b).]

(4) If $\hom(S, -)$ reflects regular epimorphisms, then S is a swell separator.

(d) In **Set** every separator is swell.

20 Monads

Algebraic constructs $(\mathbf{A}, U)$, such as **Vec**, **Grp**, **Mon**, and **Lat**, can be fully described by the following data, called the *monad* associated with $(\mathbf{A}, U)$:

1. the functor $T : \mathbf{Set} \to \mathbf{Set}$, where $T = U \circ F$ and $F : \mathbf{Set} \to \mathbf{A}$ is the associated free functor (19.4(2)),

2. the natural transformation $\eta : id_{\mathbf{Set}} \to T$ formed by universal arrows, and

3. the natural transformation $\mu : T \circ T \to T$ given by the unique homomorphism $\mu_X : T(TX) \to TX$ that extends id_{TX}.

In fact, in the above cases, there is a canonical concrete isomorphism K between $(\mathbf{A}, U)$ and the full concrete subcategory of $\mathbf{Alg}(T)$ consisting of those T-algebras $TX \xrightarrow{x} X$ that satisfy the equations $x \circ \eta_X = id_X$ and $x \circ Tx = x \circ \mu_X$. The latter subcategory is called the *Eilenberg-Moore category* of the monad (T, η, μ).

The above observation makes it possible, in the following four steps, to express the "degree of algebraic character" of *arbitrary* concrete categories that have free objects:

Step 1: With every concrete category $(\mathbf{A}, U)$ over $\mathbf{X}$ that has free objects (or, more generally, with every adjoint functor $\mathbf{A} \xrightarrow{U} \mathbf{X}$) one can associate, in an essentially unique way, an adjoint situation $(\eta, \varepsilon) : F \dashv U : \mathbf{A} \to \mathbf{X}$. (See Theorem 19.1 and Proposition 19.9.)

Step 2: With every adjoint situation $(\eta, \varepsilon) : F \dashv U : \mathbf{A} \to \mathbf{X}$ one can associate a monad $\mathbf{T} = (T, \eta, \mu)$ on $\mathbf{X}$, where $T = U \circ F : \mathbf{X} \to \mathbf{X}$. (See Proposition 20.3.)

Step 3: With every monad $\mathbf{T} = (T, \eta, \mu)$ on $\mathbf{X}$ one can associate a concrete subcategory of $\mathbf{Alg}(T)$ denoted by $(X^{\mathbf{T}}, U^{\mathbf{T}})$ and called the *category of* $\mathbf{T}$-*algebras*. (See Definition 20.4.)

Step 4: With every concrete category $(\mathbf{A}, U)$ over $\mathbf{X}$ that has free objects one can associate a distinguished concrete functor $(\mathbf{A}, U) \xrightarrow{K} (X^{\mathbf{T}}, U^{\mathbf{T}})$ into the associated category of $\mathbf{T}$-algebras called the *comparison functor* for $(\mathbf{A}, U)$. (See Proposition 20.37 and Definition 20.38.)

Concrete categories that are concretely isomorphic to a category of $\mathbf{T}$-algebras for some monad $\mathbf{T}$ have a distinct "algebraic flavor". Such categories $(\mathbf{A}, U)$ and their forgetful functors U are called *monadic*. It turns out that a concrete category $(\mathbf{A}, U)$ is monadic if and only if it has free objects and its associated comparison functor $(\mathbf{A}, U) \xrightarrow{K} (X^{\mathbf{T}}, U^{\mathbf{T}})$ is an isomorphism. Thus, for concrete categories $(\mathbf{A}, U)$ that have free objects, the associated comparison functor can be considered as a means of measuring the "algebraic character" of $(\mathbf{A}, U)$; and the associated category of $\mathbf{T}$-algebras can be considered to be the "algebraic part" of $(\mathbf{A}, U)$. In particular,

(a) every finitary variety (see Definition 16.16) is monadic,

(b) the category **TopGrp**, considered as a concrete category

  (1) over **Top**, is monadic,

  (2) over **Set**, is not monadic; the associated comparison functor is the forgetful functor **TopGrp** → **Grp**, so that the construct **Grp** may be considered as the "algebraic part" of the construct **TopGrp**,

(c) the construct **Top** is not monadic; the associated comparison functor is the forgetful functor **Top** → **Set** itself, so that the construct **Set** may be considered as the "algebraic part" of the construct **Top**; hence the construct **Top** may be considered as having a trivial "algebraic part".

Among constructs, monadicity captures the idea of "algebraicness" rather well (as will be demonstrated in §24). Unfortunately, however, the behavior of monadic categories in general is far from satisfactory. Monadic functors can fail badly to reflect nice properties of the base category (e.g., the existence of colimits or of suitable factorization structures), and they are not closed under composition. Such deficiencies are shown near the end of this section. Better behaved (and, moreover, simpler) concepts of "algebraicity" will be developed in §23.

MONADS AND ALGEBRAS

20.1 DEFINITION

A **monad** on a category $\mathbf{X}$ is a triple $\mathbf{T} = (T, \eta, \mu)$ consisting of a functor $T: \mathbf{X} \to \mathbf{X}$ and natural transformations

$$\eta: id_{\mathbf{X}} \to T \quad \text{and} \quad \mu: T \circ T \to T$$

such that the diagrams

$$\begin{array}{ccc} T \circ T \circ T & \xrightarrow{T\mu} & T \circ T \\ \mu T \downarrow & & \downarrow \mu \\ T \circ T & \xrightarrow{\mu} & T \end{array} \quad \text{and} \quad \begin{array}{ccc} T & \xrightarrow{T\eta} & T \circ T & \xleftarrow{\eta T} & T \\ & {}_{id}\searrow & \downarrow \mu & \swarrow_{id} & \\ & & T & & \end{array}$$

commute.

20.2 EXAMPLES

(1) On every category $\mathbf{X}$ there is the **trivial monad** $\mathbf{T} = (T, \eta, \mu)$ with $T = id_{\mathbf{X}}$ and $\eta = \mu = id_T$.

(2) In **Set** the **word-monad** $\mathbf{T} = (T, \eta, \mu)$ is defined as follows:

  (a) $T: \mathbf{Set} \to \mathbf{Set}$ assigns to each set X the set $TX = \bigcup_{n \in \mathbb{N}} X^n$ of tuples or "words" over X and to each function $X \xrightarrow{f} Y$ the function $TX \xrightarrow{Tf} TY$ that sends the empty word ($\in X^0$) to the empty word and for $n > 0$ sends each word $(x_1, \ldots, x_n)$ to $(f(x_1), \ldots, f(x_n))$,

(b) $\eta_X : X \to TX$ is defined by $\eta_X(x) = (x)$, i.e., η_X interprets each member of X as a one-letter word,

(c) $\mu_X : T(TX) \to TX$ is given by concatenation:

$$\mu_X((x_{11},\ldots,x_{1n_1}),(x_{21},\ldots,x_{2n_2}), \ldots,(x_{k1},\ldots,x_{kn_k})) =$$
$$(x_{11},\ldots,x_{1n_1},x_{21},\ldots,x_{2n_2},\ldots,x_{k1},\ldots,x_{kn_k}),$$

i.e., μ_X interprets each word of words in the natural way, as a word.

(3) In **Set** the **power-set monad** $\mathbf{T} = (\mathcal{P},\eta,\mu)$ is defined as follows:

(a) $\mathcal{P} : \mathbf{Set} \to \mathbf{Set}$ is the power-set functor,

(b) $\eta_X : X \to \mathcal{P}X$ is defined by $\eta_X(x) = \{x\}$,

(c) $\mu_X : \mathcal{P}(\mathcal{P}X) \to \mathcal{P}X$ is defined by $\mu_X(Z) = \bigcup Z$.

(4) Let $\mathbf{X}$ be a poset, considered as a category. A monad $\mathbf{T}$ on $\mathbf{X}$ consists of an order-preserving function $T : \mathbf{X} \to \mathbf{X}$ with $s \leq Ts$ (due to η) and $T = T \circ T$ (since $(T \circ T)s \leq Ts$ due to μ) — in other words, T is a closure operator on X.

20.3 PROPOSITION

Each adjoint situation $(\eta,\varepsilon) : F \dashv G : \mathbf{A} \to \mathbf{X}$ *gives rise to the* **associated monad** (T,η,μ) *on* $\mathbf{X}$, *defined by*

$$T = G \circ F : \mathbf{X} \to \mathbf{X} \quad \text{and} \quad \mu = G\varepsilon F : T \circ T \to T.$$

Proof:
(1). $\mu \circ T\mu = G\varepsilon F \circ GFG\varepsilon F = G(\varepsilon \circ FG\varepsilon)F = G(\varepsilon \circ \varepsilon FG)F = G\varepsilon F \circ G\varepsilon FGF = \mu \circ \mu T$, since $\varepsilon \circ FG\varepsilon = \varepsilon \circ \varepsilon FG$, in view of the fact that ε is a natural transformation. (Cf. Exercise 19J.)

(2). $\mu \circ T\eta = G\varepsilon F \circ GF\eta = G(\varepsilon F \circ F\eta) = G(id) = id$.

(3). $\mu \circ \eta T = G\varepsilon F \circ \eta GF = (G\varepsilon \circ \eta G)F = (id)F = id$. □

20.4 DEFINITION

Let $\mathbf{T} = (T,\eta,\mu)$ be a monad on $\mathbf{X}$. The full concrete subcategory of $\mathbf{Alg}(T)$ consisting of all algebras $TX \xrightarrow{x} X$ that satisfy

(1) $x \circ \eta_X = id_X$, and

(2) $x \circ Tx = x \circ \mu_X : T(TX) \to X$

is denoted by $(\mathbf{X}^{\mathbf{T}}, U^{\mathbf{T}})$ and is called the **Eilenberg-Moore category** of the monad $\mathbf{T}$, or the **category of T-algebras**.

20.5 EXAMPLES

(1) If $\mathbf{T}$ is the trivial monad on $\mathbf{X}$, then $(\mathbf{X}^{\mathbf{T}}, U^{\mathbf{T}})$ is concretely isomorphic to $(\mathbf{X}, id_{\mathbf{X}})$.

(2) If **T** is the word-monad on **X** = **Set**, then $(\mathbf{X^T}, U^\mathbf{T})$ is concretely isomorphic to the construct **Mon** of monoids. [If $TX \xrightarrow{x} X$ is a **T**-algebra, multiplication in X is defined by $a \cdot b = x((a,b))$, and if e is the value of x at the empty word, then $(X, \cdot, e)$ is a monoid.]

(3) If **T** is the power-set monad on **X** = **Set**, then $(\mathbf{X^T}, U^\mathbf{T})$ is concretely isomorphic to the construct **JCPos**. [If $TX \xrightarrow{x} X$ is a **T**-algebra, then $a \leq b$ if and only if $x(\{a,b\}) = b$ defines a partial order on X with $x(A) = \sup A$ for each $A \subseteq X$; the **T**-morphisms are precisely the join-preserving maps.] Cf. Exercise 5N(a).

(4) If $\mathbf{T} = (T, \eta, \mu)$ is a monad on a partially ordered set **X**, considered as a category, and if $E: \mathbf{Y} \to \mathbf{X}$ is the embedding of the full subcategory **Y** of **X**, whose objects are the fixed points of T, then $(\mathbf{X^T}, U^\mathbf{T})$ is concretely isomorphic to $(\mathbf{Y}, E)$.

(5) The construct **Vec** is concretely isomorphic to the construct $\mathbf{Ab^T}$ for the following monad $\mathbf{T} = (T, \eta, \mu)$ on abelian groups: for each abelian group X, let $TX = \mathbb{R} \otimes X$ be the tensor product, with morphisms $\eta_X : X \to \mathbb{R} \otimes X$ given by $x \mapsto 1 \otimes x$, and $\mu_X : \mathbb{R} \otimes \mathbb{R} \otimes X \to \mathbb{R} \otimes X$ given by $r_1 \otimes r_2 \otimes x \mapsto (r_1 r_2) \otimes x$.

20.6 REMARK

As outlined in the introduction of this section, with each concrete category $(\mathbf{A}, U)$ over **X** that has free objects (or, more generally, with every adjoint functor $\mathbf{A} \xrightarrow{U} \mathbf{X}$) one can associate a category $(\mathbf{X^T}, U^\mathbf{T})$ of **T**-algebras in three steps. Since the first of these steps does not yield a unique result, $(\mathbf{X^T}, U^\mathbf{T})$ is not uniquely determined by $(\mathbf{A}, U)$. However, any two categories of algebras that are obtained from $(\mathbf{A}, U)$ in this manner are concretely isomorphic. (See Exercise 20A.) Thus, by "abuse of language", we will sometimes designate any category of algebras that is obtained from $(\mathbf{A}, U)$ via the procedure outlined above as **the category of algebras associated with** $(\mathbf{A}, U)$ (resp. with the functor U).

20.7 PROPOSITION

Every monad $\mathbf{T} = (T, \eta, \mu)$ *on* **X** *gives rise to an* **associated adjoint situation** $(\eta, \varepsilon): F^\mathbf{T} \dashv U^\mathbf{T} : \mathbf{X^T} \to \mathbf{X}$, *where*

(1) $\mathbf{X^T}$ *and* $U^\mathbf{T}$ *are defined as in Definition 20.4,*

(2) $\mathbf{X} \xrightarrow{F^\mathbf{T}} \mathbf{X^T}$ *is defined by* $F^\mathbf{T}(X \xrightarrow{f} Y) = (TX, \mu_X) \xrightarrow{Tf} (TY, \mu_Y)$, *in particular,* (TX, μ_X) *is a free object over* X *in* $(\mathbf{X^T}, U^\mathbf{T})$,

(3) $F^\mathbf{T} U^\mathbf{T} \xrightarrow{\varepsilon} \mathrm{id}_{\mathbf{X^T}}$ *is defined by* $\varepsilon_{(X,x)} = x$.

Moreover, the monad associated with the above adjoint situation (20.3) is **T** *itself.*

Proof: Straightforward computations. □

MONADIC CATEGORIES AND FUNCTORS

20.8 DEFINITION
(1) A concrete category over $\mathbf{X}$ is called **monadic** provided that it is concretely isomorphic to $(\mathbf{X^T}, U^\mathbf{T})$ for some monad $\mathbf{T}$ on $\mathbf{X}$.

(2) A functor $\mathbf{A} \xrightarrow{U} \mathbf{X}$ is called **monadic** provided that U is faithful and $(\mathbf{A}, U)$ is monadic.

20.9 REMARK
In Proposition 20.40 it will be shown that every monadic category is concretely isomorphic to its associated category of algebras.

20.10 EXAMPLES
(1) Every isomorphism is monadic [cf. 20.5(1)].

(2) The construct **Mon** is monadic [cf. 20.5(2)]. So is every finitary variety (cf. 20.20).

(3) The construct **JCPos** is monadic [cf. 20.5(3)].

(4) An embedding of a full subcategory $\mathbf{A}$ of $\mathbf{X}$ into $\mathbf{X}$ is monadic if and only if $\mathbf{A}$ is isomorphism-closed and reflective in $\mathbf{X}$. (See Proposition 20.12, Theorem 20.17, and Exercise 20F.) A concrete category $(\mathbf{A}, U)$ over a partially ordered set $\mathbf{X}$ is monadic if and only if, up to isomorphism, $\mathbf{A} \xrightarrow{U} \mathbf{X}$ is the embedding of a full reflective subcategory of $\mathbf{X}$ [cf. 20.5(4)]. In particular, a partially ordered set, considered as a concrete category over $\mathbf{1}$, is monadic if and only if it contains precisely one element.

(5) None of the constructs **Top**, **Rel**, or **Pos** is monadic since the corresponding forgetful functors don't reflect isomorphisms (cf. 20.12).

20.11 LEMMA
The Eilenberg-Moore category of a monad $\mathbf{T} = (T, \eta, \mu)$ *is closed under the formation of mono-sources in* $\mathbf{Alg}(T)$.

Proof: Let $\mathbf{T} = (T, \eta, \mu)$ be a monad on $\mathbf{X}$, and suppose that the diagram

$$\begin{array}{ccc} TX & \xrightarrow{Tm_i} & TX_i \\ {\scriptstyle x}\downarrow & & \downarrow{\scriptstyle x_i} \\ X & \xrightarrow{m_i} & X_i \end{array}$$

commutes for each $i \in I$, where $(X \xrightarrow{m_i} X_i)_I$ is a mono-source in $\mathbf{X}$, and each (X_i, x_i) is a **T**-algebra. To show that (X, x) is a **T**-algebra, first notice that the commutativity of the diagram

$$\begin{array}{ccc} X & \xrightarrow{m_i} & X_i \\ \eta_X \downarrow & & \downarrow \eta_{X_i} \\ TX & \xrightarrow{Tm_i} & TX_i \\ x \downarrow & & \downarrow x_i \\ X & \xrightarrow{m_i} & X_i \end{array}$$

and the fact that $x_i \circ \eta_{X_i} = id_{X_i}$ imply $m_i \circ (x \circ \eta_X) = x_i \circ Tm_i \circ \eta_X = x_i \circ \eta_{X_i} \circ m_i = id_{X_i} \circ m_i = m_i \circ id_X$, and thus $x \circ \eta_X = id_X$. The commutativity of the five inner quadrangles in the diagram

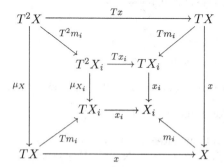

implies that $m_i \circ (x \circ \mu_X) = m_i \circ (x \circ Tx)$ and thus $x \circ \mu_X = x \circ Tx$. □

20.12 PROPOSITION

For monadic functors $\mathbf{A} \xrightarrow{U} \mathbf{X}$ the following hold:

(1) U is faithful,

(2) $(\mathbf{A}, U)$ is fibre-small,

(3) U is adjoint, i.e., $(\mathbf{A}, U)$ has free objects,

(4) $(\mathbf{A}, U)$ is uniquely transportable, hence amnestic,

(5) U creates isomorphisms, hence reflects them,

(6) U reflects epimorphisms and extremal epimorphisms,

(7) U preserves and reflects mono-sources,

(8) in $(\mathbf{A}, U)$ mono-sources are initial,

(9) U **detects wellpoweredness**, i.e., if $\mathbf{X}$ is wellpowered, then so is $\mathbf{A}$,

(10) U creates limits.

Proof: Since all of the properties mentioned above are invariant under concrete isomorphisms, we may assume that $(\mathbf{A}, U) = (\mathbf{X^T}, U^\mathbf{T})$ for some monad $\mathbf{T} = (T, \eta, \mu)$ on $\mathbf{X}$.

(1) and (2) follow immediately from the definition of monadicity.

(3) follows from Proposition 20.7.

(7) follows from (1) and (3).

(8) is immediate by straightforward computation.

(9) follows from (7) and (8).

(10). Let $D : \mathbf{I} \to \mathbf{X}^\mathbf{T}$ be a diagram and let $\mathcal{L} = (L \xrightarrow{\ell_i} D_i)_{Ob(\mathbf{I})}$ be a limit of $U^\mathbf{T} \circ D$. For each $i \in Ob(\mathbf{I})$ let $D_i = (X_i, x_i)$. Then the source $\mathcal{S} = (TL \xrightarrow{T\ell_i} TX_i \xrightarrow{x_i} X_i)_{Ob(\mathbf{I})}$ is natural for $U^\mathbf{T} \circ D$. Hence there exists a unique $\mathbf{X}$-morphism $TL \xrightarrow{y} L$ with $\mathcal{S} = \mathcal{L} \circ y$, i.e., such that each diagram

$$\begin{array}{ccc} TL & \xrightarrow{T\ell_i} & TX_i \\ y \downarrow & & \downarrow x_i \\ L & \xrightarrow{\ell_i} & X_i \end{array}$$

commutes. Thus it remains to be shown that (L, y) is a $\mathbf{T}$-algebra and that the source $\left((L, y) \xrightarrow{\ell_i} D_i\right)_{Ob(\mathbf{I})}$ is a limit of D. The former follows from the fact that $\mathcal{L}$ is a mono-source and from Lemma 20.11. The latter follows from (7) and (8) via Proposition 13.15.

(4) and (5) follow immediately from (10), by Proposition 13.36.

(6) follows immediately from (1), the preservation of monomorphisms (7) and the reflection of isomorphisms (5). □

20.13 REMARK
Even though monadic functors behave perfectly with respect to limits, their relationship to colimits is rather complex. For example, if $\mathbf{A} \xrightarrow{U} \mathbf{X}$ is a monadic functor with $\mathbf{X}$ cocomplete and if $\mathbf{A}$ has coequalizers, then $\mathbf{A}$ is cocomplete (see Exercise 20D); but $\mathbf{A}$ needn't have coequalizers (see 20.47). In fact, monadic functors are characterized by their (somewhat strange) behavior with respect to coequalizers, as will be seen in Theorem 20.17.

20.14 DEFINITION
1. A fork $A \underset{q}{\overset{p}{\rightrightarrows}} B \xrightarrow{c} C$ is called a **congruence fork** provided that (p, q) is a congruence relation of c, and c is a coequalizer of p and q.

2. A fork $A \underset{q}{\overset{p}{\rightrightarrows}} B \xrightarrow{c} C$ is called a **split fork** and c is called a **split coequalizer** of (p, q) provided that there exist morphisms s and t such that the diagram

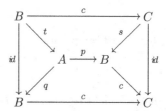

commutes.

3. A colimit $\mathcal{K}$ of a diagram $D : \mathbf{I} \to \mathbf{A}$ is called an **absolute colimit** provided that for each functor $\mathbf{A} \xrightarrow{G} \mathbf{B}$ the sink $G\mathcal{K}$ is a colimit of $G \circ D$.
   In particular, c is called an **absolute coequalizer** of p and q in $\mathbf{A}$ provided that for each functor $\mathbf{A} \xrightarrow{G} \mathbf{B}$, Gc is a coequalizer of Gp and Gq in $\mathbf{B}$.

4. A functor $\mathbf{A} \xrightarrow{G} \mathbf{B}$ is said to **create absolute colimits** provided that for each diagram $\mathbf{I} \xrightarrow{D} \mathbf{A}$ and each absolute colimit $\mathcal{K}$ of $G \circ D$ there exists a unique sink $\mathcal{C} = (D_i \xrightarrow{c_i} C)_{Ob(\mathbf{I})}$ such that $G\mathcal{C} = \mathcal{K}$ and, moreover, $\mathcal{C}$ is a (not necessarily absolute) colimit of D.

A monadic functor creating colimits

20.15 EXAMPLES

(1) If $\mathbf{A}$ is a category, f is an $\mathbf{A}$-morphism, (p, q) is a congruence relation of f in $\mathbf{A}$,

and c is a coequalizer of p and q in **A**, then $\bullet \underset{q}{\overset{p}{\rightrightarrows}} \bullet \xrightarrow{c} \bullet$ is a congruence fork in **A**. (Cf. Proposition 11.22.)

(2) A congruence fork $\bullet \underset{q}{\overset{p}{\rightrightarrows}} \bullet \xrightarrow{c} \bullet$ is a split fork if and only if c is a retraction.

(3) If $\bullet \underset{q}{\overset{p}{\rightrightarrows}} \bullet \xrightarrow{c} \bullet$ is a split fork, then c is an absolute coequalizer of p and q.

(4) If $(\eta, \varepsilon): F \dashv U: \mathbf{A} \to \mathbf{X}$ is an adjoint situation, then for each **A**-object A

$$UFUFUA \underset{U\varepsilon_{FUA}}{\overset{UFU\varepsilon_A}{\rightrightarrows}} UFUA \xrightarrow{U\varepsilon_A} UA$$

is a split fork in **B** since the diagram

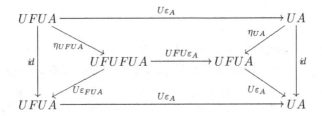

commutes.

(5) If $\mathbf{T} = (T, \eta, \mu)$ is a monad on **X**, then for each **T**-algebra (X, x)

$$T^2 X \underset{\mu_X}{\overset{Tx}{\rightrightarrows}} TX \xrightarrow{x} X$$

is a split fork in **X**. Hence, in particular, for each **X**-object X

$$T^3 X \underset{\mu_{TX}}{\overset{T\mu_X}{\rightrightarrows}} T^2 X \xrightarrow{\mu_X} TX$$

is a split fork in **X**.

20.16 PROPOSITION
Each monadic functor U creates absolute colimits.

Proof: Let $\mathbf{T} = (T, \eta, \mu)$ be a monad on **X** and let $(\mathbf{X^T}, U^\mathbf{T})$ be the associated category of **T**-algebras. Let $D: \mathbf{I} \to \mathbf{X^T}$ be a diagram and let $\mathcal{K} = (U^\mathbf{T} D_i \xrightarrow{c_i} C)_{Ob(\mathbf{I})}$ be an absolute colimit of $U^\mathbf{T} \circ D$. Then $T\mathcal{K} = (TU^\mathbf{T} D_i \xrightarrow{Tc_i} TC)_{Ob(\mathbf{I})}$ is a colimit of $T \circ U^\mathbf{T} \circ D$. If $D_i = (X_i, x_i)$ for each $i \in Ob(\mathbf{I})$, then $\mathcal{S} = (TX_i \xrightarrow{x_i} X_i \xrightarrow{c_i} C)_{Ob(\mathbf{I})}$ is a natural sink for $T \circ U^{\mathbf{T}} \circ D$. Hence there exists a unique morphism $TC \xrightarrow{c} X$ with $\mathcal{S} = c \circ T\mathcal{K}$, i.e., such that for each $i \in Ob(\mathbf{I})$ the diagram

$$\begin{array}{ccc} TX_i & \xrightarrow{Tc_i} & TC \\ {\scriptstyle x_i}\downarrow & & \downarrow{\scriptstyle c} \\ X_i & \xrightarrow{c_i} & C \end{array}$$

commutes. Thus it remains to be shown that (C, c) is a **T**-algebra and that the sink $(D_i \xrightarrow{c_i} (C, c))_{Ob(\mathbf{I})}$ is a colimit of D. The former follows from the fact that $\mathcal{K}$ and $T^2\mathcal{K}$, being colimits, are epi-sinks and from the equations

$$(c \circ \eta_C) \circ c_i = c \circ Tc_i \circ \eta_{X_i} = c_i \circ x_i \circ \eta_{X_i} = c_i \circ id_{X_i} = id_C \circ c_i$$

and

$$(c \circ Tc) \circ T^2 c_i = c \circ Tc_i \circ Tx_i = c_i \circ x_i \circ Tx_i = c_i \circ x_i \circ \mu_{X_i} = c \circ Tc_i \circ \mu_{X_i} = (c \circ \mu_C) \circ T^2 c_i.$$

For the latter, consider a natural sink $\mathcal{X} = (D_i \xrightarrow{f_i} (X, x))_{Ob(\mathbf{I})}$ for D. Then $U^{\mathbf{T}}\mathcal{X} = (X_i \xrightarrow{f_i} X)_{Ob(\mathbf{I})}$ is a natural sink for $U^{\mathbf{T}} \circ D$. Thus there exists a unique **X**-morphism $C \xrightarrow{f} X$ with $f_i = f \circ c_i$ for each $i \in Ob(\mathbf{I})$. Hence $Tf_i = Tf \circ Tc_i$ for each $i \in Ob(\mathbf{I})$ and so $x \circ Tf \circ Tc_i = x \circ Tf_i = f_i \circ x_i = f \circ c_i \circ x_i = f \circ c \circ Tc_i$. Since $T\mathcal{K}$ is an epi-sink, this implies that $x \circ Tf = f \circ c$, i.e., $(C, c) \xrightarrow{f} (X, x)$ is the unique T-homomorphism with $\mathcal{X} = f \circ \mathcal{K}$. □

20.17 CHARACTERIZATION THEOREM FOR MONADIC FUNCTORS
For any functor $\mathbf{A} \xrightarrow{U} \mathbf{X}$ the following conditions are equivalent:

(1) U is monadic,

(2) U is adjoint and creates absolute coequalizers,

(3) U is adjoint and creates split coequalizers (as defined in 20.14).

Proof: (1) ⇒ (2) follows from Propositions 20.12 and 20.16.

(2) ⇒ (3) follows from Example 20.15(3).

(3) ⇒ (1). Since U is adjoint, there is an adjoint situation $(\eta, \varepsilon) : F \dashv U : \mathbf{A} \to \mathbf{X}$. Let $\mathbf{T} = (T, \eta, \mu)$ be the associated monad and let $(\mathbf{X^T}, U^{\mathbf{T}})$ be the associated category of **T**-algebras. It suffices to construct functors $\mathbf{A} \xrightarrow{K} \mathbf{X^T}$ and $\mathbf{X^T} \xrightarrow{L} \mathbf{A}$ with $U = U^{\mathbf{T}} \circ K$, $L \circ K = id_{\mathbf{A}}$, and $K \circ L = id_{\mathbf{X^T}}$; since then U would be faithful, $(\mathbf{A}, U)$ would be a concrete category, and $(\mathbf{A}, U) \xrightarrow{K} (\mathbf{X^T}, U^{\mathbf{T}})$ would be a concrete isomorphism.

Construction of K: For each **A**-object A the pair $(UA, U\varepsilon_A)$ is a **T**-algebra [cf. 20.15(4)], and for each **A**-morphism $A \xrightarrow{f} B$, $(UA, U\varepsilon_A) \xrightarrow{Uf} (UB, U\varepsilon_B)$ is an T-homomorphism.

Thus there exists a unique functor $\mathbf{A} \xrightarrow{K} \mathbf{X}^{\mathbf{T}}$ with $K(A) = (UA, U\varepsilon_A)$ for each $\mathbf{A}$-object A.

Construction of L: For each $\mathbf{X^T}$-object (X, x) the fork

$$T^2X \underset{\mu_X}{\overset{Tx}{\rightrightarrows}} TX \xrightarrow{x} X \ =\ U(FUFX) \underset{U(\varepsilon_{FX})}{\overset{U(Fx)}{\rightrightarrows}} U(FX) \xrightarrow{x} X$$

is a split fork [cf. 20.15(5)]. Hence, by condition (3), $U(FX) \xrightarrow{x} X$ has a unique U-lift $FX \xrightarrow{\bar{x}} L(X)$ and $\bar{x}$ is a coequalizer of ε_{FX} and Fx in $\mathbf{A}$. For each T-homomorphism $(X, x) \xrightarrow{f} (Y, y)$ the following equalities hold:

$$(\bar{y} \circ Ff) \circ \varepsilon_{FX} = \bar{y} \circ \varepsilon_{FY} \circ FUFf = \bar{y} \circ Fy \circ FTf = (\bar{y} \circ Ff) \circ Fx.$$

Since $\bar{x}$ is a coequalizer of ε_{FX} and Fx in $\mathbf{A}$, there exists a unique $\mathbf{A}$-morphism $L(X) \xrightarrow{L(f)} L(Y)$ with $\bar{y} \circ Ff = L(f) \circ \bar{x}$. Since x is an epimorphism, the equalities $UL(f) \circ x = UL(f) \circ U\bar{x} = U\bar{y} \circ UFf = y \circ Tf = f \circ x$ imply that $UL(f) = f$, i.e., $U(L(X, x) \xrightarrow{L(f)} L(Y, y)) = X \xrightarrow{f} Y = U^{\mathbf{T}}((X, x) \xrightarrow{f} (Y, y))$. Thus L defines a concrete functor $\mathbf{X^T} \xrightarrow{L} \mathbf{A}$. The equation $L \circ K = id_{\mathbf{A}}$ immediately follows (via concreteness) from the fact that, for each $\mathbf{A}$-object A, $FUA \xrightarrow{\varepsilon_A} A$ U-lifts $UFUA \xrightarrow{U\varepsilon_A} UA$, so that $L \circ K(A) = A$. To show that $K \circ L = id_{\mathbf{X^T}}$, let (X, x) be an $\mathbf{T}$-algebra. Then $(K \circ L)(X, x) = (UL(X, x), U\varepsilon_{L(X,x)})$. Since $UL(X, x) = X$, it remains (in view of concreteness) to be shown that $x = \varepsilon_{L(X,x)}$. By the universality of η, the equations $U\bar{x} \circ \eta_X = x \circ \eta_X = id_X = U\varepsilon_{L(X,x)} \circ \eta_{UL(X,x)} = U\varepsilon_{L(X,x)} \circ \eta_X$ imply that $\bar{x} = \varepsilon_{L(X,x)}$ and hence that $x = U\bar{x} = U\varepsilon_{L(X,x)}$. □

20.18 PROPOSITION
Each construct of the form $\mathbf{Alg}(\Omega)$ is monadic.

Proof: By Example 8.23(6) the forgetful functor $\mathbf{Alg}(\Omega) \xrightarrow{U} \mathbf{Set}$ is adjoint, so by the above theorem it suffices to show that U creates absolute coequalizers. Consider a pair $(A, (\alpha_i)_I) \underset{q}{\overset{p}{\rightrightarrows}} (B, (\beta_i)_I)$ of Ω-homomorphisms and an absolute coequalizer $B \xrightarrow{c} C$ of p and q in $\mathbf{Set}$. For each $i \in I$, the functor $S^{n_i} : \mathbf{Set} \to \mathbf{Set}$ [3.20(10)] preserves this coequalizer. Thus $B^{n_i} \xrightarrow{c^{n_i}} C^{n_i}$ is a coequalizer of p^{n_i} and q^{n_i} in $\mathbf{Set}$. Since $(c \circ \beta_i) \circ p^{n_i} = c \circ p \circ \alpha_i = c \circ q \circ \alpha_i = (c \circ \beta_i) \circ q^{n_i}$, there exists a unique function $C^{n_i} \xrightarrow{\gamma_i} C$ with $c \circ \beta_i = \gamma_i \circ c^{n_i}$. Thus $(B, (\beta_i)_I) \xrightarrow{c} (C, (\gamma_i)_I)$ is the unique U-lift of the costructured map $(B, (\beta_i)_I) \xrightarrow{c} C$. It follows easily that $(B, (\beta_i)_I) \xrightarrow{c} (C, (\gamma_i)_I)$ is a coequalizer of $(A, (\alpha_i)_I) \underset{q}{\overset{p}{\rightrightarrows}} (B, (\beta_i)_I)$ in $\mathbf{Alg}(\Omega)$. Thus U creates absolute coequalizers. □

20.19 PROPOSITION
Let $(\mathbf{A}, U)$ be a monadic category over $\mathbf{X}$. Then each concrete full reflective subcategory of $\mathbf{A}$ that is closed under the formation of regular quotients is also monadic over $\mathbf{X}$.

Proof: Immediate from Theorem 20.17. □

20.20 PROPOSITION
Each finitary variety is a monadic construct.

Proof: Immediate from Propositions 20.18 and 20.19. □

E-MONADS AND E-MONADIC CATEGORIES AND FUNCTORS

20.21 DEFINITION
(1) A monad $\mathbf{T} = (T, \eta, \mu)$ on $\mathbf{X}$ is called an *E*-**monad** provided that $\mathbf{X}$ is an $(E, \mathbf{M})$-category for some $\mathbf{M}$ and $T[E] \subseteq E$.

RegEpi-monads in categories with regular factorizations are called **regular monads**.

(2) A concrete category $(\mathbf{A}, U)$ over $\mathbf{X}$ (or a faithful functor $\mathbf{A} \xrightarrow{U} \mathbf{X}$) is called *E*-**monadic** (resp. **regularly monadic**) provided that $(\mathbf{A}, U)$ is concretely isomorphic to $(\mathbf{X}^{\mathbf{T}}, U^{\mathbf{T}})$ for some E-monad (resp. regular monad) $\mathbf{T}$ on $\mathbf{X}$.

20.22 PROPOSITION
Every monad on **Set** *is regular.*

Proof: This follows from the facts that in **Set** every regular epimorphism is a retraction and that every functor preserves retractions. □

20.23 DEFINITION
A functor $\mathbf{A} \xrightarrow{U} \mathbf{X}$ **lifts** $(E, \mathbf{M})$-**factorizations uniquely** provided that for any source $\mathcal{S}$ in $\mathbf{A}$ and any $(E, \mathbf{M})$-factorization $U\mathcal{S} = \mathcal{M} \circ e$ in $\mathbf{X}$ there exists a unique factorization $\mathcal{S} = \hat{\mathcal{M}} \circ \hat{e}$ in $\mathbf{A}$ with $U\hat{\mathcal{M}} = \mathcal{M}$ and $U\hat{e} = e$.

20.24 PROPOSITION
If $\mathbf{X}$ *is an* $(E, \mathbf{M})$-*category and* $\mathbf{A} \xrightarrow{U} \mathbf{X}$ *is E-monadic, then U lifts* $(E, \mathbf{M})$-*factorizations uniquely.*

Proof: Assume that $(\mathbf{A}, U) = (\mathbf{X}^{\mathbf{T}}, U^{\mathbf{T}})$ for some E-monad $\mathbf{T} = (T, \eta, \mu)$ on $\mathbf{X}$. Let $\mathcal{S} = ((X, x) \xrightarrow{f_i} (X_i, x_i))_I$ be a source in $\mathbf{X}^{\mathbf{T}}$ and $(X \xrightarrow{f_i} X_i)_I = (X \xrightarrow{e} Y \xrightarrow{m_i} X_i)_I$ be an $(E, \mathbf{M})$-factorization of $U^{\mathbf{T}}\mathcal{S}$ in $\mathbf{X}$. Since $Te \in E$, there exists a unique diagonal

$TY \xrightarrow{y} Y$ in **X** that makes the diagram

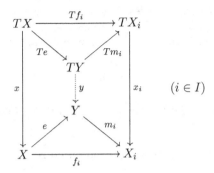

commute. That (Y, y) is a **T**-algebra follows from the fact that e and $T^2 e$ are epimorphisms [15.5(2)] and from the equations

$$y \circ \eta_Y \circ e = y \circ Te \circ \eta_X = e \circ x \circ \eta_X = e \circ id_X = id_Y \circ e$$

and

$$(y \circ Ty) \circ T^2 e = y \circ T(y \circ Te) = y \circ T(e \circ x) = y \circ Te \circ Tx = e \circ x \circ Tx = e \circ x \circ \mu_X$$
$$= y \circ Te \circ \mu_X = (y \circ \mu_Y) \circ T^2 e. \qquad \square$$

20.25 PROPOSITION

Let **A** be a full subcategory of an $(E, \mathbf{M})$-category **B**. Then the inclusion functor $\mathbf{A} \xrightarrow{U} \mathbf{B}$ is E-monadic if and only if the following conditions are satisfied:

(1) **A** is reflective in **B**,

(2) if $A \xrightarrow{e} B \xrightarrow{m} A'$ is an $(E, \mathbf{M})$-factorization of an **A**-morphism $A \xrightarrow{moe} A'$, then B belongs to **A**.

Proof: If U is E-monadic, then by Propositions 20.12(3) and 20.24, conditions (1) and (2) are satisfied. Conversely, let (1) and (2) be satisfied, and let $(\eta, \varepsilon) : F \dashv U : \mathbf{A} \to \mathbf{B}$ be an adjoint situation. If $\mathbf{T} = (T, \eta, \mu)$ is the associated monad, then $(\mathbf{A}, U)$ can be shown to be concretely isomorphic to $(\mathbf{B}^\mathbf{T}, U^\mathbf{T})$. To see this consider the concrete functor $K : (\mathbf{A}, U) \to (\mathbf{B}^\mathbf{T}, U^\mathbf{T})$ given by $KA = (A, \eta_A^{-1})$. K is an isomorphism since for each **T**-algebra (B, b) we have $b \circ \eta_B = id_B$ (where η_B is an **A**-reflection for B) and hence $\eta_B \circ b \circ \eta_B = \eta_B$ implies that $\eta_B \circ b = id$. Thus $b = \eta_B^{-1}$, and $B \in Ob(\mathbf{A})$ since, by (2), **A** is isomorphism-closed. It remains to be shown that $T[E] \subseteq E$. Consider a morphism $B \xrightarrow{e} B'$ in E. Let $TB \xrightarrow{Te} TB' = TB \xrightarrow{\hat{e}} B'' \xrightarrow{\hat{m}} TB'$ be an $(E, \mathbf{M})$-factorization of Te. Then there exists a diagonal $B \xrightarrow{d} B''$ that makes the following diagram commute:

By the universal property of η there exists a morphism $TB' \xrightarrow{\hat{d}} B''$ with $d = \hat{d} \circ \eta_{B'}$. Thus $\hat{m} \circ \hat{d} \circ \eta_{B'} = \hat{m} \circ d = \eta_{B'}$ implies that $\hat{m} \circ \hat{d} = id$. Moreover, $\hat{d} \circ \hat{m} \circ \hat{e} \circ \eta_B = \hat{d} \circ \hat{m} \circ d \circ e = \hat{d} \circ \eta_{B'} \circ e = d \circ e = \hat{e} \circ \eta_B$ implies that $\hat{d} \circ \hat{m} \circ \hat{e} = \hat{e}$. Thus, since $\hat{e}$ is an epimorphism, $\hat{d} \circ \hat{m} = id$. Hence $\hat{m}$ is an isomorphism and, consequently, $Te = \hat{m} \circ \hat{e}$ belongs to E. □

20.26 COROLLARY

*If **A** is an E-reflective subcategory of an $(E, \mathbf{M})$-category **B**, then the inclusion-functor **A** → **B** is E-monadic.*

Proof: Immediate from Theorem 16.8 and Proposition 20.25. □

20.27 EXAMPLES

As the following examples show, there exist full subcategories **A** of $(E, \mathbf{M})$-categories **B** such that the inclusion functor **A** → **B** is E-monadic, but such that **A** is neither closed under the formation of **M**-subobjects nor under the formation of E-quotient objects:

(1) **Sgr** is a (RegEpi, Mono-Source)-category, the full embedding **Grp** → **Sgr** is regularly monadic, and **Grp** is closed under the formation of regular quotients in **Sgr**. But **Grp** is not closed under the formation of subobjects in **Sgr**.

(2) **Top** is an (Epi, ExtrMono-Source)-category and the full embedding of **HComp** into **Top** is Epi-monadic. But **HComp** is neither closed under the formation of quotients nor closed under the formation of extremal subobjects in **Top**.

20.28 PROPOSITION

*If $(\mathbf{A}, U)$ is an E-monadic category over an $(E, \mathbf{M})$-category **X**, then the following hold:*

*(1) Every **A**-morphism f with $Uf \in E$ is final in $(\mathbf{A}, U)$.*

*(2) **A** is an $(U^{-1}[E], U^{-1}[\mathbf{M}])$-category.*

Proof:
(1). Assume that $(\mathbf{A}, U) = (\mathbf{X}^{\mathbf{T}}, U^{\mathbf{T}})$ for some E-monad $\mathbf{T} = (T, \eta, \mu)$ on **X**. Let $(X, x) \xrightarrow{e} (Y, y)$ be a T-homomorphism with $X \xrightarrow{e} Y$ in E, and let $Y \xrightarrow{f} Z$ be an **X**-morphism such that $(X, x) \xrightarrow{f \circ e} (Z, z)$ is a T-homomorphism. Since Te is an epimorphism, the equalities

$$(f \circ y) \circ Te = (f \circ e) \circ x = z \circ T(f \circ e) = (z \circ Tf) \circ Te$$

imply that $f \circ y = z \circ Tf$, i.e., that $(Y, y) \xrightarrow{f} (Z, z)$ is a T-homomorphism.

(2). By Proposition 20.24, **A** has $(U^{-1}[E], U^{-1}[\mathbf{M}])$-factorizations. By (1) and the fact that $U^{-1}[E] \subseteq \text{Epi}(\mathbf{A})$, **A** has the unique $(U^{-1}[E], U^{-1}[\mathbf{M}])$-diagonalization property. □

20.29 COROLLARY
If $(\mathbf{A}, U)$ is E-monadic over an E-co-wellpowered category, then $\mathbf{A}$ is $U^{-1}[E]$-co-wellpowered. □

20.30 PROPOSITION
If $(\mathbf{A}, U)$ is regularly monadic, then $\mathbf{A}$ has regular factorizations and U preserves and reflects regular and extremal epimorphisms.

Proof: By Proposition 20.28, $\mathbf{A}$ is an $(E, \text{Mono-Source})$-category, where E is the class $U^{-1}[\text{RegEpi}(\mathbf{X})]$. Thus, by Proposition 15.8(1), $E = \text{ExtrEpi}(\mathbf{A})$. It suffices to show that in $\mathbf{A}$ every extremal epimorphism $A \xrightarrow{e} B$ is regular. Since Ue is an extremal epimorphism and hence a regular epimorphism in the base category $\mathbf{X}$, it is a coequalizer of some pair $X \underset{r_2}{\overset{r_1}{\rightrightarrows}} UA$ of $\mathbf{X}$-morphisms. If $X \xrightarrow{\eta} UC$ is a universal arrow over X, then for each $i = 1, 2$ there exists an $\mathbf{A}$-morphism $C \xrightarrow{\hat{r}_i} A$ with $r_i = (U\hat{r}_i) \circ \eta$. Since Ue is a coequalizer of r_1 and r_2 in $\mathbf{X}$, the finality of e [cf. Proposition 20.28(1)] implies that e is a coequalizer of $\hat{r}_1$ and $\hat{r}_2$ in $\mathbf{A}$(cf. Exercise 8O). Hence e is a regular epimorphism in $\mathbf{A}$. □

20.31 COROLLARY
Regularly monadic functors detect extremal co-wellpoweredness. □

20.32 CHARACTERIZATION THEOREM FOR REGULARLY MONADIC FUNCTORS
A functor $\mathbf{A} \xrightarrow{U} \mathbf{X}$ is regularly monadic if and only if the following conditions hold:

(1) U is monadic,

(2) $\mathbf{X}$ has regular factorizations,

(3) U preserves regular epimorphisms.

Proof: By Proposition 20.30 every regularly monadic functor satisfies the above conditions. Conversely, assume that $(\mathbf{A}, U) = (\mathbf{X}^{\mathbf{T}}, U^{\mathbf{T}})$ for some monad $\mathbf{T} = (T, \eta, \mu)$ on $\mathbf{X}$, and that conditions (2) and (3) are satisfied. Then $F^{\mathbf{T}}$ and hence $T = U^{\mathbf{T}} \circ F^{\mathbf{T}}$ preserve regular epimorphisms. Thus $\mathbf{T}$ is a regular monad and $(\mathbf{A}, U)$ is regularly monadic. □

20.33 PROPOSITION
Regularly monadic functors detect colimits.

Proof: Let $(\mathbf{A}, U)$ be a regularly monadic category over $\mathbf{X}$, let $D: \mathbf{I} \to \mathbf{A}$ be a diagram, and let $(UD_i \xrightarrow{c_i} C)_{i \in Ob(\mathbf{I})}$ be a colimit of $U \circ D$. Consider the structured source $(C \xrightarrow{f_j} UA_j)_{j \in J}$ consisting of all structured morphisms for which each $UD_i \xrightarrow{f_j \circ c_i} UA_j$ is an $\mathbf{A}$-morphism. If $C \xrightarrow{\eta_C} UA$ is a universal arrow over C, then for each $j \in J$

there exists a unique **A**-morphism $A \xrightarrow{g_j} A_j$ with $f_j = g_j \circ \eta_C$. Let $(A \xrightarrow{g_j} A_j)_J = (A \xrightarrow{e} B \xrightarrow{m_j} A_j)_J$ be a (RegEpi, Mono-Source)-factorization.

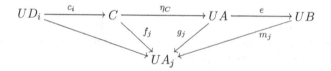

Since mono-sources are initial and each $UD_i \xrightarrow{f_j \circ c_i} UA_j$ is an **A**-morphism, each $UD_i \xrightarrow{e \circ \eta_C \circ c_i} UB$ is an **A**-morphism. It follows easily that $(D_i \xrightarrow{e \circ \eta_C \circ c_i} B)_{i \in Ob(\mathbf{I})}$ is a colimit of D. □

MONADIC CONSTRUCTS

20.34 PROPOSITION
Monadic constructs are complete, cocomplete, wellpowered, extremally co-wellpowered, and have regular factorizations.

Proof: Let $(\mathbf{A}, U)$ be a monadic construct. By Proposition 20.12, **A** is complete and wellpowered. By Proposition 20.22, U is regularly monadic. Hence Corollary 20.31 implies that **A** is extremally co-wellpowered, Proposition 20.30 implies that **A** has regular factorizations, and Proposition 20.33 implies that **A** is cocomplete. □

20.35 CHARACTERIZATION THEOREM FOR MONADIC CONSTRUCTS
For constructs $(\mathbf{A}, U)$ the following conditions are equivalent:

(1) U is monadic,

(2) U is regularly monadic,

(3) U is adjoint and creates finite limits and coequalizers of congruence relations,

(4) U is extremally co-wellpowered and creates limits and coequalizers of congruence relations.

Proof: (1) ⇔ (2) by Proposition 20.22.

(1) ⇒ (3). It suffices to show that U creates coequalizers of congruence relations. Let $A \underset{q}{\overset{p}{\rightrightarrows}} B$ be a pair of **A**-morphisms and let $UA \underset{Uq}{\overset{Up}{\rightrightarrows}} UB \xrightarrow{c} C$ be a congruence fork in **Set**. By Example 20.15(2) it is a split fork. Thus the result follows from Theorem 20.17.

Sec. 20] **Monads** 341

(3) ⇒ (1). By Theorem 20.17 it suffices to show that U creates coequalizers of split forks. Let $A \underset{q}{\overset{p}{\rightrightarrows}} B$ be a pair of **A**-morphisms and let

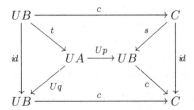

be a commutative diagram in **Set**. Let (c_1, c_2) be a congruence relation of c in **Set** and let (p_1, p_2) be a congruence relation of p in **A**. Then (Up_1, Up_2) is a congruence relation of Up in **Set**. Thus there exist unique functions u and v that make the diagrams

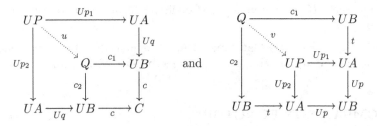

commute.

The equations $c_i \circ (u \circ v) = Uq \circ Up_i \circ v = Uq \circ t \circ c_i = c_i = c_i \circ id$ for $i = 1, 2$ imply that $u \circ v = id$. Thus, if $L \underset{\ell_2}{\overset{\ell_1}{\rightrightarrows}} UP$ is a congruence relation of u, then u is a coequalizer of ℓ_1 and ℓ_2. A straightforward computation shows that the source $(L, (\ell_i))$ can be considered as a limit of the diagram

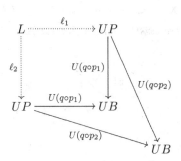

Since U creates finite limits, there is a unique lift $(\hat{L} \overset{\hat{\ell_i}}{\longrightarrow} P)$ for the structured source $(L \overset{\ell_i}{\longrightarrow} UP)$. Since $U\hat{L} \underset{U\hat{\ell_2}}{\overset{U\hat{\ell_1}}{\rightrightarrows}} UP \overset{u}{\longrightarrow} Q$ is a congruence fork, $UP \overset{u}{\longrightarrow} Q$ has a unique lift $P \xrightarrow{\hat{u}} \hat{Q}$ and $\hat{u}$ is a coequalizer of $\hat{\ell}_1$ and $\hat{\ell}_2$ in **A**. Thus the equations $c_i \circ U\hat{u} = c_i \circ u = U(q \circ p_i)$ and the faithfulness of U imply that each c_i can be lifted uniquely to an **A**-morphism $\hat{Q} \xrightarrow{\hat{c}_i} B$. Since $U\hat{Q} \underset{U\hat{c}_2}{\overset{U\hat{c}_1}{\rightrightarrows}} UB \xrightarrow{c} C$ is a congruence fork, $UB \xrightarrow{c} C$ has a unique lift $B \xrightarrow{\hat{c}} \hat{C}$ and $\hat{c}$ is a coequalizer of $\hat{c}_1$ and $\hat{c}_2$ in **A**. It remains to be shown that $\hat{c}$ is a coequalizer of p and q. Since $\hat{c} \circ p = \hat{c} \circ q$, this follows from the fact that whenever $\hat{C} \xrightarrow{f} D$ is an **A**-morphism with $f \circ p = f \circ q$, then $f \circ \hat{c}_1 \circ \hat{u} = f \circ q \circ p_1 = f \circ p \circ p_1 = f \circ p \circ p_2 = f \circ q \circ p_2 = f \circ \hat{c}_2 \circ \hat{u}$, and hence that $f \circ \hat{c}_1 = f \circ \hat{c}_2$.

(3) $\Rightarrow$ (4). In view of Proposition 18.11 this follows from (3) $\Rightarrow$ (1) and Proposition 20.34.

(4) $\Rightarrow$ (3). Since U creates limits and **Set** is strongly complete, so is **A**. Hence by Theorem 17.11(1), U has (Generating,−)-factorizations, i.e., it is adjoint. □

20.36 EXAMPLES

(1) The construct **HComp** is monadic. Condition (3) of the above theorem is easily seen to hold.

(2) Neither of the constructs **Cat**$_{of}$ and (**Ban**,O) is monadic. The corresponding forgetful functors don't preserve extremal epimorphisms [cf. 7.72(5)].

THE COMPARISON FUNCTOR

Next we will show that every monadic category is not only concretely isomorphic to *some* category of algebras, but even to its *associated* category of algebras (cf. Remark 20.6). Moreover, for every concrete category (**A**,U) that has free objects, there exists a distinguished concrete functor into its associated category of algebras; and that functor turns out to be an isomorphism if and only if (**A**,U) is monadic. (See Proposition 20.40.)

20.37 PROPOSITION

If $(\eta, \varepsilon) : F \dashv U : \mathbf{A} \to \mathbf{X}$ is an adjoint situation and $(\mathbf{X^T}, U^\mathbf{T})$ is the associated category of algebras, then there exists a unique functor $\mathbf{A} \xrightarrow{K} \mathbf{X^T}$ such that the diagram

$$\begin{array}{ccc} \mathbf{X} & \xrightarrow{F} & \mathbf{A} \\ {\scriptstyle F^\mathbf{T}}\downarrow & {\scriptstyle K} \nearrow & \downarrow {\scriptstyle U} \\ \mathbf{X^T} & \xrightarrow{U^\mathbf{T}} & \mathbf{X} \end{array}$$

commutes.

Proof: *Existence:* The functor $\mathbf{A} \xrightarrow{K} \mathbf{X^T}$, defined by

$$K(A \xrightarrow{f} B) = (UA, U\varepsilon_A) \xrightarrow{Uf} (UB, U\varepsilon_B),$$

has the desired properties.

Uniqueness: Let $\mathbf{A} \xrightarrow{\hat{K}} \mathbf{X}^{\mathbf{T}}$ be a functor satisfying $U = U^{\mathbf{T}} \circ \hat{K}$ and $F^{\mathbf{T}} = \hat{K} \circ F$. Consider an $\mathbf{A}$-morphism $A \xrightarrow{f} B$, and denote $\hat{K}(A \xrightarrow{f} B)$ by $(X, x) \xrightarrow{g} (Y, y)$. Then $U = U^{\mathbf{T}} \circ \hat{K}$ implies that $UA \xrightarrow{Uf} UB = X \xrightarrow{g} Y$. Thus $X = UA$, $Y = UB$, and $g = Uf$. It remains to be shown that $x = U\varepsilon_A$. Since $FUA \xrightarrow{\varepsilon_A} A$ is an $\mathbf{A}$-morphism, $\hat{K}(FUA \xrightarrow{\varepsilon_A} A) = F^{\mathbf{T}}UA \xrightarrow{U\varepsilon_A} \hat{K}A$ is a T-homomorphism. Thus the diagram

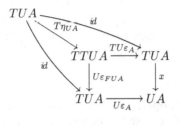

commutes.

This implies that $x = U\varepsilon_A$, so that $\hat{K} = K$. □

20.38 DEFINITION

(1) For each adjoint situation $(\eta, \varepsilon): F \dashv U : \mathbf{A} \to \mathbf{X}$, the unique functor $\mathbf{A} \xrightarrow{K} \mathbf{X}^{\mathbf{T}}$ of the above proposition is called its **comparison functor**.

(2) For each adjoint functor $\mathbf{A} \xrightarrow{U} \mathbf{X}$ (resp. each concrete category $(\mathbf{A}, U)$ that has free objects) the comparison functor of an associated adjoint situation is called a **comparison functor for U** (resp. **for $(\mathbf{A}, U)$**).

20.39 REMARKS

(1) In view of the essential uniqueness of comparison functors for U (cf. 20A) we will, by "abuse of language", usually speak of **the** comparison functor for U.

(2) Let $(\eta, \varepsilon): F \dashv U : \mathbf{A} \to \mathbf{X}$ be an adjoint situation. Then Proposition 20.37 can be interpreted as saying that the adjoint situation $(\eta^{\mathbf{T}}, \varepsilon^{\mathbf{T}}): F^{\mathbf{T}} \dashv U^{\mathbf{T}} : \mathbf{X}^{\mathbf{T}} \to \mathbf{X}$ induced from the associated monad $\mathbf{T}$ is the "largest" one with the same associated monad. In Exercise 20B it is indicated that every monad also has a "smallest realization", called its **Kleisli category**.

20.40 PROPOSITION

An adjoint functor is monadic if and only if the associated comparison functor is a concrete isomorphism.

Proof: If the comparison functor for an adjoint functor U is an isomorphism, then U is obviously monadic. Conversely, let $\mathbf{A} \xrightarrow{U} \mathbf{X}$ be a functor, let $\mathbf{T}$ be a monad on $\mathbf{X}$, and let $(\mathbf{A}, U) \xrightarrow{H} (\mathbf{X}^{\mathbf{T}}, U^{\mathbf{T}})$ be a concrete isomorphism. Let $(\eta, \varepsilon): F^{\mathbf{T}} \dashv U^{\mathbf{T}}: \mathbf{X}^{\mathbf{T}} \to \mathbf{X}$ be the adjoint situation associated with **T**. Then $(\eta, H^{-1}\varepsilon H) : H^{-1} \circ F^{\mathbf{T}} \dashv U : \mathbf{A} \to \mathbf{X}$ is an adjoint situation whose associated monad is **T**. Since the diagram

$$\begin{array}{ccc} \mathbf{X} & \xrightarrow{H^{-1} \circ F^{\mathbf{T}}} & \mathbf{A} \\ F^{\mathbf{T}} \downarrow & {}^H\!\!\nearrow & \downarrow U \\ \mathbf{X}^{\mathbf{T}} & \xrightarrow[U^{\mathbf{T}}]{} & \mathbf{X} \end{array}$$

commutes, H is the associated comparison functor for U. Thus the comparison functor is an isomorphism. □

20.41 EXAMPLES

(1) The constructs **Grp** and **TopGrp** induce the same monads on **Set**. Thus the forgetful functor **TopGrp** → **Grp** is the comparison functor for the construct **TopGrp**.

(2) Since the constructs **TfAb** and **Ab** of torsion-free abelian groups and of abelian groups induce the same monads in **Set**, the full embedding **TfAb** → **Ab** is the comparison functor for **TfAb**.

(3) The construct $\mathbf{Cat}_{of}$ is not monadic. If **Grph** (= the category of oriented graphs) is the full subconstruct of $\mathbf{Alg}(1,1)$ that consists of all objects (X, c, d) that satisfy the equations $c \circ c = d \circ c = c$ and $d \circ d = c \circ d = d$, then the concrete functor $\mathbf{Cat}_{of} \xrightarrow{K} \mathbf{Grph}$, defined by $K\mathbf{C} = (Mor(\mathbf{C}), c, d)$ (where $Mor(\mathbf{C}) \underset{d}{\overset{c}{\rightrightarrows}} Mor(\mathbf{C})$ are given by $d(A \xrightarrow{f} B) = id_A$ and $c(A \xrightarrow{f} B) = id_B$), is the comparison functor for $\mathbf{Cat}_{of}$. K is not full.

(4) The full concrete embedding of the construct (**Ban**, O) into the construct **TConv** of totally convex spaces is the comparison functor for (**Ban**, O).

(5) Since the constructs **Top** and **Set** induce the same monad on **Set**, the forgetful functor **Top** → **Set** is the comparison functor for **Top**.

20.42 THEOREM

Let $(\mathbf{A}, U) \xrightarrow{K} (\mathbf{X}^{\mathbf{T}}, U^{\mathbf{T}})$ be a comparison functor. If **A** has coequalizers, then K is adjoint.

Proof: Let $(\eta, \varepsilon) : F \dashv U : \mathbf{A} \to \mathbf{X}$ be an adjoint situation and let $\mathbf{T} = (T, \eta, \mu)$ be the associated monad. Let (X, x) be a **T**-algebra and let $FX \xrightarrow{c} C$ be a coequalizer of $FUFX \xrightarrow{Fx} FX$ and $FUFX \xrightarrow{\varepsilon FX} FX$. Then $(X, x) \xrightarrow{Uc \circ \eta_X} KC$ is a K-universal arrow as shown below:

(a). Since $(Uc \circ \eta_X) \circ x = Uc \circ UFx \circ \eta_{UFX} = Uc \circ U\varepsilon_{FX} \circ \eta_{UFX} = Uc \circ id_{UFX} = Uc \circ Uid_{FX} = Uc \circ U(\varepsilon_{FX} \circ F\eta_X) = U(c \circ \varepsilon_{FX}) \circ UF\eta_X = U(\varepsilon_C \circ FUc) \circ UF\eta_X = U\varepsilon_C \circ T(Uc \circ \eta_X)$, it follows that $(X, x) \xrightarrow{Uc \circ \eta_X} (UC, U\varepsilon_C)$ is an T-homomorphism.

(b). Let $(X,x) \xrightarrow{f} KA$ be an arbitrary K-structured arrow. Then $FX \xrightarrow{\hat{f}} A = FX \xrightarrow{\varepsilon_A \circ Ff} A$ is the unique **A**-morphism with $f = U\hat{f} \circ \eta_X$ (cf. 19J). The equations $U(\hat{f} \circ Fx) \circ \eta_{UFX} = U\hat{f} \circ UFx \circ \eta_{UFX} = U\hat{f} \circ \eta_X \circ x = f \circ x = U\varepsilon_A \circ UFf = U\hat{f} = U\hat{f} \circ U\varepsilon_{FX} \circ \eta_{UFX} = U(\hat{f} \circ \varepsilon_{FX}) \circ \eta_{UFX}$ imply that $\hat{f} \circ Fx = \hat{f} \circ \varepsilon_{FX}$. Since c is a coequalizer of Fx and ε_{FX} in **A**, there exists an **A**-morphism $C \xrightarrow{\tilde{f}} A$ with $\hat{f} = \tilde{f} \circ c$. This implies that $(X,x) \xrightarrow{f} KA = (X,x) \xrightarrow{Uc \circ \eta_X} KC \xrightarrow{K\tilde{f}} KA$, since $f = U\hat{f} \circ \eta_X = U\tilde{f} \circ Uc \circ \eta_X$ and $K\tilde{f} = U\tilde{f}$. Uniqueness of $\tilde{f}$ follows from the fact that c is an epimorphism in **A**. □

20.43 THEOREM
Let $(\eta, \varepsilon) : F \dashv U : \mathbf{A} \to \mathbf{X}$ be an adjoint situation with associated comparison functor $\mathbf{A} \xrightarrow{K} \mathbf{X}^\mathbf{T}$. Then:

(1) K is faithful if and only if U is faithful,

(2) K is full and faithful if and only if ε is a RegEpi-transformation.

Proof:
(1). Obvious.

(2). Let ε be a RegEpi-transformation. Then, by Theorem 19.14(1), U is faithful. Thus K is faithful as well. Let A and B be objects in **A** and let $KA \xrightarrow{f} KB$ be an $\mathbf{X}^\mathbf{T}$-morphism. If ε_A is a coequalizer of $C \underset{s}{\overset{r}{\rightrightarrows}} FUA$, then the equations $U\varepsilon_B \circ UFf \circ Ur = f \circ U\varepsilon_A \circ Ur = f \circ U\varepsilon_A \circ Us = U\varepsilon_B \circ UFf \circ Us$ imply that $\varepsilon_B \circ Ff \circ r = \varepsilon_B \circ Ff \circ s$ by the faithfulness of U. Thus there exists an **A**-morphism $A \xrightarrow{\hat{f}} B$ with $\varepsilon_B \circ Ff = \hat{f} \circ \varepsilon_A$. Since $U\varepsilon_A$ as a retraction is an epimorphism, the equations $K\hat{f} \circ U\varepsilon_A = U\hat{f} \circ U\varepsilon_A = U\varepsilon_B \circ UFf = f \circ U\varepsilon_A$ imply that $K\hat{f} = f$. Thus K is full. Conversely, let K be full and faithful and let A be an **A**-object. Then $\varepsilon_A \circ \varepsilon_{FUA} = \varepsilon_A \circ FU\varepsilon_A$. To see that ε_A is a coequalizer of ε_{FUA} and $FU\varepsilon_A$, let $FUA \xrightarrow{f} B$ be an **A**-morphism with $f \circ \varepsilon_{FUA} = f \circ FU\varepsilon_A$. The equalities $(Uf \circ \eta_{UA}) \circ U\varepsilon_A = Uf \circ UFU\varepsilon_A \circ \eta_{UFUA} = Uf \circ U\varepsilon_{FUA} \circ \eta_{UFUA} = Uf = Uf \circ U\varepsilon_{UFA} \circ UF\eta_{UA} = U\varepsilon_B \circ UFUf \circ UF\eta_{UA} = U\varepsilon_B \circ T(Uf \circ \eta_{UA})$ imply that $KA \xrightarrow{Uf \circ \eta_{UA}} KB$ is an T-homomorphism. By the fullness of K, there exists an **A**-morphism $A \xrightarrow{\hat{f}} B$ with $K\hat{f} = Uf \circ \eta_{UA}$. The equalities $Uf \circ \eta_{UA} = K\hat{f} = U\hat{f} = U\hat{f} \circ U\varepsilon_A \circ \eta_{UA} = U(\hat{f} \circ \varepsilon_A) \circ \eta_{UA}$ imply that $f = \hat{f} \circ \varepsilon_A$. Uniqueness follows from the fact that (due to the faithfulness of K and hence of U) ε_A is an epimorphism [cf. Theorem 19.14(1)]. □

20.44 COROLLARY
The comparison functor $(\mathbf{A}, U) \xrightarrow{K} (\mathbf{X}^\mathbf{T}, U^\mathbf{T})$ of a uniquely transportable concrete category, for which U reflects regular epimorphisms, is an isomorphism-closed full embedding. □

DEFICIENCIES OF MONADIC FUNCTORS

20.45 EXAMPLE
A composite of two regularly monadic functors need not be monadic. Consider

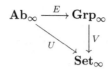

20.46 EXAMPLE
Concrete functors between monadic categories need not be adjoint. Consider

$$\mathbf{Ab}_\infty \xrightarrow{E} \mathbf{Grp}_\infty$$
$$U \searrow \quad \downarrow V$$
$$\mathbf{Set}_\infty$$

where (in each case) $\mathbf{A}_\infty$ is the full subcategory of $\mathbf{A}$ that consists of those objects of $\mathbf{A}$ that have infinite underlying sets, where U and V are the forgetful functors, and E is the inclusion functor.

20.47 EXAMPLE
Monadic functors need not detect colimits. The following yields a monadic category over **Pos** that is not cocomplete.

Denote by $\mathcal{P}^0 : \mathbf{Set} \to \mathbf{Pos}$ the power-set functor equipped with the ordering of $\mathcal{P}X$ such that $\emptyset$ is the least element, and $\mathcal{P}X - \{\emptyset\}$ is discretely ordered. Denote by $H : \mathbf{Pos} \to \mathbf{Set}$ the functor that assigns to each poset $(X, \leq)$ the set

$$H(X, \leq) = \{(x,y,z) \in X^3 \mid x < y < z\} \cup \{\alpha\} \quad (\text{where } \alpha \notin X^3)$$

and to each order-preserving function $f : (X, \leq) \to (Y, \leq)$, the map Hf with $Hf(\alpha) = \alpha$, and

$$Hf(x,y,z) = \begin{cases} (f(x), f(y), f(z)), & \text{if } f(x) \neq f(y) \neq f(z) \\ \alpha, & \text{otherwise.} \end{cases}$$

(a) The functor $T = \mathcal{P}^0 \circ H : \mathbf{Pos} \to \mathbf{Pos}$ is such that $\mathbf{Alg}(T)$ has free objects. In fact, for each poset A we have $TA = T(A + TA)$ and, hence, the free **T**-algebra over A is $(A + TA, \varphi)$, where $TA \xrightarrow{\varphi} A + TA$ is the second coproduct injection, whereas the universal arrow $A \xrightarrow{\eta} A + TA$ is the first coproduct injection. Hence, $\mathbf{Alg}(T)$ is monadic over **Pos**.

(b) $\mathbf{Alg}(T)$ does not have the coproduct of the following **T**-algebra (X, x) with itself: X is the 3-chain $a < b < c$, and $x : TX \to X$ is the constant map to a.

To prove this, for each ordinal i define T-homomorphisms $f_i, g_i : (X, x) \to (Y_i, y_i)$ as follows: Y_i is the following poset

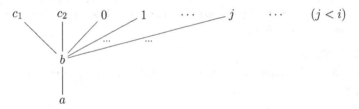

and for each $j < i$,

$$j = y_i(\{\,(a,b,k) \mid k < j\,\} \cup \{\,(a,b,c_i) \mid i = 1, 2\,\}),$$

and otherwise, y_i is constant to a. Finally, $f_i(a) = g_i(a) = a$, $f_i(b) = g_i(b) = b$, and $f_i(c) = c_1$, $g_i(c) = c_2$. It is easy to verify that f_i and g_i are homomorphisms such that whenever $(X, x) + (X, x)$ exists, the factorizing homomorphism $(X, x) + (X, x) \to (Y_i, y_i)$ is surjective. This is impossible.

20.48 EXAMPLE
Monadic categories over co-wellpowered categories need not be extremally co-wellpowered. See Exercise 16A(a).

20.49 EXAMPLE
Monadic categories over categories with regular factorizations need not have (Epi, Mono-Source)-factorizations. See Exercise 16A(b).

20.50 EXAMPLE
Monadic functors need not preserve either extremal epimorphisms or regular epimorphisms. In fact they can map regular epimorphisms into non-extremal epimorphisms [see Exercise 16A(b)] and extremal epimorphisms into non-epimorphisms [see Exercise 16A(a)].

20.51 EXAMPLE
In monadic categories, regular epimorphisms need not be final. The comparison functor $\mathbf{Cat}_{of} \xrightarrow{K} \mathbf{Grph}$ for the construct $\mathbf{Cat}_{of}$, described in Example 20.41(3), is monadic. The functor F, described in Example 7.40(6), is a regular epimorphism in $\mathbf{Cat}_{of}$ that is not final in $(\mathbf{Cat}_{of}, K)$.

20.52 EXAMPLE
Monadic functors need not reflect regular epimorphisms. As above, the comparison functor $\mathbf{Cat}_{of} \xrightarrow{K} \mathbf{Grph}$ is monadic. The functor $\mathbf{A} \xrightarrow{G \circ F} \mathbf{C}$, described in Remark 7.76(1), is not a regular epimorphism in $\mathbf{Cat}_{of}$, but $K(G \circ F)$ is a regular epimorphism in $\mathbf{Grph}$.

VARIETORS AND FREE MONADS

We now explain the role that the categories of the form $\mathbf{Alg}(T)$ play among monadic categories. They are, up to isomorphism, precisely the Eilenberg-Moore categories of free monads (introduced below).

20.53 DEFINITION
A functor $T : \mathbf{X} \to \mathbf{X}$ is called a **varietor** provided that the concrete category $\mathbf{Alg}(T)$ has free objects.

20.54 EXAMPLES
(1) $S^n : \mathbf{Set} \to \mathbf{Set}$ is a varietor. Each set generates a free S^n-algebra (i.e., an algebra with one n-ary operation; cf. 8.23(6)).

(2) If $\mathbf{X}$ has countable colimits, then $id_\mathbf{X}$ is a varietor. Each object X generates a free $id_\mathbf{X}$-algebra whose underlying object is a coproduct of countably many copies of X.

More generally, every functor $T : \mathbf{X} \to \mathbf{X}$ that preserves colimits of ω-chains is a varietor (cf. Exercise 20P).

(3) The power-set functor $\mathcal{P}$ is not a varietor. If $\mathbf{Alg}(\mathcal{P})$ would have an initial object (X, x), then $x : \mathcal{P}X \to X$ would be an isomorphism (cf. Exercise 20I). However, $\operatorname{card} \mathcal{P}X > \operatorname{card} X$.

20.55 DEFINITION
(1) Given monads $\mathbf{T} = (T, \eta, \mu)$ and $\mathbf{T}' = (T', \eta', \mu')$ over $\mathbf{X}$, a natural transformation $\tau : T \to T'$ is called a **monad morphism** (denoted by $\tau : \mathbf{T} \to \mathbf{T}'$) provided that $\eta' = \tau \circ \eta$ and $\tau \circ \mu = \mu' \circ \tau T' \circ T\tau$.

(2) A **free monad** generated by a functor $T : \mathbf{X} \to \mathbf{X}$ is a monad $\mathbf{T}^\# = (T^\#, \eta^\#, \mu^\#)$ together with a natural transformation $\lambda : T \to T^\#$ that has the following universal property: for every monad $\mathbf{T}' = (T', \eta', \mu')$ and every natural transformation $\tau : T \to T'$ there exists a unique monad morphism $\tau^\# : \mathbf{T}^\# \to \mathbf{T}'$ with $\tau = \tau^\# \circ \lambda$.

20.56 THEOREM
If $T : \mathbf{X} \to \mathbf{X}$ is a varietor, then $\mathbf{Alg}(T)$ is monadic over $\mathbf{X}$ and the associated monad is a free monad generated by T.

Proof: Let $U : \mathbf{Alg}(T) \to \mathbf{X}$ denote the forgetful functor, let $F : \mathbf{X} \to \mathbf{Alg}(T)$ be the free functor, let $T^\# = U \circ F$, and let $\mathbf{T}^\# = (T^\#, \eta^\#, \mu^\#)$ be the associated monad.

(1). U is monadic because it creates absolute coequalizers (20.17). If $(X, x) \underset{f_2}{\overset{f_1}{\rightrightarrows}} (Y, y)$ are T-homomorphisms and $X \underset{f_2}{\overset{f_1}{\rightrightarrows}} Y \overset{c}{\to} Z$ is an absolute coequalizer in $\mathbf{X}$, then Tc is a coequalizer of Tf_1 and Tf_2, and since

$$(c \circ y) \circ Tf_1 = c \circ f_1 \circ x = c \circ f_2 \circ x = (c \circ y) \circ Tf_2,$$

there is a unique $TZ \xrightarrow{z} Z$ for which $c: (Y,y) \to (Z,z)$ is a T-homomorphism. The fact that Tc is a (regular) epimorphism makes it easy to see that

$(X,x) \underset{f_2}{\overset{f_1}{\rightrightarrows}} (Y,y) \xrightarrow{c} (Z,z)$ is a coequalizer in $\mathbf{Alg}(T)$.

(2). If $FX = (X^\#, \varphi_X)$, then the morphisms

$$TX \xrightarrow{\lambda_X} T^\# X \;=\; TX \xrightarrow{T\eta_X} TX^\# \xrightarrow{\varphi_X} X^\#$$

obviously satisfy the naturality condition; i.e., they form a natural transformation $\lambda: T \to T^\#$. To verify the universal property, let (T', η', μ') be a monad, and let $\tau: T \to T'$ be a natural transformation. For each $\mathbf{X}$-object X we have a T-algebra

$$T(T'X) \xrightarrow{(\tau T')_X} (T')^2 X \xrightarrow{\mu'_X} T'X.$$

Furthermore, there exists a unique T-homomorphism $\tau_X^\# : FX \to (T'X, [\mu' \circ \tau T']_X)$ with $\tau_X^\# \circ \eta_X = \eta'_X$. It is a straightforward computation to verify that the morphisms $U\tau_X^\#$ form a natural transformation $\tau^\# : T^\# \to T'$ that satisfies all the required equalities. $\square$

20.57 COROLLARY

If $T: \mathbf{X} \to \mathbf{X}$ is a varietor, then the category $\mathbf{Alg}(T)$ is concretely isomorphic to $\mathbf{X}^{\mathbf{T}^\#}$ for a free monad $\mathbf{T}^\#$. $\square$

20.58 REMARK

The following theorem provides a partial converse to the above (in the case where $\mathbf{X}$ is strongly complete).

20.59 THEOREM

If $\mathbf{X}$ is a strongly complete category, then every functor $T: \mathbf{X} \to \mathbf{X}$ that generates a free monad is a varietor.

Proof: Let $(T^\#, \eta^\#, \mu^\#)$ together with $\lambda: T \to T^\#$ be a free monad over T. To prove that the forgetful functor $U: \mathbf{Alg}(T) \to \mathbf{X}$ is adjoint, it is sufficient to verify the solution-set condition of Theorem 18.12, since $\mathbf{Alg}(T)$ is complete and U preserves limits (13N). We will show that for each $\mathbf{X}$-object X the T-algebra

$$T(T^\# X) \xrightarrow{\lambda_{T^\# X}} (T^\#)^2 X \xrightarrow{\mu_X^\#} T^\# X$$

together with the $\mathbf{X}$-morphism $\eta_X^\# : X \to T^\# X$ forms a singleton solution set. In other words, for each T-algebra (Y,y) and each $\mathbf{X}$-morphism $f: X \to Y$ there is a T-homomorphism

$$g: (T^\# X, \mu_X^\# \circ \lambda_{T^\# X}) \to (Y,y) \quad \text{with} \quad f = g \circ \eta_X^\#.$$

Denote by $(\mathbf{B}, U')$ the full concrete subcategory of $\mathbf{Alg}(T)$ consisting of all subalgebras of products of the T-algebra (Y, y). Then $\mathbf{B}$ is closed under the formation of limits in $\mathbf{Alg}(T)$. Thus $\mathbf{B}$ is strongly complete and U' preserves limits. Moreover, (Y, y) is a coseparator of $\mathbf{B}$ (10.38), and hence, U' is adjoint (18.17). Let $F' : \mathbf{X} \to \mathbf{B}$ denote the free functor, $F'X = (X^*, \varphi_X)$, and let (T', η', μ') be the associated monad. Then the morphisms

$$TX \xrightarrow{T\eta_X} TX^* \xrightarrow{\varphi_X} X^*$$

clearly form a natural transformation $\tau : T \to T'$. By the universal property of λ there exists a unique natural transformation $\tau^\# : T^\# \to T'$ with (a) $\tau = \tau^\# \circ \lambda$, (b) $\eta' = \tau^\# \circ \eta^\#$, and (c) $\tau^\# \circ \mu^\# = \mu' \circ \tau^\# T' \circ T^\# \tau^\#$.

If $f : X \to Y$ is a morphism, then since (Y, y) lies in $\mathbf{B}$, there is a unique T-homomorphism $f^* : T'X \to (Y, y)$ with $f = f^* \circ \eta'_X$. Then $g = f^* \circ \tau_X^\# : T^\# X \to Y$ is the desired morphism. In fact, from (b) it follows that $f = g \circ \eta_X^\#$, and from (a) and (c) and the properties of monads it is easy to verify that g is a T-homomorphism (i.e., that $y \circ Ty = g \circ \mu_X^\# \circ \lambda_{T^\# x}$). $\square$

Suggestions for Further Reading

Eilenberg, S., and J. C. Moore. Adjoint functors and triples. *Illinois J. Math.* **9** (1965): 381–398.

Kleisli, H. Every standard construction is induced by a pair of adjoint functors. *Proc. Amer. Math. Soc.* **16** (1965): 544–546.

Maranda, J. M. On fundamental constructions and adjoint functors. *Canad. Math. Bull.* **9** (1966): 581–591.

Beck, J. *Triples, algebras and cohomology.* Ph. D. dissertation, Columbia University, 1967.

Manes, E. G. *A triple miscellany: Some aspects of the theory of algebras over a triple.* Ph. D. dissertation, Wesleyan University, 1967.

Linton, F. E. J. Coequalizers in categories of algebras. *Springer Lect. Notes Math.* **80** (1969): 75–90.

Duskin, J. Variations on Beck's tripleability criterion. *Springer Lect. Notes Math.* **106** (1969): 74–129.

Applegate, H., and M. Tierney. Iterated cotriples. *Springer Lect. Notes Math.* **137** (1970): 56–99.

Pumplün, D. Eine Bemerkung über Monaden und adjungierte Funktoren. *Math. Ann.* **185** (1970): 329–337.

Gabriel, P., and F. Ulmer. Lokal präsentierbare Kategorien. *Springer Lect. Notes Math.* **221** (1971): 1–200.

Paré, R. On absolute colimits. *J. Algebra* **19** (1971): 80–95.

Adámek, J. Colimits of algebras revisited. *Bull. Austral. Math. Soc.* **17** (1977): 433–450.

Mac Donald, J. L., and A. Stone. The tower and regular decomposition. *Cahiers Topol. Geom. Diff.* **23** (1982): 197–213.

Pumplün, D., and H. Röhrl. Banach spaces and totally convex spaces I. *Commun. Alg.* **12** (1984): 953–1019.

EXERCISES

20A. Associated Categories of T-Algebras Are Essentially Unique

Show that

(a) If $(\eta, \varepsilon): F \dashv U: \mathbf{A} \to \mathbf{X}$ and $(\overline{\eta}, \overline{\varepsilon}): \overline{F} \dashv U: \mathbf{A} \to \mathbf{X}$ are adjoint situations that are isomorphic in the sense of Proposition 19.9, then the associated monads $\mathbf{T} = (T, \eta, \mu)$ and $\overline{\mathbf{T}} = (\overline{T}, \overline{\eta}, \overline{\mu})$ are **isomorphic** in the following sense: there exists a natural isomorphism $T \xrightarrow{\sigma} \overline{T}$ that is a monad morphism (20.55).

(b) If monads $\mathbf{T} = (T, \eta, \mu)$ and $\overline{\mathbf{T}} = (\overline{T}, \overline{\eta}, \overline{\mu})$ in $\mathbf{X}$ are isomorphic in the above sense, then the associated categories of **T**-algebras and $\overline{\mathbf{T}}$-algebras are concretely isomorphic.

(c) If $(\mathbf{A}, U)$ is a concrete category over $\mathbf{X}$ with free objects and if $(\mathbf{A}, U) \xrightarrow{K} (X^{\mathbf{T}}, U^{\mathbf{T}})$ and $(\mathbf{A}, U) \xrightarrow{\overline{K}} (X^{\overline{\mathbf{T}}}, U^{\overline{\mathbf{T}}})$ are associated comparison functors, then there exists a concrete isomorphism $(X^{\mathbf{T}}, U^{\mathbf{T}}) \xrightarrow{H} (X^{\overline{\mathbf{T}}}, U^{\overline{\mathbf{T}}})$ with $\overline{K} = H \circ K$.

20B. The Kleisli Category of a Monad

The **Kleisli category** of a monad $\mathbf{T} = (T, \eta, \mu)$ in $\mathbf{X}$ is the following concrete category $(\mathbf{X_T}, U_{\mathbf{T}})$ over $\mathbf{X}$: $Ob(\mathbf{X_T}) = Ob(\mathbf{X})$, $\hom_{\mathbf{X_T}}(X, Y) = \hom_{\mathbf{X}}(X, TY)$, $(id_X)_{\mathbf{X_T}} = \eta_X$, and the composition of $X \xrightarrow{f} TY$ with $Y \xrightarrow{g} TZ$ in $\mathbf{X_T}$ is given by $\mu_Z \circ Tg \circ f$ (in $\mathbf{X}$). Furthermore, $U_{\mathbf{T}}(X \xrightarrow{f} TY) = TX \xrightarrow{\mu_Y \circ Tf} TY$.

(a) Prove that $U_{\mathbf{T}}$ has a co-adjoint $F_{\mathbf{T}}$ with $F_{\mathbf{T}}(X \xrightarrow{f} Y) = (X \xrightarrow{\eta_Y \circ f} TY)$, and that the associated monad of this adjoint situation is **T**.

(b) Prove that for each adjoint situation $F \dashv G: \mathbf{A} \dashv \mathbf{X}$ with the associated monad **T**, there exists a unique functor $K^*: \mathbf{X_T} \to \mathbf{A}$ with $F = K^* \circ F_{\mathbf{T}}$ and $U_{\mathbf{T}} = G \circ K^*$.

(c) Describe the Kleisli category of the power-set monad as the category of "sets and relations".

(d) Show that $U_{\mathbf{T}}$ need not detect finite limits. [Consider (c).]

* 20C. Algebraic Theories

(a) Let **T** be a monad over **Set**. Verify that in the Kleisli category $\mathbf{X_T}$ (20B) each object (set) X is a coproduct of X copies of $1 = \{0\}$ with coproduct injections $k_x : 1 \to X$ in $\mathbf{X_T}$ corresponding to $\eta_X(x) \in TX$. Prove that for each **T**-algebra A the functor $F_A : \mathbf{X_T}^{\mathrm{op}} \to \mathbf{Set}$ given by

$$F_A(X \xrightarrow{f} Y) = \hom((TX, \mu_X), A) \xrightarrow{F_A f} \hom((TY, \mu_Y), A)$$

[where $F_A f(h) = h \circ \mu_X \circ Tf$ (in **Set**)] preserves products.

(b) Conversely, prove that every product-preserving functor $F : (\mathbf{X_T})^{\mathrm{op}} \to \mathbf{Set}$ is naturally isomorphic to F_A for some **T**-algebra A. [Hint: Choose A to be $(F(1), \varphi)$, where $\varphi : TF(1) \to F(1)$ assigns to each element of $TF(1)$ represented by an $\mathbf{X_T}$-morphism $f : 1 \to F(1)$, the element of $F(1)$ that the function $Ff : F^2(1) \to F(1)$ assigns to $id \in (F(1))^{F(1)}$.] Conclude that $\mathbf{Set^T}$ is equivalent to the quasicategory of all product-preserving functors from $(\mathbf{X_T})^{\mathrm{op}}$ to **Set**.

(c) An **algebraic theory** is a category **L** whose objects are precisely all sets and such that each object X is a coproduct of X copies of 1. A **model** of **L** is a product-preserving functor from $\mathbf{L}^{\mathrm{op}}$ into **Set**. The quasicategory $\mathbf{Mod(L)}$ of models of **L** is the full subcategory of $[\mathbf{L}^{\mathrm{op}}, \mathbf{Set}]$ that has models as objects. Prove that $\mathbf{Mod(L)}$ is equivalent to a monadic construct. Moreover, show that the Kleisli category of the corresponding monad is equivalent to **L**.

(d) Verify that the above transitions between monads and algebraic theories are essentially inverse to each other (i.e., inverse up to isomorphism of monads and up to equivalence of algebraic theories).

* 20D. Cocompleteness of Monadic Categories

Let $(\mathbf{A}, U)$ be a monadic category over $\mathbf{X}$ and let $\mathbf{A}$ have coequalizers. Show that

(a) $\mathbf{A}$ has colimits over each scheme over which colimits exist in $\mathbf{X}$.

(b) If $\mathbf{X}$ is cocomplete then so is $\mathbf{A}$.

* 20E. Concrete Functors Between Monadic Categories

Let $(\mathbf{A}, U) \xrightarrow{G} (\mathbf{B}, V)$ be a concrete functor between monadic categories. Show that

(a) If $\mathbf{A}$ has coequalizers, then G is monadic.

(b) If $(\mathbf{A}, U)$ and $(\mathbf{B}, V)$ are regularly monadic, then G is regularly monadic.

20F. Idempotent Monads

A monad $\mathbf{T} = (T, \eta, \mu)$ is called **idempotent** provided that $T^2 \xrightarrow{\mu} T$ is a natural isomorphism. Show that

(a) For a monad $\mathbf{T} = (T, \eta, \mu)$ the following conditions are equivalent:

(1) **T** is idempotent,

(2) $T\eta$ is an Epi-transformation,

(3) $\mathbf{X^T} \xrightarrow{U^T} \mathbf{X}$ is full,

(4) $\mathbf{X^T} \xrightarrow{U^T} \mathbf{X}$ is an isomorphism-closed full reflective embedding.

(b) If $\mathbf{A} \xrightarrow{U} \mathbf{X}$ is an isomorphism-closed full reflective embedding, then the associated monad is idempotent.

(c) The monad associated with the forgetful functor **Top** $\to$ **Set** is idempotent.

* 20G. Monads With Rank

A monad $\mathbf{T} = (T, \eta, \mu)$ is said to have **rank** k, where k is a regular cardinal, provided that T preserves k-directed colimits, i.e., colimits of diagrams whose schemes are posets in which every subset of cardinality less than k has an upper bound.

(a) Prove that for each variety of finitary algebras the corresponding monad in **Set** has rank $\aleph_0$.

(b) Let $\mathbf{B}$ be an isomorphism-closed, full reflective subcategory of a category $\mathbf{A}$. Prove that the corresponding idempotent monad in $\mathbf{A}$ has rank k if and only if $\mathbf{B}$ is closed under the formation of k-direct colimits in $\mathbf{A}$.

(c) For each small category $\mathbf{A}$ prove that the forgetful functor $[\mathbf{A}, \mathbf{Set}] \to [Ob(\mathbf{A}), \mathbf{Set}]$ (where the set $Ob(\mathbf{A})$ is considered to be a discrete category) is monadic, and that the corresponding monad has rank $\aleph_0$.

* 20H. Locally Presentable Categories

An object A of a category $\mathbf{A}$ is called **presentable** provided that $\hom(A, -) : \mathbf{A} \to \mathbf{Set}$ preserves k-directed colimits for some regular cardinal k. A category $\mathbf{A}$ is called **locally presentable** provided that it is cocomplete and has a dense (see 12D) subcategory formed by presentable objects.

(a) Show that every object A is presentable in **Set**, **Vec**, **Pos**, and **Aut**. [Choose any regular cardinal k larger than card A.] Show that in **Top** the only presentable objects are the discrete spaces, and in **HComp** only the empty space is presentable.

(b) Show that **Set**, **Vec**, **Pos**, and **Aut** are locally presentable categories. Furthermore, show that $[\mathbf{A}, \mathbf{Set}]$ is locally presentable for each small category $\mathbf{A}$.

(c) Show that for each monad $\mathbf{T}$ with rank over a locally presentable category $\mathbf{X}$ the category $\mathbf{X^T}$ is locally presentable. In particular, if $\mathbf{Setm}$ denotes the category $\mathbf{Set}^m$ for some cardinal number m, then

$$\mathbf{Setm},\ \mathbf{Setm^{T_1}},\ (\mathbf{Setm^{T_1}})^{T_2},\ \ldots$$

are locally presentable categories for arbitrary monads $\mathbf{T}_1, \mathbf{T}_2, \ldots$ with rank.

(d) Conversely, prove that every locally presentable category $\mathbf{A}$ is equivalent to a category $(\mathbf{Setm^{T_1}})^{T_2}$ for some monads $\mathbf{T}_1, \mathbf{T}_2$ with rank; in fact, let $\mathbf{B}$ be a dense subcategory of $\mathbf{A}$ formed by presentable objects, then show that

(1) the full and faithful functor $E : \mathbf{A} \to [\mathbf{B}^{\mathrm{op}}, \mathbf{Set}]$ of Exercise 12D(c) maps $\mathbf{A}$ onto a reflective subcategory $E[\mathbf{A}]$, and the corresponding idempotent monad $\mathbf{T}_2$ on $[\mathbf{B}^{\mathrm{op}}, \mathbf{Set}]$ has rank,

(2) $[\mathbf{B}^{\mathrm{op}}, \mathbf{Set}]$ is monadic with rank $\aleph_0$ over $\mathbf{Setn} = \mathbf{Set}^n$ for $n = \mathrm{card}\,\mathbf{B}$ [20G(c)].

(e) Let us call an object A **strongly n-generated** provided that $\hom(A, -)$ preserves colimits of n-directed diagrams whose connecting morphisms are strong monomorphisms. Prove that then every quotient of A is also strongly n-generated.

In a locally n-presentable category $\mathbf{X}$ the strongly n-generated objects are precisely the quotients of n-presentable objects. And given a monad $\mathbf{T}$ with rank n prove that a $\mathbf{T}$-algebra is strongly n-presentable in $\mathbf{X}^{\mathbf{T}}$ iff it is a quotient of the free algebra $\mathbf{T}X$ for some strongly n-generated object X.

(f) In every locally presentable category $\mathbf{X}$ prove that every object is strongly n-presentable for some n. Conclude that if $\mathbf{X}$ has, for every regular cardinal n, only a set of strongly n-generated objects up to isomorphism, then $\mathbf{X}$ is co-wellpowered. Moreover, $\mathbf{X}^{\mathbf{T}}$ is also co-wellpowered for every monad $\mathbf{T}$ with rank.

(g) Conclude that every locally presentable category is complete, wellpowered, and co-wellpowered.

20I. Initial T-algebras

For an initial object (X, x) of $\mathbf{Alg}(T)$ prove that x is an isomorphism. [Hint: Use the T-algebra (TX, Tx).]

* 20J. Monadic Towers

Let $\mathbf{A}$ be cocomplete and let $\mathbf{A} \xrightarrow{U} \mathbf{X}$ be an adjoint functor. Call the associated monad $\mathbf{T}^1$ and the associated comparison functor $\mathbf{A} \xrightarrow{U^1} \mathbf{X}^{\mathbf{T}^1}$. Then $U = U^{\mathbf{T}^1} \circ U^1$. By Theorem 20.42, U^1 is adjoint. Denote the associated monad by $\mathbf{T}^2$ and the associated comparison functor by U^2. Then the diagram

$$\begin{array}{ccc} X^{\mathbf{T}^2} & \xleftarrow{U^2} & \mathbf{A} \\ {\scriptstyle U^{\mathbf{T}^2}}\downarrow & {\scriptstyle U^1}\swarrow & \downarrow{\scriptstyle U} \\ X^{\mathbf{T}^1} & \xrightarrow{U^{\mathbf{T}^1}} & \mathbf{X} \end{array}$$

commutes. This process can be iterated over all ordinals (with limit-steps obtained by forming a limit in the quasicategory $\mathbf{CAT}$).

(a) Show that $U^{\mathbf{T}^2}$ is an equivalence whenever U maps regular epimorphisms to epimorphisms.

(b) Let $\mathbf{A}_n$ (n a natural number) be the construct whose objects are pairs $(X, (\alpha_i)_{i \leq n})$, where X is a set and α_i is a partial endofunction of X defined on $x \in X$ if and only if $\alpha_j(x) = x$ for all $j < i$. Morphisms from $(X, (\alpha_i)_{i<n})$ to $(X', (\alpha'_i)_{i<n})$ are functions $f : X \to X'$ such that $f(\alpha_i(x)) = \alpha'_i(f(x))$ whenever $\alpha_i(x)$ is defined. Describe

the free algebras, comparison functors, etc. in detail and prove that the iteration described above takes precisely n steps. That is, if U_0 is the forgetful functor of $\mathbf{A}_n$ and U_{k+1} is the comparison functor associated with the kth step ($k = 0, 1, 2, \ldots$), then U_n is an equivalence but U_{n-1} is not.

(c) Show that there exists a construct for which the above iteration does not stop.

20K. The Constructs $\mathbf{Top}_0^{op}$ and Fram

Show that (cf. 5L)

(a) **Fram** is monadic.

(b) $\mathbf{Top}_0^{op}$ has free objects.

(c) The concrete functor $\mathbf{Top}_0^{op} \xrightarrow{T} \mathbf{Fram}$ described in Exercise 5L is a comparison functor.

20L. Regularly Monadic Functors Lift Regularity

Show that

(a) If $\mathbf{X}$ is regular and $\mathbf{A} \xrightarrow{U} \mathbf{X}$ is regularly monadic, then $\mathbf{A}$ is regular.

(b) Monadic constructs are regular.

20M. Extremally Monadic Functors

A functor $\mathbf{A} \xrightarrow{U} \mathbf{X}$ (resp. a concrete category $(\mathbf{A}, U)$ over $\mathbf{X}$) is called **extremally monadic** provided that $\mathbf{X}$ is an (ExtrEpi, Mono-Source)-category and U is ExtrEpi-monadic. Show that a concrete category $(\mathbf{A}, U)$ over $\mathbf{X}$ is regularly monadic if and only if it is extremally monadic and $\mathbf{X}$ has regular factorizations.

20N. When Are Order Preserving Maps Monadic?

(a) Show that a morphism in **Pos**, considered as a functor between thin categories, is monadic if and only if it is an embedding of a full reflective subcategory.

(b) For any function $A \xrightarrow{f} B$ consider $\mathcal{P}A \xrightarrow{\mathcal{P}f} \mathcal{P}B$ as a functor (cf. Exercise 6G). Show that $\mathcal{P}f$ is monadic if and only if f is surjective.

20O. Monadic Functors and Extremal Monomorphisms

Determine whether or not monadic functors preserve extremal monomorphisms.

* 20P. Varietors and Colimits of ω-chains

Let $T : \mathbf{X} \to \mathbf{X}$ be a functor with $\mathbf{X}$ having countable colimits. For each object X define a diagram $D : \mathbb{N} \to \mathbf{X}$ (where $\mathbb{N}$ is the thin category of natural numbers) by

$$D_0 = X,$$

$$D_{n+1} = X + TD_n,$$

$$D(0 \to 1) = X \to X + TX, \quad \text{the first coproduct injection,}$$

$$D(n+1 \to m+1) = X + TD_n \xrightarrow{id_X + TD(n \to m)} X + TD_m.$$

(a) If T preserves the colimit of D, prove that X generates a free T-algebra. [Hint: If $(D_n \xrightarrow{d_n} C)_\mathbb{N}$ is a colimit of D, then there is a unique $c: TC \to C$ with $c \circ Td_n = d_{n+1} \circ u_n$, where $u_n: TD_n \to X + TD_n$ is the second injection. Then $X \xrightarrow{d_0} |(C,c)|$ is universal.]

(b) Conclude that each functor $T: \mathbf{X} \to \mathbf{X}$ that preserves colimits of ω-chains (i.e., diagrams with the scheme $\mathbb{N}$) is a varietor.

(c) Find a varietor that does not preserve colimits of ω-chains.

Chapter VI

TOPOLOGICAL AND ALGEBRAIC CATEGORIES

As demonstrated in §5, the vague concept of "structure" has a formalization in the concept of a "concrete category" — the idea being that the forgetful or underlying functor "forgets" the structure in question. Many structures can be decomposed into more basic ones, which often can be classified as "topological" or "algebraic". The nature of a structure is reflected not so much in properties of its abstract category, but rather in properties of its underlying functor. For example, topological groups can be regarded:

(a) as topological structures via the forgetful functor **TopGrp** $\xrightarrow{U}$ **Grp**, i.e., as topological structures over **Grp**, or

(b) as algebraic structures via the forgetful functor **TopGrp** $\xrightarrow{V}$ **Top**, i.e., as algebraic structures over **Top**, or

(c) as topologically algebraic structures via the forgetful functor **TopGrp** $\xrightarrow{W}$ **Set**, i.e., as topologically algebraic structures over **Set**.

Since it is the underlying functor rather than the abstract category that determines the character of a structure (= concrete category), in each case we study properties of functors first. Surprisingly, each of the crucial properties under investigation (U being topological, algebraic, or topologically algebraic) implies that the functor U in question is faithful, i.e., is the forgetful functor of a concrete category.

We choose terminology such that a concrete category $(\mathbf{A}, U)$ has a certain property P if and only if its forgetful functor U has the property P. A desirable characteristic of such properties will be that they are closed under composition. Moreover, for "algebraic" properties P, it will be desirable that concrete functors between concrete categories with property P will automatically have property P as well.

Most properties under investigation can be defined either in a rigid (uniquely transportable) version or in a more flexible (closed under equivalences) one. We have found it somewhat more convenient and consistent with the current usage to choose the rigid version in the topological and algebraic cases, and the flexible version in the topologically algebraic case.

21 Topological categories

The following phenomena are typical for topological categories $(\mathbf{A}, U)$:

(1) $(\mathbf{A}, U)$ is *initially complete*, i.e., every structured source $(X \xrightarrow{f_i} UA_i)_I$ has a unique initial lift $(A \xrightarrow{f_i} A_i)_I$,

(2) $(\mathbf{A}, U)$ is *finally complete*, i.e., every structured sink $(UA_i \xrightarrow{f_i} X)_I$ has a unique final lift $(A_i \xrightarrow{f_i} A)_I$,

(3) $(\mathbf{A}, U)$ is *fibre-complete*, i.e., every fibre is a (possibly large) complete lattice (cf. Definition 5.7),

(4) $(\mathbf{A}, U)$ *has discrete structures*, i.e., every **X**-object has a discrete lift,

(5) $(\mathbf{A}, U)$ *has indiscrete structures*, i.e., every **X**-object has an indiscrete lift.

As we will see, the above conditions are not independent of each other. In fact, (1) and (2) are equivalent [Topological Duality Theorem (21.9)], and imply all the others. Moreover, (1) and (2) have many other pleasant consequences. For example, they imply that U lifts limits (and colimits) uniquely. The unique lifting of limits implies, together with (5), all of the other conditions (21.18). However, forgetful functors of algebraic categories also lift limits uniquely (see §23). Hence the very simple condition (5) in some sense may be considered to be at the heart of topology.

TOPOLOGICAL FUNCTORS

21.1 DEFINITION
A functor $\mathbf{A} \xrightarrow{G} \mathbf{B}$ is called **topological** provided that every G-structured source $(B \xrightarrow{f_i} GA_i)_I$ has a unique G-initial lift $(A \xrightarrow{\overline{f}_i} A_i)_I$.

21.2 EXAMPLES
(1) The forgetful functors of the constructs **Top**, **Unif**, **PMet**, **Rel**, and **Prost** are topological, but those of the constructs **Haus**, **Met**, **Vec**, and **Pos** are not topological.

(2) For a thin category $\mathbf{A}$, the unique functor to **1** is topological if and only if $\mathbf{A}$ is a (possibly large) complete lattice.

21.3 THEOREM
Topological functors are faithful.

Proof: Let $A \underset{s}{\overset{r}{\rightrightarrows}} A'$ be a pair of **A**-morphisms with $Gr = Gs$. Consider the source $S = (GA \xrightarrow{f_h} GA_h)_{h \in Mor(\mathbf{A})}$ with $(f_h, A_h) = (Gr, A')$ for each $h \in Mor(\mathbf{A})$. Let $\hat{S} = (\hat{A} \xrightarrow{\hat{f}_h} A')_{h \in Mor(\mathbf{A})}$ be a G-initial lift of S. Define a source $\mathcal{T} = (A \xrightarrow{g_h} A')$ by

$$g_h = \begin{cases} r, & \text{if } \hat{f}_h \circ h = s \\ s, & \text{otherwise.} \end{cases}$$

Then $G\mathcal{T} = G\hat{S} \circ id_{GA}$. By G-initiality of $\hat{S}$ there exists a morphism $A \xrightarrow{k} \hat{A}$ with $\mathcal{T} = \hat{S} \circ k$, i.e., $g_h = \hat{f}_h \circ k$ for each $h \in Mor(\mathbf{A})$. In particular, we obtain $g_k = \hat{f}_k \circ k$, which — by the definition of the g_h's — is possible only for $r = s$. □

Lifting of a source in a topological category

21.4 REMARK

We will see below that "to be topological" is a very strong and pleasant property. Much of the strength lies in the fact that the G-structured sources are allowed to be large (as can be discerned from the preceding proof). For example, the natural "forgetful" functor G from the category of modules to the category of rings is not faithful, hence not topological, even though each small G-structured source has a G-initial lift (see Exercise 21D).

21.5 PROPOSITION

If $\mathbf{A} \xrightarrow{G} \mathbf{B}$ is a functor such that every G-structured source has a G-initial lift, then the following conditions are equivalent:

(1) G *is topological,*

(2) $(\mathbf{A}, G)$ *is uniquely transportable,*

(3) $(\mathbf{A}, G)$ *is amnestic.*

Proof: (1) $\Rightarrow$ (2) is immediate from Proposition 8.14, once it is observed that G is faithful. The latter follows as in the proof of Theorem 21.3.

(2) $\Rightarrow$ (3) follows from Proposition 5.29.

(3) $\Rightarrow$ (1). If $(\overline{A} \xrightarrow{\overline{f}_i} A_i)_I$ and $(\tilde{A} \xrightarrow{\tilde{f}_i} A_i)_I$ are initial lifts of $(X \xrightarrow{f_i} GA_i)_I$, then $\overline{A} \le \tilde{A}$ and $\tilde{A} \le \overline{A}$; hence by (3), $\overline{A} = \tilde{A}$. □

21.6 PROPOSITION
If $\mathbf{A} \xrightarrow{G} \mathbf{B}$ *and* $\mathbf{B} \xrightarrow{F} \mathbf{C}$ *are topological, then so is* $\mathbf{A} \xrightarrow{F \circ G} \mathbf{C}$.

Proof: Immediate by Proposition 21.5, since amnesticity and existence of initial liftings are compositive properties. □

TOPOLOGICAL CATEGORIES

21.7 DEFINITION
A concrete category $(\mathbf{A}, U)$ is called **topological** provided that U is topological.

21.8 EXAMPLES
(1) The constructs **Top**, **Unif**, **PMet**, **Rel**, and **Prost** are topological.

(2) All functor-structured categories $\mathbf{Spa}(T)$ and all functor-costructured categories $(\mathbf{Spa}(T))^{\mathrm{op}}$ are topological.

(3) A partially ordered set, considered as concrete category over **1**, is topological if and only if it is a complete lattice.

(4) **TopGrp** is topological if it is considered as a concrete category over **Grp**, but not if it is considered as a concrete category over **Top** or over **Set**.

(5) The construct $\mathbf{Top}_1$ is not topological, even though for every structured source $(X \xrightarrow{f_i} (X_i, \tau_i))_I$ there exists a largest $\mathbf{Top}_1$-structure (= the smallest T_1-topology) τ on X making each $(X, \tau) \xrightarrow{f_i} (X_i, \tau_i)$ continuous.

The following theorem generalizes the well-known fact that each meet-complete poset is also join-complete.

21.9 TOPOLOGICAL DUALITY THEOREM
If $(\mathbf{A}, U)$ *is topological over* $\mathbf{X}$*, then* $(\mathbf{A}^{\mathrm{op}}, U^{\mathrm{op}})$ *is topological over* $\mathbf{X}^{\mathrm{op}}$ *(i.e., the existence of unique U-initial lifts of U-structured sources implies the existence of unique U-final lifts of U-structured sinks).*

Proof: Let $(\mathbf{A}, U)$ be topological. It must be shown that in $(\mathbf{A}, U)$ every structured sink $\mathcal{S} = (UA_i \xrightarrow{f_i} X)_I$ has a final lift (since uniqueness follows from amnesticity). Consider the structured source $\mathcal{T} = (X \xrightarrow{g_j} UB_j)_J$ consisting of all structured arrows (g_j, B_j) with the property that $UA_i \xrightarrow{f_i} X \xrightarrow{g_j} UB_j$ is an $\mathbf{A}$-morphism for each $i \in I$. If $(A \xrightarrow{g_j} B_j)_J$ is an initial lift of $\mathcal{T}$, then $(A_i \xrightarrow{f_i} A)_I$ is a final lift of $\mathcal{S}$. □

21.10 REMARK

The above Topological Duality Theorem implies that (as was the case for abstract and for concrete categories) there is also a **Duality Principle** available for topological categories. However, observe that, since the dual of a concrete (resp. topological) category over $\mathbf{X}$ is a concrete (resp. topological) category over $\mathbf{X}^{op}$, this does not imply a Duality Principle for (topological) concrete categories over a *fixed* category $\mathbf{X}$ (unless $\mathbf{X}^{op} = \mathbf{X}$, as for $\mathbf{X} = \mathbf{1}$). In particular, there is not a Duality Principle available for topological constructs.

21.11 PROPOSITION

Topological categories are fibre-complete. The smallest (resp. largest) member of each fibre is discrete (resp. indiscrete).

Proof: Let $(\mathbf{A}, U)$ be topological over $\mathbf{X}$, let X be an $\mathbf{X}$-object, and let $(A_i)_I$ be a family of $\mathbf{A}$-objects with $UA_i = X$. If $(A \xrightarrow{id_X} A_i)_I$ is an initial lift of $(X \xrightarrow{id_X} UA_i)_I$, then $A = \inf(A_i)$ in the fibre of X. For $I = \emptyset$, we have that A is the largest element of the fibre of X. It is an indiscrete object, since $(A, \emptyset)$ is an initial source [cf. 10.42(1)]. By duality, the smallest element of the fibre of X must be discrete. □

21.12 PROPOSITION

If $(\mathbf{A}, U)$ is topological over $\mathbf{X}$, then

(1) U is an adjoint functor; its co-adjoint $\mathbf{X} \xrightarrow{F} \mathbf{A}$ (the discrete functor) is a full embedding, satisfying $U \circ F = id_{\mathbf{X}}$.

(2) U is a co-adjoint functor; its adjoint $\mathbf{X} \xrightarrow{G} \mathbf{A}$ (the indiscrete functor) is a full embedding, satisfying $U \circ G = id_{\mathbf{X}}$.

Proof: Immediate from Proposition 21.11. □

21.13 PROPOSITION

If $(\mathbf{A}, U)$ is topological over $\mathbf{X}$, then the following hold:

(1) U preserves and reflects mono-sources and epi-sinks.

(2) An $\mathbf{A}$-morphism is an extremal (resp. regular) monomorphism if and only if it is initial and an extremal (resp. regular) $\mathbf{X}$-monomorphism.

(3) An $\mathbf{A}$-morphism is an extremal (resp. regular) epimorphism if and only if it is final and an extremal (resp. regular) $\mathbf{X}$-epimorphism.

In particular, in topological constructs, the following hold:

(4) embeddings = extremal monomorphisms = regular monomorphisms.

(5) quotient morphisms = extremal epimorphisms = regular epimorphisms.

Proof:
(1). Since U is faithful and (co)adjoint, it reflects and preserves mono-sources (epi-sinks).

(2). Let $A \xrightarrow{m} B$ be an extremal monomorphism in **A**. Then, by (1), $UA \xrightarrow{m} UB$ is a monomorphism in **X**. Let $UA \xrightarrow{m} UB = UA \xrightarrow{e} X \xrightarrow{f} UB$ be an (Epi,−)-factorization in **X**. Then $UA \xrightarrow{e} X$ has a final lift $A \xrightarrow{e} C$. Consequently, by (1), $A \xrightarrow{m} B = A \xrightarrow{e} C \xrightarrow{f} B$ is an (Epi,−)-factorization in **A**. Therefore, e is an isomorphism in **A** and hence in **X**. Thus $UA \xrightarrow{m} UB$ is an extremal monomorphism in **X**. To show initiality of $A \xrightarrow{m} B$, let $A' \xrightarrow{m} B$ be an initial lift of $UA \xrightarrow{m} UB$. Then $A \xrightarrow{m} B = A \xrightarrow{id_{UA}} A' \xrightarrow{m} B$ is an (Epi,−)-factorization. Thus $A \xrightarrow{id_{UA}} A'$ is an isomorphism; hence by amnesticity $A = A'$. So $A \xrightarrow{m} B = A' \xrightarrow{m} B$ is initial. For the converse, let $A \xrightarrow{m} B$ be an initial morphism such that $UA \xrightarrow{m} UB$ is an extremal monomorphism in **X**. Then, by (1), $A \xrightarrow{m} B$ is a monomorphism in **A**. Let $A \xrightarrow{m} B = A \xrightarrow{e} C \xrightarrow{f} B$ be an (Epi,−)-factorization in **A**. Then $UA \xrightarrow{e} UC$ is an **X**-isomorphism and $A \xrightarrow{e} C$ is initial. Hence, by Proposition 8.14, $A \xrightarrow{e} C$ is an isomorphism. Thus $A \xrightarrow{m} B$ is an extremal monomorphism in **A**.
Next, let $A \xrightarrow{m} B$ be a regular monomorphism in **A**. Then it is extremal and hence initial. By adjointness, $UA \xrightarrow{m} UB$ is a regular monomorphism in **X**. Conversely, let $A \xrightarrow{m} B$ be an initial morphism such that $UA \xrightarrow{m} UB$ is an equalizer of $UB \underset{s}{\overset{r}{\rightrightarrows}} X$ in **X**. If $B \underset{s}{\overset{r}{\rightrightarrows}} C$ is a final lift, then by Proposition 13.15, m is an equalizer of r and s in **A**.

(3) follows by duality (21.9).

(4) and (5) are immediate from (2) and (3). □

21.14 PROPOSITION

If $(\mathbf{A}, U)$ is topological over an $(E, \mathbf{M})$-category $\mathbf{X}$, then the following hold:

(1) $\mathbf{A}$ is an $(E, \mathbf{M}_{init})$-category, where $\mathbf{M}_{init}$ consists of all initial sources in $\mathbf{M}$.

(2) $\mathbf{A}$ is an $(E_{fin}, \mathbf{M})$-category, where E_{fin} consists of all final E-morphisms. □

21.15 PROPOSITION

If $(\mathbf{A}, U)$ is topological over $\mathbf{X}$, then U uniquely lifts both limits (via initiality) and colimits (via finality), and it preserves both limits and colimits.

Proof: The unique lifting follows immediately from Proposition 13.15 and its dual; the preservation from (co)adjointness of U. □

21.16 THEOREM

If $(\mathbf{A}, U)$ is topological over $\mathbf{X}$, then the following hold:

(1) $\mathbf{A}$ is (co)complete if and only if $\mathbf{X}$ is (co)complete.

(2) $\mathbf{A}$ is (co-)wellpowered if and only if $(\mathbf{A}, U)$ is fibre-small and $\mathbf{X}$ is (co-)wellpowered.

(3) $\mathbf{A}$ is extremally (co-)wellpowered if and only if $\mathbf{X}$ is extremally (co-)wellpowered.

(4) $\mathbf{A}$ is (Epi, Mono-Source)-factorizable if and only if $\mathbf{X}$ is (Epi, Mono-Source)-factorizable.

(5) $\mathbf{A}$ has regular factorizations if and only if $\mathbf{X}$ has regular factorizations.

(6) $\mathbf{A}$ has a (co)separator if and only if $\mathbf{X}$ has a (co)separator.

Proof: Let $G : \mathbf{X} \to \mathbf{A}$ be the indiscrete functor.

(1). If $\mathbf{X}$ is complete, then $\mathbf{A}$ is complete, since U lifts limits. Conversely, if $\mathbf{A}$ is complete, $D : \mathbf{I} \to \mathbf{X}$ is a small diagram, and $\mathcal{L}$ is a limit of $G \circ D : \mathbf{I} \to \mathbf{A}$, then $U\mathcal{L}$ is a limit of $U \circ G \circ D = D$.

(2). Let $(\mathbf{A}, U)$ be fibre-small and $\mathbf{X}$ be wellpowered. For any $\mathbf{A}$-object A consider a set $\mathcal{M} = \{ X_i \xrightarrow{m_i} UA \,|\, i \in I \}$ of $\mathbf{X}$-subobjects of UA such that every subobject of UA is isomorphic to some member of $\mathcal{M}$. If $\hat{\mathcal{M}}$ consists of all possible lifts $A_i \xrightarrow{m_i} A$ of members of $\mathcal{M}$, then, by fibre-smallness, $\hat{\mathcal{M}}$ is a set. By transportability each subobject of A is isomorphic to some member of $\hat{\mathcal{M}}$. Hence $\mathbf{A}$ is wellpowered.
Conversely, let $\mathbf{A}$ be wellpowered. For every $\mathbf{X}$-object X, it is clear that the class $\mathcal{A} = \{ A \in Ob(\mathbf{A}) \,|\, UA = X \}$ is a set, since otherwise $\{ A \xrightarrow{id_X} GX \,|\, A \in \mathcal{A} \}$ would by amnesticity be a proper class of pairwise non-isomorphic subobjects of GX. Thus $(\mathbf{A}, U)$ is fibre-small. If $(X_i \xrightarrow{m_i} X)_I$ would be a proper class of pairwise non-isomorphic subobjects of X, then $(GX_i \xrightarrow{m_i} GX)_I$ would be a proper class of pairwise non-isomorphic subobjects of GX in $\mathbf{A}$. Hence $\mathbf{X}$ is wellpowered.

(3). Let $\mathbf{X}$ be extremally wellpowered. For any $\mathbf{A}$-object A let $\mathcal{M} = \{ X_i \xrightarrow{m_i} UA \}_{i \in I}$ be a set of extremal subobjects of UA such that every extremal subobject of UA is isomorphic to some member of $\mathcal{M}$. For each $i \in I$ let $A_i \xrightarrow{m_i} A$ be the unique initial lift of $X_i \xrightarrow{m_i} UA$. Then every extremal subobject of A is isomorphic to some member of the set $\{ A_i \xrightarrow{m_i} A \,|\, i \in I \}$. Hence $\mathbf{A}$ is extremally wellpowered.
Conversely, let $\mathbf{A}$ be extremally wellpowered. If $(X_i \xrightarrow{m_i} X)_I$ were a proper class of pairwise non-isomorphic extremal subobjects of some $\mathbf{X}$-object X, it would follow that $(GX_i \xrightarrow{m_i} GX)_I$ would be a proper class of pairwise non-isomorphic extremal subobjects of GX in $\mathbf{A}$. Hence $\mathbf{X}$ is extremally wellpowered.

(4). If $(A \xrightarrow{f_i} A_i)_I$ is a source in $\mathbf{A}$, $(UA \xrightarrow{f_i} UA_i) = (UA \xrightarrow{e} X \xrightarrow{m_i} UA_i)$ is an (Epi, Mono-Source)-factorization in $\mathbf{X}$, and $(B \xrightarrow{m_i} A_i)_I$ is an initial lift of $(X \xrightarrow{m_i} UA_i)_I$, then $(A \xrightarrow{f_i} A_i) = (A \xrightarrow{e} B \xrightarrow{m_i} A_i)$ is an (Epi, Mono-Source)-factorization in $\mathbf{A}$. Conversely, if $(X \xrightarrow{f_i} X_i)_I$ is a source in $\mathbf{X}$ and $GX \xrightarrow{f_i} GX_i = GX \xrightarrow{e} A \xrightarrow{m_i} GX_i$ is an (Epi, Mono-Source)-factorization in $\mathbf{A}$, then $X \xrightarrow{f_i} X_i = X \xrightarrow{e} UA \xrightarrow{m_i} X_i$ is an (Epi, Mono-Source)-factorization in $\mathbf{X}$.

(5). This follows as in (4) by means of Proposition 21.13(3).

(6). If A is an **A**-separator, then UA is an **X**-separator. If X is an **X**-separator, then the discrete object A with $UA = X$ is an **A**-separator.

The "co"-parts follow by duality. □

21.17 COROLLARY
Each topological construct

(1) is complete and cocomplete,

(2) is wellpowered (resp. co-wellpowered) if and only if it is fibre-small,

(3) is an (Epi, Extremal Mono-Source)-category,

(4) has regular factorizations,

(5) has separators and coseparators. □

21.18 INTERNAL TOPOLOGICAL CHARACTERIZATION THEOREM
A concrete category $(\mathbf{A}, U)$ over $\mathbf{X}$ is topological if and only if it satisfies the following conditions:

(1) U lifts limits uniquely,

(2) $(\mathbf{A}, U)$ has indiscrete structures, i.e., every $\mathbf{X}$-object has an indiscrete lift.

Proof: By Propositions 21.11 and 21.15 topological categories satisfy (1) and (2). Conversely, let $(\mathbf{A}, U)$ satisfy (1) and (2). Then $(\mathbf{A}, U)$ is uniquely transportable. Hence, by Proposition 21.5 and by duality (21.9), it suffices to show that every structured sink $(UA_i \xrightarrow{f_i} X)_I$ has a final lift. Let B_0 be the indiscrete object with $UB_0 = X$ and let $(B_j)_J$ be the family of all **A**-objects B_j such that $UB_j = X$ and $UA_i \xrightarrow{f_i} UB_j$ is an **A**-morphism for each $i \in I$. The family $(UB_j \xrightarrow{id_X} UB_0)_J$ has an intersection $X \xrightarrow{id_X} UB_0$ in **X**. By (1) this intersection can be lifted to an intersection $B \xrightarrow{id_X} B_0$ of the family $(B_j \xrightarrow{id_X} B_0)_J$. Consequently, $\mathcal{S} = (A_i \xrightarrow{f_i} B)_I$ is a sink in **A**. To show that it is final, let $UB \xrightarrow{g} UC$ be an **X**-morphism such that each $UA_i \xrightarrow{f_i} UB \xrightarrow{g} UC$ is an **A**-morphism. Let C_0 be the indiscrete object with $UC_0 = UC$. Then the pullback in **X**

$$\begin{array}{ccc} X & \xrightarrow{g} & UC \\ {\scriptstyle id_X}\downarrow & & \downarrow{\scriptstyle id_{UC}} \\ UB & \xrightarrow{g} & UC_0 \end{array}$$

can be lifted to a pullback in **A**

$$\begin{array}{ccc} P & \xrightarrow{g} & C \\ {\scriptstyle id_X}\downarrow & & \downarrow{\scriptstyle id_{UC}} \\ B & \xrightarrow{g} & C_0 \end{array}$$

Since, for each $i \in I$, the diagram

$$\begin{array}{ccc} A_i & \xrightarrow{g \circ f_i} & C \\ f_i \downarrow & & \downarrow id_{UC} \\ B & \xrightarrow{g} & C_0 \end{array}$$

commutes, by the pullback-property each $UA_i \xrightarrow{f_i} UP$ is an **A**-morphism. Consequently, P is a member of $(B_j)_J$. This implies that $B \leq P$. Hence, by amnesticity, $B = P$. Thus $B \xrightarrow{g} C = P \xrightarrow{g} C$ is an **A**-morphism. Consequently, $(A_i \xrightarrow{f_i} B)_I$ is a final lift of $(UA_i \xrightarrow{f_i} X)_I$. □

21.19 REMARK
As we will see later, the unique lifting of limits is a widespread property shared by many reasonable forgetful functors not only in topology, but also in algebra. Hence the above theorem shows that the existence of indiscrete structures is the crucial condition that makes $(\mathbf{A}, U)$ topological.

21.20 EXAMPLE
For any construct of the form $\mathbf{Alg}(\Sigma)$ the forgetful functor $\mathbf{Alg}(\Sigma) \to \mathbf{Set}$ lifts limits and colimits uniquely. But $\mathbf{Alg}(\Sigma)$ is topological only if Σ is the empty family, i.e., only if $\mathbf{Alg}(\Sigma)$ is concretely isomorphic to the construct **Set**.

21.21 EXTERNAL TOPOLOGICAL CHARACTERIZATION THEOREM
Let $\mathbf{CAT}(\mathbf{X})$ be the quasicategory of all concrete categories and concrete functors over a fixed category $\mathbf{X}$. If $\mathcal{M}$ is the conglomerate of all full functors in $\mathbf{CAT}(\mathbf{X})$, then for each concrete category $(\mathbf{A}, U)$ over $\mathbf{X}$ the following are equivalent:

(1) $(\mathbf{A}, U)$ is topological over $\mathbf{X}$.

(2) $(\mathbf{A}, U)$ is an $\mathcal{M}$-injective object in $\mathbf{CAT}(\mathbf{X})$.

Proof: (1) ⇒ (2). Suppose that $(\mathbf{A}, U)$ is topological over $\mathbf{X}$, $(\mathbf{B}, V) \xrightarrow{E} (\mathbf{C}, W)$ belongs to $\mathcal{M}$, and $(\mathbf{B}, V) \xrightarrow{G} (\mathbf{A}, U)$ is a concrete functor over $\mathbf{X}$. For each **C**-object C consider the E-structured source $(C \xrightarrow{f_i} EB_i)_I$ consisting of all pairs (f_i, B_i) with $B_i \in Ob(\mathbf{B})$ and $C \xrightarrow{f_i} EB_i \in Mor(\mathbf{C})$. Application of W yields the U-structured source $(WC \xrightarrow{f_i} UGB_i)_I$. By hypothesis it has a U-initial lift $(A_C \xrightarrow{f_i} GB_i)_I$. Let $(\mathbf{C}, W) \xrightarrow{\hat{G}} (\mathbf{A}, U)$ be the unique concrete functor with $\hat{G}C = A_C$ for each **C**-object C. Then $G = \hat{G} \circ E$.

(2) ⇒ (1). Suppose that $(\mathbf{A}, U)$ is $\mathcal{M}$-injective in $\mathbf{CAT}(\mathbf{X})$. By Proposition 5.33 there exists an amnestic $(\mathbf{B}, V)$ and a functor $(\mathbf{A}, U) \xrightarrow{P} (\mathbf{B}, V)$ that is surjective on objects and belongs to $\mathcal{M}$. By injectivity, P is a section, hence an isomorphism. Thus $(\mathbf{A}, U)$ is

amnestic. It remains to be shown that every U-structured source $\mathcal{S} = (X \xrightarrow{f_i} UA_i)_I$ has a U-initial lift. Embed $(\mathbf{A}, U)$ as a concrete and full subcategory of a concrete category $(\mathbf{B}, V)$ by adding one object B with $VB = X$ and adding the following morphisms (for $\mathbf{A}$-objects A and $\mathbf{X}$-morphisms f):

$$UA \xrightarrow{f} VB \in Mor(\mathbf{B}) \iff UA \xrightarrow{f_i \circ f} UA_i \in Mor(\mathbf{A}) \quad \text{for each } i \in I,$$

$$VB \xrightarrow{f} UA \in Mor(\mathbf{B}) \iff \text{there exists } i \in I \text{ and an } \mathbf{A}\text{-morphism}$$
$$A_i \xrightarrow{g} A \text{ with } f = g \circ f_i,$$

$$VB \xrightarrow{f} VB \in Mor(\mathbf{B}) \iff VB \xrightarrow{f_i \circ f} UA_i \in Mor(\mathbf{B}) \quad \text{for each } i \in I.$$

By $\mathcal{M}$-injectivity, $(\mathbf{A}, U)$ is a retract of $(\mathbf{B}, V)$, i.e., the identity on $\mathbf{A}$ can be extended to a concrete functor $(\mathbf{B}, V) \xrightarrow{G} (\mathbf{A}, U)$. Then $(GB \xrightarrow{f_i} A_i)_I = (GB \xrightarrow{f_i} GA_i)_I$ is a U-initial lift of $\mathcal{S}$. □

21.22 REMARKS
(1) The above result remains true if only fibre-small or only amnestic concrete categories are considered. In the amnestic case $\mathcal{M}$ consists precisely of the concrete full embeddings [cf. 5.10(4)]. Application of the fibre-small amnestic case to $\mathbf{X} = \mathbf{1}$ yields the result that the injectives in **Pos** are precisely the complete lattices [cf. 9.3(2)].

(2) From the above proof it is easily seen that topological categories have an alternative external characterization as precisely the $\mathcal{M}$-absolute retracts in $\mathbf{CAT}(\mathbf{X})$. For a description of $\mathcal{M}$-essential extensions and $\mathcal{M}$-injective hulls in $\mathbf{CAT}(\mathbf{X})$ see Exercise 21J.

INITIALITY-PRESERVING CONCRETE FUNCTORS

An important property of abstract functors is the preservation of limits. Recall that every adjoint functor preserves limits and that, under suitable assumptions, this preservation property characterizes adjoints (see the adjoint functor theorems of §18). For concrete categories a similar condition is that of the preservation of initial sources. (See Definition 10.47.) Among other things, we will show that for Galois correspondences (F, G) the concrete functor G preserves initial sources, and that if G has a topological domain, this preservation property characterizes those concrete functors G that are part of a Galois correspondence.

21.23 PROPOSITION
Initiality-preserving concrete functors preserve indiscrete objects.

Proof: Immediate by Example 10.42(1). □

21.24 GALOIS CORRESPONDENCE THEOREM
For concrete functors $(\mathbf{A}, U) \xrightarrow{G} (\mathbf{B}, V)$ with topological domain $(\mathbf{A}, U)$ the following conditions are equivalent:

(1) G preserves initial sources,

(2) G is adjoint and has a concrete co-adjoint $(\mathbf{B}, V) \xrightarrow{F} (\mathbf{A}, U)$,

(3) there exists a (unique) $(\mathbf{B}, V) \xrightarrow{F} (\mathbf{A}, U)$ such that (F, G) is a Galois correspondence.

Proof:
(1) $\Rightarrow$ (2). For any **B**-object B consider the G-structured source $\mathcal{S}_B = (B \xrightarrow{f_i} GA_i)_I$ consisting of all pairs (f_i, A_i) with $A_i \in Ob(\mathbf{A})$ and $B \xrightarrow{f_i} GA_i \in Mor(\mathbf{B})$. Application of V yields the U-structured source $\hat{\mathcal{S}}_B = (VB \xrightarrow{f_i} UA_i)_I$. Let $(A_B \xrightarrow{f_i} A_i)_I$ be a U-initial lift of $\hat{\mathcal{S}}_B$. By (1), $(GA_B \xrightarrow{f_i} GA_i)_I$ is V-initial. Since each $VB \xrightarrow{id_{VB}} VGA_B \xrightarrow{f_i} VGA_i$ is a **B**-morphism, $VB \xrightarrow{id_{VB}} VGA_B$ must be a **B**-morphism. Thus (id_{VB}, A_B) is a G-universal arrow for B. Since it is identity-carried, the associated co-adjoint F of G satisfies $V = U \circ F$.

(2) $\Rightarrow$ (3). Let $(\mathbf{B}, V) \xrightarrow{F} (\mathbf{A}, U)$ be a concrete co-adjoint of G, induced by G-universal arrows $B \xrightarrow{\eta_B} GFB$. By Theorem 19.1 each $FB \xrightarrow{\eta_B} FGFB$ is a section in **A**; hence $VB \xrightarrow{\eta_B} UFB$ is a section in **X**. To show that it is an epimorphism in **X**, let $UFB \underset{s}{\overset{r}{\rightrightarrows}} X$ be a pair of **X**-morphisms with $r \circ \eta_B = s \circ \eta_B$. If A is an indiscrete object in $(\mathbf{A}, U)$ with $UA = X$, then $FB \underset{s}{\overset{r}{\rightrightarrows}} A$ is a pair of **A**-morphisms. Since (η_B, FB) is G-generating, the equality $Gr \circ \eta_B = Gs \circ \eta_B$ implies that $r = s$. Thus $VB \xrightarrow{\eta_B} UFB$ is an **X**-isomorphism. Let $A_B \xrightarrow{\eta_B} FB$ be a U-initial lift. Then, by Proposition 8.14, $A_B \xrightarrow{\eta_B} FB$ is an **A**-isomorphism. Hence, by Proposition 8.35, $B \xrightarrow{id_{VB}} GA_B = B \xrightarrow{\eta_B} GFB \xrightarrow{G(\eta_B)^{-1}} GA_B$ is a G-universal arrow for B. If $\hat{F}$ is the associated co-adjoint of G, then $(\hat{F}, G)$ is a Galois-correspondence. Uniqueness follows from the amnesticity of U.

(3) $\Rightarrow$ (1). Immediate from Proposition 10.49. □

21.25 REMARK

If $(\mathbf{A}, U)$ is not topological, then conditions (1) and (2) of the above theorem do not imply (3). See Exercise 21E(b).

21.26 DEFINITION

Let $\mathbf{A} \xrightarrow{U} \mathbf{X}$ and $\mathbf{B} \xrightarrow{V} \mathbf{Y}$ be functors. An adjoint situation $(\hat{\eta}, \hat{\varepsilon}): \hat{F} \dashv \hat{G}: \hat{\mathbf{A}} \to \hat{\mathbf{B}}$ is said to **lift an adjoint situation** $(\eta, \varepsilon): F \dashv G: \mathbf{X} \to \mathbf{Y}$ along U and V provided that the following conditions are satisfied:

(1) the diagrams

$$\begin{array}{ccc} \mathbf{A} & \xrightarrow{\hat{G}} & \mathbf{B} \\ U \downarrow & & \downarrow V \\ \mathbf{X} & \xrightarrow{G} & \mathbf{Y} \end{array} \quad \text{and} \quad \begin{array}{ccc} \mathbf{B} & \xrightarrow{\hat{F}} & \mathbf{A} \\ V \downarrow & & \downarrow U \\ \mathbf{Y} & \xrightarrow{F} & \mathbf{X} \end{array}$$

commute.

(2) $V\hat{\eta} = \eta V$,

(3) $U\hat{\varepsilon} = \varepsilon U$.

21.27 REMARKS

(1) In the presence of (1), conditions (2) and (3) in the above definition imply each other. They have both been included to make the symmetry apparent. If (1) and (2) hold, then (3) can be deduced from the equations $GU\hat{\varepsilon}_A \circ \eta_{GUA} = V\hat{G}\hat{\varepsilon}_A \circ \eta_{V\hat{G}A} = V\hat{G}\hat{\varepsilon}_A \circ V\hat{\eta}_{\hat{G}A} = V(\hat{G}\hat{\varepsilon}_A \circ \hat{\eta}_{\hat{G}A}) = V(id_{\hat{G}A}) = id_{V\hat{G}A} = id_{GUA} = G\varepsilon_{UA} \circ \eta_{GUA}$, since η_{GA} is G-generating.

(2) The following theorem, applied to $G = id_X$, provides a new proof of the equivalence of conditions (1) and (3) in Theorem 21.24.

21.28 TAUT LIFT THEOREM

Let $(\mathbf{A}, U)$ be a topological category over the base category $\mathbf{X}$ and $(\mathbf{B}, V)$ be a concrete category over the base category $\mathbf{Y}$. If $\mathbf{A} \xrightarrow{\hat{G}} \mathbf{B}$ is a functor and $\mathbf{X} \xrightarrow{G} \mathbf{Y}$ is an adjoint functor with $V \circ \hat{G} = G \circ U$, then the following conditions are equivalent:

(1) $\hat{G}$ sends U-initial sources into V-initial sources,

(2) every adjoint situation $(\eta, \varepsilon) : F \dashv G : \mathbf{X} \to \mathbf{Y}$ can be lifted along U and V to an adjoint situation $(\hat{\eta}, \hat{\varepsilon}) : \hat{F} \dashv \hat{G} : \mathbf{A} \to \mathbf{B}$.

Proof:
(1) $\Rightarrow$ (2). For each **B**-object B consider the $\hat{G}$-structured source $\mathcal{S}_B = (B \xrightarrow{f_i} \hat{G}A_i)$ consisting of all pairs (f_i, A_i) with $A_i \in Ob(\mathbf{A})$ and $B \xrightarrow{f_i} \hat{G}A_i \in Mor(\mathbf{B})$. For each $i \in I$ let $FVB \xrightarrow{\tilde{f}_i} UA_i$ be the unique **X**-morphism with

$$VB \xrightarrow{Vf_i} GUA_i = VB \xrightarrow{\eta_{VB}} GFVB \xrightarrow{G\tilde{f}_i} GUA_i.$$

Then the U-structured source $(FVB \xrightarrow{\tilde{f}_i} UA_i)_I$ has a U-initial lift $\mathcal{T}_B = (A_B \xrightarrow{\tilde{f}_i} A_i)_I$. By (1), $\hat{G}\mathcal{T}_B$ is V-initial. Since $V\mathcal{S}_B = V\hat{G}\mathcal{T}_B \circ \eta_{VB}$

$$\begin{array}{ccccccc}
VB & \xrightarrow{\eta_{VB}} & GFVB & = & GUA_B & = & V\hat{G}A_B \\
{\scriptstyle Vf_i}\downarrow & & \downarrow{\scriptstyle G\tilde{f}_i} & & \downarrow{\scriptstyle GU\tilde{f}_i} & & \downarrow{\scriptstyle V\hat{G}\tilde{f}_i} \\
V\hat{G}A_i & = & GUA_i & = & GUA_i & = & V\hat{G}A_i
\end{array}$$

there exists a unique **B**-morphism $B \xrightarrow{\hat{\eta}_B} \hat{G}A_B$ with $V\hat{\eta}_B = \eta_{VB}$ and $\mathcal{S}_B = \hat{G}\mathcal{T}_B \circ \hat{\eta}_B$. To show that the $\hat{G}$-structured arrow $(\hat{\eta}_B, A_B)$ is $\hat{G}$-generating and hence $\hat{G}$-universal, let $A_B \underset{s}{\overset{r}{\rightrightarrows}} A$ be a pair of **A**-morphisms with $\hat{G}r \circ \hat{\eta}_B = \hat{G}s \circ \hat{\eta}_B$. Then $GUr \circ \eta_{VB} = V\hat{G}r \circ \eta_{VB} = V(\hat{G}r \circ \hat{\eta}_B) = V(\hat{G}s \circ \hat{\eta}_B) = V\hat{G}s \circ \eta_{VB} = GUs \circ \eta_{VB}$ implies that

$Ur = Us$. Hence $r = s$ by the faithfulness of U. If $(\hat{\eta}, \hat{\varepsilon}): \hat{F} \dashv \hat{G}: \mathbf{A} \to \mathbf{B}$ is the adjoint situation determined by $\hat{G}$ and the $\hat{G}$-universal arrows $(\hat{\eta}_B, A_B)$, then $V\hat{\eta}_B = \eta_{VB}$ and $U \circ \hat{F}(B) = U(A_B) = F \circ V(B)$ for each **B**-object B. Thus for each **B**-morphism $B \xrightarrow{f} \overline{B}$ the equality $GU\hat{F}f \circ \eta_{VB} = V(\hat{G}\hat{F}f \circ \hat{\eta}_B) = V(\hat{\eta}_{\overline{B}} \circ f) = \eta_{V\overline{B}} \circ Vf = GFVf \circ \eta_{VB}$ implies that $U\hat{F}f = FVf$. Hence $U \circ \hat{F} = F \circ V$.

(2) ⇒ (1). Let $(\eta, \varepsilon): F \dashv G: \mathbf{X} \to \mathbf{Y}$ be an adjoint situation, let the adjoint situation $(\hat{\eta}, \hat{\varepsilon}): \hat{F} \dashv \hat{G}: \mathbf{A} \to \mathbf{B}$ be a lift along U and V, and let $\mathcal{S} = (A \xrightarrow{f_i} A_i)_I$ be a U-initial source. To show that $\hat{G}\mathcal{S}$ is V-initial, let $\mathcal{T} = (B \xrightarrow{g_i} \hat{G}A_i)_I$ be a **B**-source and $VB \xrightarrow{h} V\hat{G}A$ be a **Y**-morphism with $V\mathcal{T} = V\hat{G}\mathcal{S} \circ h$. For each $i \in I$ let $\hat{F}B \xrightarrow{\hat{g}_i} A_i$ be the unique **A**-morphism with $g_i = \hat{G}\hat{g}_i \circ \hat{\eta}_B$. Furthermore, let $FVB \xrightarrow{\tilde{h}} UA$ be the unique **X**-morphism with $h = G\tilde{h} \circ \eta_{VB}$.

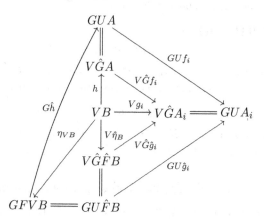

Then $G(Uf_i \circ \tilde{h}) \circ \eta_{VB} = GU\hat{g}_i \circ \eta_{VB}$ implies that $Uf_i \circ \tilde{h} = U\hat{g}_i$. Since $\mathcal{S}$ is U-initial, there exists a unique **A**-morphism $\hat{F}B \xrightarrow{h^*} A$ with $U\tilde{h} = \tilde{h}$. Thus $h^* = \hat{G}\tilde{h} \circ \hat{\eta}_B$ is a **B**-morphism with $Vh^* = V\hat{G}\tilde{h} \circ V\hat{\eta}_B = GU\tilde{h} \circ \eta_{VB} = G\tilde{h} \circ \eta_{VB} = h$. Consequently, $V(B \xrightarrow{h^*} \hat{G}A) = VB \xrightarrow{h} V\hat{G}A$. Thus $\hat{G}\mathcal{S}$ is V-initial. □

TOPOLOGICAL SUBCATEGORIES

21.29 DEFINITION
A full concrete subcategory $(\mathbf{A}, U)$ of a concrete category $(\mathbf{B}, V)$ is called **initially closed** in $(\mathbf{B}, V)$ provided that every V-initial source whose codomain is a family of **A**-objects has its domain in **A**.

DUAL NOTION: **finally closed subcategory**.

21.30 PROPOSITION
An initially closed subcategory of a topological category is topological. □

21.31 PROPOSITION
For any full concrete subcategory $(\mathbf{A}, U)$ *of a topological category* $(\mathbf{B}, V)$ *the following conditions are equivalent:*

(1) $(\mathbf{A}, U)$ *is initially closed in* $(\mathbf{B}, V)$,

(2) $(\mathbf{A}, U)$ *is concretely reflective in* $(\mathbf{B}, V)$.

Proof: $(1) \Rightarrow (2)$ follows from Theorem 21.24 applied to the inclusion functor G.

$(2) \Rightarrow (1)$. Let $\mathcal{S} = (B \xrightarrow{f_i} A_i)_I$ be a V-initial source with each A_i in $\mathbf{A}$, and let $B \xrightarrow{id_{VB}} A$ be an identity-carried $\mathbf{A}$-reflection arrow for B. Then each $A \xrightarrow{f_i} A_i$ is a morphism. Thus, by initiality of $\mathcal{S}$, it follows that $A \xrightarrow{id_{VB}} B$ is a morphism. Hence, by amnesticity of $(\mathbf{B}, V)$, $A = B$. Thus B belongs to $\mathbf{A}$. □

21.32 PROPOSITION
If a topological category $(\mathbf{A}, U)$ *is a finally dense full concrete subcategory of* $(\mathbf{B}, V)$, *then* $(\mathbf{A}, U)$ *is concretely reflective in* $(\mathbf{B}, V)$.

Proof: For each $\mathbf{B}$-object B consider the source $(B \xrightarrow{f_i} A_i)$, consisting of all morphisms $B \xrightarrow{f_i} A_i$ with A_i in $\mathbf{A}$, and let $\mathcal{S} = (A \xrightarrow{f_i} A_i)_I$ be a U-initial lift of the U-structured source $(VB \xrightarrow{f_i} UA_i)_I$. By Proposition 10.71, $\mathcal{S}$ is V-initial. Thus $VB \xrightarrow{id_{VB}} VA$ is a morphism; hence it is an identity-carried $\mathbf{A}$-reflection arrow for B. □

21.33 THEOREM
For a full concrete subcategory $(\mathbf{A}, U)$ *of a topological category* $(\mathbf{B}, V)$ *the following conditions are equivalent:*

(1) $(\mathbf{A}, U)$ *is topological,*

(2) there exists a concretely reflective subcategory $(\mathbf{C}, W)$ *of* $(\mathbf{B}, V)$ *such that* $(\mathbf{A}, U)$ *is concretely coreflective in* $(\mathbf{C}, W)$,

(3) there exists a concretely coreflective subcategory $(\mathbf{C}, W)$ *of* $(\mathbf{B}, V)$ *such that* $(\mathbf{A}, U)$ *is concretely reflective in* $(\mathbf{C}, W)$,

(4) there exists a concrete functor $(\mathbf{B}, V) \xrightarrow{R} (\mathbf{A}, U)$ *that leaves each* $\mathbf{A}$*-object fixed.*

Proof: $(1) \Rightarrow (2)$. Let $(\mathbf{C}, W)$ be the full concrete subcategory of $(\mathbf{B}, V)$ that consists of all $\mathbf{B}$-objects B for which there exists a V-initial source $(B \xrightarrow{f_i} A_i)_I$ with domain B and codomain in $\mathbf{A}$. Then $(\mathbf{C}, W)$ is initially closed in $(\mathbf{B}, V)$, hence concretely reflective in $(\mathbf{B}, V)$. Moreover, $(\mathbf{A}, U)$ is initially dense in $(\mathbf{C}, W)$, so that by the dual of Proposition 21.32 it is concretely coreflective in $(\mathbf{C}, W)$.

$(2) \Rightarrow (4)$. If $(\mathbf{B}, V) \xrightarrow{R_1} (\mathbf{C}, W)$ is a concrete reflector and $(\mathbf{C}, W) \xrightarrow{R_2} (\mathbf{A}, U)$ is a concrete coreflector, then $R = R_2 \circ R_1$ has the desired properties.

(4) ⇒ (1). Let $\mathcal{S} = (X \xrightarrow{f_i} UA_i)_I$ be a U-structured source. Considered as a V-structured source it has a V-initial lift $\mathcal{T} = (B \xrightarrow{f_i} A_i)_I$. Then $R\mathcal{T} = (RB \xrightarrow{f_i} A_i)_I$ is easily seen to be a U-initial lift of $\mathcal{S}$.

(1) ⇔ (3) follows from (1) ⇔ (2) by duality (21.9). □

FIBRE-SMALL TOPOLOGICAL CATEGORIES

21.34 PROPOSITION
For fibre-small concrete categories $(\mathbf{A}, U)$, the following conditions are equivalent:

(1) $(\mathbf{A}, U)$ is topological,

(2) every small structured source $(X \xrightarrow{f_i} UA_i)_I$ has a unique initial lift,

(3) every small structured sink $(UA_i \xrightarrow{f_i} X)_I$ has a unique final lift.

Proof: (1) ⇒ (2) is obvious.

(2) ⇒ (1). Let $(X \xrightarrow{f_i} UA_i)_I$ be a structured source. For each $i \in I$ there exists an initial lift $B_i \xrightarrow{\overline{f_i}} A_i$ of $X \xrightarrow{f_i} UA_i$. By fibre-smallness $\{ B_i \mid i \in I \}$ is a set. Thus there exists a subset J of I with $\{ B_j \mid j \in J \} = \{ B_i \mid i \in I \}$. Let $(A \xrightarrow{\overline{f_j}} A_j)_J$ be the initial lift of $(X \xrightarrow{f_j} UA_j)_J$. Then $A \leq B_j$ for each $j \in J$; hence $A \leq B_i$ for each $i \in I$. Thus each $A \xrightarrow{\overline{f_i}} A_i$ is a morphism. Consequently, by Proposition 10.46, $(A \xrightarrow{\overline{f_i}} A_i)_I$ is initial. Uniqueness is immediate.

(1) ⇔ (3) follows from (1) ⇔ (2) by duality. □

21.35 PROPOSITION
A fibre-small concrete category $(\mathbf{A}, U)$ over a category $\mathbf{X}$ with products is topological if and only if it satisfies the following conditions:

(1) $(\mathbf{A}, U)$ has concrete products,

(2) $(\mathbf{A}, U)$ has initial subobjects, i.e., every structured $\mathbf{X}$-monomorphism $X \xrightarrow{m} UA$ has a unique initial lift,

(3) $(\mathbf{A}, U)$ has indiscrete objects.

Proof: The conditions (1) – (3) are obviously necessary. To see that they are sufficient, suppose that $(X \xrightarrow{f_i} UA_i)_I$ is a small structured source. Choose an element j_0 with $j_0 \notin I$, use A_{j_0} to denote the indiscrete object with $UA_{j_0} = X$, let $f_{j_0} = id_X$, let $J = I \cup \{j_0\}$, and let $(P \xrightarrow{p_j} Aj)_J$ be a concrete product. Then $X \xrightarrow{\langle f_j \rangle} UP$ is a structured $\mathbf{X}$-section, and so has an initial lift $A \xrightarrow{\langle f_j \rangle} P$. By 10.53 and 10.45(1) the source $(A \xrightarrow{f_j} Aj)_J$ is initial. Since A_{j_0} is indiscrete, $(A \xrightarrow{f_i} A_i)_I$ is initial too. Uniqueness follows from the uniqueness requirement in condition (2). Thus Proposition 21.34 implies that $(\mathbf{A}, U)$ is topological. □

21.36 PROPOSITION
In a fibre-small topological category, a source $(A \xrightarrow{f_i} A_i)_I$ is initial if and only if there exists a subset J of I such that $(A \xrightarrow{f_j} Aj)_J$ is initial.

Proof: Immediate from the proof of (2) ⇒ (1) in Proposition 21.34. □

21.37 PROPOSITION
Let $(\mathbf{A}, U)$ be a full concrete subcategory of a fibre-small topological category $(\mathbf{B}, V)$ over a category $\mathbf{X}$ with products. Then $(\mathbf{A}, U)$ is concretely reflective in $(\mathbf{B}, V)$ if and only if it is closed under the formation of

(a) products,

(b) initial subobjects, and

(c) indiscrete objects.

Proof: By Proposition 21.31, $(\mathbf{A}, U)$ is concretely reflective in $(\mathbf{B}, V)$ if and only if it is initially closed in $(\mathbf{B}, V)$. The latter is the case if and only if $(\mathbf{A}, U)$ is closed under the formation of (a)–(c) in $(\mathbf{B}, V)$. (Cf. the proof of Proposition 21.35.) □

M-TOPOLOGICAL AND MONOTOPOLOGICAL CATEGORIES

Many familiar full subconstructs of **Top**, **Unif**, or **Rel**, defined by "separation axioms", e.g., **Top$_0$**, **Top$_1$**, **Haus**, **Tych**, **Reg$_1$**, **HUnif**, or **Pos** fail to be topological even though they are "almost" so. This leads to a natural generalization of the concept of topological categories over $\mathbf{X}$ that depends, however, on a distinguished factorization structure on $\mathbf{X}$. In particular, for constructs this leads to the concept of monotopological constructs, since **Set** has only one interesting factorization structure:

21.38 DEFINITION
A concrete category $(\mathbf{A}, U)$ over an $(E, \mathbf{M})$-category $\mathbf{X}$ is said to be **M-topological** provided that every structured source in $\mathbf{M}$ has a unique initial lift. If $\mathbf{M}$ = Mono-Sources, the term **monotopological** is used.

21.39 EXAMPLES
(1) The topological functors are precisely the Source-topological functors.

(2) The construct **Pos** is monotopological since it is closed under the formation of initial mono-sources in **Rel**.

(3) The constructs **Top$_0$**, **Top$_1$**, **Haus**, **Tych**, and **Reg$_1$** (regular T_1 topological spaces) are monotopological (since they are closed under the formation of initial mono-sources in **Top**).

(4) The construct **HUnif** of separated uniform spaces is monotopological.

(5) Let **PAlg**(Ω) denote the construct of partial algebras of type $\Omega = (n_i)$, i.e., pairs of the form $(X, \{\omega_i\})$, where X is a set and the $\omega_i : X^{n_i} \to X$ are partial functions, i.e., functions $\omega_i : D_i \to X$, whose domains D_i are subsets of X^{n_i}. Partial algebra homomorphisms $(X, \{\omega_i\}) \xrightarrow{f} (Y, \{\omega_i'\})$ are functions $f : X \to Y$ satisfying $f(\omega_i(g)) = \omega_i'(f^{n_i}(g))$ whenever $\omega_i(g)$ is defined. Then **PAlg**(Ω) is monotopological.

(6) For concrete categories over **1**, monotopological = topological.

(7) **Unif**, considered as a concrete category over **Top** is **M**-topological, where **M** is the collection of initial mono-sources in **Top**.

21.40 M-TOPOLOGICAL CHARACTERIZATION THEOREM

A concrete category $(\mathbf{A}, U)$ *over an* $(E, \mathbf{M})$-*category* $\mathbf{X}$ *is* $\mathbf{M}$-*topological if and only if* $(\mathbf{A}, U)$ *is an* E-*reflective concrete subcategory of some topological category over* $\mathbf{X}$.

Proof: If $(\mathbf{B}, V)$ is topological over $\mathbf{X}$, then by Proposition 21.14, $\mathbf{B}$ is an $(E, \mathbf{M}_{init})$-category. Thus if $(\mathbf{A}, U)$ is an E-reflective subcategory of $\mathbf{B}$, then by Theorem 16.8, $\mathbf{A}$ is closed under $\mathbf{M}_{init}$-sources in $\mathbf{B}$. Hence $(\mathbf{A}, U)$ is M-topological.

Conversely, let $(\mathbf{A}, U)$ be $\mathbf{M}$-topological. Define a concrete category $(\mathbf{A}^*, U^*)$ over $\mathbf{X}$ as follows. $\mathbf{A}^*$-objects are triples (X, e, A) with $X \in Ob(\mathbf{X})$, $A \in Ob(\mathbf{A})$, and $X \xrightarrow{e} UA \in E$. $\mathbf{A}^*$-morphisms from (X, e, A) to (X', e', A') are pairs (f, g) consisting of an $\mathbf{X}$-morphism $X \xrightarrow{f} X'$ and an $\mathbf{A}$-morphism $A \xrightarrow{g} A'$ such that the diagram

$$\begin{array}{ccc} X & \xrightarrow{f} & X' \\ {\scriptstyle e}\downarrow & & \downarrow{\scriptstyle e'} \\ UA & \xrightarrow{g} & UA' \end{array}$$

commutes. $\mathbf{A}^*$-composition is defined coordinatewise. Let $U^* : \mathbf{A}^* \to \mathbf{X}$ be defined by

$$U^*((X, e, A) \xrightarrow{(f,g)} (X', e', A')) = X \xrightarrow{f} X'.$$

Since $E \subseteq \mathrm{Epi}(\mathbf{X})$ (15.4), U^* is faithful. Hence $(\mathbf{A}^*, U^*)$ is a concrete category over $\mathbf{X}$. Moreover every U^*-structured source $\mathcal{S} = (X \xrightarrow{f_i} U^*(X_i, e_i, A_i))_I$ has an initial lift that can be constructed as follows: Let $X \xrightarrow{f_i} X_i \xrightarrow{e_i} UA_i = X \xrightarrow{e} Y \xrightarrow{m_i} UA_i$ be an $(E, \mathbf{M})$-factorization and let $(A \xrightarrow{m_i} A_i)_I$ be a U-initial lift of the U-structured $\mathbf{M}$-source $(Y \xrightarrow{m_i} UA_i)_I$. By the $(E, \mathbf{M})$-diagonalization-property, it is easily seen that the source $((X, e, A) \xrightarrow{(f_i, m_i)} (X_i, e_i, A_i))_I$ is a U^*-initial lift of $\mathcal{S}$. However, initial lifts need not be uniquely determined. By Proposition 5.33 there exists an amnestic concrete category

$(\mathbf{B}, V)$ and a surjective concrete equivalence $(\mathbf{A}^*, U^*) \xrightarrow{P} (\mathbf{B}, V)$. By Proposition 21.5, $(\mathbf{B}, V)$ is topological. Now define a concrete full embedding $(\mathbf{A}, U) \xrightarrow{E^*} (\mathbf{A}^*, U^*)$ by

$$E^*(A \xrightarrow{f} B) = (UA, id_A, A) \xrightarrow{(f,f)} (UB, id_B, B).$$

Then $E^*[\mathbf{A}]$ is E-reflective in $\mathbf{A}^*$, since for each $\mathbf{A}^*$-object (X, e, A) the $\mathbf{A}^*$-morphism $(X, e, A) \xrightarrow{(e, id_A)} E^*A$ is an $E^*[\mathbf{A}]$-reflection arrow. Thus

$$(\mathbf{A}, U) \xrightarrow{E} (\mathbf{B}, V) = (\mathbf{A}, U) \xrightarrow{E^*} (\mathbf{A}^*, U^*) \xrightarrow{P} (\mathbf{B}, V)$$

is a full concrete embedding, whose image is an E-reflective subcategory of the topological category $(\mathbf{B}, V)$. □

21.41 REMARKS
(1) Let $(\mathbf{A}, U) \xrightarrow{E} (\mathbf{B}, V)$ be the concrete embedding constructed in the above proof. Then the following hold, as can be seen easily:

  (a) E preserves $\mathbf{M}$-initial sources. Thus $E[\mathbf{A}]$ is $\mathbf{M}$-initially closed in $(\mathbf{B}, V)$.

  (b) If $\mathbf{X}$ is E-co-wellpowered and $(\mathbf{A}, U)$ is fibre-small, then $(\mathbf{B}, V)$ is fibre-small.

  (c) If $(\mathbf{A}, U)$ is topological, E needn't be an isomorphism. For example, if **Set** is considered as a monotopological construct $(\mathbf{A}, U)$ via the identity functor, then $(\mathbf{B}, V)$ is the construct of equivalence relations.

(2) Whereas $\mathbf{M}$-topological categories, i.e., E-reflective concrete subcategories of topological categories, still have a typical topological flavor, full reflective concrete subcategories of topological categories can have a completely different character. As we will see in §25 the constructs **Vec**, **Grp**, and **Rng** are examples of this kind.

21.42 THEOREM
Let $(\mathbf{A}, U)$ be a fibre-small concrete category over an E-co-wellpowered $(E, \mathbf{M})$-category with products. Then $(\mathbf{A}, U)$ is $\mathbf{M}$-topological if and only if it satisfies the following conditions:

(1) $(\mathbf{A}, U)$ has concrete products,

(2) $(\mathbf{A}, U)$ **has M-initial subobjects**, i.e., every structured $\mathbf{M}$-morphism $X \xrightarrow{m} UA$ has a unique initial lift.

Proof: If $(\mathbf{A}, U)$ is $\mathbf{M}$-topological, then it has concrete products, since $\mathbf{X}$-products, being extremal mono-sources, belong to $\mathbf{M}$.

Conversely, let $(\mathbf{A}, U)$ satisfy the conditions (1) and (2). If $\mathcal{S} = (X \xrightarrow{m_i} UA_i)_I$ is a small structured $\mathbf{M}$-source and $(\prod A_i \xrightarrow{\pi_j} A_j)_I$ is a concrete product of $(A_j)_I$, then the structured $\mathbf{M}$-morphism $X \xrightarrow{\langle m_i \rangle} UP$ has an initial lift $A \xrightarrow{\langle m_i \rangle} P$. Thus

$(A \xrightarrow{m_i} A_i)_I = (A \xrightarrow{\langle m_i \rangle} P \xrightarrow{\pi_i} A_i)_I$ is an initial lift of $\mathcal{S}$. If $\mathcal{S} = (X \xrightarrow{m_i} UA_i)_I$ is a large structured **M**-source, for each $i \in I$ let

$$X \xrightarrow{m_i} UA_i = X \xrightarrow{e_i} X_i \xrightarrow{n_i} UA_i$$

be an $(E, \mathbf{M})$-factorization and let $B_i \xrightarrow{n_i} A_i$ be an initial lift of $X_i \xrightarrow{n_i} UA_i$. Since **X** is E-co-wellpowered and $(\mathbf{A}, U)$ is fibre-small, the structured source $\mathcal{T} = (X \xrightarrow{e_i} UB_i)_I$ is representable by a small source, i.e., there exists a subset J of I such that for each $i \in I$ there is a $j(i) \in J$ and an isomorphism $h_i : B_{j(i)} \to B_i$ with $e_i = h_i \circ e_{j(i)}$. By Proposition 15.5(7), $\mathcal{T}$ belongs to **M**, and hence so does the small source $\mathcal{T}' = (X \xrightarrow{e_j} UB_j)_J$. Thus $\mathcal{T}'$ has an initial lift $(A \xrightarrow{e_j} B_j)_J$. It follows immediately that the source

$$(A \xrightarrow{m_i} A_i)_I = (A \xrightarrow{e_{j(i)}} B_{j(i)} \xrightarrow{h_i} B_i \xrightarrow{n_i} A_i)_I$$

is an initial lift of $\mathcal{S}$. □

21.43 COROLLARY

A fibre-small construct is monotopological if and only if it has concrete products and initial subobjects. □

21.44 EXAMPLES

(1) The constructs **Vec** and **Grp** have concrete products but fail to have initial subobjects.

(2) The constructs **Met** and **Met_c** have initial subspaces but fail to have concrete products.

Suggestions for Further Reading

Hušek, M. *S*-categories. *Comment. Math. Univ. Carolinae* **5** (1964): 37–46.

Gray, J. W. Fibred and cofibred categories. *Proceedings of the Conference on Categorical Algebra* (La Jolla 1965), Springer, Berlin–Heidelberg–New York, 1966, 21–83.

Antoine, P. Étude élémentaire des catégories d'ensembles structurés. *Bull. Soc. Math. Belgique* **18** (1966): 142–164 and 387–414.

Roberts, J. E. A characterization of initial functors. *J. Algebra* **8** (1968): 181–193.

Wyler, O. On the categories of general topology and topological algebra. *Archiv Math.* **22** (1971): 7–17.

Herrlich, H. Topological functors. *Gen. Topol. Appl.* **4** (1974): 125–142.

Hoffmann, R.-E. Topological functors and factorizations. *Archiv Math.* **26** (1975): 1–7.

Herrlich, H. Initial completions. *Math. Z.* **150** (1976): 101–110.

Brümmer, G. C. L., and R.-E. Hoffmann. An external characterization of topological functors. *Springer Lect. Notes Math.* **540** (1976): 136–151.

Hoffmann, R.-E. Topological functors admitting generalized Cauchy-completions. *Springer Lect. Notes Math.* **540** (1976): 286–344.

Porst, H.-E. Characterizations of Mac Neille completions and topological functors. *Bull. Austral. Math. Soc.* **18** (1978): 201–210.

Tholen, W. On Wyler's taut lift theorem. *Gen. Topol. Appl.* **8** (1978): 197–206.

Herrlich, H. Categorical topology 1971-1981. *General Topology and its Relations to Modern Analysis and Algebra V. Proceedings of the Fifth Prague Topological Symposium 1981* (ed. J. Novak), Heldermann, Berlin, 1982, 279–383.

Brümmer, G. C. L. Topological categories. *Topol. Appl.* **18** (1984): 27–41.

Nakagawa, R. Categorical Topology. *Topics in General Topology* (eds. K. Morita and J. Nagata), North Holland, Amsterdam, 1989, 563–623.

EXERCISES

21A. Existence of Indiscrete Structures
For transportable concrete categories, $(\mathbf{A}, U)$, prove that the following conditions are equivalent:

(1) $(\mathbf{A}, U)$ has indiscrete structures,

(2) U has an adjoint G with $U \circ G = id_{\mathbf{X}}$,

(3) U has an adjoint G that is a full embedding,

(4) there exists a Galois correspondence $(U, G) : (\mathbf{X}, id_{\mathbf{X}}) \to (\mathbf{A}, U)$.

21B. Fibre-Smallness
Show that in fibre-small topological categories (resp. constructs) a sink $(A_i \xrightarrow{f_i} A)_I$ is final (resp. is a final epi-sink) if and only if there exists a sub*set* J of I such that $(A_j \xrightarrow{f_j} A)_J$ is final (resp. is a final epi-sink).

21C. Characterization of Topological Categories
Show that a concrete category $(\mathbf{A}, U)$ over $\mathbf{X}$ is topological if and only if the following conditions are satisfied:

(a) $(\mathbf{A}, U)$ is fibre-complete,

(b) in $(\mathbf{A}, U)$ each structured arrow $X \xrightarrow{f} UA$ has a unique initial lift $B \xrightarrow{f} A$,

(c) in $(\mathbf{A}, U)$ each co-structured arrow $UA \xrightarrow{f} X$ has a unique final lift $A \xrightarrow{f} B$.

21D. Non-topological Functors and Categories

*(a) Consider the functor from **Mod** → **Rng** described as follows: **Mod**, the category of modules, has as objects all pairs (R, M) consisting of a ring R and a left R-module M, and as morphisms $(R, M) \xrightarrow{(f,g)} (R', M')$ all pairs (f, g) consisting of a ring homomorphism $R \xrightarrow{f} R'$ and an (abelian) group homomorphism $M \xrightarrow{g} M'$ satisfying the identity $g(r \cdot m) = f(r) \cdot g(m)$. Let **Mod** $\xrightarrow{G}$ **Rng** be the "forgetful" functor, defined by

$$G((R, M) \xrightarrow{(f,g)} (R', M')) = R \xrightarrow{f} R'.$$

Show that

(1) Each small G-structured source has a G-initial lift.

(2) There exists a functor **Rng** $\xrightarrow{F}$ **Mod** that is simultaneously an adjoint and a co-adjoint for G.

(3) G is not faithful.

(4) G is not topological.

(b) Let **A** be the full subconstruct of **Top** consisting of all discrete and all indiscrete spaces. Show that

(1) For each structured source $(X \xrightarrow{f_i} (X_i, \tau_i))_I$ there exists a largest **A**-structure τ on X making all $(X, \tau) \xrightarrow{f_i} (X_i, \tau_i)$ continuous.

(2) For each structured sink $((X_i, \tau_i) \xrightarrow{f_i} X)_I$ there exists a smallest **A**-structure τ on X making all $(X_i, \tau_i) \xrightarrow{f_i} (X, \tau)$ continuous.

(3) **A** is fibre-complete.

(4) **A** has discrete and indiscrete structures.

(5) **A** is neither complete nor cocomplete.

(6) **A** is not topological.

(c) Show that the unique functor G from $\bullet \rightrightarrows \bullet$ to $\bullet$ is not topological, even though G is faithful and each G-structured source has an initial lift.

21E. Concrete Co-adjoints and Galois Correspondences

(a) Let **X** be a category consisting of a single object X and two morphisms id_X and s with $s^2 = id_X$; let **A** be the concrete category over **X**, consisting of two objects A_0 and A_1 and the following morphism sets:

$$\hom_{\mathbf{A}}(A_i, A_j) = \begin{cases} \{id_X\}, & \text{if } i = j \\ \{s\}, & \text{if } i \neq j; \end{cases}$$

and let G and F be concrete functors from **A** to **A** over **X** defined by $GA_i = A_i$ and $FA_i = A_{1-i}$. Show that

(1) F is a co-adjoint for G.

(2) (F, G) is not a Galois correspondence between **A** and itself over **X**.

* (b) Show that a concrete functor $\mathbf{A} \xrightarrow{G} \mathbf{B}$ may have a concrete co-adjoint without being part of a Galois correspondence (F, G). [Hint: Consider the concrete embedding $\mathbf{A} \to \mathbf{B}$ constructed in Exercise 5E(d).] Cf. Theorem 21.24.

21F. Comma Categories

(a) Let $(\mathbf{A}, U)$ be topological over $\mathbf{X}$, and let B be an **A**-object. Show that $\mathbf{A} \downarrow B$ is topological over $\mathbf{X} \downarrow UB$ via the obvious forgetful functor induced by U.

(b) Consider

$$\begin{array}{ccc} \mathbf{A} \downarrow B & \xrightarrow{W} & \mathbf{A} \\ {\scriptstyle T}\downarrow & & \downarrow{\scriptstyle U} \\ \mathbf{X} \downarrow UB & \xrightarrow{V} & \mathbf{X} \end{array}$$

where V, W, T are the obvious forgetful functors. Show that

(1) If U is topological, then so is T.

(2) W preserves initiality (i.e., sends T-initial sources into U-initial sources), but in general does not preserve finality.

* 21G. Initial Completions

A full concrete embedding $(\mathbf{A}, U) \xrightarrow{E} (\mathbf{B}, V)$ is called an **initial completion** of $(\mathbf{A}, U)$ provided that $(\mathbf{B}, V)$ is topological and $E[\mathbf{A}]$ is initially dense in $(\mathbf{B}, V)$. [DUAL NOTION: **final completion**.] An initial completion $(\mathbf{A}, U) \xrightarrow{E_1} (\mathbf{B}_1, U_1)$ is said to be **smaller than** an initial completion $(\mathbf{A}, U) \xrightarrow{E_2} (\mathbf{B}_2, U_2)$ of $(\mathbf{A}, U)$ — in symbols: $E_1 \leq E_2$, provided that there exists a full concrete embedding $(\mathbf{B}_1, U_1) \xrightarrow{E} (\mathbf{B}_2, U_2)$ with $E_2 = E \circ E_1$. Show that for initial completions E_1 of $(\mathbf{A}, U)$ the following hold:

(a) The relation $\leq$ is reflexive and transitive.

(b) $E_1 \leq E_2$ and $E_2 \leq E_1$ hold simultaneously if and only if there exists a concrete isomorphism $(\mathbf{B}_1, U_1) \xrightarrow{H} (\mathbf{B}_2, U_2)$ with $E_2 = H \circ E_1$.

(c) The following conditions are equivalent:

(1) $E_1 \leq E_2$,

(2) there exists a (unique) initiality-preserving concrete functor $(\mathbf{B}_2, U_2) \xrightarrow{G} (\mathbf{B}_1, U_1)$ with $E_1 = G \circ E_2$,

(3) there exist (unique) concrete functors $(\mathbf{B}_1, U_1) \xrightarrow{E} (\mathbf{B}_2, U_2)$ and $(\mathbf{B}_2, U_2) \xrightarrow{G} (\mathbf{B}_1, U_1)$ with $E_2 = E \circ E_1$, $E_1 = G \circ E_2$ and $G \circ E = id$.

* 21H. Mac Neille Completions

A completion of $(\mathbf{A}, U)$ that is simultaneously an initial completion and a final completion is called a **Mac Neille completion** of $(\mathbf{A}, U)$. Show that

(a) [Uniqueness] If $(\mathbf{A}, U) \xrightarrow{E_1} (\mathbf{B}_1, V_1)$ and $(\mathbf{A}, U) \xrightarrow{E_2} (\mathbf{B}_2, V_2)$ are Mac Neille completions, then there exists a concrete isomorphism $(\mathbf{B}_1, V_1) \xrightarrow{H} (\mathbf{B}_2, V_2)$ with $E_2 = H \circ E_1$.

(b) [Minimality] If $(\mathbf{A}, U) \xrightarrow{E} (\mathbf{B}, V)$ is a Mac Neille completion and if $(\mathbf{A}, U) \xrightarrow{E'} (\mathbf{B}', V')$ is a full concrete embedding of $(\mathbf{A}, U)$ into a topological category $(\mathbf{B}', V')$, then there exists a full concrete embedding $(\mathbf{B}, V) \xrightarrow{H} (\mathbf{B}', V')$ with $V' = H \circ V$. In particular, Mac Neille completions of $(\mathbf{A}, U)$ are the smallest initial completions of $(\mathbf{A}, U)$.

(c) [Construction] If $(\mathbf{A}, U)$ is a full concrete subcategory of a topological category $(\mathbf{B}, V)$, if $(\mathbf{B}', V')$ is the concretely reflective hull of $\mathbf{A}$ in $(\mathbf{B}, V)$, and if $(\mathbf{B}'', V'')$ is the concretely coreflective hull of $\mathbf{A}$ in $(\mathbf{B}', V')$, then $(\mathbf{B}'', V'')$ is a Mac Neille completion of $(\mathbf{A}, U)$.

(d) [Existence] Let $(\mathbf{A}, U)$ be an amnestic concrete category over $\mathbf{X}$.
  (1) If $\mathbf{A}$ is small, then $(\mathbf{A}, U)$ has a Mac Neille completion. In particular, each poset, considered as a concrete category over $\mathbf{1}$, has a Mac Neille completion (= its familiar Dedekind-Mac Neille completion).
  (2) If $(\mathbf{A}, U)$ is **M**-topological, then it has a Mac Neille completion.

(e) [Non-existence] Let C be a proper class. Order $X = C \times \{0, 1\}$ by

$$(x, i) \leq (y, j) \Leftrightarrow \begin{cases} (x, i) = (y, j) \\ \text{or} \\ i = 0 \text{ and } j = 1 \text{ and } x \neq y. \end{cases}$$

Then $(X, \leq)$, considered as concrete category over $\mathbf{1}$, has no Mac Neille completion.

* 21I. Universal Initial Completions

An initial completion $(\mathbf{A}, U) \xrightarrow{E} (\mathbf{B}, V)$ of $(\mathbf{A}, U)$ is called a **universal initial completion** of $(\mathbf{A}, U)$ provided that the following hold:

(1) E preserves initiality,

(2) whenever $(\mathbf{C}, W)$ is topological and $(\mathbf{A}, U) \xrightarrow{F} (\mathbf{C}, W)$ is an initiality-preserving concrete functor, then there exists a unique initiality-preserving concrete functor $(\mathbf{B}, V) \xrightarrow{G} (\mathbf{C}, W)$ with $F = G \circ E$.

Show that

(a) [Uniqueness] If $(\mathbf{A}, U) \xrightarrow{E_1} (\mathbf{B}_1, V_1)$ and $(\mathbf{A}, U) \xrightarrow{E_2} (\mathbf{B}_2, V_2)$ are universal initial completions, then there exists a concrete isomorphism $(\mathbf{B}_1, V_1) \xrightarrow{H} (\mathbf{B}_2, V_2)$ with $V_2 = H \circ V_1$.

(b) [Maximality] Universal initial completions of $(\mathbf{A}, U)$ are the largest initiality-preserving initial completions of $(\mathbf{A}, U)$.

(c) Every **M**-topological category has a universal initial completion.

(d) The construct **Prox** of (not necessarily T_1) proximity spaces is a universal initial completion (but not a Mac Neille completion) of **HComp**.

(e) If **A** is small and $(\mathbf{A}, U)$ is amnestic, then $(\mathbf{A}, U)$ has a universal initial completion. In particular, if A is a finite set with $n \geq 2$ elements, ordered by equality and considered as a concrete category over **1**, then the universal initial completion of A has 2^n elements, whereas the Mac Neille completion of A has $n + 2$ elements.

(f) A proper class, ordered by equality and considered as concrete category over **1**, has a Mac Neille completion, but no universal initial completion.

* **21J. Essential Full Concrete Embeddings**

Show that in the quasicategory $\mathbf{CAT}(\mathbf{X})$ of amnestic concrete categories and concrete functors over **X**, a full concrete embedding $(\mathbf{A}, U) \xrightarrow{E} (\mathbf{B}, V)$ is

(a) essential [with respect to full concrete embeddings; cf. Definition 9.22(2)] if and only if it is initially dense and finally dense,

(b) a Mac Neille completion of $(\mathbf{A}, U)$ if and only if it is an injective hull (with respect to full concrete embeddings).

21K. M-Topological Functors

Let **M** be a conglomerate of sources in **X**. A functor $G: \mathbf{A} \to \mathbf{X}$ is called **M-topological** provided that every G-structured source in **M** has a unique G-initial lift.

Prove that if **X** is an $(E, \mathbf{M})$-category, then every **M**-topological functor $G: \mathbf{A} \to \mathbf{X}$ is faithful and uniquely transportable.

21L. Uniqueness of Forgetful Functors

Show that whenever $(\mathbf{A}, U)$ and $(\mathbf{A}, V)$ are topological constructs, then U and V are naturally isomorphic.

21M. Topological Functors and Bimorphisms

Show that if $\mathbf{A} \xrightarrow{U} \mathbf{X}$ is a topological functor, then $E = U^{-1}[\mathrm{Iso}(\mathbf{X})]$ satisfies:

(1) $E \subseteq \mathrm{Bimor}(\mathbf{A})$ (= the class of bimorphisms in **A**),

(2) if **X** is balanced, then $E = \mathrm{Bimor}(\mathbf{A})$,

(3) **A** is an $(E, -)$-category and $\mathbf{A}^{\mathrm{op}}$ is an $(E, -)$-category.

21N. Balanced Topological Categories

Show that a topological category $(\mathbf{A}, U)$ over **X** is balanced if and only if **X** is balanced and U is an isomorphism.

22 Topological structure theorems

Universal algebra can be understood as the branch of mathematics in which "nice" subconstructs of the fundamental algebraic constructs $\mathbf{Alg}(\Omega)$ of Ω-algebras are investigated. Among these are the finitary varieties and quasivarieties. These "nice" subconstructs are determined either *internally* (by means of suitable equations or implications) or *externally* (by means of certain closure properties). It turns out that an analogous situation exists in the realm of topological categories over arbitrary (!) base categories. The role of the fundamental topological categories is played by the functor-structured categories $\mathbf{Spa}(T)$ of T-spaces. "Nice" subcategories are determined either *internally* (by topological axioms) or *externally* (by suitable closure properties). In addition, duality provides a dual structure theorem.

TOPOLOGICAL AXIOMS

Functor-structured categories $\mathbf{Spa}(T)$ have been defined in Definition 5.40. By Example 21.8(2) they are fibre-small topological categories. Hence, by Proposition 21.30, each of their initially closed (= concretely reflective) full concrete subcategories is a fibre-small topological category as well. The main result of this section states that — conversely — every fibre-small topological category can be obtained in this way. Since, moreover, by Theorem 16.14 the isomorphism-closed E-reflective subcategories of an $(E, \mathbf{M})$-category $\mathbf{B}$ are precisely the E-implicational subcategories of $\mathbf{B}$, concretely reflective subcategories of topological categories $\mathbf{B}$ are precisely the identity-carried implicational subcategories of $\mathbf{B}$. This justifies the following terminology (cf. Definition 16.12):

22.1 DEFINITION
Let $(\mathbf{A}, U)$ be a concrete category over $\mathbf{X}$.

(1) Each identity-carried morphism $P \xrightarrow{p} P'$ is called a **topological axiom** in $(\mathbf{A}, U)$. An $\mathbf{A}$-object A is said to **satisfy** the axiom p provided that A is $\{p\}$-injective; that is, each $\mathbf{A}$-morphism $f : P \to A$ is also an $\mathbf{A}$-morphism $f : P' \to A$.

(2) A full subcategory $\mathbf{B}$ of $\mathbf{A}$ is said to be **definable by topological axioms** in $(\mathbf{A}, U)$ provided that it is E-implicational in $\mathbf{A}$, where E is a class of topological axioms in $(\mathbf{A}, U)$; i.e., the objects in $\mathbf{B}$ are precisely those $\mathbf{A}$-objects that satisfy each of the axioms in E.

22.2 EXAMPLES
(1) The construct **Prost** is definable by the following two topological axioms in $\mathbf{Rel} = \mathbf{Spa}(S^2)$. Cf. Example 16.13(1).

$$(A1) \quad \bullet \; \longrightarrow \; \bullet \!\circlearrowright$$

(A2)

(2) Let **Bor** denote the construct of bornological spaces, i.e., pairs (X, β), where X is a set and β is a bornology (i.e., a cover of X such that $U \in \beta$ implies $V \in \beta$ for all $V \subseteq U$ and $U_1, U_2 \in \beta$ implies $U_1 \cup U_2 \in \beta$). Morphisms $f : (X, \beta) \to (X', \beta')$ are functions $f : X \to X'$ such that $U \in \beta$ implies $f[U] \in \beta'$. **Bor** is definable by the following (proper) class of topological axioms in $\mathbf{Spa}(\mathcal{P})$:

(A_0) $(\emptyset, \emptyset) \to (\emptyset, \mathcal{P}(\emptyset))$,
(A_1) $(\{1\}, \emptyset) \to (\{1\}, \mathcal{P}\{1\})$,
$(A_{X,M,N})$ $(X, \{M, N\}) \to (X, \mathcal{P}(M \cup N))$ for any set X and any $M, N \subseteq X$.

Bor cannot be defined by a set of topological axioms in $\mathbf{Spa}(\mathcal{P})$.

(3) Let **Simp** denote the construct of simplicial complexes, i.e., pairs (X, α), where X is a set and α is a complex (i.e., a set of finite subsets of X that contains all subsets with at most one member and has the property that $U \in \alpha$ implies $V \in \alpha$ for all $V \subseteq U$). Morphisms $f : (X, \alpha) \to (X', \alpha')$ are functions $f : X \to X'$ such that $U \in \alpha$ implies $f[U] \in \alpha'$. Let $\mathcal{P}_{fin} : \mathbf{Set} \to \mathbf{Set}$ be the finite power-set functor that associates with each set X the set $\mathcal{P}_{fin}(X)$ of all finite subsets of X. Then **Simp** is definable by the following (proper) class of topological axioms in $\mathbf{Spa}(\mathcal{P}_{fin})$:

(A_0) $(\emptyset, \emptyset) \to (\emptyset, \mathcal{P}(\emptyset))$,
(A_1) $(\{1\}, \emptyset) \to (\{1\}, \mathcal{P}\{1\})$,
$(A_{X,M})$ $(X, \{M\}) \to (X, \mathcal{P}(M))$ for every set X and every finite subset M of X.

(4) Let $\mathcal{U} : \mathbf{Set} \to \mathbf{Set}$ be the ultrafilter functor defined as follows: For each set X, $\mathcal{U}(X)$ is the set of all ultrafilters on X, and for each function $f : X \to Y$, $\mathcal{U}f(\mathcal{F}) = \{M \subseteq Y \mid f^{-1}[M] \in \mathcal{F}\}$. Let $F : \mathbf{Set} \to \mathbf{Set}$ be defined by $F(X \xrightarrow{f} Y) = \mathcal{U}X \times X \xrightarrow{\mathcal{U}f \times f} \mathcal{U}Y \times Y$. Then the construct **PsTop** of pseudotopological spaces is definable by one topological axiom in $\mathbf{Spa}(F)$. An object (X, α) of $\mathbf{Spa}(F)$ is given by a set $\alpha \subseteq \mathcal{U}(X) \times X$ of pairs $(\mathcal{F}, x)$, where $\mathcal{F}$ is an ultrafilter and x is a point. [Think of $\mathcal{F} \to x$ if and only if $(\mathcal{F}, x) \in \alpha$.] Such an object is a **pseudotopological space** if and only if for each point x, α contains the pair $(\dot{x}, x)$ (where $\dot{x}$ is the trivial ultrafilter induced by x).

Thus **PsTop** is definable by the following topological axiom in $\mathbf{Spa}(F)$:

$$(\{1\}, \emptyset) \to (\{1\}, F\{1\}).$$

(5) Also **Top** is definable by topological axioms in $\mathbf{Spa}(F)$. However, a proper class of such axioms is needed.

22.3 TOPOLOGICAL STRUCTURE THEOREM

For concrete categories $(\mathbf{A}, U)$, *the following conditions are equivalent:*

(1) $(\mathbf{A}, U)$ *is fibre-small and topological,*

(2) $(\mathbf{A}, U)$ *is concretely isomorphic to an initially closed full subcategory of a functor-structured category,*

(3) $(\mathbf{A}, U)$ *is concretely isomorphic to an isomorphism-closed concretely reflective subcategory of some functor-structured category,*

(4) $(\mathbf{A}, U)$ *is concretely isomorphic to a subcategory of a functor-structured category* $\mathbf{Spa}(T)$ *that is definable by topological axioms in* $\mathbf{Spa}(T)$.

Proof:
(1) $\Rightarrow$ (2). Let $(\mathbf{A}, U)$ be a fibre-small topological category over $\mathbf{X}$. Let $T : \mathbf{X} \to \mathbf{Set}$ be the associated fibre-functor, defined on objects X by: $TX = \{ A \in Ob(\mathbf{A}) \mid UA = X \}$ and on morphisms $X \xrightarrow{f} Y$ by: $Tf(A) = B \Leftrightarrow A \xrightarrow{f} B$ is final. Then for each **A**-object A the pair $(UA, \{ B \in TUA \mid B \leq A \})$ is a T-space EA. The unique concrete functor $E : (\mathbf{A}, U) \to \mathbf{Spa}(T)$, defined on objects as just described, is easily seen to be a full embedding. Hence it remains to be shown that its image is initially closed in $\mathbf{Spa}(T)$. To see this, consider an initial source $((X, \alpha) \xrightarrow{f_i} EA_i)_I$ in $\mathbf{Spa}(T)$, and let $(A \xrightarrow{f_i} A_i)_I$ be the initial lift of the structured source $(X \xrightarrow{f_i} UA_i)_I$ in $(\mathbf{A}, U)$. Then by Example 10.42(4) the following conditions are equivalent for elements B of TX:

(a) $B \in \alpha$,

(b) $Tf_i(B) \leq A_i$ for each $i \in I$,

(c) $B \xrightarrow{f_i} A_i$ is an **A**-morphism for each $i \in I$,

(d) $B \leq A$.

Thus $(X, \alpha) = EA$.

(2) $\Rightarrow$ (1) is immediate from Proposition 21.30 and the fact that each functor-structured category is topological.

(2) $\Leftrightarrow$ (3). Proposition 21.31.

(3) $\Leftrightarrow$ (4) is immediate from Theorem 16.14, since each functor-structured category is an (Identity-Carried, Initial Source)-category. □

22.4 COROLLARY

For a construct $(\mathbf{A}, U)$ *the following conditions are equivalent:*

(1) $(\mathbf{A}, U)$ *is fibre-small and topological,*

(2) $(\mathbf{A}, U)$ *can be concretely embedded in a functor-structured construct as a full subconstruct that is closed under the formation of:*

   (a) products,

[Sec. 22]   Topological structure theorems

(b) initial subobjects, and

(c) indiscrete objects.

Proof: This follows from Theorem 22.3 and Proposition 21.37. □

22.5 REMARK
If $(\mathbf{A}, U)$ is a topological category over $\mathbf{X}$, then $(\mathbf{A}^{op}, U^{op})$ is a topological category over $\mathbf{X}^{op}$. Thus, by the above structure theorem, there exists a functor $T : \mathbf{X}^{op} \to \mathbf{Set}$, such that $(\mathbf{A}^{op}, U^{op})$ is concretely isomorphic (over $\mathbf{X}^{op}$) to a full concrete subcategory of $\mathbf{Spa}(T)$ that is definable by topological axioms. Thus, by duality, $(\mathbf{A}, U)$ is concretely isomorphic over $\mathbf{X}$ to a full concrete subcategory of the functor-costructured category $\mathbf{Spa}(T)^{op}$, definable by the dual of topological axioms, called *topological co-axioms*. Such a representation of a topological category is sometimes more natural than the one provided directly by the Topological Structure Theorem (22.3). Below we formulate these results more explicitly and provide several examples.

22.6 DEFINITION
Let $(\mathbf{A}, U)$ be a concrete category over $\mathbf{X}$.

(1) Each identity-carried morphism $P' \xrightarrow{p} P$ is called a **topological co-axiom** in $(\mathbf{A}, U)$.[71] An **A**-object A is said to **satisfy** the co-axiom p provided that A is $\{p\}$-projective; that is, each **A**-morphism $f : A \to P$ is also an **A**-morphism $f : A \to P'$.

(2) A full subcategory $\mathbf{B}$ of $\mathbf{A}$ is said to be **definable by topological co-axioms** in $\mathbf{A}$ if there exists a class of topological co-axioms in $\mathbf{A}$ such that an **A**-object A satisfies each of these co-axioms if and only if $A \in Ob(\mathbf{B})$.

22.7 EXAMPLES
(1) The construct **Sym** of symmetric relations is definable in **Rel** either

(a) by the topological axiom

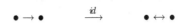

or

(b) by the topological co-axiom

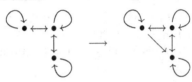

---

[71] Observe that f is a topological co-axiom if and only if it is a topological axiom. However, the concept "*A satisfies the topological co-axiom f*" is dual to the concept "*A satisfies the topological axiom f*".

(2) The construct **Top** is definable by the following (proper) class of topological co-axioms in $\mathbf{Spa}(\mathcal{Q})^{\mathrm{op}}$:

(C_1) $(\{0\},\{\{0\},\emptyset\}) \to (\{0\},\emptyset)$,

(C_2) $(\{0,1,2,3\},\{\{0,1\},\{0,2\},\{0\}\}) \to (\{0,1,2,3\},\{\{0,1\},\{0,2\}\})$,

$(C_X^{\mathcal{A}})$ $(X, \mathcal{A} \cup \{\bigcup \mathcal{A}\}) \to (X, \mathcal{A})$, for each set X and each family $\mathcal{A}$ of subsets of X.

22.8 THEOREM

For a concrete category $(\mathbf{A}, U)$ the following are equivalent:

(1) $(\mathbf{A}, U)$ is fibre-small and topological,

(2) $(\mathbf{A}, U)$ is concretely isomorphic to a finally closed full subcategory of some functor-costructured category,

(3) $(\mathbf{A}, U)$ is concretely isomorphic to a full concretely coreflective subcategory of some functor-costructured category,

(4) $(\mathbf{A}, U)$ is concretely isomorphic to a subcategory of some functor-costructured category $\mathbf{Spa}(T)^{\mathrm{op}}$ that is definable by topological co-axioms in $\mathbf{Spa}(T)^{\mathrm{op}}$. $\boxed{D}$

M-TOPOLOGICAL STRUCTURE THEOREMS

22.9 M-TOPOLOGICAL STRUCTURE THEOREM

For concrete categories $(\mathbf{A}, U)$ over an E-co-wellpowered $(E, \mathbf{M})$-category the following conditions are equivalent:

(1) $(\mathbf{A}, U)$ is fibre-small and **M**-topological,

(2) $(\mathbf{A}, U)$ is concretely isomorphic to an **M**-initially closed full subcategory of a functor-structured category,

(3) $(\mathbf{A}, U)$ is concretely isomorphic to an isomorphism-closed E-reflective subcategory of a functor-structured category,

(4) $(\mathbf{A}, U)$ is concretely isomorphic to an E-implicational subcategory of a functor-structured category.

Proof: (1) $\Rightarrow$ (2). By Theorem 21.40 and Remark 21.41(1) there exists a full concrete embedding $(\mathbf{A}, U) \xrightarrow{E} (\mathbf{B}, V)$ of $(\mathbf{A}, U)$ into a fibre-small topological category such that $E[\mathbf{A}]$ is **M**-initially closed in $(\mathbf{B}, V)$. By Theorem 22.3 there exists a full concrete embedding $(\mathbf{B}, V) \xrightarrow{F} \mathbf{Spa}(T)$ of $(\mathbf{B}, V)$ into some functor-structured category such that $F(\mathbf{B})$ is initially closed in $\mathbf{Spa}(T)$. Thus $(\mathbf{A}, U) \xrightarrow{F \circ E} \mathbf{Spa}(T)$ is a full concrete embedding such that $(F \circ E)[\mathbf{A}]$ is **M**-initially closed in $\mathbf{Spa}(T)$.

(2) $\Rightarrow$ (1) is obvious.

(2) $\Leftrightarrow$ (3) follows from Theorem 16.8 in view of Proposition 21.14(1).

(3) $\Leftrightarrow$ (4) follows from Theorem 16.14. $\square$

22.10 COROLLARY

For constructs $(\mathbf{A}, U)$ the following conditions are equivalent:

(1) $(\mathbf{A}, U)$ is fibre-small and monotopological,

(2) $(\mathbf{A}, U)$ is concretely isomorphic to a full subconstruct of a functor-structured construct that is closed under the formation of products and initial subobjects,

(3) $(\mathbf{A}, U)$ is concretely isomorphic to an implicational subconstruct of a functor-structured construct.

Proof: This follows immediately from Theorem 22.9 and the observation that in a fibre-small construct, for each initial mono-source $(A \xrightarrow{f_i} A_i)_I$, there exists a subset J of I such that $(A \xrightarrow{f_j} A_j)_J$ is an initial mono-source (cf. Proposition 21.36). □

Suggestions for Further Reading

Wyler, O. Top categories and categorical topology. *Gen. Topol. Appl.* **1** (1971): 17–28.

Menu, J., and A. Pultr. On categories determined by poset- and set-valued functors. *Comment. Math. Univ. Carolinae* **15** (1974): 665–678.

Adámek, J., H. Herrlich, and G. E. Strecker. The structure of initial completions. *Cahiers Topol. Geom. Diff.* **20** (1979): 333–352.

Herrlich, H Universal topology. *Categorical Topology. Proceedings of the International Conference, Toledo, Ohio, 1983* (ed. H. L. Bentley et. al.), Heldermann, Berlin, 1984, 223–281.

Adámek, J., and J. Reiterman. Topological categories presented by small sets of axioms. *J. Pure Appl. Algebra* **42** (1986): 1–14.

EXERCISES

22A. Characterization of Functor-Structured Categories

Show that a concrete category $(\mathbf{A}, U)$ is concretely isomorphic to some functor-structured category $\mathbf{Spa}(T)$ if and only if the following conditions are satisfied:

(1) $(\mathbf{A}, U)$ is a topological category,

(2) $(\mathbf{A}, U)$-fibres are complete atomic Boolean algebras,

(3) if $A \xrightarrow{f} B$ is final in $(\mathbf{A}, U)$, then the following hold:

  (a) if A is an atom (in its fibre), then so is B,

  (b) if B is discrete, then so is A.

22B. Topological Theories

A **topological theory** in **X** is a functor $\mathbf{X} \xrightarrow{T} \mathbf{JCPos}$. For each topological theory T in **X** denote by **Top**(T) the concrete category over **X** whose objects are pairs (X, t) with $X \in Ob(\mathbf{X})$ and $t \in T(X)$, and whose morphisms $(X,t) \xrightarrow{f} (Y,s)$ are those **X**-morphisms $X \xrightarrow{f} Y$ that satisfy $Tf(t) \leq s$.

(a) Show that if T is a topological theory in **X**, then **Top**(T) is a fibre-small topological category over **X**.

(b) Let $(\mathbf{A}, U)$ be a topological category over **X**. Consider the so-called **fibre-functor** $T : \mathbf{X} \to \mathbf{JCPos}$ that assigns to each **X**-object X the U-fibre of X, and to each **X**-morphism $X \xrightarrow{f} Y$ the unique function $TX \xrightarrow{Tf} TY$ such that $A \xrightarrow{f} Tf(A)$ is a final morphism in $(\mathbf{A}, U)$ for each $A \in TX$. Prove that T is a topological theory in **X** and that $(\mathbf{A}, U)$ is concretely isomorphic to **Top**(T).

(c) Show that if $G : \mathbf{X} \to \mathbf{Set}$ is a functor and $P : \mathbf{Set} \to \mathbf{JCPos}$ is the covariant power-set functor, then $P \circ G$ is a topological theory in **X** and **Top**$(P \circ G) = $ **Spa**(G).

* 22C. Sets of Axioms

Prove that **Top** cannot be axiomatized by a set of axioms in a functor-structured category.

22D. Co-Axiomatization

Find a co-axiomatization of **Prost** in a functor-costructured category.

23 Algebraic categories

In this section we introduce the concept of *algebraic category*, which is aimed at capturing the intuitive concept of "algebras over a base category". An attempt to describe algebraic structures, via monads, has been presented in §20. For constructs this approach has been quite successful. Unfortunately, however, besides being rather complicated, the concepts of monadic functors and monadic categories have the severe deficiencies pointed out in §20.

Here we provide the concepts of *essentially algebraic*, *algebraic*, and *regularly algebraic* functors and concrete categories. They are free from the above-mentioned deficiencies of monadic functors and categories. Moreover, as we will see in §24, the regularly algebraic functors form the compositive hull of the regularly monadic functors.

ESSENTIALLY ALGEBRAIC FUNCTORS

Each of the following concepts, which under rather mild assumptions are equivalent (see 17.13), captures the "essence of algebraicity" of a functor G:

(1) G reflects isomorphisms,

(2) G reflects limits,

(3) G reflects equalizers,

(4) G is faithful and reflects extremal epimorphisms,

(5) mono-sources are G-initial.

Each of these concepts is too weak to have many striking consequences. However, when any of them is combined with a suitably chosen factorization condition, such as (Generating, Mono-Source)-factorizability (which by itself has nothing to do with "algebraicity"), a theory emerges that is surprisingly close to intuitive notions of what concrete categories of algebras should be. Just for convenience we will also require unique transportability. This, however, is an inessential requirement.

23.1 DEFINITION

A functor is called **essentially algebraic** provided that it creates isomorphisms and is (Generating, Mono-Source)-factorizable.

23.2 PROPOSITION

For a uniquely transportable (Generating, Mono-Source)-factorizable functor $G : \mathbf{A} \to \mathbf{B}$, the following conditions are equivalent:

(1) G is essentially algebraic,

(2) G reflects isomorphisms,

(3) G reflects limits,

(4) G reflects equalizers,

(5) G is faithful and reflects extremal epimorphisms,

(6) every mono-source in **A** is G-initial.

Proof: Immediate from Propositions 17.13 and 13.36. □

23.3 PROPOSITION
Each essentially algebraic functor is a faithful adjoint functor.

Proof: Immediate from Propositions 23.2 and 18.3. □

23.4 PROPOSITION
Essentially algebraic functors are closed under composition. □

ESSENTIALLY ALGEBRAIC CATEGORIES

23.5 DEFINITION
A concrete category $(\mathbf{A}, U)$ is called **essentially algebraic** provided that U is essentially algebraic.

23.6 EXAMPLES
(1) All monadic constructs, e.g., **Alg**(Ω), **Vec**, **Grp**, and **HComp**, are essentially algebraic. However, the monadic category over **Pos** exhibited in Example 20.47 is not essentially algebraic (since it is not cocomplete, cf. Theorem 23.11).

(2) The constructs **BooSp**, **TfAb**, **Cat**$_{of}$, and (**Ban**,O) are essentially algebraic, but not monadic.

(3) Categories of topological algebras (e.g., topological groups, topological semigroups, etc.), considered as concrete categories over **Top**, are essentially algebraic.

(4) A topological category $(\mathbf{A}, U)$ is essentially algebraic only if U is an isomorphism.

(5) A partially ordered class, considered as a concrete category over **1**, is essentially algebraic if and only if it has precisely one element.

23.7 PROPOSITION
In any essentially algebraic category, the embeddings are precisely the monomorphisms.

Proof: Immediate from Proposition 23.2. □

23.8 CHARACTERIZATION THEOREM FOR ESSENTIALLY ALGEBRAIC CATEGORIES

A concrete category $(\mathbf{A}, U)$ is essentially algebraic if and only if the following conditions are satisfied:

(1) U creates isomorphisms,

(2) U is adjoint,

(3) $\mathbf{A}$ is (Epi, Mono-Source) factorizable.

Proof:
(a). If $(\mathbf{A}, U)$ is essentially algebraic, then conditions (1) and (2) are clearly satisfied (cf. 23.2 and 23.3). To show (3), let $\mathcal{S} = (A \xrightarrow{f_i} A_i)_I$ be a source in $\mathbf{A}$, and let

$$UA \xrightarrow{Uf_i} UA_i = UA \xrightarrow{g} UB \xrightarrow{Um_i} UA_i$$

be a (Generating, Mono-Source)-factorization of the corresponding structured source. Since by Proposition 23.2 each mono-source is U-initial, $A \xrightarrow{g} B$ is an $\mathbf{A}$-morphism. Hence $A \xrightarrow{f_i} A_i = A \xrightarrow{g} B \xrightarrow{m_i} A_i$ is an (Epi, Mono-Source)-factorization of $\mathcal{S}$.

(b). Let conditions (1), (2), and (3) be satisfied, and let $\mathcal{S} = (X \xrightarrow{f_i} UA_i)_I$ be a structured source. If $X \xrightarrow{u} UA$ is a universal arrow, then for each $i \in I$ there exists an $\mathbf{A}$-morphism $A \xrightarrow{g_i} A_i$ with $f_i = (Ug_i) \circ u$. If $A \xrightarrow{g_i} A_i = A \xrightarrow{e} B \xrightarrow{m_i} A_i$ is an (Epi, Mono-Source)-factorization, then $X \xrightarrow{f_i} UA_i = X \xrightarrow{(Ue) \circ u} UB \xrightarrow{Um_i} UA_i$ is a (Generating, Mono-Source)-factorization of $\mathcal{S}$. □

23.9 REMARKS

(1) Condition (3) in the above theorem is surprising since it is a requirement on $\mathbf{A}$ and not on U. In particular, the identity functor on a category $\mathbf{A}$ is essentially algebraic if and only if $\mathbf{A}$ has (Epi, Mono-Source)-factorizations.

(2) Monadic categories satisfy conditions (1) and (2) of the above theorem, but may fail to satisfy (3). Hence a monadic category $(\mathbf{A}, U)$ is essentially algebraic if and only if $\mathbf{A}$ has (Epi, Mono-Source)-factorizations. In particular, regularly monadic categories are essentially algebraic.

23.10 COROLLARY

If $(\mathbf{A}, U)$ is essentially algebraic, then $\mathbf{A}$ has coequalizers.

Proof: Immediate since (Epi, Mono-Source)-factorizable categories have coequalizers. (Cf. Proposition 15.7 and Theorem 15.10.) □

23.11 THEOREM

Every essentially algebraic functor

(1) detects colimits,

(2) *preserves and creates limits.*

Proof: Let $(\mathbf{A}, U)$ be an essentially algebraic category.

(1). Let $D: \mathbf{I} \to \mathbf{A}$ be a diagram, and let $(UD_i \xrightarrow{c_i} K)_I$ be a colimit of $U \circ D$. Consider the structured source $\mathcal{S} = (K \xrightarrow{f_j} UA_j)_J$ consisting of those structured arrows (f_j, A_j) with the property that $UD_i \xrightarrow{f_j \circ c_i} UA_j$ is an $\mathbf{A}$-morphism for each $i \in I$. Let $K \xrightarrow{f_j} UA_j = K \xrightarrow{g} UA \xrightarrow{Um_j} UA_j$ be a (Generating, Mono-Source)-factorization of $\mathcal{S}$. Since mono-sources in $\mathbf{A}$ are initial, each $UD_i \xrightarrow{g \circ c_i} UA$ is an $\mathbf{A}$-morphism. Hence $(D_i \xrightarrow{g \circ c_i} A)_I$ is a colimit of D.

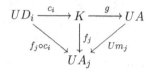

(2). As an adjoint functor, U preserves limits. To show creation, let $D: \mathbf{I} \to \mathbf{A}$ be a diagram and let $\mathcal{L} = (L \xrightarrow{\ell_i} UD_i)$ be a limit of the composite functor $U \circ D$. If $L \xrightarrow{\ell_i} UD_i = L \xrightarrow{g} UA \xrightarrow{Um_i} UD_i$ is a (Generating, Mono-Source)-factorization, then $\mathcal{M} = (A \xrightarrow{m_i} D_i)_I$ is a natural source for D. Let $\mathcal{S} = (B \xrightarrow{f_i} D_i)_I$ be an arbitrary natural source for D. Then $U\mathcal{S}$ is a natural source for $U \circ D$. Consequently, there exists a morphism $UB \xrightarrow{f} L$ with $U\mathcal{S} = \mathcal{L} \circ f$.

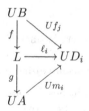

Since $\mathcal{M}$ is a mono-source and, hence, is initial, $UB \xrightarrow{g \circ f} UA$ is an $\mathbf{A}$-morphism with $\mathcal{S} = \mathcal{M} \circ (g \circ f)$. Thus $\mathcal{M}$ is a limit of D. Since U preserves limits, $U\mathcal{M}$ is a limit of $U \circ D$. Hence $L \xrightarrow{g} UA$ is an isomorphism. Therefore, transportability of U implies that U lifts limits. Hence, by Proposition 13.25, U creates limits. □

23.12 PROPOSITION

If $(\mathbf{A}, U)$ is essentially algebraic over $\mathbf{X}$, then the following hold:

(1) If $\mathbf{X}$ is (strongly) complete, then $\mathbf{A}$ is (strongly) complete.

(2) If $\mathbf{X}$ has coproducts, then $\mathbf{A}$ is cocomplete.

(3) If $\mathbf{X}$ is wellpowered, then $\mathbf{A}$ is wellpowered.

Proof: (1) follows from Theorem 23.11 and the fact that U, being adjoint, preserves monomorphisms.

(2) follows from Theorem 23.11 and Corollary 23.10.

(3) follows from the fact that monomorphisms are initial and U preserves and reflects them. □

23.13 COROLLARY
Every essentially algebraic construct is complete, cocomplete, and wellpowered. □

23.14 PROPOSITION
For essentially algebraic categories, the following conditions are equivalent:

(1) **A** *is extremally co-wellpowered,*

(2) U *is extremally co-wellpowered,*

(3) U *is concretely co-wellpowered.*

Proof: (1) $\Rightarrow$ (2) follows from Proposition 18.11.

(2) $\Leftrightarrow$ (3) is immediate by Proposition 23.7.

(3) $\Rightarrow$ (1) is immediate by Remark 8.18(3). □

23.15 THEOREM
If $(\mathbf{A}, U)$ *is a concretely co-wellpowered category over a strongly complete category* **X**, *then the following conditions are equivalent:*

(1) $(\mathbf{A}, U)$ *is essentially algebraic,*

(2) U *creates limits.*

Proof: (1) $\Rightarrow$ (2) follows from Theorem 23.11.

(2) $\Rightarrow$ (1). By Theorem 13.19, **A** is strongly complete and U preserves strong limits. Thus (1) follows from Theorem 17.11(2). □

23.16 COROLLARY
Concretely co-wellpowered monadic categories over strongly complete categories are essentially algebraic. □

23.17 THEOREM
Concrete functors between essentially algebraic categories are essentially algebraic.

Proof: Let $(\mathbf{A}, U) \xrightarrow{G} (\mathbf{B}, V)$ be a concrete functor between essentially algebraic categories over **X**. G reflects isomorphisms since V preserves and U reflects them. G is faithful and amnestic since U is. G is transportable since U is transportable, V is amnestic, and **B**-monomorphisms are V-initial. Hence G creates isomorphisms. Let $\mathcal{S} = (B \xrightarrow{f_i} GA_i)_I$ be a G-structured source. Consider a (Generating, Mono-Source)-factorization
$$VB \xrightarrow{Vf_i} UA_i \;=\; VB \xrightarrow{g} UA \xrightarrow{Um_i} UA_i$$
of the U-structured source $(VB, (Vf_i, A_i)_I)$. Since U preserves and V reflects mono-sources, $(GA \xrightarrow{Gm_i} GA_i)_I$ is a mono-source, and hence is V-initial. Thus there exists a **B**-morphism $B \xrightarrow{\hat{g}} GA$ with $V\hat{g} = g$ and $f_i = Gm_i \circ \hat{g}$ for each $i \in I$. The G-structured morphism $(\hat{g}, A)$ is generating, since the U-structured morphism $(V\hat{g}, A)$ is generating. Hence
$$B \xrightarrow{f_i} GA_i = B \xrightarrow{\hat{g}} GA \xrightarrow{Gm_i} GA_i$$
is a (Generating, Mono-Source)-factorization of the G-structured source $\mathcal{S}$. □

23.18 COROLLARY
The forgetful functors **Rng** $\to$ **Ab** *and* **Rng** $\to$ **Mon** *are essentially algebraic functors; in particular, they are adjoint functors.* □

ALGEBRAIC FUNCTORS AND CATEGORIES

Since (Epi, Mono-Source)-factorizable categories are (Extremal Epi, Mono-Source) categories (15.10), there is a certain symmetry between mono-sources on the one hand and extremal epimorphisms on the other. Unfortunately, essentially algebraic functors, as nice as they are otherwise, may not interact well with this symmetry. In particular,

(1) they preserve mono-sources, but may fail to preserve extremal epimorphisms, e.g., **Ban** $\xrightarrow{O}$ **Set** or **Cat**$_{of}$ $\xrightarrow{U}$ **Set**, cf. Example 7.72(5),

(2) they detect wellpoweredness, but may fail to detect extremal co-wellpoweredness, cf. Exercise 23I,

(3) mono-sources must be initial, but extremal epimorphisms may fail to be final, e.g, **Ban** $\xrightarrow{O}$ **Set** or **Cat**$_{of}$ $\xrightarrow{U}$ **Set**, cf. Example 8.11(4).

The following condition, which is quite often satisfied, restores a proper interaction with this symmetry.

23.19 DEFINITION
(1) A functor is called **algebraic** provided that it is essentially algebraic and preserves extremal epimorphisms.

(2) A concrete category $(\mathbf{A}, U)$ is called **algebraic** provided that U is algebraic.

23.20 EXAMPLES

(1) Each regularly monadic category and, in particular, each monadic construct is algebraic (cf. 20.30).

(2) Every full subconstruct of a monadic construct that is closed under the formation of products and subobjects is algebraic. In particular, **BooSp** and **TfAb** are algebraic. Neither, however, is monadic.

(3) The constructs **Cat**$_{of}$ and (**Ban**, O) are essentially algebraic, but not algebraic [cf. 7.72(5)].

(4) **Cat**$_{of}$, considered as a concrete category over **Alg**$(1,1)$ via the functor that forgets composition, is essentially algebraic and monadic but is not algebraic.

(5) **TopGrp**, considered as a concrete category over **Top**, is algebraic. The category of Hausdorff semigroups, considered as a concrete category over **Haus**, is essentially algebraic and monadic, but is not algebraic.

(6) The embedding **Reg** $\hookrightarrow$ **Top** of the full subcategory of **Top** that consists of all regular spaces (the T_1-condition not being required) is essentially algebraic and monadic, but is not algebraic.

23.21 PROPOSITION
Algebraic functors are closed under composition. □

23.22 PROPOSITION
Each concrete functor between algebraic categories is algebraic.

Proof: This follows from Theorem 23.17 and the fact that (essentially) algebraic functors reflect extremal epimorphisms. □

23.23 PROPOSITION
In algebraic categories every extremal epimorphism is final.

Proof: Suppose that $A \xrightarrow{e} B$ is an extremal epimorphism in an algebraic category $(\mathbf{A}, U)$ over $\mathbf{X}$. Let $UB \xrightarrow{f} UC$ be an **X**-morphism and $A \xrightarrow{g} C$ be an **A**-morphism with $Ug = f \circ Ue$. If

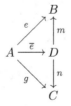

is an (ExtrEpi, Mono-Source)-factorization of the 2-source $B \xleftarrow{e} A \xrightarrow{g} C$ (which exists in view of Theorems 23.8 and 15.10), then $Un \circ U\bar{e} = Ug = f \circ Ue = f \circ Um \circ U\bar{e}$. Since U is algebraic, $U\bar{e}$ is an (extremal) epimorphism in **X**, which implies that $Un =$

$f \circ Um$. Since (Um, Un) is a mono-source, Um is a monomorphism in **X**, so that m is a monomorphism in **A**. Since e is an extremal epimorphism, m must be an isomorphism. Thus $B \xrightarrow{n \circ m^{-1}} C$ is the unique **A**-morphism with $U(n \circ m^{-1}) = f$. □

23.24 COROLLARY
Algebraic functors reflect regular epimorphisms.

Proof: This follows from Proposition 23.23, as in the proof of Proposition 20.30. □

23.25 REMARK
Algebraic functors need not preserve regular epimorphisms. See Exercise 23J.

23.26 COROLLARY
In any algebraic category over a balanced category, the quotient morphisms are precisely the extremal epimorphisms. □

23.27 COROLLARY
*If $(\mathbf{A}, U)$ is algebraic over an extremally co-wellpowered category **X**, then **A** is extremally co-wellpowered.* □

23.28 PROPOSITION
*If $(\mathbf{A}, U)$ is algebraic over **X**, then the following hold:*

*(1) If extremal epimorphisms are closed under the formation of pullbacks (resp. products) in **X**, then they are closed under the formation of pullbacks (resp. products) in **A**.*

*(2) If **X** has enough extremal projectives, then so does **A**.*

Proof:
(1). This follows from the fact that U preserves limits and preserves and reflects extremal epimorphisms.

(2). Let A be an **A**-object. Then in **X** there exists an extremal epimorphism $P \xrightarrow{e} UA$ with extremally projective domain P. If $P \xrightarrow{u} U\hat{P}$ is a universal arrow, then there exists a unique **A**-morphism $\hat{P} \xrightarrow{\hat{e}} A$ with $e = U\hat{e} \circ u$. Since e is an extremal epimorphism, so is $U\hat{e}$. Since U reflects extremal epimorphisms, $\hat{e}$ is an extremal epimorphism in **A**. It remains to show that $\hat{P}$ is extremally projective. Let $B \xrightarrow{g} C$ be an extremal epimorphism in **A** and $\hat{P} \xrightarrow{f} C$ be an arbitrary **A**-morphism. Since Ug is an extremal epimorphism in **X**, there exists some **X**-morphism $P \xrightarrow{k} UB$ such that the square

$$\begin{array}{ccc} P & \xrightarrow{u} & U\hat{P} \\ {\scriptstyle k}\downarrow & & \downarrow{\scriptstyle Uf} \\ UB & \xrightarrow{Ug} & UC \end{array}$$

commutes. If $\hat{P} \xrightarrow{\hat{k}} B$ is the unique **A**-morphism with $k = U\hat{k} \circ u$, then $f = g \circ \hat{k}$. □

23.29 COROLLARY

Any algebraic construct is complete, cocomplete, wellpowered, and extremally co-wellpowered. It has enough extremal projectives, and its extremal epimorphisms are closed under composition and the formation of pullbacks and products. □

23.30 CHARACTERIZATION THEOREM FOR ALGEBRAIC CATEGORIES

A concrete category $(\mathbf{A}, U)$ *is algebraic if and only if the following conditions are satisfied:*

(1) $\mathbf{A}$ *is (Epi, Mono-Source)-factorizable,*

(2) U *is adjoint,*

(3) U *is uniquely transportable,*

(4) U *preserves and reflects extremal epimorphisms.*

Proof: The necessity of these conditions has already been established. Conversely, it need only be shown that for any concrete category $(\mathbf{A}, U)$ with the above properties, U must reflect isomorphisms. This follows immediately from the fact that U reflects extremal epimorphisms and monomorphisms. □

23.31 THEOREM

For a concrete category $(\mathbf{A}, U)$ *over an (Extremal Epi, Mono-Source)-category* $\mathbf{X}$ *the following conditions are equivalent:*

(1) $(\mathbf{A}, U)$ *is algebraic,*

(2) U *is an adjoint functor and lifts (Extremal Epi, Mono-Source)-factorizations uniquely,*

(3) U *is an adjoint functor and* **creates** *(***Extremal Epi, Mono-Source***)-***factorizations***, i.e., U lifts these factorizations uniquely and reflects extremal epimorphisms and mono-sources.*

Proof: (1) ⇒ (3). Since, by Theorem 23.30 and Theorem 15.10, $\mathbf{A}$ has (Extremal Epi, Mono-Source)-factorizations and U preserves these and is transportable, U lifts such factorizations. Since, moreover, U reflects extremal epimorphisms and mono-sources and is amnestic, the liftings are uniquely determined.

(3) ⇒ (2). Trivial.

(2) ⇒ (1). Since U, being faithful, reflects epimorphisms and mono-sources, $\mathbf{A}$ has (Epi, Mono-Source)-factorizations. Observe that U reflects identities, i.e., any $\mathbf{A}$-morphism $A \xrightarrow{f} B$ with $Uf = id_X$ must be id_A, since the (Extremal Epi, Mono-Source)-factorization $UA \xrightarrow{Uf} UB = UA \xrightarrow{id_X} X \xrightarrow{id_X} UB$ has the lifts $A \xrightarrow{f} B \xrightarrow{id_B} B$ and $A \xrightarrow{id_A} A \xrightarrow{f} B$. Next consider the class E of $\mathbf{A}$-morphisms that are mapped by U into extremal epimorphisms. To show that any element $A \xrightarrow{e} B$ of E is final, let

$UB \xrightarrow{f} UC$ be an **X**-morphism and let $A \xrightarrow{g} C$ be an **A**-morphism with $Ug = f \circ Ue$. Then the (Extremal Epi, Mono-Source)-factorization in **X**

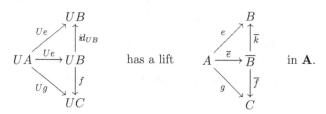

Since $U\bar{k} = id_{UB}$, we conclude that $\bar{k} = id_B$; in particular, $\bar{B} = B$. Hence $B \xrightarrow{\bar{f}} C$ is an **A**-morphism with $U\bar{f} = f$. This finality-property implies that **A** is an $(E, \mathbf{M})$-category, where **M** is the conglomerate of all **A**-sources that are mapped via U to mono-sources in **X**. Since U is faithful and adjoint, **M** consists precisely of the mono-sources in **A**. Thus, by Proposition 15.8(1), E consists precisely of the extremal epimorphisms in **A**. Hence U preserves and reflects extremal epimorphisms. It remains to be shown that U is transportable. If $X \xrightarrow{h} UA$ is a structured isomorphism, then the (Extremal Epi, Mono-Source)-factorization $UA \xrightarrow{U(id_A)} UA = UA \xrightarrow{h^{-1}} X \xrightarrow{h} UA$ has a lift $A \xrightarrow{\bar{f}} B \xrightarrow{\bar{g}} A$. Hence $B \xrightarrow{\bar{g}} A$ is a morphism with $U\bar{g} = h$. Since U reflects extremal epimorphisms and monomorphisms, $\bar{g}$ is both, i.e., an isomorphism. □

23.32 COROLLARY

A full concrete subcategory $(\mathbf{A}, U)$ *of an algebraic category* $(\mathbf{B}, V)$ *is algebraic if and only if* **A** *is reflective in* **B** *and contains with every source its (Extremal Epi, Mono-Source)-factorization in* **B**.

Proof: By Propositions 23.21 and 23.22, $(\mathbf{A}, U)$ is algebraic if and only if the inclusion functor $\mathbf{A} \to \mathbf{B}$ is algebraic. Since **B** has (Epi, Mono-Source)-factorizations and hence is an (Extremal Epi, Mono-Source)-category, the desired result follows immediately from Theorem 23.31. □

23.33 COROLLARY

Each extremally epireflective concrete subcategory of an algebraic category is algebraic. □

23.34 EXAMPLES

(1) The constructs **Grp** and **Sgr** are algebraic, but **Grp** is not closed under the formation of subobjects in **Sgr**.

(2) **HComp** and **Top**, considered as concrete categories over **Top**, are algebraic, but **HComp** is closed neither under the formation of subobjects nor under the formation of extremal quotient objects in **Top**.

(3) Epireflective full concrete subcategories of algebraic categories need not be algebraic. See Exercise 16A.

REGULARLY ALGEBRAIC FUNCTORS AND CATEGORIES

Algebraic functors whose codomains have regular factorizations are particularly well-behaved. They are closely related to regularly monadic functors — in fact, they form the compositive hull of the regularly monadic functors (see Theorem 24.2). This suggests the following terminology:

23.35 DEFINITION
(1) A functor $\mathbf{A} \xrightarrow{U} \mathbf{X}$ is called **regularly algebraic** provided that U is algebraic and $\mathbf{X}$ has regular factorizations.

(2) A concrete category $(\mathbf{A}, U)$ is called **regularly algebraic** provided that U is regularly algebraic.

23.36 PROPOSITION
Regularly monadic functors are regularly algebraic.

Proof: Immediate from Proposition 20.24 and Theorem 23.31. □

23.37 COROLLARY
Monadic constructs are regularly algebraic. □

23.38 CHARACTERIZATION THEOREM FOR REGULARLY ALGEBRAIC CATEGORIES
A concrete category $(\mathbf{A}, U)$ over $\mathbf{X}$ is regularly algebraic if and only if the following conditions are satisfied:

(1) $\mathbf{A}$ has coequalizers,

(2) $\mathbf{X}$ has regular factorizations,

(3) U is adjoint,

(4) U is uniquely transportable,

(5) U preserves and reflects extremal epimorphisms.

Proof: By Theorem 23.30 it suffices to show that conditions (1) – (5) imply that $\mathbf{A}$ is (Epi, Mono-Source)-factorizable. Let $(A \xrightarrow{f_i} A_i)_I$ be a source in $\mathbf{A}$ and let

$$UA \xrightarrow{Uf_i} UA_i = UA \xrightarrow{e} X \xrightarrow{m_i} UA_i$$

be a (RegEpi, Mono-Source)-factorization in $\mathbf{X}$. Then e is a coequalizer of some pair $Y \underset{p_2}{\overset{p_1}{\rightrightarrows}} UA$ of morphisms in $\mathbf{X}$. Let $Y \xrightarrow{\eta_Y} UB$ be a universal arrow for Y. Then there exist $\mathbf{A}$-morphisms $B \xrightarrow{\hat{p}_j} A$ with $p_j = (U\hat{p}_j) \circ \eta_Y$ for $j = 1, 2$. Let $A \xrightarrow{c} C$ be a coequalizer of $\hat{p}_1$ and $\hat{p}_2$ in $\mathbf{A}$. For each $i \in I$ the equalities $U(f_i \circ \hat{p}_1) \circ \eta_Y = Uf_i \circ p_1 = m_i \circ e \circ p_1 = m_i \circ e \circ p_2 = Uf_i \circ p_2 = U(f_i \circ \hat{p}_2) \circ \eta_Y$ imply that $f_i \circ \hat{p}_1 = f_i \circ \hat{p}_2$.

Thus for each $i \in I$ there exists a unique **A**-morphism $C \xrightarrow{n_i} A_i$ with $f_i = n_i \circ c$. It remains to be shown that $(C \xrightarrow{n_i} A_i)_I$ is a mono-source in **A**. Since $Uc \circ p_1 = U(c \circ \hat{p}_1) \circ \eta_Y = U(c \circ \hat{p}_2) \circ \eta_Y = Uc \circ p_2$, there exists a unique **X**-morphism $X \xrightarrow{f} UC$ with $Uc = f \circ e$. Since, by (5), Uc is an extremal (and hence a regular) epimorphism in **X** and $(X \xrightarrow{m_i} A_i)_I$ is a mono-source in **X**, there exists a unique diagonal d in the following diagram:

$$\begin{array}{ccc} UA & \xrightarrow{Uc} & UC \\ {\scriptstyle e}\downarrow & {\scriptstyle d}\nearrow & \downarrow{\scriptstyle Un_i} \\ X & \xrightarrow{m_i} & UA_i \end{array}$$

Thus $f \circ d \circ Uc = f \circ e = Uc$ implies that $f \circ d = id_{UC}$ and $d \circ f \circ e = d \circ Uc = e$ implies that $d \circ f = id_X$. Hence d is an isomorphism and $U(C \xrightarrow{n_i} A_i)_I = (UC \xrightarrow{d} X \xrightarrow{m_i} UA_i)_I$ is a mono-source. By faithfulness, $(C \xrightarrow{n_i} A_i)_I$ is a mono-source. □

23.39 COROLLARY

If $(\mathbf{A}, U)$ is a regularly algebraic category, then **A** has regular factorizations and U preserves and reflects regular epimorphisms.

Proof: In the proof of Theorem 23.38 regular factorizations have been constructed. Since in categories with regular factorizations every extremal epimorphism is regular (cf. 14.14), U preserves and reflects regular epimorphisms. □

23.40 COROLLARY

Concrete functors between regularly algebraic categories are regularly algebraic. □

23.41 THEOREM

A construct $(\mathbf{A}, U)$ is monadic if and only if it is algebraic and U reflects congruence relations.

Proof: (1). Let $(\mathbf{A}, U)$ be a monadic construct. By Corollary 23.37, $(\mathbf{A}, U)$ is algebraic. That U reflects congruence relations follows easily from the facts that U creates coequalizers of congruence relations (20.35) and that every point-separating 2-source in **A** is a mono-source and, hence, is initial.

(2). Let $(\mathbf{A}, U)$ be an algebraic construct and let U reflect congruence relations. By Theorem 20.35 it suffices to show that U creates coequalizers of congruence relations. Let $A \overset{p}{\underset{q}{\rightrightarrows}} B$ be a pair of **A**-morphisms and let $UA \overset{Up}{\underset{Uq}{\rightrightarrows}} UB \xrightarrow{c} C$ be a congruence fork in **Set**. Let $B \xrightarrow{\hat{c}} \hat{C}$ be a coequalizer of p and q in **A**. Since U reflects congruence relations, Proposition 11.22(1) implies that $A \overset{p}{\underset{q}{\rightrightarrows}} B \xrightarrow{\hat{c}} \hat{C}$ is a congruence fork in **A**. Since $U\hat{c}$ is an extremal (and hence a regular epimorphism) in **Set** and since

U preserves pullbacks, Proposition 11.22(2) implies that $UA \underset{Uq}{\overset{Up}{\rightrightarrows}} UB \xrightarrow{U\hat{c}} U\hat{C}$ is a congruence fork in **Set**. Thus there is an isomorphism $C \xrightarrow{h} U\hat{C}$ with $U\hat{c} = h \circ c$. By transportability, h lifts to an isomorphism $\tilde{C} \xrightarrow{\tilde{h}} \hat{C}$ in **A**. Thus

$$B \xrightarrow{\tilde{c}} \tilde{C} = B \xrightarrow{\hat{c}} \hat{C} \xrightarrow{\tilde{h}^{-1}} \tilde{C}$$

is a lift of $UB \xrightarrow{c} C$ to a coequalizer of p and q in **A**. Let $B \xrightarrow{c'} C'$ be an arbitrary lift of $UB \xrightarrow{c} C$. Then faithfulness implies that $c' \circ p = c' \circ q$. Thus there exists an **A**-morphism $\tilde{C} \xrightarrow{k} C'$ with $c' = k \circ \tilde{c}$. Application of U yields $c = Uk \circ c$. Thus $Uk = id_C$. Amnesticity yields $\tilde{C} \xrightarrow{k} C' = \tilde{C}' \xrightarrow{id} \tilde{C}$. Therefore U creates coequalizers of congruence relations. □

23.42 REMARK
The following diagram exhibits some of the results on the "algebraic character" of functors $\mathbf{A} \xrightarrow{U} \mathbf{X}$:

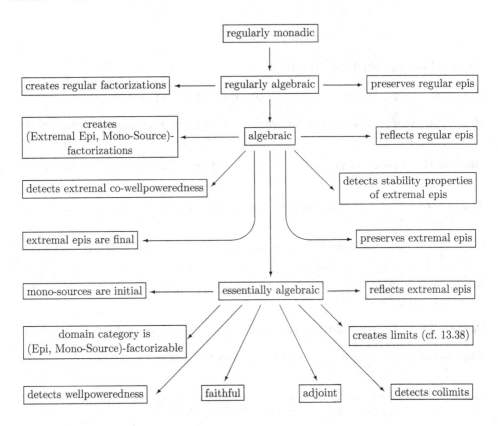

An algebraic construct $(\mathbf{A}, U)$ is monadic iff U reflects congruence relations

Suggestions for Further Reading

Dubuc, E. Adjoint triangles. *Springer Lect. Notes Math.* **61** (1968): 69–91.

Huq, S. A. An interpolation theorem for adjoint functors. *Proc. Amer. Math. Soc.* **25** (1970): 880–883.

Herrlich, H. Regular categories and regular functors. *Canad. J. Math.* **26** (1974): 709–720.

Tholen, W. Adjungierte Dreiecke, Colimites und Kan-Erweiterungen. *Math. Ann.* **217** (1975): 121–129.

Bargenda, H., and G. Richter. Varietal hulls of functors. *Quaest. Math.* **4** (1980): 121–158.

Richter, G. Algebraic categories of topological spaces. *Springer Lect. Notes Math.* **962** (1982): 263–271.

Herrlich, H. Essentially algebraic categories. *Quaest. Math.* **9** (1986): 245–262.

Bargenda, H. *Algebraische Hüllen rechtsadjungierter Funktoren.* Ph. D. dissertation, Bremen University, 1987.

Im, G. B., and G. M. Kelly. Adjoint triangle theorems for conservative functors. *Bull. Austral. Math. Soc.* **36** (1987): 133–136.

EXERCISES

23A. Weakly Algebraic Functors

A functor $G: \mathbf{A} \to \mathbf{B}$ is called **weakly algebraic** provided that it is an adjoint and mono-sources in $\mathbf{A}$ are G-initial. Show that

(a) G is weakly algebraic if and only if G is an adjoint functor that reflects swell epimorphisms. (Cf. Exercise 15A).

(b) Weakly algebraic functors

   (1) reflect extremal epimorphisms, swell epimorphisms, epimorphisms, and isomorphisms,

   (2) are faithful,

   (3) preserve and reflect mono-sources,

   (4) preserve and reflect limits,

   (5) detect wellpoweredness.

(c)

   (1) Every monadic functor is weakly algebraic.

   (2) Every essentially algebraic functor is weakly algebraic.

   (3) Every full reflective embedding is weakly algebraic.

   (4) Weakly algebraic embeddings need not be full. [$\mathbf{A}$ = pointed sets and all maps, $\mathbf{B} = \mathbf{pSet}$.]

   (5) A poset, considered as a concrete category over **1**, is weakly algebraic if and only if it contains exactly one element.

(d) $G: \mathbf{A} \to \mathbf{B}$ is essentially algebraic if and only if G is uniquely transportable and weakly algebraic and $\mathbf{A}$ is (Epi, Mono-Source)-factorizable.

(e) Weakly algebraic functors are closed under composition.

(f) Concrete functors between weakly algebraic categories need not be weakly algebraic. (Cf. Example 20.46.)

(g) If $G: \mathbf{A} \to \mathbf{B}$ is weakly algebraic and $\mathbf{B}$ has (Epi, Mono-Source)-factorizations, then each co-adjoint for G preserves extremal epimorphisms.

23B. Essentially Algebraic Embeddings

(a) Prove that the embedding $E: \mathbf{A} \to \mathbf{B}$ of a full subcategory is essentially algebraic if and only if the following two conditions hold:

   (1) E is monadic; i.e., $\mathbf{A}$ is reflective and isomorphism-closed in $\mathbf{B}$.

   (2) $\mathbf{A}$ is (Epi, Mono-Source) factorizable.

(b) Show that embeddings of extremally epireflective, full, isomorphism-closed subcategories into (Epi, Mono-Source)-factorizable categories are essentially algebraic.

(c) Prove that embeddings of epireflective, full, isomorphism-closed subcategories into (Epi, Mono-Source)-factorizable categories need not be essentially algebraic. [See Exercise 16A(b).]

23C. Essentially Algebraic vs. Monadic Functors

Prove that a monadic functor with an extremally co-wellpowered domain and strongly complete codomain is essentially algebraic. [Cf. Theorem 23.16 and Proposition 18.11.]

23D. Algebraic Categories

(a) Let $(\mathbf{A}, U)$ be a concrete category over an $(E, \mathbf{M})$-category and let U lift $(E, \mathbf{M})$-factorizations uniquely. Show that

　(1) each **A**-morphism f with $Uf \in E$ is a final epimorphism,

　(2) **A** is an $(U^{-1}[E], U^{-1}[\mathbf{M}])$-category.

(b) Prove that algebraic functors reflect regular epimorphisms. [Hint: Proposition 23.23.]

23E. Regularly Algebraic Functors and Categories

Show that

(a) If **B** has regular factorizations, then the following are equivalent for any functor $G: \mathbf{A} \to \mathbf{B}$:

　(1) G is regularly algebraic,

　(2) G is essentially algebraic and preserves regular epimorphisms,

　(3) G is essentially algebraic and preserves swell epimorphisms,

　(4) G is adjoint and creates regular factorizations,

　(5) **A** has regular factorizations and G is adjoint, uniquely transportable, and preserves and reflects regular epimorphisms.

(b) Regularly algebraic functors are closed under composition.

(c) Regular epireflective full isomorphism-closed subcategories of regularly algebraic categories are regularly algebraic.

(d) If **B** has regular factorizations and $G: \mathbf{A} \to \mathbf{B}$ is an embedding of a full, reflective subcategory, then the following are equivalent:

　(1) G is regularly algebraic,

　(2) **A** contains with any source its regular factorization in **B**,

　(3) **A** contains with any morphism its regular factorization in **B**,

(4) **A** contains any **B**-object that is simultaneously a subobject of some **A**-object and a regular quotient object of some **A**-object,

(5) **A** is isomorphism-closed in **B**, and G preserves regular epimorphisms.

(e) If **B** has regular factorizations, then the following conditions are equivalent for any functor $G: \mathbf{A} \to \mathbf{B}$:

(1) G is regularly monadic,

(2) G is regularly algebraic and monadic.

* 23F. Varietal Functors and Varietal Categories

A functor $G: \mathbf{A} \to \mathbf{B}$ is called **varietal** provided that G is regularly algebraic, **B** has pullbacks, and G reflects congruence relations. A concrete category $(\mathbf{A}, U)$ is called **varietal** provided that U is varietal. Show that

(a) A concrete category $(\mathbf{A}, U)$ over a category with pullbacks and regular factorizations is varietal if and only if $(\mathbf{A}, U)$ is algebraic and U creates coequalizers of congruence relations.

(b) A regularly algebraic category $(\mathbf{A}, U)$ over a category with pullbacks is monadic if and only if U creates those coequalizers of congruence relations that are retractions.

(c) Every varietal functor is regularly monadic.

(d) A construct is varietal if and only if it is monadic.

(e) The full embeddings **TfAb** $\hookrightarrow$ **Ab** and **BooSp** $\hookrightarrow$ **Top** are regularly monadic, but are not varietal.

(f) Varietal functors are closed under composition.

(g) Concrete functors between varietal categories are varietal.

(h) If $\mathbf{A} \xrightarrow{G} \mathbf{B}$ is varietal, then **A** has pullbacks and regular factorizations.

23G. Extremal Epimorphisms and Quotients

Prove that for algebraic constructs, quotients = extremal epimorphisms = swell epimorphisms = strict epimorphisms = regular epimorphisms.

23H. Extremal Projectives

Show that if $\mathbf{A} \xrightarrow{G} \mathbf{B}$ is algebraic with co-adjoint F and if P is extremally projective in **B**, then FP is extremally projective in **A**.

* 23I. Extremal Co-wellpoweredness

Show that the following (legitimate quasi-)construct $(\mathbf{A}, U)$ is essentially algebraic but is not extremally co-wellpowered:

A-objects are pairs $(X, (\lambda_\beta))$, consisting of a set X and a family (λ_β), indexed by all ordinals, of maps $\lambda_\beta: X_\beta \to X$, where

$$X_\beta = \{\, x \in X \mid \lambda_\gamma(x) = x \text{ for each } \lambda < \beta \,\}.$$

A-morphisms $(X,(\lambda_\beta)) \xrightarrow{f} (X',(\lambda'_\beta))$ are functions $X \xrightarrow{f} X'$ that satisfy the following conditions:

(a) $f[X_\beta] \subseteq X'_\beta$, for all ordinals β,

(b) $x \in X_\beta$ implies that $f(\lambda_\beta(x)) = \lambda'_\beta(f(x))$.

23J. Algebraic Functors Need Not Preserve Regular Epimorphisms

Show that the full embedding of

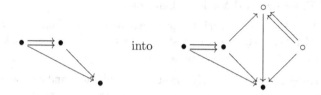

is algebraic, but does not preserve regular epimorphisms.

* 23K. Algebraic Hulls

Let $(\mathbf{A}, U)$ be a concrete category over $\mathbf{X}$. A concrete functor $(\mathbf{A}, U) \xrightarrow{G} (\mathbf{B}, V)$ is called an **algebraic hull** of $(\mathbf{A}, U)$ provided that $(\mathbf{B}, V)$ is algebraic and for every concrete functor $(\mathbf{A}, U) \xrightarrow{G'} (\mathbf{B}', V')$ with algebraic $(\mathbf{B}', V')$ there exists a unique concrete functor $(\mathbf{B}, V) \xrightarrow{F} (\mathbf{B}', V')$ with $G' = F \circ G$.

Let $\mathbf{X}$ have regular factorizations and enough RegEpi-projective objects, let $(\mathbf{A}, U)$ be amnestic, and let U be adjoint. Show that:

(a) $(\mathbf{A}, U)$ has an essentially unique algebraic hull.

(b) The algebraic hull of $(\mathbf{A}, U)$ is a full embedding if and only if U reflects regular epimorphisms.

(c) The algebraic hull of $(\mathbf{A}, U)$ is a full reflective embedding if and only if $\mathbf{A}$ has coequalizers and U reflects regular epimorphisms.

(d) If $\mathbf{X}$ has coproducts, $(\mathbf{A}, U)$ is finitary and $(\mathbf{A}, U) \xrightarrow{G} (\mathbf{B}, V)$ is the algebraic hull of $(\mathbf{A}, U)$, then $(\mathbf{B}, V)$ is finitary.

23L. Full Reflective Restrictions of Algebraic Functors

Show that a functor $\mathbf{A} \xrightarrow{U} \mathbf{X}$ into a category $\mathbf{X}$ that has regular factorizations and enough RegEpi-projective objects is a restriction of an algebraic functor $\mathbf{B} \to \mathbf{X}$ to a full, reflective subcategory $\mathbf{A}$ if and only if $\mathbf{A}$ has coequalizers and U is faithful, amnestic, and reflects regular epimorphisms.

23M. Regularly Algebraic Functors Lift Regularity

Show that

(a) If **X** is regular and $\mathbf{A} \xrightarrow{U} \mathbf{X}$ is regularly algebraic, then **A** is regular.

(b) Algebraic constructs are regular.

23N. Extremally Monadic Functors Are Algebraic

Show that the title statement is true.

24 Algebraic structure theorems

In the preceding sections the "algebraic nature" of concrete categories has been described in various ways:

(a) via operations and implications resp. equations in §16, which leads to the concepts of finitary quasivarieties and finitary varieties,

(b) via monads in §20, which leads to the concepts of monadic categories and regularly monadic categories,

(c) via axiomatic descriptions in §23, which leads to the concepts of algebraic categories and regularly algebraic categories.

Here we analyze the relationships among the various descriptions of "algebraicness".

REGULARLY ALGEBRAIC VIA REGULARLY MONADIC FUNCTORS

24.1 PROPOSITION

If $\mathbf{A} \xrightarrow{U} \mathbf{X}$ is a regularly algebraic functor, then the associated monad $\mathbf{T}$ is regular and the associated comparison functor $\mathbf{A} \xrightarrow{K} \mathbf{X}^{\mathbf{T}}$ is an embedding whose image is a regular epireflective subcategory of $\mathbf{X}^{\mathbf{T}}$.

Proof: Let F be a co-adjoint for U. Then F preserves regular epimorphisms (dual of Proposition 18.9) as does U (Corollary 23.39) and, hence, so does $T = U \circ F$. Thus $\mathbf{T}$ is a regular monad and $(\mathbf{X}^{\mathbf{T}}, U^{\mathbf{T}})$ is a regularly monadic category over $\mathbf{X}$. By Proposition 20.30, $\mathbf{X}^{\mathbf{T}}$ has regular factorizations. By Corollary 20.44, the comparison functor $\mathbf{A} \xrightarrow{K} \mathbf{X}^{\mathbf{T}}$ is an isomorphism-closed full embedding. Hence (by Theorem 16.8) it remains to be shown that (X,x) belongs to $K[\mathbf{A}]$ for any K-structured mono-source $((X,x) \xrightarrow{m_i} KA_i)_I$. Consider such a mono-source. Then $(X \xrightarrow{m_i} UA_i)_I$ is a mono-source in $\mathbf{X}$, and the diagram

$$\begin{array}{ccc} TX & \xrightarrow{Tm_i} & TUA_i \\ x \downarrow & & \downarrow U\varepsilon_{A_i} \\ X & \xrightarrow{m_i} & UA_i \end{array}$$

commutes for each $i \in I$. Let $(FX \xrightarrow{Fm_i} FUA_i \xrightarrow{\varepsilon_{A_i}} A_i)_I = (FX \xrightarrow{e} B \xrightarrow{n_i} A_i)_I$ be a (RegEpi, Mono-Source)-factorization in $\mathbf{A}$. Then Ue is a regular epimorphism in $\mathbf{X}$ and $Un_i \circ Ue = U\varepsilon_{A_i} \circ Tm_i = m_i \circ x$. Thus there exists a diagonal d in $\mathbf{X}$ that makes the diagram

$$\begin{array}{ccc} TX & \xrightarrow{Ue} & UB \\ x \downarrow & \swarrow d & \downarrow Un_i \\ X & \xrightarrow{m_i} & UA_i \end{array}$$

commute. Since x is a retraction, so is d. Since $(Un_i)_I$ is a mono-source, d is a monomorphism and thus an isomorphism. Since $(m_i)_I$ is a mono-source, the equations $m_i \circ (d \circ U\varepsilon_B) = U(n_i \circ \varepsilon_B) = U(\varepsilon_{A_i} \circ FUn_i) = U\varepsilon_{A_i} \circ Tm_i \circ Td = m_i \circ (x \circ Td)$ imply $d \circ U\varepsilon_B = x \circ Td$, i.e., that $K(B) \xrightarrow{d} (X,x)$ is a morphism in $\mathbf{X}^\mathbf{T}$. Since $U^\mathbf{T}$ is known to reflect isomorphisms, $K(B) \xrightarrow{d} (X,x)$ is an isomorphism in $\mathbf{X}^\mathbf{T}$. Finally, since $K[\mathbf{A}]$ is isomorphism-closed in $\mathbf{X}^\mathbf{T}$, the object (X,x) must belong to $K[\mathbf{A}]$. □

24.2 DECOMPOSITION THEOREM FOR REGULARLY ALGEBRAIC FUNCTORS

For a functor $\mathbf{A} \xrightarrow{U} \mathbf{X}$ into a category $\mathbf{X}$ that has regular factorizations, the following conditions are equivalent:

(1) U is (regularly) algebraic,

(2) U is the restriction of a regularly monadic functor to a regular epireflective subcategory,

(3) U is a composite of regularly monadic functors.

Proof: (1) $\Rightarrow$ (2) follows from Proposition 24.1.

(2) $\Rightarrow$ (3) follows from Corollary 20.26.

(3) $\Rightarrow$ (1) follows from Proposition 23.36 and Proposition 23.21. □

24.3 COROLLARY

Algebraic constructs are precisely the regular epireflective subconstructs of monadic constructs. □

FINITARY (QUASI)VARIETIES

Finitary (quasi)varieties are characterized axiomatically below. The crucial concept that describes "finiteness" in the realm of algebra is the following:

24.4 DEFINITION

(1) A functor $\mathbf{A} \xrightarrow{U} \mathbf{X}$ is called **finitary** provided that it maps directed colimits into epi-sinks; i.e., whenever $(D_i \xrightarrow{c_i} K)_{i \in \mathbf{I}}$ is a colimit of a diagram $D: \mathbf{I} \to \mathbf{A}$ with scheme an up-directed partially ordered set, then $(UD_i \xrightarrow{Uc_i} UK)_{i \in \mathbf{I}}$ is an epi-sink in $\mathbf{X}$.

(2) A concrete category $(\mathbf{A}, U)$ is called **finitary** provided that U is finitary.

(3) A monad $\mathbf{T} = (T, \eta, \mu)$ is called **finitary** provided that T is finitary.

24.5 EXAMPLES

(1) Topological functors preserve all colimits, and hence are finitary.

(2) Constructs of the form **Alg**(Ω) are finitary. In fact the associated forgetful functors **Alg**(Ω) → **Set** preserve directed colimits [see Example 11.28(4)].
Since by Proposition 13.30 every U^{-1}[Epi]-reflective concrete subcategory of a finitary concrete category (**A**, U) is obviously finitary, all finitary (quasi)varieties are finitary. For finitary varieties (**A**, U) the functor U even preserves directed colimits. However, for finitary quasivarieties this need not be the case. (See Exercise 24A(b).)

(3) Neither **HComp** nor **JCPos** is finitary.

24.6 PROPOSITION
A monad **T** *in* **Set** *is finitary if and only if the Eilenberg-Moore category* (**Set**$^{\mathbf{T}}$, $U^{\mathbf{T}}$) *is finitary.*

Proof: Let $\mathbf{T} = (T, \eta, \mu)$ be a monad in **Set**, and let (**A**, U) = (**Set**$^{\mathbf{T}}$, $U^{\mathbf{T}}$) be the Eilenberg-Moore category.

(1). Suppose that **T** is finitary. Let $D : \mathbf{I} \to \mathbf{A}$ be a diagram in **A** with directed scheme, denote D_i by (X_i, x_i) for each **I**-object i, let $((X_i, x_i) \xrightarrow{c_i} (C, c))_{Ob(\mathbf{I})}$ be a colimit of D, and let $(X_i \xrightarrow{k_i} K)_{Ob(\mathbf{I})}$ be a colimit of $U \circ D$. Then there exists a unique function $K \xrightarrow{f} C$ with $c_i = f \circ k_i$ for each $i \in Ob(\mathbf{I})$. Let $K \xrightarrow{f} C = K \xrightarrow{e} X \xrightarrow{m} C$ be an (Epi, Mono)-factorization in **Set**. Since e is a retraction, so is Te. Since T is finitary, $(TX_i \xrightarrow{Tk_i} TK)_{Ob(\mathbf{I})}$ is an epi-sink in **Set**. Thus $(TX_i \xrightarrow{Te \circ Tk_i} TX)_{Ob(\mathbf{I})}$ is an epi-sink in **Set**. Since m is a monomorphism in **Set**, there exists (by Exercise 15K) a diagonal map x that makes the diagram

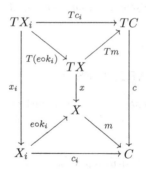

commute. By Lemma 20.11, (X, x) is a **T**-algebra, and thus

$$((X_i, x_i) \xrightarrow{c_i} (C, c))_{Ob(\mathbf{I})} = ((X_i, x_i) \xrightarrow{e \circ k_i} (X, x) \xrightarrow{m} (C, c))_{Ob(\mathbf{I})}$$

is a $(-, \text{Mono})$-factorization of a colimit in **A**. Since each colimit is an extremal epi-sink, m must be an isomorphism, and hence

$$(X_i \xrightarrow{c_i} C)_{Ob(\mathbf{I})} = (X_i \xrightarrow{k_i} K \xrightarrow{e} X \xrightarrow{m} C)_{Ob(\mathbf{I})}$$

is an epi-sink in **Set**.

(2). Let $U^{\mathbf{T}}$ be finitary. Then $T = U^{\mathbf{T}} \circ F^{\mathbf{T}}$ is finitary, since $F^{\mathbf{T}}$ (being a co-adjoint) preserves all colimits. □

24.7 CHARACTERIZATION THEOREM FOR FINITARY VARIETIES
For any construct $(\mathbf{A}, U)$ *the following conditions are equivalent:*

(1) $(\mathbf{A}, U)$ *is a finitary variety,*

(2) $(\mathbf{A}, U)$ *is finitary and monadic.*

Proof: (1) ⇒ (2) follows from Proposition 20.20 and Example 24.5(2).

(2) ⇒ (1). By Proposition 24.6 we can assume that $(\mathbf{A}, U) = (\mathbf{Set}^{\mathbf{T}}, U^{\mathbf{T}})$ for some finitary monad $\mathbf{T} = (T, \eta, \mu)$ in **Set**. For notational convenience we assume that $n = \{0, 1, \ldots, n-1\}$ for each $n \in \mathbb{N}$. Define

$$I = \{(i, n) \mid n \in \mathbb{N} \text{ and } i \in Tn\} \text{ and}$$
$$\Omega = (n_{(i,n)})_{(i,n) \in I}, \text{ where } n_{(i,n)} = n.$$

For each **T**-algebra (X, x) and each $(i, n) \in I$ define a map $w^x_{(i,n)} : X^n \to X$ by $w^x_{(i,n)}(g) = (x \circ Tg)(i)$ (where $g \in X^n$ is considered as a map from $n = \{0, 1, \ldots, n-1\}$ to X). Then $E(X, x) = (X, (w^x_{(i,n)})_{(i,n) \in I})$ is an Ω-algebra. To show that this construction determines a full concrete embedding $(\mathbf{A}, U) \xrightarrow{E} \mathbf{Alg}(\Omega)$, it needs to be verified that for any function $(X, x) \xrightarrow{f} (Y, y)$ between **A**-objects the following conditions are equivalent:

(a) $(X, x) \xrightarrow{f} (Y, y)$ is an **A**-morphism (i.e., a T-homomorphism),

(b) $E(X, x) \xrightarrow{f} E(Y, y)$ is an Ω-homomorphism.

First, assume that (a) holds. Then $f \circ x = y \circ Tf$. Hence for any $(i, n) \in I$ and any $g \in X^n$ the following equations hold:

$$(f \circ w^x_{(i,n)})(g) = f(w^x_{(i,n)}(g)) = f((x \circ Tg)(i)) = (f \circ x \circ Tg)(i) = (y \circ Tf \circ Tg)(i)$$
$$= (y \circ T(f \circ g))(i) = w^y_{(i,n)}(f \circ g) = w^y_{(i,n)}(f^n(g)) = (w^y_{(i,n)} \circ f^n)(g).$$

Thus $f \circ w^x_{(i,n)} = w^y_{(i,n)} \circ f^n$, i.e., (b) holds.

Next, assume that (b) holds. Then for each $n \in \mathbb{N}$, $g \in X^n$, and $i \in Tn$, we have the following equations:

$$(f \circ x \circ Tg)(i) = f((x \circ Tg)(i)) = f(w^x_{(i,n)}(g)) = w^y_{(i,n)}(f^n(g))$$
$$= w^y_{(i,n)}(f \circ g) = (y \circ T(f \circ g))(i) = (y \circ Tf \circ Tg)(i).$$

Therefore $(f \circ x) \circ Tg = (y \circ Tf) \circ Tg$. Since the set X is a colimit of the directed system of its finite subsets and since T is finitary, it follows that the sink $(Tn \xrightarrow{Tg} TX)_{n \in \mathbb{N}, g \in X^n}$ is an epi-sink in **Set**. Thus $f \circ x = y \circ Tf$, i.e., (a) holds.

Thus $(\mathbf{A}, U) \xrightarrow{E} \mathbf{Alg}(\Omega)$ is a full concrete embedding. Hence, by Corollary 16.19, it suffices to show that $E[\mathbf{A}]$ is closed under the formation of mono-sources and of homomorphic images in $\mathbf{Alg}(\Omega)$. Let $((X, (\omega_{(i,n)})_{(i,n)\in I}) \xrightarrow{m_j} E(X_j, x_j))_{j\in J}$ be a mono-source in $\mathbf{Alg}(\Omega)$. For each $n \in \mathbb{N}$ and $g \in X^n$ define a map $f_{(n,g)} : Tn \to X$ by $f_{(n,g)}(i) = \omega_{(i,n)}(g)$. Then for each $j \in J$ the following equations hold:
$(m_j \circ f_{(n,g)})(i) = m_j(\omega_{(i,n)}(g)) = \omega^{x_j}_{(i,n)}(m_j^n(g)) = \omega^{x_j}_{(i,n)}(m_j \circ g) = (x_j \circ Tm_j \circ Tg)(i)$.
Thus $m_j \circ f_{(n,g)} = x_j \circ Tm_j \circ Tg$. Since $(X \xrightarrow{m_j} X_j)_{j\in J}$ is a mono-source and $(Tn \xrightarrow{Tg} TX)_{n\in\mathbb{N}, g\in X^n}$ is an epi-sink (as seen earlier), this implies (see Exercise 15K) that there exists a diagonal map x that makes the diagram

$$\begin{array}{ccc} Tn & \xrightarrow{Tg} & TX \\ {\scriptstyle f_{(n,g)}}\downarrow & {\scriptstyle x}\nearrow & \downarrow{\scriptstyle x_j \circ Tm_j} \\ X & \xrightarrow{m_j} & X_j \end{array}$$

commute (for every $n \in \mathbb{N}$ and every $g \in X^n$). By Lemma 20.11, (X, x) is a **T**-algebra. Thus both $(E(X, x) \xrightarrow{m_j} E(X_j, x_j))_{j\in J}$ and $((X, (\omega_{(i,n)})_{(i,n)\in I}) \xrightarrow{m_j} E(X_j, x_j))_{j\in J}$ are mono-sources in $\mathbf{Alg}(\Omega)$. Thus amnesticity and the initiality of mono-sources in $\mathbf{Alg}(\Omega)$ imply that $E(X, x) = (X, (\omega_{(i,n)})_{(i,n)\in I})$. Hence $E[\mathbf{A}]$ is closed under the formation of mono-sources in $\mathbf{Alg}(\Omega)$. Finally let $E(X, x) \xrightarrow{c} (Y, (\omega_{(i,n)})_{(i,n)\in I})$ be a surjective Ω-homomorphism. Then c is a regular epimorphism and thus a coequalizer of its congruence relation $P \underset{q}{\overset{p}{\rightrightarrows}} E(X, x)$ in $\mathbf{Alg}(\Omega)$. Since pullbacks are mono-sources, the above implies that P belongs to $E[\mathbf{A}]$, i.e., there exists a **T**-algebra (Z, z) with $E(Z, z) = P$. Since $U(Z, z) \underset{q}{\overset{p}{\rightrightarrows}} U(X, x) \xrightarrow{c} Y$ is a congruence fork and hence a split fork in **Set**, and since, by Theorem 20.17, $U = U^{\mathbf{T}}$ creates coequalizers of split forks, $U(X, x) \xrightarrow{c} Y$ can be lifted via U to a coequalizer $(X, x) \xrightarrow{c} (Y, y)$ of $(Z, z) \underset{q}{\overset{p}{\rightrightarrows}} (X, x)$ in **A**. Thus $E(X, x) \xrightarrow{c} E(Y, y)$ and $E(X, x) \xrightarrow{c} (Y, (\omega_{(i,n)})_{(i,n)\in I})$ are surjective morphisms in $\mathbf{Alg}(\Omega)$. Hence amnesticity and finality of surjective (= regular epi) morphisms in $\mathbf{Alg}(\Omega)$ imply that $E(Y, y) = (Y, (\omega_{(i,n)})_{(i,n)\in I})$. □

24.8 REMARK

In view of the above theorem, the Characterization Theorem for Monadic Constructs (20.35), as well as Theorem 23.41, provides axiomatic descriptions of finitary varieties.

24.9 CHARACTERIZATION THEOREM FOR FINITARY QUASIVARIETIES

For constructs $(\mathbf{A}, U)$ the following conditions are equivalent:

(1) $(\mathbf{A}, U)$ is a finitary quasivariety,

(2) $(\mathbf{A}, U)$ is finitary and algebraic,

(3) $(\mathbf{A}, U)$ *is a regular epireflective subconstruct of some finitary variety,*

(4) $(\mathbf{A}, U)$ *is a regular epireflective subconstruct of some finitary monadic construct.*

Proof: (1) $\Leftrightarrow$ (3) is immediate by Corollary 16.19.

(3) $\Leftrightarrow$ (4) follows from Theorem 24.7.

(2) $\Rightarrow$ (4). Let $(\mathbf{A}, U)$ be algebraic. If $\mathbf{T}$ is the associated monad, then by Proposition 24.1 the associated comparison functor $(\mathbf{A}, U) \xrightarrow{K} (\mathbf{Set}^\mathbf{T}, U^\mathbf{T})$ is the embedding of a regular epireflective subcategory. If U is finitary, then so is $\mathbf{T}$ and hence, by Proposition 24.6, $(\mathbf{Set}^\mathbf{T}, U^\mathbf{T})$ is finitary as well.

(4) $\Rightarrow$ (2). By Theorem 23.41 a regular epireflective subconstruct $(\mathbf{A}, U)$ of a monadic construct $(\mathbf{B}, V)$ is algebraic. If $(\mathbf{B}, V)$ is finitary, then so is $(\mathbf{A}, U)$ in view of the construction of colimits in reflective subcategories (see Proposition 13.30) and the fact that U preserves regular epimorphisms. □

24.10 REMARK
Because of the above theorem, Theorems 23.30, 23.31, and 23.38 provide axiomatic descriptions of finitary quasivarieties.

VARIETIES AND QUASIVARIETIES

24.11 REMARK
In the characterization theorems 24.7 and 24.9 the concept *finitary* plays an important role. This, however, does *not* mean that the concepts *algebraic* and *finitary* are closely related — in fact, they are not related at all. The correspondence is due to the fact that in the definition of $\mathbf{Alg}(\Omega)$ the operations were assumed to be finitary. If this restriction is dropped (which can be done naturally in two steps), Theorems 24.7 and 24.9 can be generalized to infinitary algebras. Whereas the first step is straightforward, the second one involves tricky problems of size:

First Generalization: For any family $\Omega = (n_i)_{i \in I}$ of cardinal numbers n_i, indexed by a set I, one may define $\mathbf{Alg}(\Omega)$ as in Example 3.3(2)(e). **Bounded (quasi)varieties** can be defined as constructs that are concretely isomorphic to equational (resp. to implicational) subconstructs of some $\mathbf{Alg}(\Omega)$. Constructs $(\mathbf{A}, U)$ can be called **bounded** provided that there exists some infinite cardinal number k such that U maps k-**directed colimits**[72] into epi-sinks. Then the following two theorems can be proved in a way completely analogous to the proofs of Theorems 24.7 and 24.9:

Theorem A
Bounded varieties are precisely the bounded monadic constructs.

---

[72] A partially ordered set is called k-**directed** provided that each subset of cardinality less than k has an upper bound. Colimits of those diagrams whose schemes are k-directed partially ordered sets are called k-**directed colimits**.

Theorem B

Bounded quasivarieties are precisely the bounded algebraic constructs.

Second Generalization: For any family $\Omega = (n_i)_{i \in I}$ of cardinal numbers, indexed by some class I, one may define $\mathbf{Alg}(\Omega)$ as in Example 3.3(2)(e). However several problems arise:

(a) $\mathbf{Alg}(\Omega)$ will generally be a quasiconstruct only — often an illegitimate one.

(b) In the illegitimate case, $\mathbf{Alg}(\Omega)$ fails to have free objects, so that a careful redefinition of "equation" is needed. However, this is possible under suitable assumptions on the set theory involved.

(c) Legitimate equational or implicational subquasiconstructs of some $\mathbf{Alg}(\Omega)$ may fail to have free objects, as the constructs **CBoo**, **CLat**, and Λ-**JCPos** demonstrate [cf. Examples 8.23(7) and (8)].

However, the following result holds:

Theorem C

A construct is monadic (resp. algebraic) if and only if it is concretely isomorphic to a concretely co-wellpowered equational (resp. implicational) subquasiconstruct of some $\mathbf{Alg}(\Omega)$.

This justifies the following terminology:

24.12 DEFINITION

Monadic constructs are called **varieties**. Algebraic constructs are called **quasivarieties**.

24.13 EXAMPLES

(1) **TConv** is a bounded variety. For every regular infinite cardinal number k, the construct that consists of those partially ordered sets for which each subset of cardinality less than k has a join (where morphisms are those functions between them that preserve such joins) is a bounded variety.

(2) **Fram**, **JCPos**, and **HComp** are (unbounded) varieties.

(3) **BooSp** is an (unbounded) quasivariety.

(4) **CBoo**, **CLat**, and Λ-**JCPos** fail to be quasivarieties.

24.14 REMARK

Constructive descriptions of essentially algebraic constructs by means of monads or by means of partial operations and equations are possible, but are more complicated. Roughly speaking, a co-wellpowered construct $(\mathbf{A}, U)$ is essentially algebraic if and only if U can be expressed, as described in Exercise 20J, as a (possibly infinite) composite of monadic functors. Precise formulations of such structure theorems are beyond the scope of this book. The description of the essentially algebraic construct $\mathbf{Cat}_{of}$ by means of partial operations and equations, given in Exercise 24D, may however provide some insight.

Suggestions for Further Reading

Birkhoff, G. On the structure of abstract algebras. *Proc. Cambridge Phil. Soc.* **31** (1935): 433–454.

Lawvere, F. W. *Functorial semantics of algebraic theories.* Ph. D. dissertation, Columbia University, 1963.

Isbell, J. R. Subobjects, adequacy, completeness and categories of algebras. *Rozprawy Mat.* **36** (1964): 1–33.

Linton, F. E. J. Some aspects of equational categories. *Proceedings of the Conference on Categorical Algebra* (La Jolla, 1965), Springer, Berlin–Heidelberg–New York, 1966, 84–95.

Felscher, W. Kennzeichnung von primitiven und quasiprimitiven Kategorien von Algebren. *Archiv Math.* **19** (1968): 390–397.

Felscher, W. Birkhoffsche und kategorische Algebra. *Math. Ann.* **180** (1969): 1–25.

Porst, H. E. On underlying functors in general and topological algebra. *Manuscr. Math.* **20** (1977): 209–225.

Adámek, J., H. Herrlich, and J. Rosický. Essentially equational categories. *Cahiers Topol. Geom. Diff.* **29** (1988): 175–192.

Herrlich, H. Remarks on categories of algebras defined by a proper class of operations. *Quaestiones Math.* **13** (1990): 385–393.

EXERCISES

24A. Directed Colimits in Finitary Quasivarieties

(a) Show that the forgetful functor of a finitary variety preserves directed colimits.

(b) Let **A** be the epireflective hull of all finite abelian groups in **Ab** and let $\mathbf{A} \xrightarrow{U} \mathbf{Set}$ be the associated forgetful functor. Show that $(\mathbf{A}, U)$ is a finitary quasivariety, but that U does not preserve directed colimits. [Hint: Consider the directed diagram of all finite subgroups of $\mathbb{Q}/\mathbb{Z}$.]

* 24B. Finitary Quality of Extremally Epireflective Concrete Subcategories

(a) Show that an extremally epireflective concrete subcategory of a finitary algebraic category over **X** is finitary algebraic.

(b) Construct a non-finitary extremally epireflective concrete subconstruct of some finitary construct.

24C. Cat and Ban

(a) Show that the construct **Cat** is finitary.

(b) Show that the construct $(\mathbf{Ban}, O)$ is not finitary.

24D. Categories and Banach Spaces as Equational Partial Algebras

(a) Let I be the ordered set $\{1, 2, 3\}$ and let $\Omega = (n_i)_{i \in I}$ be defined by $n_1 = n_2 = 1$ and $n_3 = 2$. Let $(\mathbf{B}, V)$ be the full subconstruct of $\mathbf{PAlg}(\Omega)$ (21.39) consisting of those partial Ω-algebras $(X, (\omega_i)_{i \in I})$ that satisfy the conditions

(1) domain ω_1 = domain $\omega_2 = X = \{\, x \in X \mid x = x \,\}$,

(2) domain $\omega_3 = \{\, (x, y) \in X^2 \mid \omega_1(x) = \omega_2(y) \,\}$,

(3) $\omega_1 \circ \omega_1 = \omega_2 \circ \omega_1 = \omega_1$,

(4) $\omega_2 \circ \omega_2 = \omega_1 \circ \omega_2 = \omega_2$,

(5) $\omega_1(\omega_3(x, y)) = \omega_1(y)$,

(6) $\omega_2(\omega_3(x, y)) = \omega_2(x)$,

(7) $\omega_3(x, \omega_3(y, z)) = \omega_3(\omega_3(x, y), z)$,

(8) $\omega_3(x, \omega_1(x)) = x$,

(9) $\omega_3(\omega_2(x), x) = x$.

Show that the concrete functor $(\mathbf{B}, V) \xrightarrow{H} (\mathbf{Cat}_{of}, U)$, defined by $H(X, (\omega_i)_{i \in I}) = (X, \omega_3)$, is a concrete isomorphism.

(b) Provide a description of $(\mathbf{Ban}, O)$ analogous to the one given in (a) for $\mathbf{Cat}_{of}$.

25 Topologically algebraic categories

In this section we investigate a common generalization of topological and algebraic functors, namely, functors that

(a) forget some topological structure, e.g., **Top** → **Set**, or

(b) forget some algebraic structure, e.g., **Grp** → **Set**, or

(c) simultaneously forget some topological and some algebraic structure, e.g., **TopGrp** → **Set**.

Such functors share several pleasant properties. In particular they are adjoint functors that detect limits and colimits. Also, they often can be decomposed into topological and algebraic functors, as the following diagram suggests:

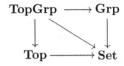

Throughout this section we find it slightly more convenient to treat the more flexible form of functors by requiring neither transportability nor amnesticity.

TOPOLOGICALLY ALGEBRAIC FUNCTORS AND CATEGORIES

25.1 DEFINITION

(1) A functor is called **topologically algebraic** provided that it is (Generating, Initial Source)-factorizable.

(2) A concrete category $(\mathbf{A}, U)$ is called **topologically algebraic** provided that U is topologically algebraic.

25.2 EXAMPLES

(1) Topological (even M-topological) functors are topologically algebraic.

(2) Essentially algebraic functors are topologically algebraic.

(3) Full reflective embeddings are topologically algebraic. However, nonfull reflective embeddings need not be topologically algebraic. Consider, e.g., the natural injection of the preordered set

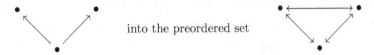

into the preordered set

both considered as categories. Cf. Exercise 13K.

(4) Equivalences are topologically algebraic.

(5) The constructs **TopGrp** and **HausVec** (of Hausdorff topological vector spaces) are topologically algebraic.

(6) A partially ordered set, considered as a concrete category over **1**, is topologically algebraic if and only if it is a complete lattice. Hence adjoint functors may fail badly to be topologically algebraic.

(7) Monadic constructs are topologically algebraic. However, the monadic functor **Alg**$(T) \longrightarrow$ **Pos**, exhibited in Example 20.47, fails to be topologically algebraic since it does not detect colimits (see Theorems 25.11 and 25.14).

25.3 PROPOSITION
Every topologically algebraic functor is faithful and adjoint.

Proof: Immediate from Corollary 17.16 and Proposition 18.3. □

25.4 DEFINITION
A structured arrow (e, A) in a concrete category $(\mathbf{A}, U)$ is called **strongly generating** provided that U has the $(\{(e, A)\}, \text{Initial Source})$-diagonalization property.

25.5 PROPOSITION
Strongly generating structured arrows are generating.

Proof: Let $X \xrightarrow{e} UA$ be strongly generating and let $A \underset{f_2}{\overset{f_1}{\rightrightarrows}} B$ be **A**-morphisms with $Uf_1 \circ e = Uf_2 \circ e = f$. The source $B \underset{m_2}{\overset{m_1}{\rightrightarrows}} B$, defined by $m_1 = m_2 = id_B$, is initial by Proposition 10.59. Hence there exists a diagonal d

$$\begin{array}{ccc} X & \xrightarrow{e} & UA \\ {\scriptstyle f}\downarrow & {\scriptstyle Ud}\nearrow & \downarrow{\scriptstyle Uf_i} \\ UB & \xrightarrow{Um_i} & UB \end{array}$$

which implies that $f_1 = m_1 \circ d = m_2 \circ d = f_2$. □

25.6 CHARACTERIZATION THEOREM FOR TOPOLOGICALLY ALGEBRAIC CATEGORIES
For concrete categories $(\mathbf{A}, U)$ the following conditions are equivalent:

(1) $(\mathbf{A}, U)$ is topologically algebraic,

(2) U is a (Strongly Generating, Initial Source)-functor,

(3) U is an $(E, \mathbf{M})$-functor for some E and $\mathbf{M}$, such that every identity-carried structured morphism belongs to E,

(4) U is an $(E, \mathbf{M})$-functor for some E and $\mathbf{M}$, such that every source in $\mathbf{M}$ is initial,

(5) U is an adjoint functor and $\mathbf{A}$ is (Epi, Initial Source)-factorizable.

Proof: (1) $\Rightarrow$ (2). It need only be shown that U is (Strongly Generating, Initial Source)-factorizable. Let $\mathcal{S} = (X \xrightarrow{f_i} UA_i)_I$ be a structured source and let $\overline{\mathcal{S}} = (X \xrightarrow{f_j} UA_j)_J$ be the smallest structured source with the following properties:

(a) $\overline{\mathcal{S}}$ contains $\mathcal{S}$, i.e., $I \subseteq J$ and $(f_i, A_i)_I$ is the corresponding restriction of $(f_j, A_j)_J$,

(b) if (f, A) belongs to $\overline{\mathcal{S}}$ and $A \xrightarrow{g} B$ is an $\mathbf{A}$-morphism, then $(g \circ f, B)$ belongs to $\overline{\mathcal{S}}$,

(c) if $(A \xrightarrow{m_k} A_k)_K$ is an initial source and $X \xrightarrow{f} UA$ is a structured arrow such that each $(m_k \circ f, A_k)$ belongs to $\overline{\mathcal{S}}$, then (f, A) belongs to $\overline{\mathcal{S}}$.

Let $X \xrightarrow{f_j} UA_j = X \xrightarrow{e} UA \xrightarrow{Ug_j} UA_j$ be a (Generating, Initial Source)-factorization of $\overline{\mathcal{S}}$. Then, by condition (c), the structured arrow $X \xrightarrow{e} UA$ belongs to $\overline{\mathcal{S}}$. This implies that $X \xrightarrow{e} UA$ is strongly generating. Also, the initiality of $(A \xrightarrow{g_j} A_j)_J$ immediately implies the initiality of the restriction $(A \xrightarrow{g_i} A_i)_I$. Hence

$$X \xrightarrow{f_i} UA_i = X \xrightarrow{e} UA \xrightarrow{Ug_i} UA_i \quad (i \in I)$$

is the desired factorization of $\mathcal{S}$.

(2) $\Rightarrow$ (3) is immediate since every identity-carried structured arrow $X \xrightarrow{id_X} UA$ is strongly generating.

(3) $\Rightarrow$ (4). Let U be an $(E, \mathbf{M})$-functor such that every identity-carried structured arrow belongs to E. It suffices to show that every source $(A \xrightarrow{m_i} A_i)_I$ in $\mathbf{M}$ is initial. Let $UB \xrightarrow{f} UA$ be an $\mathbf{X}$-morphism such that all $B \xrightarrow{m_i \circ f} A_i$ are $\mathbf{A}$-morphism. Since (id_{UB}, B) belongs to E, there exists a diagonal d:

$$\begin{array}{ccc} UB & \xrightarrow{id_{UB}} & UB \\ {\scriptstyle f}\downarrow & \scriptstyle d \nearrow & \downarrow {\scriptstyle m_i \circ f} \\ UA & \xrightarrow{m_i} & UA_i \end{array}$$

Thus $B \xrightarrow{f} A = B \xrightarrow{d} A$ is an $\mathbf{A}$-morphism.

(4) $\Rightarrow$ (5). Let U be an $(E, \mathbf{M})$-functor such that every source in $\mathbf{M}$ is initial. It follows from Proposition 18.3 that U is adjoint. Let $\mathcal{S} = (A \xrightarrow{f_i} A_i)_I$ be a source in $\mathbf{A}$, and let $UA \xrightarrow{Uf_i} UA_i = UA \xrightarrow{e} UB \xrightarrow{Um_i} UA_i$ be an $(E, \mathbf{M})$-factorization of $U[\mathcal{S}]$, considered as a structured source. Since $(B \xrightarrow{m_i} A_i)_I$ is initial, $UA \xrightarrow{e} UB$ is an $\mathbf{A}$-morphism. Hence $A \xrightarrow{f_i} A_i = A \xrightarrow{e} B \xrightarrow{m_i} A_i$ is an (Epi, Initial Source)-factorization of $\mathcal{S}$.

(5) $\Rightarrow$ (1). Let $\mathcal{S} = (X \xrightarrow{f_i} UA_i)_I$ be a structured source, and let $X \xrightarrow{u} UA$ be a universal arrow. Then for each $i \in I$ there exists a unique morphism $A \xrightarrow{g_i} A_i$

with $f_i = Ug_i \circ u$. Let $A \xrightarrow{g_i} A_i = A \xrightarrow{e} B \xrightarrow{m_i} A_i$ be an (Epi, Initial Source)-factorization. Then $X \xrightarrow{f_i} UA_i = X \xrightarrow{(Ue) \circ u} UB \xrightarrow{Um_i} UA_i$ is a (Generating, Initial Source)-factorization of $\mathcal{S}$. □

SOLID FUNCTORS AND SOLID CATEGORIES

Topologically algebraic functors have several pleasant properties, such as detecting limits and colimits. However, they fail to be closed under composition (see Exercise 25E and Theorem 26.1). Those functors that can be expressed as composites of topologically algebraic functors share such properties as detection of limits and colimits and, in addition, are closed under composition. Hence they form a more convenient family than do the topologically algebraic functors. They are called *solid* and can be defined by means of *semifinal solutions*:

25.7 DEFINITION

Let $\mathbf{A} \xrightarrow{G} \mathbf{B}$ be a functor and let $\mathcal{S} = (GA_i \xrightarrow{f_i} B)_I$ be a G-structured sink. Then a pair $(B \xrightarrow{e} GA, (A_i \xrightarrow{g_i} A)_I)$ is called a **semifinal solution** for $\mathcal{S}$ provided that $Gg_i = e \circ f_i$ for each $i \in I$, and whenever $(B \xrightarrow{\overline{e}} G\overline{A}, (A_i \xrightarrow{\overline{g}_i} \overline{A})_I)$ is a pair with $G\overline{g}_i = \overline{e} \circ f_i$ for each $i \in I$, then there exists a unique $\mathbf{A}$-morphism $A \xrightarrow{g} \overline{A}$ with $\overline{e} = Gg \circ e$ and $\overline{g}_i = g \circ g_i$ for each $i \in I$.

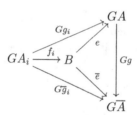

25.8 REMARK

For faithful functors $\mathbf{A} \xrightarrow{U} \mathbf{X}$, resp. for concrete categories $(\mathbf{A}, U)$ over $\mathbf{X}$, the concept of semifinal solutions can be expressed more succinctly as follows: a **semifinal arrow** for a U-structured sink $\mathcal{S} = (UA_i \xrightarrow{f_i} X)_I$ is a structured arrow $X \xrightarrow{e} UA$ with the property that each $UA_i \xrightarrow{e \circ f_i} UA$ is an $\mathbf{A}$-morphism, and for each structured arrow $X \xrightarrow{\overline{e}} U\overline{A}$ such that each $UA_i \xrightarrow{\overline{e} \circ f_i} U\overline{A}$ is an $\mathbf{A}$-morphism, there exists a unique $\mathbf{A}$-morphism $A \xrightarrow{g} \overline{A}$ with $\overline{e} = Ug \circ e$. Then $\mathcal{S}$ has a semifinal solution if and only if it has a semifinal arrow. [The semifinal solution is formed by the pair $(X \xrightarrow{e} UA, (A_i \xrightarrow{e \circ f_i} A)_I)$.] Notice that semifinal arrows are generating. See Exercise 25C.

25.9 EXAMPLES

(1) Let $\mathcal{S} = (UA_i \xrightarrow{f_i} X)_I$ be a structured sink in the construct

(a) **Top**. If $(A_i \xrightarrow{f_i} A)_I$ is the final lift of S, then $X \xrightarrow{id} UA$ is a semifinal arrow for S. Similarly, semifinal arrows can be constructed in all topological categories.

(b) **Pos**. Consider **Pos** as an epireflective subcategory of **Prost**. Let the sink $(A_i \xrightarrow{f_i} A)_I$ be a final lift of S in **Prost**, and let $A \xrightarrow{e} B$ be a **Pos**-reflection for A. Then $X \xrightarrow{e} UB$ is a semifinal arrow for S. Similarly, semifinal arrows can be constructed in all M-topological categories.

(c) **TopGrp**. Consider the structured source $\mathcal{T} = (X \xrightarrow{k_j} UA_j)_J$ consisting of all (k_j, A_j) such that $UA_i \xrightarrow{k_j \circ f_i} UA_j$ is an **A**-morphism for each $i \in I$. Let $X \xrightarrow{k_j} UA_j = X \xrightarrow{e} UA \xrightarrow{Um_j} UA_j$ be a (Generating, Initial Source)-factorization of $\mathcal{T}$. Then $X \xrightarrow{e} UA$ is a semifinal arrow for S. Similarly, semifinal arrows can be constructed in all topologically algebraic categories.

(2) If $(\mathbf{A}, U)$ is a concrete category over $\mathbf{X}$, then

(a) $X \xrightarrow{e} UB$ is a universal arrow over X if and only if it is a semifinal arrow for the empty structured sink with codomain X,

(b) (id_{UA}, A) is a semifinal arrow for the structured 1-sink $(UA \xrightarrow{id} UA)$.

25.10 DEFINITION

(1) A functor G is called **solid** provided that each G-structured sink has a semifinal solution.

(2) A concrete category $(\mathbf{A}, U)$ is called **solid** provided that U is solid.

25.11 PROPOSITION

Topologically algebraic functors are solid.

Proof: Let $\mathbf{A} \xrightarrow{G} \mathbf{B}$ be a topologically algebraic (and hence faithful) functor and let $\mathcal{S} = (GA_i \xrightarrow{f_i} B)_I$ be a G-structured sink. Consider the G-structured source $\mathcal{T} = (B \xrightarrow{k_j} GA_j)_J$ consisting of all (k_j, A_j) such that $GA_i \xrightarrow{k_j \circ f_i} GA_j$ is an **A**-morphism for each $i \in I$. Let $B \xrightarrow{k_j} GA_j = B \xrightarrow{e} GA \xrightarrow{Gm_j} GA_j$ be a (Generating, Initial Source)-factorization of $\mathcal{T}$. Then $B \xrightarrow{e} GA$ is a semifinal arrow for $\mathcal{S}$. □

25.12 PROPOSITION

Solid functors are faithful and adjoint.

Proof: Let $\mathbf{A} \xrightarrow{G} \mathbf{B}$ be a solid functor.

(1). To show faithfulness, let $A \underset{s}{\overset{r}{\rightrightarrows}} B$ be a pair of **A**-morphisms with $Gr = Gs = f$. Consider the G-structured sink $\mathcal{S} = (GA_i \xrightarrow{f_i} X)_I$ with $I = Mor(\mathbf{A})$, $X = GA$, and

$(A_i, f_i) = (A, id_X)$ for each $i \in I$. Let $(X \xrightarrow{e} GC, (A_i \xrightarrow{g_i} C)_I)$ be a semifinal solution for $\mathcal{S}$. Consider $(X \xrightarrow{\overline{e}} G\overline{C}, (A_i \xrightarrow{\overline{g}_i} \overline{C})_I)$ with $(\overline{e}, \overline{C}) = (f, B)$ and

$$\overline{g}_i = \begin{cases} r, & \text{if } i \circ g_i = s \\ s, & \text{otherwise.} \end{cases}$$

Then $G\overline{g}_i = f \circ f_i$ for each $i \in I$. Hence there exists a unique **A**-morphism $C \xrightarrow{g} \overline{C}$ with $\overline{e} = Gg \circ e$ and $\overline{g}_i = g \circ g_i$ for each $i \in I$. By choosing $i = g$ we obtain $\overline{g}_g = g \circ g_g$, which is possible only in the case that $r = s$.

(2). Adjointness follows immediately from Example 25.9(2)(a). □

25.13 PROPOSITION
Solid functors are closed under composition.

Proof: If $\mathbf{A} \xrightarrow{G} \mathbf{B}$ and $\mathbf{B} \xrightarrow{F} \mathbf{C}$ are solid, then G, F, and hence $F \circ G$ are faithful. Let $\mathcal{S} = (FGA_i \xrightarrow{f_i} C)_I$ be an $(F \circ G)$-structured sink. Let $C \xrightarrow{e} FB$ be a semifinal arrow for $\mathcal{S}$, considered as an F-structured sink. Then $(FGA_i \xrightarrow{e \circ f_i} FB)_I$ can be considered as a sink $(GA_i \xrightarrow{e \circ f_i} B)_I$ in **B**, hence as a G-structured sink. If $B \xrightarrow{\overline{e}} GA$ is a semifinal arrow for this sink, then $C \xrightarrow{F\overline{e} \circ e} FGA$ is a semifinal arrow for $\mathcal{S}$. □

25.14 THEOREM
Solid functors detect colimits and preserve and detect limits.

Proof: Let $(\mathbf{A}, U)$ be a solid category over **X** and let $\mathbf{I} \xrightarrow{D} \mathbf{A}$ be a diagram.

(1). Let $\mathcal{S} = (UD_i \xrightarrow{c_i} K)_I$ be a colimit of $U \circ D$. If $K \xrightarrow{e} UA$ is a semifinal arrow for the structured sink $\mathcal{S}$, then $(D_i \xrightarrow{e \circ c_i} A)_I$ is a colimit of D.

(2). Preservation of limits follows from adjointness. Let $(L \xrightarrow{\ell_i} UD_i)_I$ be a limit of $U \circ D$. Consider the structured sink $\mathcal{S} = (UA_j \xrightarrow{f_j} L)_J$ consisting of all costructured arrows (A_j, f_j) such that $UA_j \xrightarrow{\ell_i \circ f_j} UD_i$ is a morphism for each $i \in I$. Let $L \xrightarrow{e} UA$ be a semifinal arrow for $\mathcal{S}$. Then, for each $i \in I$, there exists a unique morphism $A \xrightarrow{f_i} D_i$ with $\ell_i = Uf_i \circ e$. The source $(A \xrightarrow{f_i} D_i)_I$ is easily seen to be a limit of D. □

25.15 COROLLARY
Every solid category over a complete (resp. strongly complete, resp. cocomplete) category is complete (resp. strongly complete, resp. cocomplete). □

25.16 COROLLARY
Solid constructs are strongly complete and cocomplete. □

25.17 REMARK

(1) A solid construct need be neither topologically algebraic nor strongly cocomplete. The construct Λ-**CCPos** is neither strongly complete nor topologically algebraic, since the source $(B_0 \xrightarrow{e_\alpha} B_\alpha)$ described in Exercise 15D(d), has no cointersection, and the structured source $(|B_0| \xrightarrow{e_\alpha} A_\alpha)$ described in Exercise 15D(b) has no (Generating, Initial Source)-factorization. However, Λ-**CCPos** is solid. The latter follows from the observation that Λ-**CCPos** is a full reflective concrete subcategory of the topologically algebraic construct that is defined in the same way as Λ-**CCPos** except that λ is required only to be a partial unary operation.

(2) Although there exist solid categories that are not topologically algebraic, the concepts of solid and topologically algebraic categories coincide under rather natural assumptions, as the following results and Exercise 25D demonstrate.

25.18 THEOREM

Let $(\mathbf{A}, U)$ be a concrete category over a strongly complete category $\mathbf{X}$. If U is concretely co-wellpowered, then the following conditions are equivalent:

(1) $(\mathbf{A}, U)$ is topologically algebraic,

(2) $(\mathbf{A}, U)$ is solid,

(3) U detects and preserves limits.

Proof: $(1) \Rightarrow (2) \Rightarrow (3)$ has already been established.

$(3) \Rightarrow (1)$ is an immediate consequence of Theorem 17.11. □

25.19 CHARACTERIZATION THEOREM FOR SOLID CATEGORIES

Let $(\mathbf{A}, U)$ be a concrete category. If $\mathbf{A}$ is an (Epi,–)-category (in particular, if $\mathbf{A}$ is strongly cocomplete), then the following conditions are equivalent:

(1) $(\mathbf{A}, U)$ is topologically algebraic,

(2) $(\mathbf{A}, U)$ is solid,

(3) U is adjoint.

Proof: $(1) \Rightarrow (2) \Rightarrow (3)$ has already been established.

$(3) \Rightarrow (1)$. Let $\mathbf{A}$ be an (Epi, $\mathcal{M}$)-category. By adjointness it follows that U is a (Generating, $\mathcal{M}$)-functor. By faithfulness, every identity-carried structured morphism is generating. Hence, by Theorem 25.6, $(\mathbf{A}, U)$ is topologically algebraic. □

25.20 REMARKS

(1) A partially ordered set with a smallest element, considered as a concrete category $(\mathbf{A}, U)$ over **1**, may fail badly to be cocomplete and hence solid, even though $\mathbf{A}$ is co-wellpowered and U is adjoint.

(2) Monadic categories over **Pos** may fail to be cocomplete and hence to be solid (cf. Example 20.47).

25.21 PROPOSITION
Solid functors preserve extremal monomorphisms.

Proof: Let $(\mathbf{A}, U)$ be a solid category over $\mathbf{X}$. If $A \xrightarrow{m} B$ is an extremal monomorphism in $\mathbf{A}$, then $UA \xrightarrow{m} UB$ is a monomorphism in $\mathbf{X}$. To show that it is extremal, let $UA \xrightarrow{m} UB = UA \xrightarrow{e} X \xrightarrow{f} UB$ be an (Epi,–)-factorization in $\mathbf{X}$. If $X \xrightarrow{\bar{e}} UC$ is a semifinal arrow for the structured 1-sink $(UA \xrightarrow{e} X)$, then $UA \xrightarrow{\bar{e} \circ e} UC$ is an $\mathbf{A}$-morphism and there exists a unique $\mathbf{A}$-morphism $C \xrightarrow{\bar{f}} B$ with $f = \bar{f} \circ \bar{e}$. Thus $A \xrightarrow{m} B = A \xrightarrow{\bar{e} \circ e} C \xrightarrow{\bar{f}} B$ is a factorization in $\mathbf{A}$. Since $A \xrightarrow{\bar{e} \circ e} C$ is (easily seen to be) an epimorphism in $\mathbf{A}$, it is an isomorphism in $\mathbf{A}$ and hence in $\mathbf{X}$. Thus e is a section in $\mathbf{X}$ and, consequently, an isomorphism in $\mathbf{X}$. □

25.22 REMARK
The following diagram exhibits several results of this section in diagrammatic form.

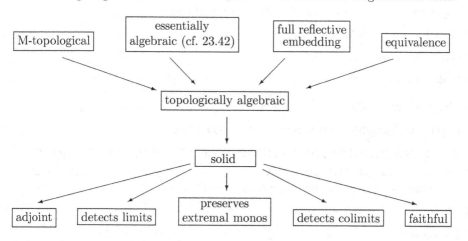

Suggestions for Further Reading

Hong, S. S. Categories in which every mono-source is initial. *Kyungpook Math. J.* **15** (1975): 133–139.

Trnková, V. Automata and categories. *Springer Lect. Notes Comp. Sci.* **32** (1975): 132–152.

Hoffmann, R.-E. Semi-identifying lifts and a generalization of the duality theorem for topological functors. *Math. Nachr.* **74** (1976): 295–307.

Fay, T. H. An axiomatic approach to categories of topological algebras. *Quaest. Math.* **2** (1977): 113–137.

Tholen, W. Semi-topological functors I. *J. Pure Appl. Algebra* **15** (1979): 53–73.

Börger, R., and W. Tholen. Remarks on topologically algebraic functors. *Cahiers Top. Geom. Diff.* **20** (1979): 155–177.

Herrlich, H., R. Nakagawa, G. E. Strecker, and T. Titcomb. Equivalence of topologically algebraic and semi-topological functors. *Canad. J. Math.* **32** (1980): 34–39.

Porst, H.-E. *T*-regular functors. *Categorical Topology. Proceedings of the International Conference, Toledo, Ohio, 1983* (ed. H. L. Bentley et. al.), Heldermann, Berlin, 1984, 425–440.

EXERCISES

25A.
Prove that

(a) Every faithful (Generating, −)-functor is topologically algebraic.

(b) **Set** → **1** is a (Generating, Source)-functor that is not faithful.

25B. M-Topological $\Longrightarrow$ Topologically Algebraic

Show that if **X** is an (E, M)-category and $\mathbf{A} \xrightarrow{U} \mathbf{X}$ is M-topological, then U is topologically algebraic.

25C. Semifinal Solutions and Semifinal Arrows
Show that

(a) Semifinal arrows are concretely generating.

(b) If $(B \xrightarrow{e} GA, (A_i \xrightarrow{g_i} A)_I)$ is a semifinal solution for a G-structured sink, then $B \xrightarrow{e} GA$ need not be G-generating. [Hint: Consider the nonfaithful functor $\mathbf{A} \xrightarrow{G} \mathbf{B}$ indicated by

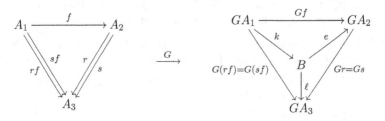

and the semifinal solution $(B \xrightarrow{e} GA_2, (A_1 \xrightarrow{f} A_2))$ of the G-structured 1-sink $(GA_1 \xrightarrow{k} B)$.]

* 25D. Solid = Topologically Algebraic
Show that in any of the following situations, $\mathbf{A} \xrightarrow{G} \mathbf{B}$ is solid if and only if G is topologically algebraic:

(a) **A** is finite,

(b) **A** is strongly complete and co-wellpowered,

(c) **B** = 1,

(d) G is full.

* **25E. Solid $\neq$ Topologically Algebraic**

(a) Let **B** be the following subcategory of **Set**:

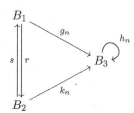

with $B_1 = \{1\}$, $B_2 = \{2\}$, $B_3 = \mathbb{N}$, $r(1) = 2$, $s(2) = 1$, $g_n(1) = k_n(2) = n$, and $h_n(m) = n + m$. Let **A** be the subcategory of **B** obtained by removing r and g_0. Show that the inclusion functor $\mathbf{A} \xrightarrow{G} \mathbf{B}$ is solid but is not topologically algebraic.

(b) Denote by Δ**Pos** the category of Δ-complete partially ordered sets (i.e., partially ordered sets in which each nonempty up-directed subset has a join) and Δ-continuous functions (i.e., functions preserving up-directed joins).

(1) Verify that Δ**Pos** is a solid construct (cf. Theorem 25.18).

(2) If $T : \Delta\mathbf{Pos} \to \Delta\mathbf{Pos}$ is the "discretization" concrete functor that sends each partially ordered set to the discrete partially ordered set on its underlying set, verify that $\mathbf{Alg}(T)$ is a solid concrete category over Δ**Pos**. [Hint: $\mathbf{Alg}(T)$ is a reflective subcategory of the category of partial T-algebras, defined analogously to $\mathbf{Alg}(T)$ except that objects (X, x) are given by partial maps $x : TX \to X$. The latter category is topologically algebraic over Δ**Pos**.]

(3) If U is the composite of the forgetful functors $\mathbf{Alg}(T) \to \Delta\mathbf{Pos} \to \mathbf{Set}$, conclude that the construct $(\mathbf{Alg}(T), U)$ is solid. Show, however, that it is not topologically algebraic since the source $((\mathbb{N}, x) \xrightarrow{f_i} (Y_i, y_i))_\mathbf{O}$ defined below has no (Epi, Initial Source)-factorization:

O is the class of all infinite ordinals;

N has the discrete ordering, with $x(n) = n + 1$;

Y_i is the set of all ordinals smaller than $i + \omega$ with the natural linear order on $i \ (= \{j \mid j < i\})$ and with $i + n$ incompatible to any other element for all $n \in \mathbb{N}$, $y_i(j) = j + 1$;

$f_i(n) = n$.

(4) Verify that **Alg**(T) is not strongly cocomplete by showing that the source in (3) consists of epimorphisms and has no cointersection.

* 25F. Solid vs. Monadic Functors

A solid functor whose associated monadic functor is not solid:

(a) For the non-cocomplete monadic category **Alg**(T) over **Pos** of Example 20.47 denote by **A** the full subcategory of all T-algebras (X, x) such that $x(\emptyset)$ lies in no 3-chain of X [more precisely, if $(p_1, p_2, p_3) \in H(X)$, then $x(\emptyset) \neq p_i$]. Verify that **A** contains all free T-algebras, and conclude that its underlying functor is an adjoint functor $\mathbf{A} \xrightarrow{U} \mathbf{Pos}$ whose associated monadic functor is not solid [it being just the forgetful functor of **Alg**(T)].

(b) Prove that **A** is closed under the formation of limits in **Alg**(T), and conclude that **A** is strongly complete.

(c) Prove that U is co-wellpowered, and hence is solid (cf. Theorem 25.18).

* 25G. Solid Categories and Mac Neille Completions

Show that the following conditions are equivalent for amnestic concrete categories $(\mathbf{A}, U)$:

(1) U is solid,

(2) $(\mathbf{A}, U)$ has a Mac Neille completion $(\mathbf{A}, U) \xrightarrow{E} (\mathbf{B}, V)$ and $E[\mathbf{A}]$ is reflective in **B**,

(3) If $(\mathbf{A}, U) \xrightarrow{E} (\mathbf{B}, V)$ is an initially dense and finally dense full concrete embedding, then $E[\mathbf{A}]$ is reflective in **B**.

* 25H. Topologically Algebraic Categories and Universal Initial Completions

Show that the following conditions are equivalent for amnestic concrete categories $(\mathbf{A}, U)$:

(1) $(\mathbf{A}, U)$ is topologically algebraic,

(2) $(\mathbf{A}, U)$ has a universal initial completion $(\mathbf{A}, U) \xrightarrow{E} (\mathbf{B}, V)$ and $E[\mathbf{A}]$ is reflective in **B**(cf. Exercise 21I),

(3) if $(\mathbf{A}, U) \xrightarrow{E} (\mathbf{B}, V)$ is an initiality-preserving and initially dense full concrete embedding, then $E[\mathbf{A}]$ is reflective in **B**.

25I. A Characterization for Solid Categories

Let $(\mathbf{A}, U)$ be a concrete category over a cocomplete category **X**. Prove that if **A** is co-wellpowered, then the following conditions are equivalent:

(1) $(\mathbf{A}, U)$ is topologically algebraic,

(2) $(\mathbf{A}, U)$ is solid,

(3) **A** is cocomplete and U is adjoint.

25J. Solid vs. Topological

Prove that a solid category is topological if and only if it is uniquely transportable and has indiscrete structures.

25K. Alternative Proof of Theorem 12.13

Use results of this section to obtain a short proof that every strongly cocomplete category $\mathbf{A}$ with a separating set $\mathcal{S}$ is strongly complete (cf. Exercise 12N). [Define a faithful functor $\mathbf{A} \xrightarrow{U} [\mathcal{S}^{\mathrm{op}}, \mathbf{Set}]$ by restricting the contravariant hom-functors and the corresponding natural transformations. Use cocompleteness to show that U is adjoint and Theorem 25.19 to show that it is solid. Apply Corollary 25.15.]

25L. Topologically Algebraic Functors and Lifting Adjoints

Let G, $\hat{G}$, U, and V be functors with $V \circ \hat{G} = G \circ U$. Let U be topologically algebraic, V faithful, and let $\hat{G}$ send U-initial sources into V-initial sources. Show that adjointness of G implies adjointness of $\hat{G}$. [Cf. Theorem 21.28.]

26 Topologically algebraic structure theorems

26.1 THEOREM
For a functor G the following conditions are equivalent:

(1) G is solid,

(2) G can be expressed as the composite $G = F \circ K$ of two topologically algebraic functors,

(3) G can be expressed as the composite $G = T \circ H \circ E$ of a topological functor T, an equivalence H, and a full reflective embedding E,

(4) G belongs to the smallest conglomerate of functors that contains all topologically algebraic functors and is closed under composition,

(5) G belongs to the smallest conglomerate of functors that contains all topological functors, all equivalences, all full reflective embeddings, and is closed under composition.

Proof: (1) $\Rightarrow$ (3). If $\mathbf{A} \xrightarrow{G} \mathbf{X}$ is a solid functor, consider $(\mathbf{A}, G)$ as a concrete category over $\mathbf{X}$, and let $(\mathbf{B}, U)$ be the following concrete category over $\mathbf{X}$: objects are pairs $(X, (e, A))$, consisting of an $\mathbf{X}$-object X and a semifinal G-structured arrow $X \xrightarrow{e} GA$; morphisms $(X, (e, A)) \xrightarrow{f} (X', (e', A'))$ are those $\mathbf{X}$-morphisms $X \xrightarrow{f} X'$ for which there exists a (necessarily unique) $\mathbf{A}$-morphism $A \xrightarrow{g} A'$ with $e' \circ f = Gg \circ e$. The concrete functor $(\mathbf{A}, G) \xrightarrow{E} (\mathbf{B}, U)$ defined by $E(A) = (GA, (id_{GA}, A))$ is a full embedding. Moreover, for each $\mathbf{B}$-object $(X, (e, A))$, the E-structured arrow $(X, (e, A)) \xrightarrow{e} EA$ is E-universal. Hence $\mathbf{A} \xrightarrow{E} \mathbf{B}$ is a full reflective embedding. In $(\mathbf{B}, U)$ every structured source $\mathcal{S} = (X \xrightarrow{f_i} U(X_i, (e_i, A_i)))_I$ has a (not necessarily unique) initial lift. This can be seen as follows: in $(\mathbf{A}, G)$ consider the structured sink $\mathcal{T} = (GA_k \xrightarrow{g_k} X)_K$ that consists of all pairs (A_k, g_k) such that $GA_k \xrightarrow{e_i \circ f_i \circ g_k} GA_i$ is an $\mathbf{A}$-morphism for each $i \in I$. Let the G-structured arrow $X \xrightarrow{e} GA$ be a semifinal arrow for $\mathcal{T}$. It follows that $((X, (e, A)) \xrightarrow{f_i} (X_i, (e_i, A_i)))_I$ is a U-initial lift of $\mathcal{S}$. Hence $(\mathbf{B}, U)$ may fail to be topological only by not being amnestic. By Proposition 5.33 there exists an amnestic concrete category $(\mathbf{C}, T)$ over $\mathbf{X}$ and a surjective concrete equivalence $(\mathbf{B}, U) \xrightarrow{H} (\mathbf{C}, T)$. Obviously, T is topological. Hence $G = T \circ H \circ E$ is the desired factorization.

(3) $\Rightarrow$ (5) is obvious.

(5) $\Rightarrow$ (4) is immediate from the fact that topological functors, equivalences, and full reflective embeddings are topologically algebraic.

(4) $\Rightarrow$ (1) is immediate from the fact that topologically algebraic functors are solid and solid functors are closed under composition.

(3) $\Rightarrow$ (2) and (2) $\Rightarrow$ (4) are obvious. □

26.2 COROLLARY

A concrete category is solid and amnestic if and only if it is a full reflective concrete subcategory of a topological category.

Proof: In the proof of Theorem 26.1 amnesticity of G implies that $\mathbf{A} \xrightarrow{H \circ E} \mathbf{C}$ is a full reflective embedding. □

26.3 DECOMPOSITION THEOREM FOR SOLID FUNCTORS

For any functor G between (Epi, Mono-Source)-factorizable categories, the following conditions are equivalent:

(1) G is solid and uniquely transportable,

(2) G can be expressed as a composite $G = T \circ A$ of a topological functor T and an essentially algebraic functor A,

(3) G belongs to the smallest conglomerate of functors that contains all topological and all essentially algebraic functors and is closed under composition.

Proof: (1) ⇒ (2). By Theorem 26.1 G can be expressed as a composite $G = T \circ H \circ E$ of a topological functor T, an equivalence H, and a full reflective embedding E. In the case that G is uniquely transportable, the above constructed $A = H \circ E$ is an isomorphism-closed full reflective embedding. Hence, by Theorem 23.8, it is an essentially algebraic functor.

(2) ⇒ (3) ⇒ (1) is obvious. □

26.4 THEOREM

For any functor $\mathbf{A} \xrightarrow{G} \mathbf{X}$ into a category $\mathbf{X}$ that has regular factorizations the following conditions are equivalent:

(1) G is solid, uniquely transportable, and preserves regular epimorphisms,

(2) G can be expressed as a composite $G = T \circ R$ of a topological functor T and a regularly monadic functor R,

(3) G belongs to the smallest conglomerate of functors that contains all topological and all regularly monadic functors and is closed under composition.

Proof: (1) ⇒ (2). In view of Theorem 20.32 it suffices to show that the full reflective embedding $\mathbf{A} \xrightarrow{E} \mathbf{B}$ constructed in the proof of Theorem 26.1 preserves regular epimorphisms. Let $B \xrightarrow{c} C$ be a regular epimorphism in $\mathbf{A}$, and thus a coequalizer of some pair $A \underset{q}{\overset{p}{\rightrightarrows}} B$ of $\mathbf{A}$-morphisms. Then $GB \xrightarrow{Gc} GC$ is a regular epimorphism in $\mathbf{X}$, and so is a coequalizer of some pair $X \underset{s}{\overset{r}{\rightrightarrows}} GB$ of $\mathbf{X}$-morphisms. Consider the G-structured sink $\mathcal{S} = (GA_i \xrightarrow{f_i} X)_I$ that consists of those G-costructured arrows $GA_i \xrightarrow{f_i} X$, for which $GA_i \xrightarrow{r \circ f_i} GB$ and $GA_i \xrightarrow{s \circ f_i} GB$ are $\mathbf{A}$-morphisms, and

Sec. 26] Topologically algebraic structure theorems

let $X \xrightarrow{e} G\overline{A}$ be a semifinal arrow for $\mathcal{S}$. Then there exist **A**-morphisms $\overline{A} \xrightarrow{\overline{r}} B$ and $\overline{A} \xrightarrow{\overline{s}} B$ with $r = G\overline{r} \circ e$ and $s = G\overline{s} \circ e$. Thus $(X, (e, \overline{A})) \underset{s}{\overset{r}{\rightrightarrows}} E(B)$ is a pair of **B**-morphisms. It suffices to show that $E(B \xrightarrow{c} C) = E(B) \xrightarrow{Gc} E(C)$ is a coequalizer of this pair in **B**. Obviously, $Gc \circ r = Gc \circ s$. Let $E(B) \xrightarrow{f} (\overline{X}, (\overline{e}, \overline{B}))$ be a **B**-morphism with $f \circ r = f \circ s$. Since Gc is a coequalizer of r and s in **X**, there exists a unique **X**-morphism $GC \xrightarrow{\overline{f}} \overline{X}$ with $f = \overline{f} \circ Gc$. Hence it suffices to show that $E(C) \xrightarrow{\overline{f}} (\overline{X}, (\overline{e}, \overline{B}))$ is a **B**-morphism. Since $E(B) \xrightarrow{f} (\overline{X}, (\overline{e}, \overline{B}))$ is a **B**-morphism, there exists an **A**-morphism $B \xrightarrow{g} \overline{B}$ with $\overline{e} \circ f = Gg \circ id_{GB}$. Thus $G(g \circ p) = \overline{e} \circ f \circ Gp = \overline{e} \circ \overline{f} \circ Gc \circ Gp = \overline{e} \circ \overline{f} \circ Gc \circ Gq = \overline{e} \circ f \circ Gq = G(g \circ q)$. Hence $g \circ p = g \circ q$. Since c is a coequalizer of p and q in **A**, there exists an **A**-morphism $C \xrightarrow{\overline{g}} \overline{B}$ with $g = \overline{g} \circ c$. Thus the equations $G\overline{g} \circ Gc = Gg = \overline{e} \circ f = \overline{e} \circ \overline{f} \circ Gc$ and the fact that Gc is an epimorphism in **X** imply that $G\overline{g} = \overline{e} \circ \overline{f}$, i.e., that $E(C) \xrightarrow{\overline{f}} (\overline{X}, (\overline{e}, \overline{B}))$ is a **B**-morphism.

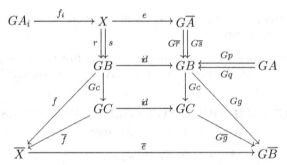

(2) $\Rightarrow$ (3) $\Rightarrow$ (1) obvious. □

Next we wish to characterize those concrete categories that can be embedded as full reflective concrete subcategories of fibre-small topological categories. By Theorem 22.3 this will provide a characterization of full reflective subcategories of functor-structured categories. By the above theorem they must be solid and, obviously, fibre-small. But fibre-smallness is not enough. We now introduce a slightly stronger condition than fibre-smallness, and show that for concrete categories that have this property, solidness is equivalent to being a reflective subcategory of some functor-structured category.

26.5 DEFINITION

(1) Structured arrows $X \xrightarrow{f_1} UA_1$ and $X \xrightarrow{f_2} UA_2$ in a concrete category $(\mathbf{A}, U)$ are said to be **equivalent** if for each costructured arrow $UB \xrightarrow{g} X$ the following holds:

$B \xrightarrow{f_1 \circ g} A_1$ is an **A**-morphism if and only if $B \xrightarrow{f_2 \circ g} A_2$ is an **A**-morphism.

(2) A concrete category over **X** is said to be **strongly fibre-small** provided that for each **X**-object X there exists no proper class of pairwise non-equivalent structured arrows with domain X.

26.6 EXAMPLES

(1) Each strongly fibre-small amnestic concrete category is fibre-small.

(2) A topological category is strongly fibre-small if and only if it is fibre-small. This follows immediately from the fact that each structured arrow $X \xrightarrow{f} UA$ is equivalent (via its initial lift) to a structured arrow of the form $X \xrightarrow{id_X} UA'$.

(3) Each fibre-small hereditary (i.e., monomorphisms have initial lifts) construct (e.g., each of **Pos**, **Top**, **Haus**, and **Met**) is strongly fibre-small: for each structured arrow $X \xrightarrow{f} UA$ consider the factorization of f as a surjective function $X \xrightarrow{f'} UA'$ followed by an initial monomorphism $A' \to A$. Then two such arrows $X \xrightarrow{f_i} UA_i$, $i = 1, 2$, are equivalent if and only if $X \xrightarrow{f'_i} UA'_i$ are equivalent.

(4) Each fibre-small uniquely transportable construct that has (surjective, injective)-factorizations is strongly fibre-small. Thus **Vec**, **Grp**, **HComp**, **Ban**, **Boo**, Σ-**Seq**, etc. are strongly fibre-small constructs.

(5) For concrete categories over $\mathbf{X} = \mathbf{1}$, smallness, fibre-smallness, and strong fibre-smallness are equivalent.

(6) Let **A** be a large discrete category and let $U \colon \mathbf{A} \to \mathbf{Set}$ be an injective functor with $\mathrm{card}(UA) = 1$ for all **A**-objects A. Then $(\mathbf{A}, U)$ is fibre-small, but not strongly fibre-small.

26.7 STRUCTURE THEOREM FOR SOLID CATEGORIES

For a concrete category $(\mathbf{A}, U)$ the following conditions are equivalent:

(1) $(\mathbf{A}, U)$ is amnestic, solid, and strongly fibre-small,

(2) $(\mathbf{A}, U)$ is a full reflective concrete subcategory of a fibre-small topological category,

(3) $(\mathbf{A}, U)$ is concretely isomorphic to a full reflective subcategory of some functor-structured category.

Proof: (1) $\Rightarrow$ (2). Let $(\mathbf{A}, G)$ be a solid category and let $(\mathbf{A}, G) \xrightarrow{E} (\mathbf{B}, U)$ and $(\mathbf{B}, U) \xrightarrow{H} (\mathbf{C}, T)$ be the concrete functors constructed in the proof of Theorem 26.1. If G is amnestic, then $H \circ E$ is a full reflective embedding. Thus it suffices to show that strong fibre-smallness of $(\mathbf{A}, G)$ implies fibre-smallness of $(\mathbf{C}, T)$. For this it is sufficient to show that for any two **B**-objects $(X, (e, A))$ and $(X, (e', A'))$ that are equivalent when considered as G-structured arrows $X \xrightarrow{e} GA$ and $X \xrightarrow{e'} GA'$, the equality $H(X, (e, A)) = H(X, (e', A'))$ holds. This follows immediately from the fact that for any such pair $(X, (e, A)) \xrightarrow{id_X} (X, (e', A'))$ is a **B**-isomorphism.

(2) $\Rightarrow$ (3) follows from Theorem 22.3.

(3) $\Rightarrow$ (1) follows immediately from Corollary 26.2, Example 26.6(2), and the fact that every full concrete subcategory of a strongly fibre-small category is strongly fibre-small. □

26.8 COROLLARY

For a co-wellpowered concrete category $(\mathbf{A}, U)$ over a cocomplete base category, the following are equivalent:

(1) $(\mathbf{A}, U)$ is cocomplete, amnestic, and has free objects,

(2) $(\mathbf{A}, U)$ is concretely isomorphic to a full reflective subcategory of a functor-structured category.

Proof: (1) ⇒ (2). By Proposition 18.11, U is co-wellpowered. By Corollary 15.17, $\mathbf{A}$ is an (Epi, ExtrMono-Source)-category, so that by Theorem 25.19, U is topologically algebraic, and thus has (Generating, InitialSource)-factorizations. These two facts immediately imply that $(\mathbf{A}, U)$ is strongly fibre-small. Thus (2) follows from Theorem 26.7.

(2) ⇒ (1) is immediate from the fact that full reflective embeddings detect colimits (cf. 13.32). □

Suggestions for Further Reading

Hoffmann, R.-E. Note on semi-topological functors. *Math. Z.* **160** (1978): 69–74.

Rosický, J. Concrete categories and infinitary languages. *J. Pure Appl. Algebra* **22** (1981): 309–338.

Rosický, J. Semi-initial completions. *J. Pure Appl. Algebra* **40** (1986): 177–183.

Herrlich, H., Mossakowski, T., and G. E. Strecker. Algebra ∪ Topology. *Category Theory at Work* (eds. H. Herrlich and H.-E. Porst), Heldermann Verlag 1991, 137–148.

EXERCISES

* **26A. Decomposition of Functors and Factorization Structures**

Let $\mathbf{A} \xrightarrow{G} \mathbf{B}$ be a uniquely transportable, solid functor, and let $\mathbf{B}$ have regular factorizations. Show that:

(a) $\mathbf{A}$ need not have (Epi, Mono-Source)-factorizations. [Cf. Exercise 16A(b).]

(b) There need not exist a decomposition $G = T \circ A$ with T topological and A essentially algebraic.

26B. Composites are Topologically Algebraic

Show that each composite $G = A \circ T$ of a monotopological functor T and an essentially algebraic functor A is topologically algebraic.

26C. Topological-Algebraic via Algebraic-Topological Decompositions

(a) Consider the conditions:

(1) there exist a topological functor T and an essentially algebraic functor A with $G = A \circ T$,

(2) there exist a topological functor T and an essentially algebraic functor A with $G = T \circ A$.

Show that:

  (i) (1) implies (2).

  (ii) If $(\mathbf{A}, U)$ is a non-topological, monotopological construct and there exists an **A**-object A such that UA has at least 2 points, then U satisfies (2) but not (1).

  (iii) The forgetful functor for Λ-**CCPos** satisfies (2) but not (1). Cf. Exercise 26B.

(b) Let $(\mathbf{A}, U)$ be a concrete category over a category **X** with regular factorizations. Let U be uniquely transportable, solid, and preserve regular epimorphisms. Let E be the class of those **A**-morphisms whose underlying morphisms are regular epimorphisms in **X**. Show that the following conditions are equivalent:

(1) **A** is an $(E, \text{InitMono-Source})$-category,

(2) U can be expressed as a composite $U = A \circ T$ of a monotopological functor T and a regularly monadic functor A,

(3) the comparison functor for U is monotopological.

26D. Strong Fibre-Smallness

(a) Verify that a construct is strongly fibre-small if and only if it can be fully embedded into a fibre-small topological construct. Under what assumptions on **X** does the same result hold for concrete categories over **X**?

(b) Show that in Corollary 26.8 it is not possible to delete the (co-)wellpoweredness hypothesis. [Hint: Consider **Ord** as a concrete category over itself.]

Chapter VII

CARTESIAN CLOSEDNESS AND PARTIAL MORPHISMS

27 Cartesian closed categories

In some constructs, we can form objects B^A by structuring the set of all morphisms from A to B in such a way that the evaluation map

$$ev: A \times B^A \to B \quad \text{given by:} \quad (x, f) \mapsto f(x)$$

becomes a co-universal arrow. For example, in **Set** we have the set B^A of all functions from A to B, in **Pos** we have the poset B^A of all order-preserving functions from A to B (ordered pointwise), etc. We will study this phenomenon first for abstract categories, then for concrete ones, and finally for constructs.

If a category **A** has finite products, then for each **A**-object A we can define a functor $(A \times -): \mathbf{A} \to \mathbf{A}$ by $(B \xrightarrow{f} B') \mapsto (A \times B \xrightarrow{id_A \times f} A \times B')$. (Since products are not unique, such a rule usually gives a large collection of pairwise naturally isomorphic endofunctors of **A**. By an "abuse of language" we will, however, speak about *the* functor $(A \times -)$, meaning that a certain product $A \times B$ has been chosen for each B.)

27.1 DEFINITION
A category **A** is called **cartesian closed** provided that it has finite products and for each **A**-object A the functor $(A \times -): \mathbf{A} \to \mathbf{A}$ is co-adjoint.

27.2 NOTATION
The essential uniqueness of products and of co-universal arrows allows us to introduce the following standard notation for cartesian closed categories: "The" adjoint functor for $(A \times -)$ is denoted on objects by $B \mapsto B^A$, and "the" associated co-universal arrows are denoted by

$$ev: A \times B^A \to B.$$

Thus, a category with finite products is cartesian closed if and only if for each pair (A, B) of objects there exists an object B^A and a morphism $ev: A \times B^A \to B$ with the following universal property: for each morphism $f: A \times C \to B$ there exists a unique morphism $\hat{f}: C \to B^A$ such that

$$\begin{array}{c} A \times C \\ {\scriptstyle id_A \times \hat{f}} \downarrow \quad \searrow{\scriptstyle f} \\ A \times B^A \xrightarrow[ev]{} B \end{array}$$

commutes. We call the objects B^A **power objects**, the morphisms $ev: A \times B^A \to A$ **evaluation morphisms**, and the $\hat{f}$, associated with f, the **exponential morphism for** f.

A power object in a cartesian closed category

27.3 EXAMPLES

(1) The following categories are cartesian closed:

(a) **Set**: B^A is the set of all functions from A to B; ev is the usual evaluation map; $(\hat{f}(c))(a) = f(a,c)$.

(b) **Rel**: $(B,\sigma)^{(A,\rho)}$ is (B^A, τ) with $f\tau g \Leftrightarrow [a\rho a' \Rightarrow f(a)\sigma g(a')]$; ev is the usual evaluation map; $(\hat{f}(c))(a) = f(a,c)$.

(c) **Pos**: $(B, \leq)^{(A,\preceq)}$ is the set of all order-preserving functions from A to B, ordered as follows: $f \leq g \Leftrightarrow f(a) \leq g(a)$ for all $a \in A$; ev is the usual evaluation map; $(\hat{f}(c))(a) = f(a,c)$.

(d) **Alg(1)** (algebras with one unary operation): B^A is the set of all homomorphisms from $A \times \mathbb{N}$ to B [where $\mathbb{N}$ denotes the algebra of all natural numbers with the successor operation], together with the unary operation that sends each $h: A\times\mathbb{N} \to B$ to $h': A\times\mathbb{N} \to B$, where $h'(a,n) = h(a,n+1)$; $ev(a,f) = f(a,0)$; and for each homomorphism $f: A \times C \to B$, $\hat{f}(c)$ sends (a,n) to $f(a,\gamma^n(c))$, where γ is the unary operation for C.

(e) **Cat**: $\mathbf{B}^{\mathbf{A}}$ is the functor category $[\mathbf{A}, \mathbf{B}]$; $ev: \mathbf{A} \times \mathbf{B}^{\mathbf{A}} \to \mathbf{B}$ is defined on objects by $ev(A, F) = FA$ and on morphisms by $ev(h, \tau) = \tau_{A'} \circ Fh$ [where $A \xrightarrow{h} A'$]; for $F: \mathbf{A} \times \mathbf{C} \to \mathbf{B}$ we have $\hat{F}C: \mathbf{A} \to \mathbf{B}$ defined by

$$\hat{F}C(A \xrightarrow{h} A') = F(A,C) \xrightarrow{F(h,id)} F(A',C).$$

(2) **Top** is not cartesian closed (since $(\mathbb{Q} \times -)$: **Top** $\to$ **Top** does not preserve quotients, and hence does not preserve coequalizers). However, **Top** has cartesian closed supercategories [e.g., **Conv** and **PsTop**, where the power objects carry the structure of "continuous convergence" (see Exercise 27G)] as well as cartesian closed subcategories [e.g., the category **kTop** of (compact Hausdorff)-generated topological spaces, where the power objects are formed by the **kTop**-coreflections of the compact-open topologies].

(3) A poset A, considered as a category, is cartesian closed if and only if A has finite meets and for each pair (a, b) of elements, the set $\{x \in A \mid a \wedge x \le b\}$ has a largest member. In particular, a complete lattice is cartesian closed if and only if it satisfies the distributive law: $a \wedge \bigvee b_i = \bigvee (a \wedge b_i)$; i.e., if and only if it is a frame.

27.4 CHARACTERIZATION THEOREM FOR CARTESIAN CLOSED CATEGORIES

Let **A** be a cocomplete and co-wellpowered category that has a separator. Then **A** is cartesian closed if and only if it has finite products, and for each **A**-object A the functor $(A \times -)$ preserves colimits.

Proof: Immediate from the Special Adjoint Functor Theorem (18.17). □

27.5 DEFINITION

Let **A** be a cartesian closed category. For each object C

(1) "the" **covariant exponential functor for** C, denoted by $(-)^C : \mathbf{A} \to \mathbf{A}$, is "the" adjoint functor for $(C \times -)$ and is defined (for an **A**-morphism $A \xrightarrow{f} B$) by:

$$(-)^C(A \xrightarrow{f} B) = A^C \xrightarrow{f^C} B^C,$$

where f^C is the unique **A**-morphism that makes the diagram

$$\begin{array}{ccc} C \times A^C & \xrightarrow{ev} & A \\ {\scriptstyle id \times f^C} \downarrow & & \downarrow {\scriptstyle f} \\ C \times B^C & \xrightarrow{ev} & B \end{array}$$

commute;

(2) "the" **contravariant exponential functor for** C, denoted by $C^{(-)} : \mathbf{A}^{op} \to \mathbf{A}$, is defined (for an **A**-morphism $A \xrightarrow{f} B$) by

$$C^{(-)}(A \xrightarrow{f} B) = C^B \xrightarrow{C^f} C^A,$$

where C^f is the unique **A**-morphism that makes the diagram

$$\begin{array}{ccc} A \times C^B & \xrightarrow{f \times id} & B \times C^B \\ {\scriptstyle id \times C^f} \downarrow & & \downarrow {\scriptstyle ev} \\ A \times C^A & \xrightarrow{ev} & C \end{array}$$

commute.

27.6 REMARK
Observe that exponential functors are not determined uniquely, but that any two covariant (resp. contravariant) exponential functors for an **A**-object C are naturally isomorphic.

27.7 PROPOSITION
In a cartesian closed category **A**, *every contravariant exponential functor* $C^{(-)}$ *is an adjoint functor and has its own dual* $(C^{(-)})^{\mathrm{op}}$ *as a co-adjoint.*

Proof: Let A be an **A**-object and let $s \colon C^A \times A \to A \times C^A$ be the isomorphism that is determined by interchanging the projections. Then the unique **A**-morphism $u_A \colon A \to C^{(C^A)}$ that makes the diagram

$$\begin{array}{ccc} C^A \times A & \xrightarrow{s} & A \times C^A \\ {\scriptstyle id \times u_A} \downarrow & & \downarrow {\scriptstyle ev} \\ C^A \times C^{(C^A)} & \xrightarrow{ev} & C \end{array}$$

commute is a $C^{(-)}$-universal arrow for A. To see this let $f \colon A \to C^B$ be a $C^{(-)}$-structured arrow, let $\bar{s} \colon A \times B \to B \times A$ be the isomorphism determined by interchanging the projections, and let $\bar{f} \colon B \to C^A$ be the unique **A**-morphism that makes the diagram

$$\begin{array}{ccc} A \times B & \xrightarrow{(id \times f) \circ \bar{s}} & B \times C^B \\ {\scriptstyle id \times \bar{f}} \downarrow & & \downarrow {\scriptstyle ev} \\ A \times C^A & \xrightarrow{ev} & C \end{array}$$

commute. Then $id \times \bar{f} = s \circ (\bar{f} \times id) \circ \bar{s}$ implies that in the following diagram $\boxed{1}$ commutes:

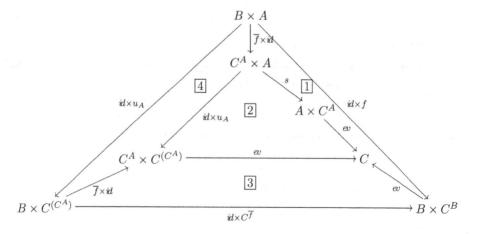

$\boxed{2}$ commutes by the definition of u, $\boxed{3}$ commutes by the definition of $C^{(-)}$, and $\boxed{4}$ commutes trivially. Thus $f = C^{(\bar{f} \circ u_A)}$. That $\bar{f}$ is determined by this equality follows immediately by retracing the steps from $C^{\bar{f}}$ via $\bar{f}$ to f in the above diagram. Thus $u_A \colon A \to C^{(C^A)}$ is a $C^{(-)}$-universal arrow. In order to see that $(C^{(-)})^{\mathrm{op}} \colon \mathbf{A} \to \mathbf{A}^{\mathrm{op}}$ is a co-adjoint for $C^{(-)}$ it need only be shown that for any $\mathbf{A}$-morphism $f \colon A \to B$ the equality $u_B \circ f = C^{(C^{(-)})^{\mathrm{op}}(f)} \circ u_A$ holds. This follows immediately from the commutativity of the following diagram, where s, $\bar{s}$, and $\tilde{s}$ are isomorphisms determined by interchanging the projections

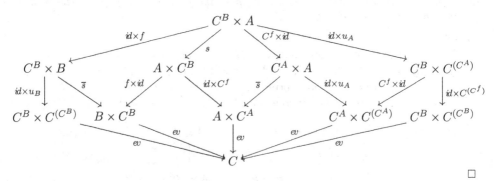

27.8 PROPOSITION
In a cartesian closed category the following hold:

(1) **First Exponential Law**: $A^{B \times C} \cong (A^B)^C$,

(2) **Second Exponential Law**: $(\prod A_i)^B \cong \prod (A_i^B)$,

(3) **Third Exponential Law**: $A^{\amalg B_i} \cong \prod (A^{B_i})$,

(4) **Distributive Law**: $A \times \amalg B_i \cong \amalg (A \times B_i)$,

(5) *Finite products of epimorphisms are epimorphisms.*

Proof:
(1). This follows from the definition of covariant exponential functors (27.5) and the fact that, up to natural equivalence, the functor $((B \times C) \times -)$ is the composite of $(B \times -)$ with $(C \times -)$.

(2). $(-)^B$ preserves products since it is an adjoint functor.

(3). $A^{(-)} \colon \mathbf{A}^{\mathrm{op}} \to \mathbf{A}$ preserves products since it is an adjoint functor.

(4). $(A \times -)$ preserves coproducts since it is a co-adjoint functor.

(5). This follows from the fact that the functors $(A \times -)$, as co-adjoints, preserve epimorphisms and the fact that

$$A_1 \times A_2 \xrightarrow{f_1 \times f_2} B_1 \times B_2 = A_1 \times A_2 \xrightarrow{f_1 \times id} B_1 \times A_2 \xrightarrow{id \times f_2} B_1 \times B_2. \quad \square$$

CARTESIAN CLOSED SUBCATEGORIES

27.9 PROPOSITION
Let **A** be an isomorphism-closed full subcategory of a cartesian closed category **B**.

(1) If **A** is reflective in **B** and the **A**-reflector preserves finite products, then **A** is closed under the formation of finite products and powers in **B**, and hence is cartesian closed.

(2) If **A** is coreflective in **B** and is closed under the formation of finite products in **B**, then **A** is cartesian closed.

Proof:

(1). Let A and B be **A**-objects and let $r \colon B^A \to \overline{A}$ be an **A**-reflection arrow for B^A. Then by assumption $id \times r \colon A \times B^A \to A \times \overline{A}$ is an **A**-reflection arrow for $A \times B^A$. Thus there exists a unique morphism $e \colon A \times \overline{A} \to B$ with $ev = e \circ (id_A \times r)$. Furthermore, there exists a unique morphism $s \colon \overline{A} \to B^A$ with $e = ev \circ (id_A \times s)$.

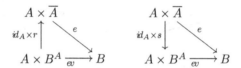

This immediately implies that $s \circ r = id_{B^A}$. Hence $r \circ s \circ r = id_{\overline{A}} \circ r$, so that $r \circ s = id_{\overline{A}}$. Consequently, $r \colon B^A \to \overline{A}$ is an isomorphism; i.e., B^A belongs to **A**.

(2). If A and B are **A**-objects, $ev \colon A \times B^A \to B$ is an evaluation morphism in **B**, and $c \colon \overline{A} \to B^A$ is an **A**-coreflection arrow for B^A, then $ev \circ (id_A \times c) \colon A \times \overline{A} \to B$ is easily seen to be an evaluation morphism in **A**. □

27.10 EXAMPLES

(1) The full subcategory **Sym** of **Rel** is cartesian closed (and is closed under the formation of power objects in **Rel**) since it is a reflective subcategory of **Rel** and the reflector preserves products. In contrast, **Pos**, although it is cartesian closed, is not closed under the formation of power objects in **Rel**. Also, the full (reflective) subcategory that consists of all transitive relations is not cartesian closed: for $A = (\{0,1\}, \{0,1\})$ the functor $(A \times -)$ does not preserve the quotient of $A+A$ that identifies 1 of the first copy of A with 0 of the second copy.

(2) **Top** is not cartesian closed even though it is a reflective subcategory of **Conv**, which is cartesian closed.

CARTESIAN CLOSED CONCRETE CATEGORIES

27.11 DEFINITION
A concrete category $(\mathbf{A}, U)$ over **X** is called **concretely cartesian closed** provided that the following hold:

(1) **A** and **X** are cartesian closed,

(2) U preserves finite products, power objects, and evaluation; in particular, whenever $A \times B^A \xrightarrow{ev} B$ is an evaluation in **A**, then
$$U(A \times B^A \xrightarrow{ev} B) = UA \times UB^{UA} \xrightarrow{ev} UB$$
is an evaluation in **X**.

27.12 EXAMPLES
(1) **Rel** is a concretely cartesian closed construct. Cf. Example 27.3(1)(b).

(2) **Pos** and **Alg**(1) are constructs that are cartesian closed but not concretely so. Cf. Examples 27.3(1)(c) and (d).

(3) A partially ordered set, considered as concrete category over **1**, is cartesian closed if and only if it is concretely so.

(4) **Spa**($\mathcal{P}$) is concretely cartesian closed. More generally, a functor-structured category is cartesian closed if and only if it is concretely so (cf. Exercise 27I).

27.13 REMARK
For topological categories, concrete cartesian closedness is well understood, as the following results indicate.

27.14 PROPOSITION
*If $(\mathbf{A}, U)$ is a topological category over **X** and if **A** is cartesian closed, then so is **X**.*

Proof: **X** is isomorphic to the full subcategory of **A** consisting of all indiscrete objects, which by Proposition 27.9(1) is cartesian closed. □

27.15 CHARACTERIZATION THEOREM FOR CONCRETELY CARTESIAN CLOSED TOPOLOGICAL CATEGORIES
*For a topological category $(\mathbf{A}, U)$ over a cartesian closed category **X** the following are equivalent:*

(1) $(\mathbf{A}, U)$ is concretely cartesian closed,

*(2) **A** is cartesian closed and every **A**-morphism with a discrete codomain has a discrete domain,*

*(3) for each **A**-object A the functor $(A \times -)$ preserves final sinks.*

Proof: (1) $\Rightarrow$ (3). Let $\mathcal{S} = (B_i \xrightarrow{f_i} B)_I$ be a final sink. To show that $A \times \mathcal{S} = (A \times B_i \xrightarrow{id_A \times f_i} A \times B)_I$ is final, let $U(A \times B) \xrightarrow{f} UC$ be an **X**-morphism such that all $A \times B_i \xrightarrow{f \circ (id_A \times f_i)} C$ are **A**-morphisms. Let $A \times C^A \xrightarrow{ev} C$ be an evaluation in **A**. Then by (1), $UA \times UC^{UA} \xrightarrow{ev} UC$ is an evaluation in **X**. Thus in **A**, for each $i \in I$, there exists a unique morphism $B_i \xrightarrow{g_i} C^A$ with $f \circ (id_A \times f_i) = ev \circ (id_A \times g_i)$, and in **X** there exists a unique morphism $UB \xrightarrow{g} UC^{UA}$ with $f = ev \circ (id_{UA} \times g)$. Hence $ev \circ (id_A \times (g \circ f_i)) = ev \circ ((id_A \times g) \circ (id_A \times f_i)) = f \circ (id_A \times f_i) = ev \circ (id_A \times g_i)$ implies that $g \circ f_i = g_i$, so that since $\mathcal{S}$ is final, $UB \xrightarrow{g} U(C^A)$ is an **A**-morphism. Consequently, $U(A \times B) \xrightarrow{f} UC = U(A \times B) \xrightarrow{id_A \times g} U(A \times C^A) \xrightarrow{ev} UC$ is an **A**-morphism as well.

(3) $\Rightarrow$ (1). Let A and B be **A**-objects and let $UA \times UB^{UA} \xrightarrow{ev} UB$ be an evaluation morphism in **X**. Consider the structured sink $\mathcal{S} = (UA_i \xrightarrow{f_i} UB^{UA})_I$ consisting of all **X**-morphisms f_i such that $U(A \times A_i) \xrightarrow{ev \circ (id_A \times f_i)} UB$ is an **A**-morphism. Let $(A_i \xrightarrow{f_i} B^A)_I$ be a final lift of $\mathcal{S}$. Then, by (3), the sink $(A \times A_i \xrightarrow{id_A \times f_i} A \times B^A)_I$ is final too. Thus $U(A \times B^A) \xrightarrow{ev} UB$ is an **A**-morphism. It is easily seen to be the desired evaluation in **A**.

(1) $\Rightarrow$ (2). Let $A \xrightarrow{f} D$ be an **A**-morphism with discrete codomain D. To show that A is discrete, consider an **X**-morphism $UA \xrightarrow{g} UB$. Let $(A \times D, (\pi_A, \pi_D))$ be a product. Since an object C is discrete if and only if the empty sink with codomain C is final, (3) implies that $A \times D$ is discrete. Hence $U(A \times D) \xrightarrow{g \circ \pi_A} UB$ is an **A**-morphism. Consequently, $UA \xrightarrow{g} UB = UA \xrightarrow{\langle id_A, f \rangle} U(A \times D) \xrightarrow{g \circ \pi_A} UB$ is an **A**-morphism.

(2) $\Rightarrow$ (1). Let $A \times B^A \xrightarrow{ev} B$ be an evaluation in **A**. To establish that its image $UA \times U(B^A) \xrightarrow{ev} UB$ is an evaluation in **X**, let $UA \times X \xrightarrow{f} UB$ be an **X**-morphism and let D be the discrete object with $UD = X$. By condition (2), $A \times D$ is discrete since the projection $A \times D \xrightarrow{\pi_D} D$ is an **A**-morphism. Thus $U(A \times D) \xrightarrow{f} UB$ is an **A**-morphism. Hence there exists a unique **A**-morphism $D \xrightarrow{g} B^A$ with $f = e \circ (id_A \times g)$. Thus, by discreteness of D, $X \xrightarrow{g} U(B^A)$ is the unique **X**-morphism with $f = e \circ (id_{UA} \times g)$. □

CARTESIAN CLOSED CONSTRUCTS

Relatively few of the familiar cartesian closed constructs are concretely cartesian closed. More frequently, the underlying sets $U(B^A)$ of powers in cartesian closed constructs $(\mathbf{A}, U)$ are not of the form UB^{UA}, but rather of the form $\hom_{\mathbf{A}}(A, B)$. As we will see, this phenomenon is closely related to the fact that the terminal objects are often discrete, or, equivalently, that all constant maps between **A**-objects are **A**-morphisms.

27.16 PROPOSITION
Every concretely cartesian closed amnestic construct with discrete terminal object is, up to concrete isomorphism, a full subconstruct of **Set**.

Proof: In Theorem 27.15 the implication (1) $\Rightarrow$ (2) holds without the assumption that $(\mathbf{A}, U)$ is topological. Thus the discreteness of the terminal object implies that every object is discrete. □

27.17 DEFINITION
A construct $(\mathbf{A}, U)$ is said to **have function spaces** provided that the following hold:

(1) $(\mathbf{A}, U)$ has finite concrete products,

(2) $\mathbf{A}$ is cartesian closed and the evaluation morphisms $A \times B^A \xrightarrow{ev} B$ can be chosen in such a way that $U(B^A) = \hom_{\mathbf{A}}(A, B)$ and ev is the restriction of the canonical evaluation map in **Set**.

27.18 PROPOSITION
Let $(\mathbf{A}, U)$ be a construct with finite concrete products. If $\mathbf{A}$ is cartesian closed, then the following conditions are equivalent:

(1) $(\mathbf{A}, U)$ has function spaces,

(2) terminal $\mathbf{A}$-objects are discrete,

(3) each constant function[73] between $\mathbf{A}$-objects is an $\mathbf{A}$-morphism.

Proof: (1) $\Rightarrow$ (2). Let T be a terminal object in $\mathbf{A}$ and let $UT \xrightarrow{f} UA$ be a function. Then f is a constant function, whose value will be denoted by a. Consider the evaluation $T \times A^T \xrightarrow{ev} A$ as specified by Definition 27.17. Since the projection $T \times A \xrightarrow{\pi} A$ is a morphism, $A \xrightarrow{\hat{\pi}} A^T$ is a morphism. Hence $\hat{\pi}(a) \in \hom(T, A)$. Since $\hat{\pi}(a)$ agrees with f, it follows that $UT \xrightarrow{f} UA$ is a morphism. Thus T is discrete.

(2) $\Rightarrow$ (1). Let $A \times B^A \xrightarrow{ev} B$ be an evaluation morphism in $\mathbf{A}$ and let T be a terminal $\mathbf{A}$-object. Since empty products are concrete, UT is a singleton set, say, $\{t\}$. Since T is terminal, it is clear that the function $h : \hom_{\mathbf{A}}(A, B) \to \hom_{\mathbf{A}}(A \times T, B)$ defined by $h(A \xrightarrow{f} B) = A \times T \xrightarrow{\pi_A} A \xrightarrow{f} B$ is a bijection. Also, we have the bijection $k : \hom_{\mathbf{A}}(A \times T, B) \to \hom_{\mathbf{A}}(T, B^A)$ that associates with any morphism $A \times T \xrightarrow{f} B$ the unique morphism $T \xrightarrow{k(f)} B^A$ with $f = ev \circ (id_A \times k(f))$. The function $\ell : \hom_{\mathbf{A}}(T, B^A) \to U(B^A)$ defined by $\ell(f) = f(t)$ is a bijection since T is terminal, hence discrete by (2). Let $C \xrightarrow{\ell \circ k \circ h} B^A$ be an initial lift of the bijection $\hom_{\mathbf{A}}(A, B) \xrightarrow{\ell \circ k \circ h} U(B^A)$ and set $\overline{ev} = ev \circ (id_A \times (\ell \circ k \circ h))$. Then $A \times C \xrightarrow{\overline{ev}} B$ is an evaluation with $UC = \hom_{\mathbf{A}}(A, B)$. A simple calculation shows that $\overline{ev}(a, f) = f(a)$ for $a \in UA$ and $f \in \hom_{\mathbf{A}}(A, B)$. Thus $(\mathbf{A}, U)$ has function spaces.

(2) $\Leftrightarrow$ (3) is obvious. $\square$

27.19 REMARK
In topological constructs, terminal objects are discrete if and only if the fibres of each one-element set contain exactly one element. To avoid uninteresting pathologies we will require a little more:

27.20 DEFINITION
A construct is called **well-fibred** provided that it is fibre-small and for each set with at most one element, the corresponding fibre has exactly one element.

---
[73] A function is called **constant** provided that it factors through a one-element set.

27.21 EXAMPLES

The following topological constructs are well-fibred and have function spaces: **Conv** of convergence spaces, **PsTop** of pseudotopological spaces, **sTop** of sequential topological spaces, **kTop** of (compact Hausdorff)-generated topological spaces, **Bor** of bornological spaces, **Simp** of simplicial complexes, **Prost** of preordered sets, and **Rere** of reflexive relations.

27.22 THEOREM
For well-fibred topological constructs the following conditions are equivalent:

(1) **A** *is cartesian closed,*

(2) $(\mathbf{A}, U)$ *has function spaces,*

(3) for each **A**-*object A the functor* $(A \times -)$ *preserves final epi-sinks,*

(4) for each **A**-*object A the functor* $(A \times -)$ *preserves colimits,*

(5) for each **A**-*object A the functor* $(A \times -)$ *preserves (a) coproducts and (b) quotients.*

Proof:
(1) $\Leftrightarrow$ (2). Immediate from Proposition 27.18.

(1) $\Leftrightarrow$ (4). Theorem 27.4.

(4) $\Rightarrow$ (5) follows from the fact that quotient maps are regular epimorphisms. (Cf. Proposition 21.13.)

(5) $\Rightarrow$ (3) follows from the facts that: (a) a sink $(B_i \xrightarrow{f_i} B)_I$ is a final epi-sink if and only if there exists a sub*set* J of I such that the small sink $(B_j \xrightarrow{f_j} B)_J$ is a final epi-sink, and (b) a small sink $(B_j \xrightarrow{f_j} B)_J$ is a final epi-sink if and only if $\coprod_J B_j \xrightarrow{[f_j]} B$ is a quotient map.

(3) $\Rightarrow$ (4) follows from concreteness of both colimits and finite products in $(\mathbf{A}, U)$ and the fact that the functor $(UA \times -) : \mathbf{Set} \to \mathbf{Set}$ preserves colimits. $\square$

27.23 EXAMPLES
(1) **Top** is a well-fibred topological construct that satisfies the above condition (5)(a), but not (5)(b).

(2) **Unif** is a well-fibred topological construct that satisfies the above condition (5)(b), but not (5)(a).

(3) **Alg**(1) and **Cat** [considered as a construct via the functor $U : \mathbf{Cat} \to \mathbf{Set}$, defined by $U(\mathbf{A} \xrightarrow{F} \mathbf{B}) = Mor(\mathbf{A}) \xrightarrow{F_M} Mor(\mathbf{B})$] are cartesian closed well-fibred non-topological constructs that have finite concrete products, but don't have function spaces.

27.24 PROPOSITION
For cartesian closed, well-fibred topological constructs the following hold:

(1) products with discrete factors A are coproducts:
$$A \times B \cong {}^{|A|}B = \coprod_{x \in |A|} B,$$

(2) power objects with discrete exponents A are powers:
$$B^A \cong B^{|A|} = \prod_{x \in |A|} B.$$

Proof: This follows immediately from $A \cong \coprod_{x \in X} T$, where T is a terminal object, and from Proposition 27.8(3) and (4). □

Suggestions for Further Reading

Eilenberg, S., and G. M. Kelly. Closed categories. *Proceedings of the Conference on Categorical Algebra* (La Jolla, 1965), Springer, Berlin–Heidelberg–New York, 1966, 421–562.

Day, B. J., and G. M. Kelly. On topological quotient maps preserved by pullbacks or products. *Proc. Cambridge Phil. Soc.* **67** (1970): 553–558.

Day, B. A reflection theorem for closed categories. *J. Pure Appl. Algebra* **2** (1972): 1–11.

Wyler, O. Convenient categories for topology. *Gen. Topol. Appl.* **3** (1973): 225–242.

Herrlich, H. Cartesian closed topological categories. *Math. Coll. Univ. Cape Town* **9** (1974): 1–16.

Nel, L. D. Initially structured categories and cartesian closedness. *Canad. J. Math.* **27** (1975): 1361–1377.

Isbell, J. R. Function spaces and adjoints. *Math. Scand.* **36** (1975): 317–339.

Herrlich, H., and L. D. Nel. Cartesian closed topological hulls. *Proc. Amer. Math. Soc.* **62** (1977): 215–222.

Wyler, O. Function spaces in topological categories. *Springer Lect. Notes Math.* **719** (1979): 411–420.

Frölicher, A. Catégories cartésienne fermées engendrées par des monoids. *Cahiers Topol. Geom. Diff.* **21** (1980): 367–375.

Schwarz, F. Powers and exponential objects in initially structured categories and applications to categories of limit spaces. *Quaest. Math.* **6** (1983): 227–254.

Adámek, J., and H. Herrlich. Cartesian closed categories, quasitopoi and topological universes. *Comment. Math. Univ. Carolinae* **27** (1986): 235–257.

Herrlich, H., and G. E. Strecker. Cartesian closed topological hulls as injective hulls. *Quaest. Math.* **9** (1986): 263–280.

Isbell, J. R. General function spaces, products and continuous lattices. *Math. Proc. Cambridge Phil. Soc.* **100** (1986): 193–205.

EXERCISES

27A. Cartesian Closed Categories With Zero-Objects
Let **A** be cartesian closed. Show that

(a) If I is an initial object, then $A \times I \cong I$ for each **A**-object A.

(b) The following are equivalent:

  (1) **A** is pointed,

  (2) **A** has a zero-object,

  (3) **A** is equivalent to **1**.

27B. ΔPos is Cartesian Closed
Show that the category Δ**Pos** is cartesian closed. [Analogously to the situation in **Pos**, B^A is the pointwise-ordered set of all Δ-continuous functions from A to B.]

27C. Coreflective Hulls and Cartesian Closedness
Let $(\mathbf{A}, U)$ be a cartesian closed topological construct and let **B** be a full subcategory of **A** that is closed under the formation of finite products. Show that the bicoreflective hull of **B** in **A** is cartesian closed.

27D. Cartesian Closed Constructs
Let T be a terminal object in a cartesian closed topological construct $(\mathbf{A}, U)$, and for each **A**-object A let A^* denote the discrete object with $UA^* = UA$. Show that

(a) Power objects B^A and evaluation morphisms $A \times B^A \xrightarrow{ev} B$ can be chosen such that
$$U(B^A) = \{UA \xrightarrow{f} UB \mid A \times T^* \xrightarrow{\pi_A} A \xrightarrow{f} B \text{ is an } \mathbf{A}\text{-morphism}\}$$
$$ev(a, f) = f(a).$$

(b) $(\mathbf{A}, U)$ has function spaces if and only if $A \times T^* \xrightarrow{\pi_A} A$ is an isomorphism for each **A**-object A.

(c) $(\mathbf{A}, U)$ is concretely cartesian closed if and only if $A \times T^* = (A \times T)^*$ for each **A**-object A.

(d) $(\mathbf{A}, U)$ is concretely cartesian closed and has function spaces if and only if U is an isomorphism.

27E. Composition as a Morphism

In cartesian closed constructs describe explicitly the unique morphism $comp: B^A \times C^B \to C^A$ that makes the diagram

$$
\begin{array}{ccc}
A \times (B^A \times C^B) & \xrightarrow{ev \times id_{C^B}} & B \times C^B \\
{\scriptstyle id_A \times comp} \downarrow & & \downarrow {\scriptstyle ev} \\
A \times C^A & \xrightarrow{ev} & C
\end{array}
$$

commute.

* 27F. (Concretely) Cartesian Closed Topological Categories as Injective Objects

Let **X** be a cartesian closed category and let $\mathbf{CAT_p(X)}$ be the quasicategory whose objects are the amnestic concrete categories over **X** with finite concrete products, and whose morphisms are the concrete functors over **X** that preserve finite products.

(a) Show that the injective objects in $\mathbf{CAT_p(X)}$ (with respect to full embeddings) are precisely the concretely cartesian closed topological categories over **X**.

Let $\mathbf{CONST_p}$ be the quasicategory whose objects are the amnestic well-fibred constructs with finite concrete products, and whose morphisms are the concrete functors that preserve finite products.

(b) Show that the injective objects in $\mathbf{CONST_p}$ (with respect to full embeddings) are precisely the well-fibred topological constructs that have function spaces.

* 27G. (Concretely) Cartesian Closed Topological Hulls

Let **X** be a cartesian closed category. A morphism $(\mathbf{A}, U) \xrightarrow{E} (\mathbf{B}, V)$ in $\mathbf{CAT_p(X)}$ is called a **concretely cartesian closed topological hull** (shortly: a **CCCT hull**) of $(\mathbf{A}, U)$ provided that the following conditions are satisfied:

(1) E is a full embedding,

(2) $E[\mathbf{A}]$ is finally dense in $(\mathbf{B}, V)$,

(3) $\{ EA^{E\overline{A}} \mid A, \overline{A} \in Ob(\mathbf{A}) \}$ is initially dense in $(\mathbf{B}, V)$,

(4) $(\mathbf{B}, V)$ is a concretely cartesian closed topological category.

(a) Show that the injective hulls in $\mathbf{CAT_p(X)}$ are precisely the CCCT hulls.

A morphism $(\mathbf{A}, U) \xrightarrow{E} (\mathbf{B}, V)$ in $\mathbf{CONST_p}$ is called a **cartesian closed topological hull** (shortly: a **CCT hull**) of $(\mathbf{A}, U)$ provided the above conditions (1), (2), (3), and the following (4*) hold:

(4*) $(\mathbf{B}, V)$ is a cartesian closed topological category.

Show that

(b) The injective hulls in $\mathbf{CONST_p}$ are precisely the CCT hulls.

(c) The concrete embedding **PrTop** $\hookrightarrow$ **PsTop** is a cartesian closed topological hull of **PrTop**.

27H. Well-Fibred Topological Constructs

Show that

(a) In well-fibred topological constructs extremal subobjects of discrete objects are discrete. [Cf. Exercise 8M(a).]

(b) A functor-structured construct $\mathbf{Spa}(T)$ is well-fibred only if T is the constant functor, defined by $T(X \xrightarrow{f} Y) = \emptyset \xrightarrow{id_\emptyset} \emptyset$, i.e., only if $\mathbf{Spa}(T)$ is concretely isomorphic to the construct **Set**.

27I. Cartesian Closed Functor-Structured Categories

(a) Prove that if $\mathbf{Spa}(T)$ is cartesian closed, then it is concretely cartesian closed.

(b) Prove that $\mathbf{Spa}(T)$ is (concretely) cartesian closed whenever **X** is cartesian closed and T **weakly preserves pullbacks**, i.e., for each 2-sink $\bullet \xrightarrow{f} \bullet \xleftarrow{g} \bullet$ the factorizing morphism of the T-image of the pullback of (f,g) through the pullback of (Tf, Tg) is a retraction.

(c) Verify that the **Set**-functors S^n and $\mathcal{P}$ weakly preserve pullbacks.

27J. Re: Theorem 27.15

Prove that the equivalence of 27.15(1) and 27.15(3) follows from the dual of the Galois Correspondence Theorem (21.24).

28 Partial morphisms, quasitopoi, and topological universes

A partial morphism from A to B is a morphism from a subobject of A to B. Sometimes it is possible to "represent" partial morphisms into B by ordinary morphisms into a suitable extension B^* of B in such a way that partial morphisms from A into B are precisely the pullbacks of ordinary morphisms from A to B^* along a distinguished embedding $B \xrightarrow{m_B} B^*$. Since there is no satisfactory categorical concept of subobject or of embedding, partial morphisms will be defined with respect to an arbitrary class M of morphisms, acting as "embeddings of subobjects". The most interesting cases will be those where M consists of all extremal monomorphisms resp. of all monomorphisms. As in §27 we first will consider abstract categories, then topological categories, and, finally, topological constructs.

REPRESENTATIONS OF PARTIAL MORPHISMS

28.1 DEFINITION
Let M be a class of morphisms in **A**.

(1) A 2-source $(A \xleftarrow{m} \bullet \xrightarrow{f} B)$ with $m \in M$ is called an M-**partial morphism** from A into B. (Extremal) Mono-partial morphisms are called (**extremal**) **partial morphisms**.

(2) An M-morphism $B \xrightarrow{m_B} B^*$ is said to **represent** M-partial morphisms into B provided that the following two conditions are satisfied:

(a) for every morphism $\bullet \xrightarrow{f} B^*$ there exists a pullback

$$\begin{array}{ccc} \circ & \xrightarrow{\overline{m}} & \bullet \\ \overline{f} \downarrow & & \downarrow f \\ B & \xrightarrow{m_B} & B^* \end{array}$$

and every such $\overline{m}$ belongs to M,

(b) for every M-partial morphism $(\bullet \xleftarrow{m} \circ \xrightarrow{f} B)$ there exists a unique morphism $\bullet \xrightarrow{f^*} B^*$ such that

$$\begin{array}{ccc} \circ & \xrightarrow{m} & \bullet \\ f \downarrow & & \downarrow f^* \\ B & \xrightarrow{m_B} & B^* \end{array}$$

is a pullback.

451

(3) **A** is said to have **representable M-partial morphisms** provided that for each **A**-object B there exists some $B \xrightarrow{m_B} B^*$ that represents M-partial morphisms into B.

28.2 EXAMPLES

(1) **Set** has representable (extremal) partial morphisms. For each set B the one-point extension $B \hookrightarrow B \uplus \{\infty\}$ represents both types of partial morphisms into B: given $A_0 \subseteq A$ and $A_0 \xrightarrow{f} B$, define $f^* : A \to B \uplus \{\infty\}$ by

$$f^*(x) = \begin{cases} f(x), & \text{for } x \in A_0 \\ \infty, & \text{otherwise.} \end{cases}$$

(2) The category **Alg**(1) of algebras with one unary operation has representable (extremal) partial morphisms. For each object (B, β) the extension $(B, \beta) \xrightarrow{m} ((B \times \mathbb{N}) \cup \{\infty\}, \beta^*)$, defined by $m(b) = (b, 0)$ and

$$\beta^*(x) = \begin{cases} \infty, & \text{if } x = \infty \\ (b, n), & \text{if } x = (b, n+1) \\ (\beta(b), 0), & \text{if } x = (b, 0), \end{cases}$$

represents (extremal) partial morphisms into (B, β).

(3) **Rel** has representable extremal partial morphisms. For each object (X, ρ) the one-point extension

$$(X, \rho) \hookrightarrow (X \uplus \{\infty\}, \rho \cup (X \times \{\infty\}) \cup (\{\infty\} \times X) \cup \{(\infty, \infty)\})$$

represents extremal partial morphisms into (X, ρ). However, **Rel** does not have representable partial morphisms.

(4) **Top** has neither representable extremal partial morphisms nor representable partial morphisms. However, for the class M of all closed [resp. open] embeddings, it does have representable M-partial morphisms. The one-point extension

$$(X, \tau) \hookrightarrow (X \uplus \{\infty\}, \{\emptyset\} \cup \{A \uplus \{\infty\} \mid A \in \tau\})$$

$$[\text{resp. } (X, \tau) \hookrightarrow (X \uplus \{\infty\}, \tau \cup \{X \uplus \{\infty\}\})]$$

represents M-partial morphisms into (X, τ).

(5) Every category has representable Iso-partial morphisms, since these are essentially just morphisms. In particular, every partially ordered set, considered as a category, has representable extremal partial morphisms. However, such a category has representable partial morphisms only if the order relation is equality.

28.3 PROPOSITION

*If **A** has representable M-partial morphisms, then the following hold:*

Sec. 28] Partial morphisms, quasitopoi, and topological universes   453

(1) $\text{Iso}(\mathbf{A}) \subseteq M \subseteq \text{RegMono}(\mathbf{A})$.

(2) Pullbacks of M-morphisms exist and belong to M.

Proof:
(1). Let $A \xrightarrow{f} B$ be a morphism and let $B \xrightarrow{m_B} B^*$ represent M-partial morphisms into B. If f belongs to M, then there exists a morphism $B \xrightarrow{g} B^*$ such that

$$\begin{array}{ccc} A & \xrightarrow{f} & B \\ {\scriptstyle f}\downarrow & & \downarrow{\scriptstyle g} \\ B & \xrightarrow{m_B} & B^* \end{array}$$

is a pullback. This implies that f is an equalizer of g and m_B, and hence is a regular monomorphism.

If f is an isomorphism, then the square

$$\begin{array}{ccc} A & \xrightarrow{f} & B \\ {\scriptstyle f}\downarrow & & \downarrow{\scriptstyle m_B} \\ B & \xrightarrow{m_B} & B^* \end{array}$$

is a pullback since m_B, by the above, is a monomorphism. Thus $f \in M$.

(2). Let $(A \xrightarrow{m} B \xleftarrow{f} C)$ be a 2-sink with $m \in M$, and let $A \xrightarrow{m_A} A^*$ represent M-partial morphisms into A. Then there exist pullbacks

$$\begin{array}{ccc} A & \xrightarrow{m} & B \\ {\scriptstyle id_A}\downarrow & & \downarrow{\scriptstyle g} \\ A & \xrightarrow{m_A} & A^* \end{array} \quad \text{and} \quad \begin{array}{ccc} \bullet & \xrightarrow{\overline{m}} & C \\ {\scriptstyle \overline{f}}\downarrow & & \downarrow{\scriptstyle g \circ f} \\ A & \xrightarrow{m_A} & A^* \end{array}$$

with $\overline{m} \in M$. Thus there exists a unique morphism $\bullet \xrightarrow{k} A$ such that

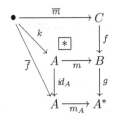

commutes. By Proposition 11.10(2) the square designated by $\boxed{*}$ is a pullback. □

28.4 REMARK

If $\mathbf{A}$ has representable M-partial morphisms, then M need not be closed under composition, as the following example shows: $\mathbf{A}$ is the category of sets and injective maps, and M consists of all injective maps $f \colon A \to B$ such that $B \setminus f[A]$ has at most one element.

28.5 PROPOSITION

If **A** has finite products and representable M-partial morphisms, where M is a family that contains all sections, then the following hold:

(1) **A** is finitely complete.

(2) $M = RegMono(\mathbf{A})$.

Proof:

(1). In view of Theorem 12.4 it suffices to show that any pair of **A**-morphisms $A \underset{g}{\overset{f}{\rightrightarrows}} B$ has an equalizer. Since $\langle id_A, f \rangle \colon A \to A \times B$ is a section, it belongs to M. Hence by Proposition 28.3 there exists a pullback

$$\begin{array}{ccc} \bullet & \xrightarrow{m} & A \\ \downarrow & & \downarrow {\scriptstyle \langle id_A, g \rangle} \\ A & \xrightarrow[\langle id_A, f \rangle]{} & A \times B \end{array}$$

with $m \in M$. By Proposition 11.14, m is an equalizer of f and g.

(2). By Proposition 28.3(1) it suffices to show that every regular monomorphism belongs to M. Let $A \xrightarrow{e} B$ be an equalizer of $B \underset{g}{\overset{f}{\rightrightarrows}} C$. Then

$$\begin{array}{ccc} A & \xrightarrow{e} & B \\ {\scriptstyle e} \downarrow & & \downarrow {\scriptstyle \langle id_B, g \rangle} \\ B & \xrightarrow[\langle id_B, f \rangle]{} & B \times C \end{array}$$

is a pullback. Since $\langle id_B, f \rangle$ is a section, and so belongs to M, by Proposition 28.3(2), e belongs to M. □

28.6 COROLLARY

If **A** has finite products and representable (extremal) partial morphisms, then the following hold:

(1) **A** is finitely complete.

(2) $(Extr)Mono(\mathbf{A}) = RegMono(\mathbf{A})$.

(3) In **A** regular monomorphisms are closed under composition.

Proof: (1) and (2) follow from Proposition 28.5, and (3) follows from (2) and Proposition 7.62(1). □

QUASITOPOI

28.7 DEFINITION
Let M be a class of morphisms in a category **A**. Then **A** is called an M-**topos** provided that it:

(1) has representable M-partial morphisms,

(2) is cartesian closed, and

(3) is finitely cocomplete.

Mono-topoi are called **topoi**, and ExtrMono-topoi are called **quasitopoi**.

28.8 PROPOSITION
Topoi are precisely the balanced quasitopoi.

Proof: Immediate from Proposition 28.5(2) and Proposition 7.67. □

28.9 EXAMPLES
(1) **Set**, **Alg**(1), the category of finite sets, and all functor categories of the form [**A**, **Set**], where **A** is small, are topoi.

(2) **Rel**, **Conv**, and **PsTop** are quasitopoi, but are not topoi.

(3) A linearly ordered set, considered as a category, is

   (a) cartesian closed if and only if it has a largest element,

   (b) a quasitopos if and only if it has a largest element and a smallest element,

   (c) a topos if and only if it has precisely one element.

(4) **Cat** and **kTop** are cartesian closed and cocomplete, hence Iso-topoi; but they are not quasitopoi.

(5) **pSet**, and **PrTop** (equivalently: **Clos**, see 5N) are complete and cocomplete and their extremal partial morphisms are representable, but they fail to be cartesian closed.

28.10 PROPOSITION
Each quasitopos is (Epi, RegMono)-structured.

Proof: Immediate from Proposition 28.6(3) and the dual of Proposition 14.22. □

28.11 REMARKS
Quasitopoi have a rich inner structure:

(1) There is an "extremal-subobject classifier", i.e., a morphism $T \to \Omega$ that represents extremal partial morphisms into a terminal object T.

Cartesian closed categories with subobject classifiers are automatically topoi. But cartesian closed categories with extremal-subobject classifiers may fail badly to be quasitopoi. For example,

(a) **Prost, Cat**, and **kTop** are cartesian closed, cocomplete, and have extremal-sub-object classifiers, but fail to be quasitopoi,

(b) the ordered sets

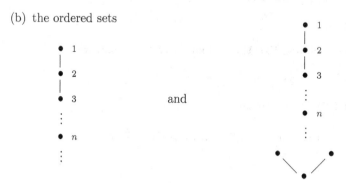

and

considered as categories, are cartesian closed and have representable extremal partial morphisms, but fail to be finitely cocomplete.

(2) For each object A there is an "elementhood-morphism" ε_A, defined by the pullback

(3) For each object A there is a "singleton-morphism" $\sigma_A \colon A \to \Omega^A$, defined by the diagram

$$A \xrightarrow{\langle id_A, id_A \rangle} A \times A \xrightarrow{id_A \times \sigma_A} A \times \Omega^A$$

with pullback to $T \to \Omega$ and ev.

(4) For each morphism $A \xrightarrow{f} B$ there is a "preimage-morphism" $f^* \colon B \to \Omega^A$, defined by the diagram

$$A \xrightarrow{\langle id_A, f \rangle} A \times B \xrightarrow{id_A \times f^*} A \times \Omega^A$$

with pullback to $T \to \Omega$ and ev.

Sec. 28] Partial morphisms, quasitopoi, and topological universes   457

(5) For each morphism $A \xrightarrow{f} B$ there is a "power-set-morphism" $\Omega^B \xrightarrow{\Omega f} \Omega^A$, defined by the diagram

$$\begin{array}{ccc} A \times \Omega^B & \xrightarrow{f \times id} & B \times \Omega^B \\ {\scriptstyle id_A \times \Omega f} \downarrow & & \downarrow {\scriptstyle ev} \\ A \times \Omega^A & \xrightarrow{ev} & \Omega \end{array}$$

(6) For each morphism $A \xrightarrow{f} B$ there is a "pullback along f" functor $(\mathbf{A} \downarrow B) \to (\mathbf{A} \downarrow A)$ between the corresponding comma categories that is simultaneously an adjoint and a co-adjoint functor.

CONCRETE QUASITOPOI

If a topological category $(\mathbf{A}, U)$ has representable partial morphisms, then, by Proposition 28.3(1), $\mathbf{A}$ is balanced, so that U is an isomorphism (cf. Exercise 21N). Thus for topological categories the concepts of representability of partial morphisms, and in particular of topoi, are not interesting. However, the concepts of representability of extremal partial morphisms (and in particular of quasitopoi) are fruitful.

28.12 PROPOSITION
Let $(\mathbf{A}, U)$ be a topological category over $\mathbf{X}$, let M be a class of morphisms in $\mathbf{X}$, and let M_{init} be the class of all initial $\mathbf{A}$-morphisms m with $Um \in M$. If $\mathbf{A}$ has representable M_{init}-partial morphisms, then $\mathbf{X}$ has representable M-partial morphisms. U preserves these representations if and only if M_{init}-subobjects of discrete objects are discrete.

Proof:
(1). To show that $\mathbf{X}$ has representable M-partial morphisms, consider the full subcategory $\mathbf{B}$ of $\mathbf{A}$ that consists of all indiscrete objects. Since U restricts to an isomorphism from $\mathbf{B}$ to $\mathbf{X}$, and since $\mathbf{B}$ is closed under the formation of pullbacks in $\mathbf{A}$, it suffices to show that $\mathbf{B}$ is closed under representations $B \xrightarrow{m_B} B^*$ of M_{init}-partial morphisms in $\mathbf{A}$, i.e., that $B \in \mathbf{B}$ implies that $B^* \in \mathbf{B}$. For this purpose let $UA \xrightarrow{f} UB^*$ be an $\mathbf{X}$-morphism. Let D be the discrete object with $UD = UA$, and let

$$\begin{array}{ccc} P & \xrightarrow{\hat{m}} & D \\ {\scriptstyle \hat{f}} \downarrow & & \downarrow {\scriptstyle f} \\ B & \xrightarrow{m_B} & B^* \end{array}$$

be a pullback in $\mathbf{A}$. Then $\hat{m} \in M_{init}$. Since $B \xrightarrow{m_B} B^*$ represents M_{init}-partial morphisms into B and since D is discrete, $UD \xrightarrow{f} UB^*$ is the only $\mathbf{X}$-morphism g

that makes

$$\begin{array}{ccc} UP & \xrightarrow{\hat{m}} & UD \\ \hat{f}\downarrow & & \downarrow g \\ UB & \xrightarrow{m_B} & UB^* \end{array} \qquad (*)$$

a pullback in **X**. Let $C \xrightarrow{\hat{m}} A$ be an initial lift of $UP \xrightarrow{\hat{m}} UA$. Then $C \xrightarrow{\hat{m}} A$ belongs to M_{init}. Moreover, $UC \xrightarrow{\hat{f}} UB$ is an **A**-morphism, since B is indiscrete. Hence $A \xleftarrow{\hat{m}} C \xrightarrow{\hat{f}} B$ is an M_{init}-partial morphism from A to B. Thus there exists an **A**-morphism $A \xrightarrow{g} B^*$ that makes

$$\begin{array}{ccc} C & \xrightarrow{\hat{m}} & A \\ \hat{f}\downarrow & & \downarrow g \\ B & \xrightarrow{m_B} & B^* \end{array}$$

a pullback in **A**. Hence

$$\begin{array}{ccc} UC & \xrightarrow{\hat{m}} & UA \\ \hat{f}\downarrow & & \downarrow g \\ UB & \xrightarrow{m_B} & UB^* \end{array}$$

a pullback in **X**. By $(*)$ this implies that $g = f$. Hence $UA \xrightarrow{f} UB^*$ is an **A**-morphism.

(2). That U preserves representations if and only if M_{init}-subobjects of discrete objects are discrete follows by a straightforward calculation. □

28.13 DEFINITION

(1) Let $B \xrightarrow{m} A$ be a morphism and let $\mathcal{S} = (A_i \xrightarrow{f_i} A)$ be a sink. If for each $i \in I$ the diagram

$$\begin{array}{ccc} B_i & \xrightarrow{m_i} & A_i \\ \hat{f_i}\downarrow & & \downarrow f_i \\ B & \xrightarrow{m} & A \end{array}$$

is a pullback, then the sink $(B_i \xrightarrow{\hat{f_i}} B)_I$ is called a **pullback of $\mathcal{S}$ along** m.

(2) Let M be a class of morphisms and let $\mathcal{C}$ be a conglomerate of sinks in a category **A**. $\mathcal{C}$ is called **stable under pullbacks along** M provided that every pullback of a sink in $\mathcal{C}$ along a M-morphism is a member of $\mathcal{C}$.
In particular, $\mathcal{C}$ is called **pullback-stable** provided that $\mathcal{C}$ is stable under pullbacks along $Mor(\mathbf{A})$. In the case that M is a class of monomorphisms, $\mathcal{C}$ is called **reducible** provided that it is stable under pullbacks along M. When M is the class of all extremal monomorphisms, we say that $\mathcal{C}$ is **extremally reducible**.

28.14 REMARK
In view of the fact that diagrams of the form

$$\begin{array}{ccc} A \times B & \xrightarrow{\pi_B} & B \\ {\scriptstyle id_A \times f} \downarrow & & \downarrow {\scriptstyle f} \\ A \times C & \xrightarrow{\pi_C} & C \end{array}$$

are pullbacks (see Exercise 11D), Theorem 27.15 can be restated as follows:

If **X** is cartesian closed, then for each topological category $(\mathbf{A}, U)$ over **X** the following conditions are equivalent:

(1) $(\mathbf{A}, U)$ is concretely cartesian closed,

(2) final sinks in $(\mathbf{A}, U)$ are stable under pullbacks along projections.

28.15 THEOREM
If **X** has representable M-partial morphisms, then for each topological category $(\mathbf{A}, U)$ over **X** the following conditions are equivalent:

(1) **A** has representable M_{init}-partial morphisms and U preserves these representations,

(2) final sinks in $(\mathbf{A}, U)$ are M_{init}-reducible.

Proof: (1) $\Rightarrow$ (2). Let $\mathcal{S} = (A_i \xrightarrow{f_i} A)_I$ be a final sink and let $B \xrightarrow{m} A$ belong to M_{init}. Then for each $i \in I$ there exists a pullback

$$\begin{array}{ccc} B_i & \xrightarrow{m_i} & A_i \\ {\scriptstyle \hat{f}_i} \downarrow & & \downarrow {\scriptstyle f_i} \\ B & \xrightarrow{m} & A \end{array}$$

with $m_i \in M_{\text{init}}$. To show that the sink $\mathcal{T} = (B_i \xrightarrow{\hat{f}_i} B)_I$ is final, let $UB \xrightarrow{g} UC$ be an **X**-morphism such that all $UB_i \xrightarrow{g \circ \hat{f}_i} UC$ are **A**-morphisms. Let $C \xrightarrow{m_C} C^*$ represent M_{init}-partial morphisms into C. Then, by (1), $UC \xrightarrow{m_C} UC^*$ represents M-partial morphisms into UC. Hence there exists a unique **X**-morphism $UA \xrightarrow{\hat{g}} UC^*$ such that

$$\begin{array}{ccc} UB & \xrightarrow{m} & UA \\ {\scriptstyle g} \downarrow & & \downarrow {\scriptstyle \hat{g}} \\ UC & \xrightarrow{m_C} & UC^* \end{array}$$

is a pullback in **X**, and for each $i \in I$ there is a unique **A**-morphism $A_i \xrightarrow{g_i} C^*$ such that

$$\begin{array}{ccc} B_i & \xrightarrow{m_i} & A_i \\ {\scriptstyle g \circ \hat{f}_i} \downarrow & & \downarrow {\scriptstyle g_i} \\ C & \xrightarrow{m_C} & C^* \end{array}$$

is a pullback in **A**. Thus, for each $i \in I$,

$$UB_i \xrightarrow{m_i} UA_i \qquad UB_i \xrightarrow{m_i} UA_i$$
$$g\circ \hat{f}_i \downarrow \quad \downarrow g_i \quad \text{and} \quad g\circ \hat{f}_i \downarrow \quad \downarrow \hat{g}\circ f_i$$
$$UC \xrightarrow{m_C} UC^* \qquad UC \xrightarrow{m_C} UC^*$$

are pullbacks in **X**. Hence $UA_i \xrightarrow{g_i} UC^* = UA_i \xrightarrow{f_i} UA \xrightarrow{\hat{g}} UC^*$. Since $\mathcal{S}$ is final, $UA \xrightarrow{\hat{g}} UC^*$ is an **A**-morphism. Thus $UB \xrightarrow{g} UC \xrightarrow{m_C} UC^* = UB \xrightarrow{m} UA \xrightarrow{\hat{g}} UC^*$ is an **A**-morphism. Since $C \xrightarrow{m_C} C^*$ is initial, $UB \xrightarrow{g} UC$ is an **A**-morphism.

(2) $\Rightarrow$ (1). Let B be a **B**-object and let $UB \xrightarrow{m_B} X$ represent M-partial morphisms into UB. Consider the family $(A_i \xleftarrow{m_i} B_i \xrightarrow{f_i} B)_I$ of all M_{init}-partial morphisms into B. Then for each $i \in I$ there exists a unique **X**-morphism $UA_i \xrightarrow{g_i} X$ such that

$$UB_i \xrightarrow{m_i} UA_i$$
$$f_i \downarrow \quad \downarrow g_i$$
$$UB \xrightarrow{m_B} X$$

is a pullback in **X**. Let $(A_i \xrightarrow{g_i} B^*)_I$ be a final lift of the structured sink $(UA_i \xrightarrow{g_i} X)_I$. In order to prove that $B \xrightarrow{m_B} B^*$ represents M_{init}-partial morphisms into B it suffices to show that $UB \xrightarrow{m_B} UB^*$ belongs to M_{init}. Since

$$UB \xrightarrow{id_B} UB$$
$$id_{UB} \downarrow \quad \downarrow m_B$$
$$UB \xrightarrow{m_B} X$$

is a pullback, there exists an $i \in I$ with $(A_i, g_i) = (B, m_B)$. Thus $UB \xrightarrow{m_B} UB^* = UA_i \xrightarrow{g_i} UB^*$ is an **A**-morphism. If $A \xrightarrow{m_B} B^*$ is an initial lift of $UB \xrightarrow{m_B} UB^*$, then $B \leq A$ (cf. 5.4). So it remains to be shown that $A \leq B$. For each $i \in I$ the diagram

$$B_i \xrightarrow{m_i} A_i$$
$$f_i \downarrow \quad \downarrow g_i$$
$$A \xrightarrow{m_B} B^*$$

is a pullback in **A** since it is a pullback in **X** and $B_i \xrightarrow{m_i} A_i$ is initial. Hence, by (2), the sink $(B_i \xrightarrow{f_i} A)_I$ is final. Thus $UB_i \xrightarrow{f_i} UB = UB_i \xrightarrow{f_i} UA \xrightarrow{id_{UA}} UB$ implies that $UA \xrightarrow{id_{UA}} UB$ is an **A**-morphism, i.e., that $A \leq B$. □

28.16 DEFINITION

A concrete category is called **universally topological** provided that it is topological and final sinks are pullback-stable.

28.17 EXAMPLES

(1) **Rel** is a universally topological construct.

(2) If a functor $T : \mathbf{X} \to \mathbf{Set}$ weakly preserves pullbacks [cf. Exercise 27I], then $\mathbf{Spa}(T)$ is universally topological over $\mathbf{X}$. In particular, $\mathbf{Spa}(\mathcal{P})$ is universally topological.

(3) Frames are (cartesian closed and hence) universally topological over **1**.

(4) The construct of compactly generated topological spaces is not universally topological (although it is cartesian closed and topological). In fact, a "non-concretely" cartesian closed construct cannot be universally topological:

28.18 THEOREM
For topological categories $(\mathbf{A}, U)$ *over a quasitopos* $\mathbf{X}$ *the following conditions are equivalent:*

(1) $(\mathbf{A}, U)$ *is universally topological,*

(2) $(\mathbf{A}, U)$ *is a concrete quasitopos, i.e.,* $\mathbf{A}$ *is a quasitopos and* U *preserves evaluation, power objects, and representations of extremal partial morphisms.*

Proof: (1) $\Rightarrow$ (2). Immediate from Remark 28.14, Theorem 28.15, and Proposition 28.12.

(2) $\Rightarrow$ (1). By the above-mentioned results, (2) implies that in $(\mathbf{A}, U)$ final sinks are stable under pullbacks along projections and along extremal monomorphisms. Since each $\mathbf{A}$-morphism $A \xrightarrow{f} B$ can be expressed as $A \xrightarrow{f} B = A \xrightarrow{\langle id_A, f \rangle} A \times B \xrightarrow{\pi_B} B$, since $\langle id_A, f \rangle$ is a section and hence is an extremal monomorphism, and since pullbacks can be composed [cf. Proposition 11.10(1)], this implies that $(\mathbf{A}, U)$ is universally topological.□

TOPOLOGICAL UNIVERSES

28.19 THEOREM
For well-fibred topological constructs, the following conditions are equivalent:

(1) extremal partial morphisms are representable,

(2) final sinks are extremally reducible,

(3) final epi-sinks are extremally reducible,

(4) coproducts and quotients are extremally reducible.

Proof: (1) $\Leftrightarrow$ (2) is immediate from Theorem 28.15. Cf. Exercise 28B(a).

(2) $\Rightarrow$ (3) is obvious.

(3) ⇒ (2). Let O (resp. T) be the unique object with underlying set $\emptyset$ (resp. $\{0\}$). If $\mathcal{S} = (A_i \xrightarrow{f_i} A)_I$ is a final sink, enlarge $\mathcal{S}$ to a final epi-sink $\mathcal{T} = (A_j \xrightarrow{f_j} A)_J$ by adding all morphisms from T to A. Let $B \xrightarrow{m} A$ be an extremal monomorphism and let

$$\begin{array}{ccc} B_j & \xrightarrow{\hat{m}_j} & A_j \\ \hat{f}_j \downarrow & & \downarrow f_j \\ B & \xrightarrow{m} & A \end{array}$$

be a pullback for each $j \in J$. Then $(B_j \xrightarrow{\hat{f}_j} B)_J$ is a final epi-sink. Since for each $j \in J \setminus I$ the morphism $\hat{f}_j$ has a discrete domain (namely, O or an object isomorphic to T), the restricted sink $(B_i \xrightarrow{\hat{f}_i} B)_I$ is final as well.

(3) ⇔ (4) as in Theorem 27.22. □

28.20 EXAMPLES

(1) Extremal partial morphisms are representable in each of the following well-fibred topological constructs: **Conv**, **PsTop**, **PrTop** (= pretopological spaces = closure spaces), **Bor**, **Simp**, and **Rere**.

(2) Extremal partial morphisms are not representable for any of the following well-fibred topological constructs: **Top**, **Unif**, **PMet**, **sTop**, **kTop**, and **Prost**.

28.21 DEFINITION

A well-fibred topological construct $(\mathbf{A}, U)$ for which $\mathbf{A}$ is a quasitopos is called a **topological universe**.

A topological universe

28.22 THEOREM
For a well-fibred topological construct $(\mathbf{A}, U)$ the following conditions are equivalent:

(1) $(\mathbf{A}, U)$ is a topological universe,

(2) $(\mathbf{A}, U)$ has function-spaces and representable extremal partial morphisms,

(3) in $(\mathbf{A}, U)$ final epi-sinks are pullback-stable.

Proof: (1) ⇔ (2) is immediate from Theorem 27.22.

(2) ⇔ (3) can be established as in the proof of Theorem 28.18. □

28.23 EXAMPLES
Conv, **PsTop**, **Simp**, **Bor**, and **Rere** are topological universes.

Suggestions for Further Reading

Lawvere, F. W. Introduction. *Springer Lect. Notes Math.* **274** (1972): 1–12.

Freyd, P. Aspects of topoi. *Bull. Austral. Math. Soc.* **7** (1972): 1–76 and 467–480.

Paré, R. Co-limits in topoi. *Bull. Amer. Math. Soc.* **80** (1974): 556–561.

Wyler, O. Are there topoi in topology? *Springer Lect. Notes Math.* **540** (1976): 700–719.

Penon, J. Sur les quasi-topos. *Cahiers Topol. Geom. Diff.* **18** (1977): 181–218.

Booth, P. I., and R. Brown. Spaces of partial maps, fibred mapping spaces and the compact-open topology. *Gen. Topol. Appl.* **8** (1978): 181–195.

Dubuc, E. J. Concrete quasitopoi. *Springer Lect. Notes Math.* **753** (1979): 239–254.

Lambek, J., and P. Scott. Algebraic aspects of topos theory. *Cahiers Topol. Geom. Diff.* **22** (1981): 129–140.

Nel, L. D. Topological universes and smooth Gelfand-Naimark duality. *Contemp. Math.* **30** (1984): 244–276.

Dyckhoff, R., and W. Tholen. Exponentiable morphisms, partial products and pullback complements. *J. Pure Appl. Algebra* **49** (1987): 103–116.

Herrlich, H. On the representability of partial morphisms in **Top** and in related constructs. *Springer Lect. Notes Math.* **1348** (1988): 143–153.

Schwarz, F. Description of the topological universe hull. *Springer Lect. Notes Comp. Sci.* **393** (1989): 325–332.

EXERCISES

28A. Injectivity in Topological Constructs

Let $(\mathbf{A}, U)$ be a well-fibred topological construct with representable extremal partial morphisms. Show that in $(\mathbf{A}, U)$

(a) An object B is injective if and only if there exists some $A \xrightarrow{m} B$ that represents extremal partial morphisms into A.

(b) $A \xrightarrow{m} B$ is an injective hull if and only if either

   (1) A is injective and m is an isomorphism, or

   (2) A is not injective and $A \xrightarrow{m} B$ represents extremal partial morphisms into A.

28B. Extensional Topological Constructs

A well-fibred topological construct $(\mathbf{A}, U)$ is called **extensional** provided that it has representable extremal partial morphisms.

(a) Show that if $(\mathbf{A}, U)$ is extensional then U preserves representations of extremal partial morphisms. [Cf. Proposition 28.12 and Exercise 27H(a).]

(b) For each $\mathbf{A}$-object B denote by B^* the final lift of the structured sink consisting of all $A \xrightarrow{a} |B| \uplus \infty$ whose restriction a' to the initial subobject A' of A on the set $a^{-1}[|B|]$ is an $\mathbf{A}$-morphism $A' \xrightarrow{a'} B$. If the inclusion map $B \hookrightarrow B^*$ is an initial (= extremal) monomorphism, we call B^* the **canonical one-point extension** of B. Prove that a well-fibred topological construct is extensional if and only if every $\mathbf{A}$-object has a canonical one-point extension.

* (c) Let $\mathbf{CONST_s}$ be the quasicategory, whose objects are the well-fibred constructs in which every injective structured arrow has a unique initial lift, and whose morphisms are the concrete functors that preserve embeddings. Show that the injective objects in $\mathbf{CONST_s}$ (with respect to full embeddings) are precisely the extensional topological constructs.

* 28C. Extensional Topological Hulls

A morphism $(\mathbf{A}, U) \xrightarrow{E} (\mathbf{B}, V)$ in $\mathbf{CONST_s}$ is called an **extensional topological hull** of $(\mathbf{A}, U)$ provided that the following conditions are satisfied:

(1) E is a full embedding,

(2) $E[\mathbf{A}]$ is finally dense in $(\mathbf{B}, V)$,

(3) the image under E of

$$\{ B_A \mid A \in Ob(\mathbf{A}) \text{ and } A \xrightarrow{m_A} B_A \text{ represents extremal partial morphisms into } A \}$$

is initially dense in $(\mathbf{B}, V)$,

(4) $(\mathbf{B}, V)$ is an extensional topological construct.

Show that

(a) Injective hulls in **CONST**$_s$ are precisely the extensional topological hulls.

(b) Every object in $(\mathbf{A}, U)$ has an injective hull.

(c) The concrete embedding **Top** $\hookrightarrow$ **PrTop** is an extensional topological hull of **Top**.

28D. Posets as Quasitopoi

Let **A** be a poset, considered as a category. Show that:

(a) **A** is a topos if and only if **A** contains precisely one element.

(b) If **A** is complete, then **A** is a quasitopos if and only if **A** is a frame.

28E. A Characterization of Topoi

(a) Show that **A** is a topos if and only if **A** is cartesian closed and has representable partial morphisms.

(b) Show that the category **A** of nonempty sets has the following properties:

  (1) **A** is cartesian closed,

  (2) for each **A**-object B there exists a monomorphism $B \xrightarrow{m_B} B^*$ such that for every partial morphism $\bullet \xleftarrow{m} \circ \xrightarrow{f} B$ there exists a unique morphism $\bullet \xrightarrow{f^*} B$ such that

$$\begin{array}{ccc} \circ & \xrightarrow{m} & \bullet \\ f \downarrow & & \downarrow f^* \\ B & \xrightarrow{m_B} & B^* \end{array}$$

  is a pullback,

  (3) in **A** partial morphisms are not representable.

28F. Regular Monomorphisms in (Quasi)Topoi

Show that:

(a) In a quasitopos pushouts of regular monomorphisms are monomorphisms.

(b) In a topos monomorphisms and regular monomorphisms are pushout-stable.

(c) In a (quasi)topos every (extremal) monomorphism is regular.

* 28G. Universally Topological Categories as Injective Objects

Let **X** be a quasitopos and let **CAT**$_{\mathbf{ps}}$ be the quasicategory whose objects are the concrete categories over **X** with finite concrete products and unique initial lifts of regular monomorphic structured arrows, and whose morphisms are the concrete functors that preserve finite products and those embeddings that are carried by regular monomorphisms.

(a) Show that the injective objects in $\mathbf{CAT_{ps}}$ (with respect to full embeddings) are precisely the universally topological categories.

(b) Describe injective hulls in $\mathbf{CAT_{ps}(X)}$.

(c) Show that the concrete embedding $\mathbf{Top} \hookrightarrow \mathbf{PsTop}$ is an injective hull in the quasi-category $\mathbf{CAT_{ps}(Set)}$.

28H. Relational Categories

For every small concrete category $(\mathbf{A}, U)$ over $\mathbf{X}$ the concrete category $\mathbf{Rel}(\mathbf{A}, U)$ of $(\mathbf{A}, U)$-**ary relations** over $\mathbf{X}$ is $\mathbf{Spa}(F_{(\mathbf{A},U)})$, where $F_{(\mathbf{A},U)}: \mathbf{X} \to \mathbf{Set}$ is the functor, defined by

$$F_{(\mathbf{A},U)}(X) = \{(A, a) \mid A \in Ob(\mathbf{A}) \text{ and } UA \xrightarrow{a} X \in Mor(\mathbf{X})\}$$

and $F_{(\mathbf{A},U)}(f)(A, a) = (A, f \circ a)$. A concrete category that is concretely isomorphic to some $\mathbf{Rel}(\mathbf{A}, U)$ is called **relational**. Show that

(a) **Rel** is a relational construct.

(b) Every relational category is functor-structured.

(c) If a functor-costructured construct $(\mathbf{A}, U)$ is relational, then U is an isomorphism.

(d) **Top** is not concretely isomorphic to a full subconstruct of some relational construct.

(e) Relational categories are universally topological.

28I. Functor-Costructured Constructs

Prove that for functor-costructured constructs $(\mathbf{A}, U)$, the following conditions are equivalent:

(1) $(\mathbf{A}, U)$ is relational,

(2) $(\mathbf{A}, U)$ is functor-structured,

(3) $(\mathbf{A}, U)$ is universally topological,

(4) $(\mathbf{A}, U)$ is cartesian closed and there exists only one **A**-object A with $UA = \emptyset$,

(5) $(\mathbf{A}, U)$ is well-fibred,

(6) U is an isomorphism.

28J. Products of Final Morphisms

Prove that

(a) In relational categories, products of final morphisms are final.

(b) In universally topological categories, finite products of final morphisms are final.

(c) If $T: \mathbf{Set} \to \mathbf{Set}$ is defined by $T(X)$ is the set of all finite subsets of X, and $Tf(Y) = f[Y]$, then $\mathbf{Spa}(T)$ is universally topological, but in $\mathbf{Spa}(T)$ countable products of final morphisms can fail to be final.

Sec. 28]   Partial morphisms, quasitopoi, and topological universes   467

(d) In concretely cartesian closed topological categories, finite products of final maps are final.

(e) In a well-fibred topological construct $(\mathbf{A}, U)$, finite products of final maps are final if and only if U is an isomorphism.

(f) In a topological construct with function spaces, finite products of quotient maps are quotient maps.

(g) A topological full subconstruct $(\mathbf{A}, U)$ of **Top** has function spaces if and only if in $(\mathbf{A}, U)$ finite products of quotient maps are quotient maps.

28K. Functor-Structured Categories

Prove that $\mathbf{Spa}(T)$ is universally topological if and only if the base category is a quasitopos and T weakly preserves pullbacks (cf. Exercise 27I).

28L. Re: Theorems 27.22 and 28.19

Let $\mathbf{A}$ (resp. $\mathbf{B}$) be the full subconstruct of **Rel** (resp. **Rere**) that consists of all objects (X, ρ) that have no proper cycles; i.e., $x_1 \rho x_2$, $x_2 \rho x_3$, $\cdots$, $x_{n-1} \rho x_n$ and $x_1 = x_n$ imply that $x_1 = x_2 = \cdots = x_n$. Show that the following hold:

(a) $\mathbf{A}$ (resp. $\mathbf{B}$) is closed under the formation of mono-sources in **Rel** (resp. **Rere**),

(b) $\mathbf{A}$ and $\mathbf{B}$ are monotopological constructs,

(c) $\mathbf{A}$ and $\mathbf{B}$ are wellpowered and co-wellpowered (epimorphisms are surjective),

(d) $\mathbf{A}$ and $\mathbf{B}$ are locally presentable,

(e) $\mathbf{B}$ is well-fibred; $\mathbf{A}$ is not,

(f) in $\mathbf{A}$ final sinks are pullback stable; in $\mathbf{B}$ final epi-sinks are pullback stable,

(g) in $\mathbf{B}$ all functors of the form $(B \times -)$ preserve colimits; in $\mathbf{A}$ some functors of the form $(A \times -)$ don't preserve coequalizers,

(h) $\mathbf{B}$ is cartesian closed; $\mathbf{A}$ is not,

(i) neither $\mathbf{A}$ nor $\mathbf{B}$ has representable extremal partial morphisms; neither even has an extremal-subobject classifier,

(j) neither $\mathbf{A}$ nor $\mathbf{B}$ is a quasitopos.

28M. Re: Theorem 28.19

Show that for a topological construct $\mathbf{A}$ the following are equivalent:

(1) in $\mathbf{A}$ final sinks are extremally reducible,

(2) in $\mathbf{A}$
  (i) final epi-sinks are extremally reducible, and
  (ii) subobjects of discrete objects are discrete.

Observe that condition (2)(ii) is automatically satisfied whenever the fibre of the empty set has exactly one element.

BIBLIOGRAPHY

Adámek, J. *Theory of Mathematical Structures.* Reidel, Dordrecht–Boston–Lancaster, 1983.

Adámek, J., and J. Rosicky. *Locally Presentable and Accessible Categories.* Cambridge University Press, 1993.
Corrections under http://www.iti.cs.tu-bs.de/~adamek/corrections.ps.

Adámek, J., and V. Trnková. *Automata and Algebras in Categories.* Kluwer, Dordrecht, 1989.

Arbib, M. A. *Theories of Abstract Automata.* Prentice-Hall, Englewood Cliffs, NJ, 1969.

Arbib, M. A., and E. G. Manes. *Arrows, Structures and Functors. The Categorical Imperative.* Academic Press, New York, 1975.

Barr, M. *∗-Autonomous Categories.* Lecture Notes in Mathematics, Vol. 752. Springer, Berlin–Heidelberg–New York, 1979.

Barr, M., P. A. Grillet, and D. H. van Osdol. *Exact Categories and Categories of Sheaves.* Lecture Notes in Mathematics, Vol. 236. Springer, Berlin–Heidelberg–New York, 1971.

Barr, M., and C. Wells. *Category Theory for Computer Science.* Third edition, CRM Montreal, 1999.

Toposes, Triples and Theories. Springer, Berlin–Heidelberg–New York, 1985. Revised and corrected version:
http://www.cwru.edu/artsci/math/wells/pub/ttt.html

Bell, J. L. *Toposes and Local Set Theories — An Introduction.* Oxford Logic Guides, Vol. 14. Clarendon Press, New York–Oxford, 1988.

Bénabou, J. *Introduction to Bicategories.* Lecture Notes in Mathematics, Vol. 47. Springer, Berlin–Heidelberg–New York, 1967.

Structures Algébriques dans les Catégories. Cahiers Topol. Géom. Diff. **10** (1968): 1–126.

Blyth, T. S. *Categories.* Wiley, New York, 1986.

Borceaux, F. *Handbook of Categorical Algebra.* Three volumes, Cambridge University Press, 1994.

Borceaux, F., and G. Janelidze. *Galois Theories.* Cambridge University Press, 2001.

Brinkmann, H.-B., and D. Puppe. *Kategorien und Funktoren.* Lecture Notes in Mathematics, Vol. 18. Springer, Berlin–Heidelberg–New York, 1966.

Abelsche und exakte Kategorien, Korrespondenzen. Lecture Notes in Mathematics, Vol. 96. Springer, Berlin–Heidelberg–New York, 1969.

BIBLIOGRAPHY

Bucur, I., and A. Deleanu (with the collaboration of P. J. Hilton and N. Popescu). *Introduction to the Theory of Categories and Functors.* Wiley, New York, 1968.

Budach, L. *Quotientenfunktoren und Erweiterungstheorie.* Mathematische Forschungsberichte, Vol. 22. Deutscher Verlag der Wissenschaften, Berlin, 1967.

Budach, L., and H.-J. Hoehnke. *Automaten und Funktoren.* Akademie Verlag, Berlin, 1975.

Budach, L., and H.-P. Holzapfel. *Localizations and Grothendieck Categories,* Deutscher Verlag der Wissenschaften, Berlin, 1975.

Burmeister, P., and H. Reichel. *Introduction to Theory and Application of Partial Algebras.* Akademie Verlag, Berlin, 1986.

Castellini, G. *Categorical Closure Operators.* Birkhäuser, Boston–Basel–Berlin, 2003.

Diers, Y. *Categories of Commutative Algebras.* Oxford University Press, Oxford, 1992.

Dubuc, E. *Kan Extensions in Enriched Category Theory.* Lecture Notes in Mathematics, Vol. 145. Springer, Berlin–Heidelberg–New York, 1970.

Ehresmann, C. *Catégories et Structures.* Dunod, Paris, 1965.

Prolongements Universels d'un Foncteur par Adjonction de Limites. Diss. Math. Rozprawy Mat., Vol. 64, 1969.

Oevres Complétes et Commentées I-IV (ed. by A. C. Ehresmann). Cahiers Topol. Géom. Diff., Vols. 20-24, (1979–83): Supplements.

Ehrig, H., K. D. Kiermeier, H.-J. Kreowski, and W. Kühnel. *Universal Theory of Automata.* Teubner, Stuttgart–Leipzig, 1974.

Eilenberg, S. *Automata, Languages and Machines Vol.I.* Academic Press, New York, 1974.

Eilenberg, S., and S. Mac Lane. *Collected Works.* Academic Press, New York, 1986.

Freyd, P. *Abelian Categories. An Introduction to the Theory of Functors.* Harper & Row, New York, 1964.

Freyd, P., and A. Scedrov. *Categories, Allegories.* North-Holland, Amsterdam, 1990.

Frölicher A., and A. Kriegl. *Linear Spaces and Differentiation Theory.* Wiley, New York, 1988.

Gabriel, P., and F. Ulmer. *Lokal präsentierbare Kategorien.* Lecture Notes in Mathematics, Vol. 221. Springer, Berlin–Heidelberg–New York, 1971.

Goldblatt, R. *Topoi.* North Holland, Amsterdam, 1984.

Gorey, E. $Cat^e gor\ y$. Adama Books, 1973.

Grätzer, G. *Universal Algebra.* Van Nostrand, New York, 1968.

BIBLIOGRAPHY

Grothendieck, A. *Catégories Cofibrées Additives et Complexe Cotangent Relatif.* Lecture Notes in Mathematics, Vol. 79. Springer, Berlin–Heidelberg–New York, 1968.

Hasse, M., and L. Michler. *Theorie der Kategorien.* Mathematische Monographien, Vol. 7. Deutscher Verlag der Wissenschaften, Berlin, 1966.

Henkin, L., J. D. Monk, and A. Tarski. *Cylindric Algebras Part II.* North Holland, Amsterdam, 1985.

Herrlich, H. *Topologische Reflexionen und Coreflexionen.* Lecture Notes in Mathematics, Vol. 78. Springer, Berlin–Heidelberg–New York, 1968.

Herrlich, H., and G. E. Strecker. *Category Theory.* Allyn Bacon, Boston, 1973, 2nd ed. Heldermann, Berlin, 1979.

Isbell, J. R. *Subobjects, Adequacy, Completeness and Categories of Algebras.* Diss. Math. Rozprawy Mat., Vol. 36, 1964.

Johnstone, P. T. *Topos Theory.* Academic Press, New York, 1977.

Stone Spaces. Cambridge Univ. Press, Cambridge–New York–Melbourne, 1982.

Sketches of an Elephant. A Topos Theory Compendium. Two volumes, Oxford University Press, Oxford, 2002.

Kelly, G. M. *Basic Concepts of Enriched Category Theory.* Cambridge Univ. Press, Cambridge–New York–Melbourne, 1982.

Kuroš, A. G., A. H. Livšic, E. G. Šul'geifer, and M. S. Calenko. *Zur Theorie der Kategorien,* Mathematische Forschungsberichte, Vol. 15. Deutscher Verlag der Wissenschaften, Berlin, 1963.

Lambek, J. *Completions of Categories.* Lecture Notes in Mathematics, Vol. 24. Springer, Berlin–Heidelberg–New York, 1966. tablecat.tex

Lambek, J., and P. J. Scott. *Introduction to Higher-Order Categorical Logic.* Cambridge Univ. Press, Cambridge–New York–Melbourne, 1986.

Leinster, Tom. *Higher Operads, Higher Categories,* Cambridge University Press, 2003. Available from http://www.arxiv.org/abs/math.CT/0305049

Mac Lane, S. *Categories for the Working Mathematician.* Springer, Berlin–Heidelberg–New York, 2nd ed., 1997.

Mac Lane, S., and I. Moerdijk. *Sheaves in Geometry and Logic.* Springer, Berlin–Heidelberg–New York, 1992.

Mal'cev, A. I. *Algebraic Systems.* Die Grundlehren der mathematischen Wissenschaften, Vol. 192. Springer, Berlin–Heidelberg–New York, 1972.

McLarty, C. *Elementary Categories, Elementary Toposes.* Oxford University Press, Oxford, 1992.

Makkai, M., and R. Paré. *Accessible Categories: The Foundations of Categorical Model Theory.* American Mathematical Society, 1989.

Manes, E. G. *Algebraic Theories.* Springer, Berlin–Heidelberg–New York, 1976.

Michor, P. W. *Functors and Categories of Banach Spaces.* Lecture Notes in Mathematics, Vol. 651. Springer, Berlin–Heidelberg–New York, 1978.

Mitchell, B. *Theory of Categories.* Pure and Applied Mathematics **17**. Academic Press, New York, 1965.

Morita, K., and J. Nagata (editors). *Topics in General Topology.* North Holland, Amsterdam, 1989.

Pareigis, B. *Categories and Functors.* Academic Press, New York, 1970.

Popescu, N. *Abelian Categories with Applications to Rings and Modules.* Academic Press, New York, 1973.

Preuss, G. *Theory of Topological Structures.* Reidel, Dordrecht–Boston–Lancaster, 1988.

Pultr, A., and V. Trnková. *Combinatorial, Algebraic and Topological Representations of Groups, Semigroups and Categories.* North Holland, Amsterdam, 1980.

Richter, G. *Kategorielle Algebra.* Akademie Verlag, Berlin, 1979.

Rivano, N. S. *Catégories Tannakiennes.* Lecture Notes in Mathematics, Vol. 265. Springer, Berlin–Heidelberg–New York, 1972.

Rosenthal, K. I. *The Theory of Quantaloids.* Pitman Research Notes in Mathematics 348, Longman, Harlow, 1996.

Salicrup, G. *Categorical Topology – The Complete Work of Graciela Salicrup* (ed. by H. Herrlich and C. Prieto). Aportaciones Matemática, Vol. 2, 1988.

Schubert, H. *Categories.* Springer, Berlin–Heidelberg–New York, 1972.

Semadeni, Z. *Projectivity, Injectivity, and Duality.* Diss. Math. Rozprawy Mat., Vol. 35, 1963.

Banach Spaces of Continuous Functions. Polish Scientific Publishers, Warsaw, 1971.

Semadeni, Z., and A. Wiweger. *Einführung in die Theorie der Kategorien und Funktoren.* Teubner, Stuttgart–Leipzig, 1979.

Słominsky, J. *The Theory of Abstract Algebras with Infinitary Operations.* Diss. Math. Rozprawy Mat., Vol. 18, 1959.

Taylor, P. *Practical Foundations of Mathematics.* Cambridge University Press, 1999.

Wyler, O. *Lecture Notes on Topoi and Quasitopoi.* World Scientific, Singapore-New Jersey-London-Hong Kong, 1991.

TABLES

FUNCTORS AND MORPHISMS: PRESERVATION PROPERTIES

	monos	extr. monos	reg. monos	epis	extr. epis	reg. epis
topological	+	+	+	+	+	+
monadic	+	?	+	−	−	−
reg. monadic	+	+	+	−	+	+
reg. algebraic	+	+	+	−	+	+
algebraic	+	+	+	−	+	−
ess. algebraic	+	+	+	−	−	−
solid	+	+	+	−	−	−
full refl. emb.	+	+	+	−	−	−
adjoint	+	−	+	−	−	−
faithful	−	−	−	−	−	−

FUNCTORS AND MORPHISMS: REFLECTION PROPERTIES

	monos	extr. monos	reg. monos	epis	extr. epis	reg. epis	isos
topological	+	−	−	+	−	−	−
monadic	+	−	−	+	+	−	+
reg. monadic	+	−	−	+	+	+	+
reg. algebraic	+	−	−	+	+	+	+
algebraic	+	−	−	+	+	+	+
ess. algebraic	+	−	−	+	+	−	+
solid	+	−	−	+	−	−	−
full refl. emb.	+	−	−	+	+	+	+
adjoint	−	−	−	−	−	−	−
faithful	+	−	−	+	−	−	−

FUNCTORS AND LIMITS

	creates	lifts uniquely	preserves	reflects
topological	−	+	+	−
monadic	+	+	+	+
ess. algebraic	+	+	+	+
solid, uniquely transp.	−	+	+	−
adjoint	−	−	+	−

FUNCTORS AND COLIMITS

	lifts uniquely	preserves	detects
topological	+	+	+
monadic	−	−	−
solid	−	−	+

STABILITY PROPERTIES OF SPECIAL EPIMORPHISMS

	composition	pushouts	cointersections
isomorphisms	+	+	+
retractions	+	−	−
reg. epis	−	+	−
strict epis	−	+	+
swell epis	+	+	+
strong epis	+	+	+
extr. epis	−	−	−
epis	+	+	+

TABLE OF CATEGORIES

($\mathbf{A} \downarrow K$), 3K (objects over K)

Ab, 3.26 (abelian groups) 3.39, 4.17, 7.18, 7.69, 8.23, 9.3, 9.13, 9.17, 9.24, 9.28, 9A, 10G, 11O, 12.11, 14F, 16.13, 24A

AbTop, 16.13 (abelian topological groups)

AbTor, 10.20 (abelian torsion groups) 10.55, 13.2

Alg(1), 27.3 (algebras with one unary operator) 4K, 5O, 9.3, 27.12, 28.2, 28.9

Alg(S^2), 5.38 (binary algebras)

Alg(T), 5.37 (T-algebras) 5G, 10U, 13N, 20.56, 20I

Alg(Σ), 3.26 (Σ-algebras) 8.41, 18.2, 21.20

Alg(Ω), 3.3 (Ω-algebras) 5O, 6I, 7.2, 7.40, 7.69, 7.72, 7O, 8.23, 8H, 9.28, 10.67, 10U, 16.19, 20.18, 24.5

Aut, 3.3 (automata) 4K, 5.2, 7.15, 13.13, 15.3, 20H

Aut$_r$, 8.31 (reachable automata) 18.2, 19.12

Ban, 3.3 (Banach spaces with linear contractions) 7.9, 7.18, 7.40, 7.50, 7.58, 9.3, 10.20, 10.31, 10J, 26.6

Ban$_b$, 3.3 (Banach spaces with bounded linear maps) 5.6, 7.9, 7.18, 7.40, 7.50, 7.58, 8.8, 8.23, 10.31, 10J, 13.18

(**Ban**, O), 5.2 (Banach spaces as a construct with unit ball functor) 6.11, 6E, 7.72, 8.8, 8.11, 8.23, 8K, 10.55, 10J, 20.36, 20.41, 23.20, 24C

(**Ban**, U), 5.2 (Banach spaces as a construct with usual forgetful functor) 6.11, 6E, 8.8, 8.23, 8K 10.55, 10J, 13.2, 18.2

Beh, 8.31 (behaviors) 18.2, 19.12

BiComp, 4F (bicompact spaces) 16.10

BiTop, 4F (bitopological spaces) 16.10

Boo, 3.20 (boolean algebras) 3.26, 3.39, 7.2, 7.18, 9.3, 9.13, 9A, 26.6

BooRng, 3.26 (boolean rings)

BooSpa, 3.39 (boolean spaces) 9.3, 9.18, 9.28, 10C, 12M, 23.6, 24.13

Bor, 22.2 (bornological spaces) 27.21, 28.20, 28.23

Cat, 3.47 (small categories); 5I, 7.40, 7.72, 7.76, 7S, 8.11, 8.23, 8C, 11J, 14.2, 14.23, 15.3, 15M, 24C, 27.3, 27.23, 28.9, 28.11

TABLE OF CATEGORIES

CAT, 3.50 (categories) 7.18, 7.33

CAT(X), 5.15 (concrete categories over **X**) 7.5, 9.24, 21.21, 21.22, 21J

Cat$_{of}$, 3.55 (object-free small categories); 5I, 20.36, 20.41 20.51, 20.52, 23.6, 23.20, 24D

CAT$_{of}$, 3.55 (object-free categories) 13.18

CBoo, 8E (complete boolean algebras) 8F, 8.23, 10S, 12.6, 12I, 18.13, 18G

Class, 15.3 (classes)

CLat, 4.3 (complete lattices with join- and meet-preserving maps) 8.23, 8G, 10S, 12.6, 12I, 18.13

Clos 5N (closure spaces)

Comp, 5A (compact spaces)

CONST, 5.15 (constructs)

Conv, 5N (convergence spaces) 27.3, 27.21, 28.9, 28.20, 28.23

ΔPos, 25E (Δ-complete partially ordered sets) 27B

DivAb, 7.33(5) (divisible abelian groups)

DLat, 9.3 (distributive lattices)

DRail, 8.8 (derailed spaces) 11H

$(F \downarrow G)$, 3K (comma category F over G)

FHaus, 7.65 (functionally Hausdorff spaces) 7I, 7J, 8.8

Field, 9.8 (fields) 9.17, 10I, 12.6, 16.13

Fram, 5L (frames) 8G, 20K, 24.13

Grp, 3.3 (groups) 4.17, 7.18, 7.58, 7C, 7H, 8.8, 9.3, 10.67, 14F, 16N, 17.5, 20.27, 23.34, 26.6

Grph, 20.41 (oriented graphs) 20.51, 20.52

Haus, 4.3 (Hausdorff spaces) 5O, 7.18, 7.40, 7.58, 8.8, 9.3, 9E, 11J, 13.2, 21.39

HausVec, 25.2 (Hausdorff topological vector spaces)

HComp, 3.39 (compact Hausdorff spaces) 4.17, 4D, 7.58, 9.3, 9.18, 9.28, 9A, 9D, 10.18, 10.67, 10C, 12K, 20.27, 20.36, 21I, 23.13, 23.34, 24.5, 26.6

HComp^{op}, 5M (as a construct)

hTop, 5J(d) (topological spaces and homotopy equivalence classes of maps) 7.33, 7.40

HUnif, 21.39 (separated uniform spaces)

JCPos, 4.3(3) (complete lattices and join-preserving maps) 4.17, 5N, 8.23, 8G, 9.3, 20.5, 22B, 24.5, 24.13

TABLE OF CATEGORIES

JPos, 4.17(10) (posets and join-preserving maps) 8G, 12.6

$(K \downarrow \mathbf{A})$, 3K (objects under K)

kTop, 27.3 ((compact Hausdorff)-generated topological spaces) 7.52, 27.21, 28.9, 28.11, 28.20

Λ-CCPos, 15D (chain complete λ-posets and chain join-preserving λ-homomorphisms) 15.26, 15J, 18.16, 18B, 25.17, 26C

Lat, 4.3 (lattices) 7.40, 7.72, 9.13, 9H

Loc 5L (locales)

M-Act, 3.26 (M-actions)

Mat, 3.3 (matrices (as morphisms)) 3.35

MCPos, 5N (complete lattices and meet-preserving maps)

Met, 3.3 (metric spaces and non-expansive maps) 8.8, 9.3, 9.13, 9.17, 9.24, 10.31, 10.70

Met$_c$, 3.3 (metric spaces and continuous maps) 3.35, 5.6, 5.34, 10.31

Met$_u$, 3.3 (metric spaces and uniformly continuous maps) 10.31

Mod, 21D (modules)

Mod-R, 3.26 (right R-modules)

Mon, 4.3 (monoids) 4.17, 8.23, 14F, 20.5

Neigh, 5N (neighborhood spaces)

1, 3.3 (terminal category (with only one morphism))

Ord, 7.83 (ordinal numbers) 7.90, 12.6, 15.8, 18.13, 18.16

PAlg(Ω), 21.39 (partial algebras of type Ω)

pGrp, 3B (pointed groups)

PMet, 8.2 (pseudometric spaces) 8.11, 10.31, 10.42, 21.8

Pos, 4.3 (partially ordered sets = posets) 4.17, 6I, 7.18, 8.8, 8.23, 9.3, 9.13, 9.17, 10.39, 10.55, 10.64, 10.65, 10.66, 10.67, 10.68, 10.70, 11.12, 12.6 12.11, 13.2, 16.13, 20H, 21.22, 21.39, 25.9, 27.3, 27.10, 27.12

Prost, 4.3 (preordered sets) 4.17, 21.8, 22.2, 27.21, 28.11, 28.20

Prox, 21I (proximity spaces)

PrTop, 5N (pretopological spaces) 27G, 28.9, 28.20, 28C

pSet, 3B (pointed sets) 14F, 16.13, 23A, 28.9

PsTop, 4F (pseudotopological spaces) 22.2, 27.3, 27.21, 27G, 28.9, 28.20, 28.23, 28G

pTop, 3B (pointed spaces)

Reg, 23.20 (regular topological spaces)

Reg$_1$ 21.39 (regular T_1 topological spaces)

Rel, 3.3 (relations) 4.7, 4E, 5K, 5.12, 7.69, 8.8, 8.11, 13I, 18.2, 21.8, 21.37, 27.3, 28.2, 28.9, 28L

Rere, 27.21 (reflexive relations) 15L, 27.21, 28.20, 28.23, 28L

R-**Mod**, 3.26 (left *R*-modules)

Rng, 5.11 (rings) 7.2, 7.18, 7.40, 7.58, 8.8, 8.23, 21D, 23.18

Set, 3.3 (sets) 3.39, 4D, 4E, 6F, 6I, 7.18, 7.40, 7.48, 7.69, 9.3, 9.13, 9.17, 9A, 10.6, 10.20, 10.28, 10.64, 10.67, 11.4, 11.28, 11K, 11P, 12.11, 12E, 13.2, 13B, 13I, 14.2, 14B, 15.3, 15K, 15J, 16.5, 18.2, 18A, 18E, 19G, 20.22, 20H, 27.3, 28.2

Set^{op}, 5.2(4) (as a construct)

Set^2, (arrow category of **Set**, i.e., functions) 9E, 15N, 16I

Sgr, 4.3 (semigroups) 4.7, 6I, 7.18, 7.40, 7.58, 7.60, 8.23, 8B, 14.23, 14I, 16N

Σ-Seq, 3.3 (Σ-acceptors) 3.35, 4.17, 4.26, 4K, 7.2, 7.33, 7.58, 7.69, 8.8, 8.23, 8P, 9I, 10.20, 10.64, 10.70, 16.13, 26.6

Σ-Seq$_0$, 10.18 (observable Σ-acceptors with no initial state)

Simp, 22.2 (simplicial complexes) 27.21, 28.20, 28.23

SLat, 4.3(3) (semi-lattices)

Spa(*T*), 5.40 (*T*-spaces) 5H, 6C, 6I, 7.2, 7R, 8.41, 10.42, 10.68, 13.13, 14.23, 18.2, 21.8, 22A, 22B, 27H, 27I, 28.17, 28H, 28J, 28K

sTop, 27.21 (sequential topological spaces) 4.26, 7.52, 27.21, 28.20

Sym, 4.26 (symmetric relations) 4E, 22.7, 27.10

TConv, 20.41 (totally convex spaces) 24.13

TfAb, 4.17 (torsion-free abelian groups) 7.40, 16N, 20.41, 20.45, 23.6

Top, 3.3 (topological spaces) 4E, 5.12, 5D, 7.18, 7.58, 7.60, 7.69, 7L, 8.8, 8.11, 8.23, 8.41, 9.3, 9.24, 9.26, 9E, 10.18, 10.42, 10.64, 10.66, 10.67, 10.70, 12.11, 13I, 13J, 14.2, 15.22, 15L, 16.25, 16D, 16L, 17.5, 18.2, 21.8, 21D, 22.7, 22C, 27.3, 27.23, 28.2

Top^{op}, 3.20

Top$_0$, 4.17 (T_0 topological spaces) 5L, 7.18, 9.3, 9.24, 21.39

Top$_0$^{op}, 5L (as a construct) 20K

Top$_1$, 7.73 (T_1 topological spaces) 8.5, 9.3, 9F, 16.13, 21.8, 21.39

Top_m, 3.35 (metrizable topological spaces) 5.34

TopGrp, 5.2 (topological groups) 5A, 8.8, 18.2, 20.41, 21.8, 23.6, 23.20, 25.2, 25.9

TopVec, 5.2 (topological vector spaces)

Tych, 4.3 (Tychonoff spaces) 7.18, 9.3, 9.13, 9.21, 10.18, 10C, 14.2, 16F, 21.39

Unif, 9.3 (uniform spaces) 21.8, 21.39, 27.23

Vec, 3.3 (vector spaces) 3.20, 4H, 6I, 7.40, 7.58, 8.8, 8.23, 9.3, 9.13, 9.17, 9.28, 10.67, 10.70, 12.11, 13I, 14F, 16.5, 20.5, 20H, 26.6

TABLE OF SYMBOLS

$\{x \in X \mid P(x)\}$, 2.1

$\mathcal{P}(X)$, 2.1 (power set)

$\{X, Y\}$, 2.1

(X, Y), 2.1 (ordered pair)

$X \cup Y$, 2.1 (binary union)

$X \cap Y$, 2.1 (binary intersection)

$X \times Y$, 2.1 (binary cartesian product)

$X \setminus Y$, 2.1 (relative complement)

Y^X, 2.1

$\{X_i \mid i \in I\}$, 2.1

$\bigcup_{i \in I} X_i$, 2.1 (union)

$\bigcap_{i \in I} X_i$, 2.1 (intersection)

$\prod_{i \in I} X_i$, 2.1 (cartesian product)

$\biguplus_{i \in I} X_i$, 2.1 (disjoint union)

$\mathbb{N}$, 2.1 (natural numbers)

$\mathbb{Z}$, 2.1 (integers)

$\mathbb{Q}$, 2.1 (rational numbers)

$\mathbb{R}$, 2.1 (real numbers)

$\mathbb{C}$, 2.1 (complex numbers)

$\hom(A, B)$, 3.1 (hom-set)

id, 3.1 (identity)

$\circ$, 3.1 (composition)

$A \xrightarrow{f} B$, 3.1 (morphism)

$Ob(\mathbf{A})$, 3.2 (class of $\mathbf{A}$-objects)

$Mor(\mathbf{A})$, 3.2 (class of $\mathbf{A}$-morphisms)

$dom(f)$, 3.2 (domain of f)

$cod(f)$, 3.2 (codomain of f)

TABLE OF SYMBOLS

1, 3.3(4) (terminal category)

A × **B**, 3.3(4) (product of categories)

A^{op}, 3.5 (dual category)

□ (end of proof)

$\boxed{D}$, 3.7 (proof by duality)

$\boxed{A}$, 8.25 (footnote) (proof by analogy)

f^{-1}, 3.12 (inverse of f)

$\mathcal{P}$, 3.20(8) (covariant power-set functor)

$\mathcal{Q}$, 3.20(9) (contravariant power-set functor)

S^n, 3.20(10) (nth power functor)

F^n, 3.23 (nth composite of F)

F^{op}, 3.41 (dual functor)

$(F \downarrow G)$, 3K (comma category F over G)

$(\mathbf{A} \downarrow K)$, 3K (category of objects over K)

$(K \downarrow \mathbf{A})$, 3K (category of objects under K)

$\mathbf{A}^2$ 3K (arrow category of **A**)

$|\ |$, 5.3 (underlying functor)

$A \leq B$, 5.4 (order in a fibre)

$F \leq G$, 5.18 (order of concrete functors)

τ^{op}, 6.3 (dual natural transformation)

$[\mathbf{A}, \mathbf{B}]$, 6.15 (functor quasicategory)

$(\hat{F}, \hat{G}) \circ (F, G)$, 6.27 (composition of Galois correspondences)

$(A, m) \leq (A', m')$, 7.79 (order of subobjects)

$(e, B) \geq (e', B')$, 7.85 (order of quotient objects)

$X \xrightarrow{f} |A| = (f, A)$, 8.15 (structured arrow)

$(A \xrightarrow{f_i} A_i)_I = (A, f_i)_I = (A, f_i) = (A, (f_i)_{i \in I})$, 10.1, 10.2 (source)

$(\prod A_i, \pi_j)_{j \in I}$, 10.23 (product of objects)

$\langle f_i \rangle$, 10.23 (morphism induced by product)

$A \times B$, 10.23 (binary product)

Πf_i, $f \times g$ 10.34 (product of morphisms)

TABLE OF SYMBOLS

A^I, 10.37 (I-th power of A)

$(\mu_j, \coprod A_i)_{j \in I}$, 10.63 (coproduct of objects)

$[f_i]$, 10.63 (morphism induced by coproduct)

$\coprod f_i$, 10.63 (coproduct of morphisms)

$A + B$, $f + g$, 10.63 (binary coproduct)

$^I A$, 10.63 (I-th copower of A)

Δ_A, 10V (diagonal morphism of A)

$(X \xrightarrow{f_i} GA_i)_{i \in I}$, 17.2 (structured source)

$ev: A \times B^A \to B$, 18.2, 27.2 (evaluation)

$(\eta, \varepsilon): F \dashv G: \mathbf{A} \to \mathbf{B}$, 19.3 (adjoint situation)

$\mathbf{T} = (T, \eta, \mu)$, 20.1 (monad)

$(\mathbf{X^T}, U^\mathbf{T})$, 20.4 (category of $\mathbf{T}$-algebras)

ConGen = ConGen($\mathbf{A}$) = the class of concretely generating structured arrows (in $\mathbf{A}$)

Emb = Emb($\mathbf{A}$) = the class of embeddings (in $\mathbf{A}$)

Epi = Epi($\mathbf{A}$) = the class of epimorphisms (in $\mathbf{A}$)

ExtrEpi = ExtrEpi($\mathbf{A}$) = the class of extremal epimorphisms (in $\mathbf{A}$)

ExtrGen = ExtrGen($\mathbf{A}$) = the class of extremally generating structured arrows (in $\mathbf{A}$)

ExtrMono = ExtrMono($\mathbf{A}$) = the class of extremal monomorphisms (in $\mathbf{A}$)

ExtrMono-Source = ExtrMono-Source($\mathbf{A}$) = the conglomerate of extremal mono-sources (in $\mathbf{A}$)

Final = Final($\mathbf{A}$) = the class of final morphisms (in $\mathbf{A}$)

Gen = Gen($\mathbf{A}$) = the class of generating structured arrows (in $\mathbf{A}$)

Init = Init($\mathbf{A}$) = the class of initial morphisms (in $\mathbf{A}$)

Iso = Iso($\mathbf{A}$) = the class of isomorphisms (in $\mathbf{A}$)

Mono = Mono($\mathbf{A}$) = the class of monomorphisms (in $\mathbf{A}$)

Mono-Source = Mono-Source($\mathbf{A}$) = the conglomerate of mono-sources (in $\mathbf{A}$)

Mor = Mor($\mathbf{A}$) = the class of morphisms (in $\mathbf{A}$)

NormalEpi = NormalEpi($\mathbf{A}$) = the class of normal epimorphisms (in $\mathbf{A}$)

NormalMono = NormalMono($\mathbf{A}$) = the class of normal monomorphisms (in $\mathbf{A}$)

Quot = Quot($\mathbf{A}$) = the class of quotient morphisms (in $\mathbf{A}$)

RegEpi = RegEpi(**A**) = the class of regular epimorphisms (in **A**)

RegMono = RegMono(**A**) = the class of regular monomorphisms (in **A**)

Retr = Retr(**A**) = the class of retractions (in **A**)

Sect = Sect(**A**) = the class of sections (in **A**)

Source = Source(**A**) = the conglomerate of sources (in **A**)

StableEpi = StableEpi(**A**) − the class of stable epimorphisms (in **A**)

StrictMono = StrictMono(**A**) = the class of strict monomorphisms (in **A**)

StrongMono = StrongMono(**A**) = the class of strong monomorphisms (in **A**)

Surj = Surj(**A**) = the class of surjective morphisms (in **A**)

SwellEpi = SwellEpi(**A**) = the class of swell morphisms (in **A**)

INDEX

Absolute coequalizer, 20.14
Absolute colimit, 20.14
Absolute retract, 9.6
   vs. enough injectives, 9.10
Adjoint, for a functor, 19.10
Adjoint functor, 18.1 ff
   characterizations, 18.3
   characterization theorems, 18.12, 18.14, 18.17, 18.19
   comparison functor for, 20.38, 20.42
   composition of, 18.5
   monadic functor is, 20.12
   between posets, 18H
   preserves mono-sources and limits, 18.6, 18.9
   smallness conditions for, 18B
   theorem, concrete, 18.19
   theorem, first, 18.12
   theorem, special, 18.17
   vs. adjoint situation, 19.1, 19.4, 19.8
   vs. algebraic category, 23.31
   vs. co-adjoint functor, 18A, 19.1
   vs. colimit, 18D
   vs. co-wellpoweredness, 18.11, 18.14, 18.19
   vs. completeness, 18.12, 18.14, 18.17, 18.19
   vs. essentially algebraic functor, 23.8
   vs. exponential functor, 27.7
   vs. extremal monomorphism, 18J
   vs. free object, 18.19
   vs. full, faithful functor, 19I
   vs. (Generating, **M**)-functor, 18.4
   vs. monadic category, 20.46
   vs. monadic functor, 20.12, 20.17
   vs. reflection of epimorphisms, 18I, 19B
   vs. regularly algebraic category, 23.38
   vs. representable functor, 18C
   vs. solid functor, 25.12, 25.19
   vs. topologically algebraic functor, 25.3, 25.6, 25.19
   vs. wellpoweredness, 18.19
Adjoint sequence, 19F
Adjoint situation, 19.3 ff
   alternative description, 19A
   associated with a monad, 20.7
   composition of, 19.13
   consequences of, 19.14
   duality for, 19.6
   gives rise to a monad, 20.3
   induced by (co-)adjoint functor, 19.7
   lifting of, 21.26, 21.28
   uniqueness, 19.9
   vs. adjoint functor, 19.1, 19.4, 19.8
   vs. equivalence functor, 19.8, 19H
   vs. free object, 19.4
   vs. Galois correspondence, 19.8
   vs. indiscrete structure, 21A
   vs. monad, 20A
   vs. reflective subcategory, 19.4
   vs. universal arrow, 19.7
Algebra, partial binary, 3.52
Algebraic category, 23.19 ff, 23D
   characterization theorem for, 23.30, 23.31
   concrete functor between, 23.22
   implies extremal epimorphisms are final, 23.23
   vs. adjoint functor, 23.31
   vs. (Epi, Mono-Source)-factorizable category, 23.30, 23.31
   vs. extremal co-wellpoweredness, 23.27
   vs. extremal epimorphism, 23.30
   vs. unique lift of (ExtrEpi, Mono-Source)-factorizations, 23.31
Algebraic construct, = regular epireflective subconstruct of monadic construct, 24.3

INDEX

bounded, vs. bounded quasivariety, 24.11
  vs. monadic construct, 23.41
Algebraic functor, 23.19 ff
  closed under composition, 23.21
  need not preserve regular epimorphisms, 23.25, 23J
  reflects regular epimorphisms, 23.24
  restrictions of, 23L
  vs. composite of regular monadic functors, 24.2
  vs. regular epimorphism, 23J
  vs. regular factorization, 24.2
  vs. uniquely transportable functor, 23.30
Algebraic hull of a concrete category, 23K
Algebraic subcategory, conditions for, 23.32, 23.33
  vs. extremally epireflective subcategory, 23.33
Algebraic theory, 20C
Algebraic-topological decompositions, 26C
Algebraic-type functors, relationships among, 23.42
Amnesticity, vs. lifting of limits, 13.21
Amnestic concrete category, 5.4 ff
  vs. Galois correspondence, 6.29–6.36
  vs. topological functor, 21.5
  vs. transportable concrete category, 5.29, 5.30
Amnestic functor, 3.27, 5.6, 13.21, 13.25, 13.28
Amnestic modification, of a concrete category, 5.6, 5.33, 5.34, 5F
Arrow,
  costructured, = costructured arrow, 8.40
  co-universal, = co-universal costructured arrow, 8.40
  reflection, = reflection arrow, 4.16
  structured, = structured arrow, 8.15

universal, = universal structured arrow, 8.22
Arrow category, 3K, 6.17
Associativity, composition of morphisms has, 3.1, 3.53, 3C
Axiom, topological, 22.6
Axiom of choice, 2.3, 9A
Axiom of replacement, 2.2

Balanced category, 7.49 ff
  topological, scarcity of, 21N
  vs. extremal monomorphism, 7.67
  vs. mono-source, 10.12
Base category, 5.1 ff
Bicoreflective subcategory,
  vs. coreflective subcategory, 16.4
  vs. separator, 16.4
Bimorphism, 7.49
  vs. topological functor, 21M
Bireflective subcategory, 16.1
  monoreflective subcategory is, 16.3
Birkhoff-theorem, 16G
Boolean ring, 3.26, footnote

Cartesian closed category, 27.1 ff
  characterization theorem, 27.4
  vs. coreflective hull, 27C
  vs. products of epimorphisms, 27.8
  vs. zero object, 27A
Cartesian closed construct, 27.16 ff, 27D
  vs. well-fibred topological construct, 27.22
Cartesian closed subcategory, 27.9
  vs. (co)reflective subcategory, 27.9
Cartesian closed topological category,
  implies base category is cartesian closed, 27.14
  as injective object, 27F
Cartesian closed topological construct,
  vs. powers with discrete exponents (resp. factors), 27.24

Cartesian closed topological hull, 27G

Cartesian product of sets, 2.1

Categorical statement involving functors, dual of, 3.40, 3.42

Category, 3.1 ff
algebraic, = algebraic category, 23.19 ff
of all categories, can't be formed because of set-theoretical restrictions, 3.48
with all products must be thin, 10.32
alternative definition, 3C
arrow, 3K
base, = base category, 5.1 ff
cartesian closed, = cartesian closed category, 27.1
(co)complete, = (co)complete category, 12.2
comma, 3K, 5.38, 21F
compact, 18K
of concrete categories, 5.15, 5C
concretely cartesian closed, = concretely cartesian closed category, 27.11
concretizable, 5J
co-wellpowered, = co-wellpowered category, 7.87
discrete, 3.26
dual, = opposite category, 3.5
Eilenberg-Moore, 20.4
$(E, \mathbf{M})$, = $(E, \mathbf{M})$-category, 15.1 ff
empty, 3.3(4)
essentially algebraic, = essentially algebraic category, 23.5 ff
$(E, -)$-structured, 14H
exact, 14F
extremally co-wellpowered, = extremally co-wellpowered category, 7.87
extremally wellpowered, = extremally wellpowered category, 7.82
finitary, 24.4, 24.7
finitely (co)complete, 12.2
free, 3A
functor-costructured, 5.43
functor-structured, = functor-structured category, 5.40
graph of, 3A
isomorphic, 3.24
Kleisli, 20.39, 20B
large, 3.44
locally presentable, 20H
monadic, = monadic category, 20.8 ff
non-co-wellpowered, 7L
object-free, = object-free category, 3.53
of objects over (resp. under) an object, 3K
opposite, = opposite category, 3.5
pointed, 3B, 7B
product, 3.3(4)
is a quasicategory, 3.51
regular co-wellpowered, 7.87–7.89
regular wellpowered, 7.82, 7.88, 7.89
regularly algebraic, = regularly algebraic category, 23.35 ff
skeleton of a, 4.12–4.15, 4I
small, 3.44–3.45
of small categories, 3.47
— is a large category, 3.48
— vs. concrete category, 5I
solid, = solid category, 25.10 ff
strongly (co)complete, = strongly (co)com- plete category, 12.2
sub-, see Subcategory
terminal, 3.3(4)
thin, 3.26, 3.29, 3G
topological, = topological category, 21.7 ff
topologically algebraic, = topologically algebraic category, 25.1 ff
total, 6I
of type 2, 3C
universal, 4J
universally topological, 28.16, 28.18, 28G
wellpowered, = wellpowered category,

7.82

Category theory, "object-free version", 3.55

Choice, axiom of, 2.3, 9A

Class, 2.2
  of all sets, = universe, 2.2
  as a category, (is not a construct), 3.3(4)
  large, = proper class, 2.2
  preordered, 3.3(4)
  proper, 2.2
  small, = set, 2.2

Closed under the formation of intersections, 11.26

Closed under the formation of **M**-sources, 16.7
  and E-quotients, vs. E-equational subcategory, 16.17
  vs. E-reflective, 16.8

Closed under the formation of products, and extremal subobjects, vs. (epi)reflective subcategory, 16.9, 16.10
  vs. E-equational subcategory, 16.17

Closed under the formation of pullbacks, 11.17

Closure space, 5N

Co-, see Dual concept

Co-adjoint, for a functor, 19.10

Co-adjoint functor, 18.1 ff
  vs. adjoint functor, 18A, 19.1, 19.2, 19.11
  vs. colimit, 18D
  vs. contravariant exponential functor, 27.7

Coarser than, (preorder) relation on concrete functors, 5.18

Co-axiom, topological, 22.6

Cocomplete category, 12.2 ff
  almost implies complete, 12.7 ff
  with a small colimit-dense subcategory is complete, 12.12

  vs. cartesian closed category, 27.4
  vs. colimit-dense subcategory, 12.12
  vs. complete category, 12.13, 25K
  vs. copower, 12H
  vs. essentially algebraic category, 23.12, 23.13
  vs. solid category, 25.15, 25.16
  vs. topological category, 21.16, 21.17

Cocomplete subcategory, vs. reflective subcategory, 12K

Cocone, = natural sink, 11.27

Codomain,
  of a function, 2.1
  of a morphism, 3.2
  of a source, 10.1
  of a sink, 10.62
  of a structured source, 17.1

Coequalizer, 7.68 ff
  absolute, 20.14
  as a colimit, 11.28
  existence, 12.1, 15F
  preservation, 13.1
  split, 20.14
  uniqueness, 7.70
  vs. completeness, 12J
  vs. epimorphism, 7M
  vs. forgetful functor, 7.73
  vs. isomorphism, 7.70
  vs. monomorphism, 7.70
  vs. pushout square, 11.33

Coessential morphism, w.r.t. a class of morphisms, 9.27

Coessential quotient map, 9.27

Cointersection, 11.28
  existence, 12.1
  vs. $(E, \mathbf{M})$-category, 15.14, 15.15, 15.16

Colimit, 11.27 ff
  absolute, 20.14
  concrete, 13.12
  creation, 13.17
  detected by regularly monadic functor, 20.33

detection, = detection of colimit, 13.22
directed, = directed colimit, 11.28
domain of, 11.27
lifting of, 13.17
— vs. topological functor, 21.15
preservation, 13.1
— vs. topological functor, 21.15
reflection, 13.22
uniqueness, 11.29
vs. (co-)adjoint functor, 18D
vs. coproduct, 11.28
vs. monadic functor, 20.13
vs. reflection arrow, 13.30
Colimit-closed subcategory, vs. reflective subcategory, 13.29
Colimit-dense full embedding, preserves limits, 13L
Colimit-dense subcategory, 12.10
embedding of preserves limits, 13.11
vs. (co)completeness, 12.12
Comma category, 5.38, 3K, 21F
Commuting triangle and square, 3.4
Compact category, 18K
Comparison functor, 20.37 ff
vs. adjoint functor, 20.42
vs. faithful functor, 20.43
vs. full functor, 20.43, 20.44
Complete category, 12.2 ff
characterization, 12.3
and wellpowered category, is strongly complete, 12.5
vs. adjoint functor, 18.12, 18.14, 18.17, 18.19
vs. cocomplete category, 12.13, 25K
vs. coequalizer, 12J
vs. colimit-dense subcategory, 12.12
vs. essentially algebraic category, 23.12, 23.13
vs. solid category, 25.15
vs. strongly complete category, 12.5, 12I
vs. topological category, 21.16, 21.17

Complete lattice, vs. has products, 10.32
Completion,
of abstract categories, 12L
regarded as reflection, 4.17
universal, 12M
Complex numbers, 2.1
Composite,
of (concrete) functors, 3.23, 5.14
— vs. natural transformation, 6.3
of epimorphisms, is epimorphism, 7.41
of essential embeddings, 9.14
of functions, 2.1
of isomorphisms, is isomorphism, 3.14
of monomorphisms, is monomorphism, 7.34
of morphisms, 3.1
of natural transformations, 6.13
of sources, 10.3
— vs. $(E, \mathbf{M})$-category, 15.5
Composition, as a morphism, 27E
Composition of morphisms, 3.1 ff
is associative, 3.1
in a quasicategory, 3.49
vs. $(E, \mathbf{M})$-category, 15.14, 15.15
vs. (E, M)-structured category, 14.6
Concept, (self-)dual, 3.7
Concrete adjoint functor, vs. concrete coadjoint functor, 21.24
Concrete category, 5.1 ff
amnestic, = amnestic concrete category, 5.4 ff
amnestic modification, 5.6, 5.33, 5.34
$(\mathbf{A}, U)$ has property P means $\mathbf{A}$ (or U) has property P, 5.3
"concretely isomorphic" is stronger than "isomorphic", 5.12
duality principle for, 5.20
fibre-complete, 5.7, 5.42
fibre-discrete, 5.7, 5.8, 5.39
monadic, 20.8 ff
over **1**, vs. concrete functors, 5.11
over **Set**, = construct, 3.3(2), 5.1 ff
solid, = solid concrete category,

INDEX 489

25.10 ff
strongly fibre-small, 26.5
topological, = topological concrete category, 21.7 ff
(uniquely) transportable, 5.28, 5.35, 5.36
vs. **Cat**, 5I
Concrete co-adjoint functor,
vs. concrete adjoint functor, 21.24
vs. Galois correspondence, 21E
Concrete colimit, 13.12
Concrete coproduct, 13.12
Concrete coreflector, i.e., coreflector induced by identity-carried coreflection arrows, 5.22
Concrete co-wellpoweredness, 8.19
vs. extremal co-wellpoweredness, 8E
Concrete embedding, essential, 21J
Concrete equivalence, 5.13
Concrete functor, 5.9 ff
between algebraic categories is algebraic, 23.22
composite, 5.14
between constructs, 5D
between essentially algebraic categories is essentially algebraic, 23.17
existence of a concrete natural transformation between, 6.24
between regularly algebraic categories is regularly algebraic, 23.40
specified by its values on objects, 5.10, 5.11
vs. embedding functor and equivalent objects, 5.10
vs. monadic category, 20E
Concrete generation of an object, 8.15
vs. extremal epimorphism, 8.18
vs. (extremal) generation, 8.16
vs. universal arrow, 8.24
Concrete isomorphism, 5.12, 5N
Concrete limit, 13.12

characterized, 13.15
reflection of, 13C
two step construction, 13.16
vs. concrete colimit, 13.10
Concrete limit dense, 12M
Concretely cartesian closed category, 27.11
vs. discrete objects, 27.15
vs. topological category, 27.15
Concretely cartesian closed construct, vs. subconstruct of **Set**, 27.16
Concretely complete, 12M
Concretely coreflective subcategory, 5.22
vs. topological subcategory, 21.33
Concretely co-wellpowered concrete category, 8.19
Concretely co-wellpowered functor, 8.37
vs. (ConGen, Initial Mono-Source) functor, 17.11
Concretely equivalent, is not symmetric, 5.13
Concretely generating structured arrow, see Concrete generation of an object
Concretely reflective concrete subcategory, 5.22
of amnestic category is full, 5.24
of non-amnestic concrete category need not be full, 5.25, 4.21
vs. finally dense subcategory, 21.32
vs. initial source, 10.50
vs. initially closed subcategory, 21.31
vs. reflector, 5.26, 5.27, 5.31, 5.32
vs. topological (sub)category, 21.32, 21.33
Concrete natural transformation, 6.23
vs. concrete functors, 6.24
Concrete product, 10.52
as a concrete limit, 13.12
preserved by composition, 10.56
vs. has **M**-initial subobjects, 21.42
vs. initial source, 10.53
vs. monotopological construct, 21.42

Concrete quasitopos, vs. universally topological category, 28.18

Concrete reflector, = reflector induced by identity-carried reflection arrows, 5.22, 5E
   vs. concrete functors that are reflectors, 5.22

Concrete subcategory, 5.21 ff
   concretely (co)reflective, = concretely (co)reflective subcategory, 5.22

Concretizable category, 5J, 10L
   vs. separating set, 7Q

Cone, = natural source, 11.3

(ConGen, Initial Mono-Source)-factorizable category, 17J

(ConGen, Initial Mono-Source)-functor, vs. concretely co-wellpowered, 17.11

Conglomerate, 2.3
   of all classes, 2.3
   codability by, 2.3
   (il)legitimate, 2.3
   of morphisms between two objects in a quasicategory, 3.49
   of objects in a quasicategory, 3.49
   small, 2.3

Congruence fork, 20.14

Congruence relation, 11.20
   coequalizer of, vs. monadic construct, 20.35
   vs. equalizer, 11.20
   vs. monomorphism, 11.20, 11R
   vs. pulation square, 11.33

Constant functions are morphisms, vs. has function spaces, 27.18

Constant functor, 3.20(2)

Constant morphism, 7A, 10W

Construct, 3.3(2), 5.1 ff
   bounded, 24.11
   concretely co-wellpowered, 8D
   monadic, = monadic construct, 20.34 ff
   must be regular wellpowered and regular co-wellpowered, 7.88
   not determinated by object class, 3.3(2)

Contraction, 3.3(3)

Contravariant exponential functor, 27.5
   vs. (co-)adjoint functor, 27.7

Contravariant hom-functor, 3.20(5)

Contravariant power-set functor, 3.20(9)

Copower, of an object, 10.63
   non-existence, 10S
   vs. cocompleteness, 12H
   vs. free object, 10R

Coproduct, 10.63 ff
   concrete, 13.12
   existence, 10I, 12.1, 12G
   of functors, 10U
   preservation of, 13.1
   vs. colimit, 11.28

Coreflection, Galois, 6.26

Coreflection arrow, 4.25 ff

Coreflective hull, vs. cartesian-closed category, 27C

Coreflective modification, of a concrete category, 5.22, 5K

Coreflective subcategory, 4.27, 16.1 ff
   vs. bicoreflective subcategory, 16.4
   vs. cartesian closed subcategory, 27.9

Coreflector for a coreflective subcategory, 4.27

Coreflector, concrete, 5.22

Coseparating set, 18L

Coseparator, 7.16
   category with, is regular wellpowered and regular co-wellpowered, 7.89
   extremal, 10.17
   is extremal in balanced category, 10.18
   vs. faithful hom-functor, 7.17
   vs. power of an object, 10.38
   vs. topological category, 21.16, 21.17

Costructured arrow, 8.40
Costructured sink, 17.4
Co-unit of an adjunction, 19.3, 19J
Co-universal arrow, 8.40
    vs. universal arrow, 19.1, 19.2
Covariant exponential functor, 27.5
Covariant hom-functor, 3.20(4)
Covariant power-set functor, 3.20(8)
Cover, projective, 9.27
Co-wellpowered category, 7.87 ff
    concrete, 8.19
    vs. adjoint functor, 18.11, 18.14, 18.19
    vs. cartesian closed category, 27.4
    vs. $(E, \mathbf{M})$-category, 15.25, 15.26
    vs. epireflective hulls, 16C
    vs. essentially algebraic category, 23.14
    vs. extensions of factorization structures, 15.20, 15.21
    vs. monadic category, 20.29
    vs. topological category, 21.16, 21.17
    vs. wellpoweredness, 12.13
Co-wellpowered functor, 8.37
    concrete, 8.37
    extremal, 8.37
    implies domain is co-wellpowered, 8.38
    vs. adjoint functor, 18.11, 18.14
    vs. essentially algebraic category, 23.14
Creation of absolute coequalizers, vs. monadic functor, 20.17
Creation of absolute colimits, 20.14
    monadic functor does, 20.16
Creation of colimits, 13.17
Creation of (Extremal Epi, Mono-Source)- factorizations, 23.31
Creation of finite limits, vs. monadic construct, 20.35
Creation of isomorphisms, 13.35, 13M
    vs. creation of limits, 13.36
    vs. essentially algebraic functor, 23.8

    vs. monadic functor, 20.12
    vs. reflection of identities & isomorphisms, 13.36
Creation of limits, 13.17, 13N
    monadic functor does, 20.12
    vs. creation of isomorphisms, 13.36
    vs. essentially algebraic functor, 23.15
    vs. lifts limits uniquely and reflects limits, 13.20, 13.25
    vs. reflection of isomorphisms, 13.25
Creation of split coequalizers, vs. monadic functor, 20.17

Decomposition of functors, 3N
    vs. factorization structure, 26A
Decomposition of Galois correspondence, 6.35
Decomposition theorems for solid functors, 26.3, 26.4
Dense subcategory, 12D
Detection of colimits, 13.22
    vs. reflective subcategory, 13.32
    vs. solid functor, 25.14
Detection of limits, 13.22
    vs. lifting of limits, 13.34
    vs. solid functor, 25.14
Detection and preservation of limits,
    vs. solid functor, 25.18, 25.19
    vs. topologically algebraic functor, 25.18, 25.19
Detection of wellpoweredness, vs. monadic functor, 20.12
Diagonal, 14.1, 14B, 15.1, 15K, 17.3, 17D
Diagonal morphism, 10V, 10W
Diagonalization property,
    causes E and $\mathbf{M}$ to determine each other, 15.5
    $(E, \mathbf{M})$, = $(E, \mathbf{M})$-diagonalization property, 15.1
    w.r.t. a functor, 17.3
Diagram, = functor, 11.1 ff

limit of, 11.3
Directed colimit, 11.28, 11O
  vs. finitary quasivariety, 24A
Discrete category, 3.26
Discrete functor, vs. topological functor, 21.12
Discrete object, 8.1, 8M
  must be smallest element in the fibre, 8.4
  vs. concretely cartesian closed subcategory, 27.15
  vs. topological category, 21.11
Discrete quasicategory, 3.51
Discrete space functor, is a full embedding, 3.29
Discrete terminal object, vs. has function spaces, 27.18
Disjoint union, of a family of sets, 2.1
Dispersed factorization structure, 15L
Distributive law for cartesian closed category, 27.8
Domain,
  of a diagram, = scheme, 11.1
  of a function, 2.1
  of a morphism, 3.2
  of a sink, 10.62
  of a source, 10.1
  of a structured source, 17.1
Dominion, 14J
Down-directed poset, 11.4
Dual of a categorical statement involving functors, 3.40, 3.42
Dual category, = opposite category, 3.5
Dual concept, of a "categorical concept", 3.7
Dual functor, = opposite functor, 3.41
Duality principle,
  for categories, 3.7, 3E
  for concrete categories, 5.20
  for topological category, 21.10

Duality theories, vs. representability, 10N
Dually equivalent categories, 3.37, 3.38
Dual property, 3.7
Dual statement, of a "categorical statement", 3.7

$(E,-)$-category, 15B, 15D
E-equation, 16.16
E-equational subcategory, 16.16
  vs. closed under the formation of $\mathbf{M}$-sources and E-quotients, 16.17
  vs. closed under the formation of products, 16.17
$(E,-)$-functor, 17.4
Eilenberg-Moore category, 20.4
E-implicational subcategory, 16.12
  vs. E-reflective subcategory, 16.14
Elementhood-morphism, 28.11
Embeddable, fully, 4.6
Embedding functor, 3.27
  closed under composition and first factor, 3.30
  finally dense, 10.72
  is (up to isomorphism) inclusion of subcategory, 4.5
  vs. faithful functor, 3.28, 4.5
  vs. full functor, 3.29
  vs. preservation of limit, 13.11
Embedding morphism, 8.6 ff
  in **Cat**, 8C
  composition, 8.9
  essential, 9.12
  first factor of, 8.9
  vs. monomorphism, 8.7
  vs. regular monomorphism, 8.7, 8A
  vs. section, 8.7
Embedding, Yoneda, 6.19, 6J
$(E,\mathbf{M})$-category, 15.1
  conditions on E, 15.14, 15.15
  consequences of, 15.5
  implies $E \subseteq$ Epi, 15.4

values for E and $\mathbf{M}$, 15.8
vs. co-wellpowered category, 15.25, 15.26
vs. $(E, \mathbf{M})$-factorization, 15.10
vs. $(E, \mathbf{M})$-functor, 17.4
vs. extremally co-wellpowered category, 15.25, 15.26
vs. mono-source, 15.6–15.9
vs. regular epimorphism, 15.7
vs. strongly complete category, 15.25, 15.26
vs. topological category, 21.14

(E, M)-diagonalization property, 14.1

$(E, \mathbf{M})$-diagonalization property, 15.1

(E, M)-factorization, 14.1
vs. (E, M)-structured category, 14.7

$(E, \mathbf{M})$-factorization, 15.1
uniqueness, 15.5
vs. $(E, \mathbf{M})$-category, 15.10
vs. faithful functor, 17.15, 17.16
vs. topological category, 21.16, 21.17

$(E, \mathbf{M})$-functor, = adjoint functor, 17.3, 18.3, 18.4
implies $E \subseteq$ Gen, 17.6
implies factorizations are essentially unique, 17.7
implies $\mathbf{M}$ determines E, 17.7
$\mathbf{M}$ need not be closed under composition nor determined by E, 17.8
not every functor is, 17.4
properties of, 17E
vs. $(E, \mathbf{M})$-category, 17.4
vs. has $(E, \mathbf{M})$-factorizations, 17.10
vs. topologically algebraic functor, 25.6

E-monad, 20.21 ff

E-monadic functor, 20.21 ff
lifts $(E, \mathbf{M})$-factorizations uniquely, 20.24
vs. reflective subcategory, 20.25

$(E, \text{Mono-Source})$-category, vs. (Epi, $\mathbf{M}$)-category, 15.11

$(E, \text{Mono-Source})$-functor, implies $E \subseteq$ ExtrGen, 17.9

(E, Mono)-structured category, vs. regular epimorphism, 14.14

Empty category, 3.3(4)

Empty source, 10.2
vs. product, 10.20

(E, M)-structured category, 14.1 ff
consequences of, 14.6, 14.9, 14.11
duality for, 14.3
relationship to limits, 14.15 ff
relationship to special morphisms, 14.10 ff
uniqueness of factorizations, 14.4
vs. composition of morphisms, 14.6
vs. (E, M)-factorization property, 14.7
vs. (extremal) epimorphisms, 14.10–14.14
vs. extremal monomorphisms, 14.10
vs. isomorphisms, 14.5

Enough injectives, 9.9
vs. absolute retract, 9.10
vs. injective hull, 9D

(Epi, ExtrMono-Source)-category, vs. (ExtrEpi, Mono-Source)-category, 15C
characterization, 15.16
strongly complete category is, 15.17

(Epi, ExtrMono)-structured category,
vs. equalizer, 14.19
vs. intersection, 14.19

(Epi, Initial Source)-factorizable category, vs. topologically algebraic category, 25.6

(Epi, $\mathbf{M}$)-category, characterization, 15.16
vs. $(E, \text{Mono-Source})$-category, 15.11

(Epi, Mono-Source)-factorizations, vs. monadic category, 20.49

(Epi, Mono-Source)-factorizable category,

vs. algebraic category, 23.30, 23.31
vs. essentially algebraic category, 23.8,

23.9
(Epi, Mono-Source)-factorizations, imply (ExtrEpi, Mono-Source)-category, 15.10
Epimorphism, 7.38 ff
  closed under composition, 7.41
  equals implication, 16.12
  extremal, = extremal epimorphism, 7.74
  as extremal monomorphism is isomorphism, 7.66
  for groups, 7H
  preserved and reflected by equivalence functor, 7.47
  products of, 10D
  reflected by faithful functor, 7.44
  regular, = regular epimorphism, 7.71
  as section is isomorphism, 7.43
  split, = retraction, 7.24
  stable, 11J
  swell, 7.76, 15A
  types, 7.76
  vs. coequalizer, 7M
  vs. $(E, \mathbf{M})$-category, 15.8, 15.14, 15.15
  vs. (E, M)-structured category, 14.10
  vs. equalizer, 7.54
  vs. generating structured arrows, 8.16, 8.18, 8.36
  vs. hom-functor, 19C
  vs. quotient morphism, 8.12
  vs. regular epimorphism, 7O
  vs. retraction, 7.42
Epireflective hulls, vs. co-wellpoweredness, 16C
Epireflective subcategory, 16.1 ff
  with bad behavior, 16A
  regular, 16I
  vs. closure under the formation of products and extremal subobjects, 16.9, 16.10
  vs. extremal mono-source, 16H
  vs. strongly limit-closed subcategory, 16L

Epi-sink, 10.63
  vs. pullback, 11I
Epi-transformation, 6.5, 7R
  vs. adjoint situation, 19.14
Equalizer, 7.51 ff
  existence, 12.1
  preservation of, 13.1, 13.3
  and product, vs. pullback square, 11.11, 11.14
  reflection of, 17.13
  uniqueness, 7.53
  vs. congruence relation, 11.20
  vs. (Epi, ExtrMono)-structured category, 14.19
  vs. epimorphism and isomorphism, 7.54
  vs. product and pullback square, 10.36, 11.14, 11S
  vs. regular monomorphism, 13.6
Equation, = regular implication with free domain, 16.16
Equational subcategory, 16.16
  of $\mathbf{Alg}(\Omega)$, char. 16.19
Equational subconstruct, vs. epireflective subconstruct, 16.18
Equivalence,
  natural, = natural isomorphism, 6.5
  of quasicategories, 3.51
  of "standard" and "object-free" versions of category theory, 3.55
Equivalence functor, = full, faithful, and iso-morphism-dense functor, 3.33, 3H
  concrete, 5.13
  is topologically algebraic, 25.2
  preserves and reflects special morphisms, 7.47
  properties of, 3.36
  vs. adjoint situation, 19.8, 19H
  vs. natural isomorphism, 6.8
Equivalence relation on a conglomerate, 2.3
Equivalent categories, 3.33

INDEX

dually, 3.37, 3.38
   vs. isomorphic categories, 3.35

Equivalent objects, in a concrete category, 5.4

E-reflective hull, = smallest E-reflective subcategory, 16.21
   characterization of members, 16.22

E-reflective subcategory, 16.1
   intersection of, 16.20
   vs. closure under the formation of **M**-sources, 16.8
   vs. E-implicational subcategory, 16.14
   vs. E-monadic functor, 20.25

Essential embedding, 9.12
   properties of, 9.14

Essential extension, 9.11, 9H
   types of, 9.20
   vs. injective object, 9.15

Essentially algebraic category, 23.5 ff
   characterization, 23.8
   concrete functor between, 23.17
   has coequalizers, 23.10
   implies embeddings = monomorphisms, 23.7
   vs. completeness & cocompleteness, 23.12, 23.13
   vs. co-wellpoweredness, 23.14
   vs. monadic category, 23.16
   vs. wellpoweredness, 23.12, 23.13

Essentially algebraic embedding, 23B

Essentially algebraic functor, 23.1 ff
   closed under composition, 23.4
   detects colimits, 23.11
   equivalent conditions, 23.2
   is faithful adjoint functor, 23.3
   preserves and creates limits, 23.11
   is topologically algebraic, 25.2
   vs. creates limits, 23.15
   vs. monadic functor, 23C
   vs. solid functor, 26.3

Essential morphism, w.r.t. a class of morphisms, 9.22

Essential uniqueness, of universal arrow, 8.25, 8.35

$(E, -)$-structured category, 14H

Evaluation morphism, 27.2

Exact category, 14F

Exponential functors, 27.5

Exponential law for cartesian closed category, 27.8

Exponential morphism, 27.2

Extension, injective, 9.11

Extensional topological construct & hull, 28B, 28C

Extension of a factorization structure, 15.19 ff
   equivalent conditions for, 15.20
   vs. has products, 15.19, 15.21

Extension of an object, 8.6

Extremal coseparator, 10.17
   characterization, 10B, 10C

Extremal co-wellpoweredness, 7.87 ff, 23I
   detected by regularly monadic functor, 20.31
   vs. concrete co-wellpoweredness, 8E
   vs. monadic category, 20.48

Extremal epimorphism, 7.74 ff
   is final morphism in algebraic category, 23.23
   preservation & reflection of, vs. regularly monadic category, 20.30
   preservation & reflection of, vs. regularly algebraic category, 23.38
   reflection of, 17.13
   vs. concrete generation, 8.18
   vs. $(E, \mathbf{M})$-category, 15.8
   vs. (E, M)-structured category, 14.12, 14.13, 14.14
   vs. extremally generating structured arrows, 8.16, 8.18, 8.36
   vs. final morphism, 8.11(5)
   vs. pullbacks and products, 23.28, 23.29
   vs. quotient morphism in algebraic

category, 23.26, 23G
  vs. regular epimorphism, 7.75, 21.13
  vs. topological category, 21.13
Extremal epi-sink, 10.63
Extremal generation of an object, 8.15
  vs. concrete generation, 8.16
  vs. extremal epimorphism, 8.16, 8.18
Extremally co-wellpowered category, 7.87
  vs. algebraic category, 23.27
  vs. $(E, \mathbf{M})$-category, 15.25, 15.26
Extremally co-wellpowered functor, 8.37
Extremally epireflective subcategory, 16.1 ff
  characterization, 16M
  vs. algebraic subcategory, 23.33
Extremally generating structured arrow, 8.15
  with respect to a functor, 8.30
Extremally monadic construct, is topologically algebraic, 25.2
Extremally monadic functor, 20M, 23N
Extremally projective object, 23H
Extremally reducible conglomerate of sinks, 28.13
Extremally wellpowered category, 7.82 ff
  vs. wellpowered category, 7.83
Extremal monomorphism, 7.61 ff
  closure under composition and intersections, 14.18
  composite need not be extremal, 7N
  as epimorphism is isomorphism, 7.66
  preserved by solid functors, 25.21
  products of, 10D
  vs. adjoint functor, 18J
  vs. balanced category, 7.67
  vs. (E, M)-diagonalization property, 14.18
  vs. (E, M)-structured category, 14.10
  vs. extremal mono-source, 10.26
  vs. monadic functor, 20O
  vs. regular monomorphism, 7.62, 7.63, 12B, 14.20, 14I, 21.13

  vs. topological category, 21.13
Extremal mono-source, 10.11 ff
  characterization, 10A
  not preserved by composition, 10.14, 7N
  preserved by first factor, 10.13
  vs. $(E, \mathbf{M})$-category, 15.8
  vs. epireflective subcategory, 16H
  vs. extremal monomorphism, 10.26
  vs. product, 10.21
  vs. pullback square, 11.9
  vs. subsource, 10.15
Extremal partial morphism, 28.1
Extremal quotient object, 7.84
Extremal reducibility of final sinks, 28.19
Extremal separator, 10.63
  characterization, 10B, 10C
Extremal subobject, 7.77
Extremal-subobject classifier, 28.11
(ExtrEpi, Mono-Source)-category, vs. (Epi, ExtrMono-Source)-category, 15C
(ExtrGen, Mono)-factorization, 17I
(ExtrGen, Mono-Source)-functor, 17K
  vs. preservation of strong limits, 17.11, 17H

Factorization lemma, 14.16
Factorization structure for morphisms, 14.1 ff, 14A
  extension of, to a factorization structure for sources, 15.24
Factorization structure, 15.1 ff
  dispersed, 15J
  for empty sources, 15.2, 15G
  extensions of, 15.19 ff
  w.r.t. a functor, 17.3
  — existence, 17C
  — two methods, 17.5
  inheritance, 16B
  for sinks, 15.2
  for small sources, 15J

INDEX 497

for 2-sources, 15I
  vs. decomposition of functors, 26A
  vs. monadic category, 20.28

Faithful functor, 3.27 ff, 7G
  characterization, 8N
  closed under composition, 3.30
  is isomorphism iff full and bijective on objects, 3.28
  means epimorphisms are generating, 8.32
  monadic functor is, 20.12
  reflects epimorphisms, 7.44
  reflects monomorphisms, 7.37
  reflects mono-sources, 10.7
  topological functor is, 21.3
  vs. adjoint situation, 19.14
  vs. comparison functor, 20.43
  vs. concrete functor, 5.10
  vs. embedding functor, 3.28, 3.29, 4.5
  vs. $(E, \mathbf{M})$-factorization, 17.15, 17.16
  vs. essentially algebraic functor, 23.2
  vs. inclusions of subcategories, 4.5
  vs. initial source, 10.59
  vs. reflection of equalizers, 13.24, 17.13, 17.14
  vs. reflection of (extremal) epimorphisms, 19.14
  vs. reflection of products, 10.60
  vs. reflection of special morphisms, 17G
  vs. solid functor, 25.12
  vs. thin category, 3.29, 3G
  vs. topologically algebraic functor, 25.3

Fibre of a base object, 5.4, 5A
  largest element need not be discrete, 8.5
  order on, 5.4

Fibre-complete concrete category, 5.7
  functor-structured categories are, 5.42

Fibre-discrete concrete category, means fibres are ordered by equality, 5.7
  categories of T-algebras are, 5.39

concrete category is, iff forgetful functor reflects identities, 5.8

Fibre-small concrete category, 5.4, 5.6
  monadic category is, 20.12

Fibre-small topological category, 21.34 ff, 21B

Final lift, vs. topological category, 21.34

Finally closed subcategory, 21.29
  vs. concretely coreflective subcategory, 22.8

Finally dense embedding, 10.72

Finally dense subcategory, 10.69
  vs. concretely reflective subcategory, 21.32
  vs. preserves initial sources, 10.71
  vs. topological category, 21.32

Final morphism, 8.10 ff
  composition of, 8.13
  is extremal epimorphism in algebraic category, 23.23
  first factor of, 8.13
  products of, 28J
  vs. extremal epimorphism, 8.11(5)
  vs. isomorphism, 8.14
  vs. monadic category, 20.28
  vs. regular epimorphism, 8.11(4), 8O, 20.51

Final quotient object, 8.10

Final sink, 10.63 ff
  reducibility of, vs. representable partial morphisms, 28.15

Finer than, (preorder) relation on concrete functors, 5.18

Finitary category, 24.4
  = finitary variety, if monadic, 24.7

Finitary functor, 24.4

Finitary monad, 24.4

Finitary quasivariety, 16.12
  axiomatic descriptions, 23.30, 23.31, 23.38, 24.9, 24.10,
  characterization theorem, 24.9

= finitary and algebraic category, 24.9
= regular epireflective subconstruct of some finitary monadic construct, 24.9
vs. directed colimit, 24A
vs. finitary variety, 16K, 16N, 24.9

Finitary variety, 16.16
axiomatic descriptions, 20.35, 23.40, 24.7, 24.8
characterization theorem for, 24.7
is monadic, 20.20
vs. finitary quasivariety, 16K, 16N, 24.9
vs. monadic category, 24.7

Finitely (co)complete category, 12.2
characterization, 12.4

Finitely complete category, vs. has finite products and representable M-partial morphisms, 28.5, 28.6

Finite product, 10.29
vs. product and projective limit, 11B

Forgetful functor, 3.20(3), 5.1 ff
vs. coequalizer, 7.73
vs. regular epimorphism, 7.73
vs. topological construct, 21L

Fork, congruence, 20.14
split, 20.14

Frame, 5L, 8G

Free automata, 8P

Free category, 3A

Free functor, 19.4

Free monad, 20.55
vs. varietor, 20.56, 20.57, 20.58

Free object, 8.22, 8F–8L
existence, 18.15
retract of, 9.29
vs. adjoint functor, 18.19
vs. adjoint situation, 19.4
vs. copower, 10R
vs. initial object, 8.23
vs. monomorphism, 8.28, 8.29
vs. preservation and reflection of mono-sources, 18.10
vs. projective object, 9.29, 9.30
vs. representable forgetful functor, 8.23
vs. wellpoweredness, 18.10

Full embeddability of familiar constructs, 4.7, 4K

Full embedding functor, 4.6
vs. comparison functor, 20.44
vs. inclusion of a full subcategory, 4.5, 4.6

Full faithful functor,
is isomorphism iff bijective on objects, 3.28
properties of, 3.31
reflects isomorphisms, 3.32
vs. adjoint situation, 19.14

Full functor, 3.27 ff
closed under composition, 3.30
second factor of is not always full, 3.30
vs. comparison functor, 20.43

Full reflective embedding,
is topologically algebraic, 25.2
vs. solid functor, 26.1

Full reflective subcategory,
characterization, 4.20

Full subcategory, 4.1 ff, 4C
isomorphism-closed, 4.9, 4B
isomorphism-dense, 4.9
vs. concretely reflective subcategory of amnestic category, 5.24

Function,
between classes, 2.2
between conglomerates, 2.3
between sets, 2.1

Functor, 3.17 ff
adjoint, = adjoint functor, 18.1 ff
algebraic, = algebraic functor, 23.19 ff
co-adjoint, = co-adjoint functor, 18.1 ff
comparison, = comparison functor, 20.37 ff

composite of, 3.23
concretely co-wellpowered, 8.37
constant functor, 3.20(2)
contravariant hom-functor, 3.20(5)
contravariant power-set functor, 3.20(9)
coproduct of, 10U
coreflector for a coreflective subcategory, 4.27
covariant hom-functor, 3.20(4)
covariant power-set functor, 3.20(8)
co-wellpowered, implies domain co-well- powered, 8.37, 8.38
decomposition of, 3N
discrete space, 3.29
dual, = opposite functor, 3.41
duality functor for vector spaces, 3.20(12)
$(E, -)$, = $(E, -)$-functor, 17.4
embedding, = embedding functor, 3.27
is embedding iff faithful and injective on objects, 3.28
$(E, \mathbf{M})$, = $(E, \mathbf{M})$-functor, 17.3 ff
E-monadic, = E-monadic functor, 20.21 ff
equivalence, = equivalence functor, 3.33
essentially algebraic, = essentially algebraic functor, 23.1 ff
exponential, 27.5
extremally co-wellpowered, 8.37
faithful, = faithful functor, 3.27
as family of functions between morphism classes, 3.19
finitary, 24.4
forgetful functor, = underlying or forgetful functor, 3.20(3), 5.1
free, 19.4
full, = full functor, 3.27
fundamental group, 3.22
identity, 3.20(1)
inclusion, 4.4
indiscrete space, 3.29, 21.12
inverse, 3.25
isomorphism, 3.24
is isomorphism iff full, faithful, and bijective on objects, 3.28
isomorphism-dense, 3.33
monadic, = monadic functor, 20.8 ff
M-topological, 21K
naturally isomorphic to id_A, 6D
need not reflect isomorphisms, 3.22
notation, 3.18
nth power functor, 3.20(10)
object-free, 3.55
object-part determined by the morphism-parts, 3.19
opposite, = opposite functor, 3.41
preserves isomorphisms, 3.21
between quasicategories, 3.51
regularly algebraic, = regularly algebraic functor, 23.35 ff
regularly monadic, = regularly monadic functor, 20.21 ff
representable, = representable functor, 6.9
solid, = solid functor, 25.10 ff
Stone-functor, 3.20(11)
topologically algebraic, = topologically algebraic functor, 25.1 ff
topological, = topological functor, 21.1 ff
underlying, = forgetful functor, 3.20(3), 5.1
vs. reflection of identities, 3D
Functor-costructured category, 5.43
Functor quasicategory, 6.15, 6H
full embedding of any category into, 6.20
Functors, relationships among, 25.22
Functor-structured category, 5.40, 6C, 28K
cartesian closed, 27I
characterization, 22A
reflective subcategory of, vs. solid

strongly fibre-small category, 26.7

Galois adjoint, = residual functor, 6.25
Galois adjunction, 19D
Galois co-adjoint, = residuated functor, 6.25
Galois connection, 6.26, 6G
    composition of, 6.27(1)
    dual of, 6.27(2)
Galois coreflection, 6.26, 6.35
Galois correspondence, 6.25, 19E
    between amnestic concrete categories, 6.29–6.36
    decomposition of, 6.35
    equivalently described, 6.28
    for constructs, 6.26
    theorem, 21.24
    vs. adjoint situation, 19.8
    vs. concrete co-adjoint, 21E
    vs. initial source preservation, 10.49, 21.24
Galois isomorphism, 6.26, 6.35
Galois reflection, 6.26
    characterization, 6.34
    vs. decomposition of Galois correspondence, 6.35
(Generating, −)-factorizable functor, composites of, 17B
(Generating, −)-factorization, for 2-sources implies preservation of mono-sources, 17.12
(Generating, M)-functor, vs. adjoint functor, 18.4
(Generating, Mono-Source)-factorizations of 2-sources, vs. reflection of isomorphisms, 17.13
Generating structured arrow, 8.15 ff
    characterization, 17A
    concretely, 8.15
    extremally, 8.15
    strongly generating structured arrow is, 25.5
    w.r.t. a functor, 8.30
Generation,
    vs. concrete generation, 8.16
    vs. epimorphism, 8.16, 8.18
Graph of a category, 3A

Has a separator, vs. cartesian closed category, 27.4
Has coequalizers, 12.1
    vs. regularly algebraic category, 23.38
Has concrete (co)limits, 13.12
Has concrete products, 10.54
Has (E, M)-factorizations, vs. (E, M)-functor, 17.10
    w.r.t. a functor, 17.3
    vs. adjoint functor, 18.3
Has enough injectives, 9.9, 9.22
Has equalizers, 12.1
    and (finite) products, means (finitely) complete, 12.3, 12.4
Has (finite) (co)intersections, 12.1
Has (finite) (co)products, 10.29, 12.1
Has finite intersections,
    and finite products, means finitely complete, 12.4
    and products, means complete, 12.3
Has finite products, and representable M-partial morphisms, vs. finitely complete, 28.5, 28.6
Has free objects, 8.26
Has function spaces, 27.17
    vs. constant functions are morphisms, 27.18
    vs. discrete terminal objects, 27.18
    vs. topological category, 27.22
Has limits, vs. (E, M)-structured, 14.17
Has M-initial subobjects, vs. concrete products, 21.42
Has products, 10.29
    of all sizes, implies thin category, 10.32

INDEX

and equalizers (resp. intersections),
   means complete, 12.3
   vs. extension of factorization
      structure, 15.19, 15.21
Has pullbacks, 12.1
   and a terminal object, means finitely
      complete, 12.4
Has pushouts, 12.1
Has regular factorizations, 15.12
   vs. regularly monadic functor, 20.32
Has representable M-partial morphisms,
   28.1
   consequences of, 28.3
Has small concrete colimits,
   characterized, 13.14
Hewitt, E. see Duality principle
Hom-functor, 3.20(4), 3.20(5)
   preserves limits, 13.7
   vs. (co)separator, 7.12, 7.17
   vs. epimorphism, 19C
   vs. limit, 13H
   vs. product, 10E, 10F
   vs. quasicategory, 3.51
   vs. set-valued functor, 6.18
Hull,
   algebraic, 23K
   E-reflective, = E-reflective hull, 16.21
   injective, 9.16

Idempotent monad, 20F
Identity-carried **A**-morphism, 5.3
Identity-carried reflection arrow, see
   Concretely reflective concrete
   subcategory
Identity functor, 3.20(1), 5.14
Identity morphism, 3.1 ff
   in a category, as unit in the
      corresponding object-free category,
      3.54
   reflection, 3D
   vs. isomorphism, 3.13
   vs. object, 3.19

   vs. subcategory, 4A
Identity natural transformation, 6.6
Illegitimate conglomerate, 2.3
Illegitimate quasicategory, 6.16
Image of the indexing function, 2.1
Image, inverse 11.19
Implication, = epimorphism, 16.12
   satisfaction of, 16.12
Implicational subcategory, 16.12
Inclusion functor, 4.4
Indiscrete object, 8.3
   must be largest element in the fibre,
      8.4
   preservation, 21.23
   vs. initial source, 10.42
   vs. initial subobjects, 21.35
   vs. topological category, 21.11, 21.35
Indiscrete space functor,
   is a full embedding, 3.29
   vs. topological functor, 21.12
Indiscrete structure, vs. adjoint situation,
   21A
   vs. topological functor, 21.18, 21.19,
      21.20
Inductive limit, 11.28
Initial completion, 21G
Initiality-preserving concrete functor,
   preserves indiscrete objects, 21.23
Initial lift, vs. topological category, 21.34
Initially closed subcategory, 21.29
   of a topological category is
      topological, 21.30
   vs. concretely reflective subcategory,
      21.31, 22.3, 22.4
Initially dense subcategory, 10.69
Initial morphism, 8.6
   composition of, 8.9
   first factor of, 8.9
   universally, 10P
   vs. injective morphism, 8B
   vs. isomorphism, 8.14

502 INDEX

Initial object, 7.1
  uniqueness, 7.3
  vs. free object, 8.23
  vs. limit, 11A
Initial source, 10.41
  composition of, 10O
  domain of, 10.43
  preservation, 10.47
  — vs. Galois correspondence, 21.24
  preserved by composition, 10.45
  preserved by first factor, 10.45
  vs. concrete product, 10.53
  vs. concretely reflective subcategory, 10.50
  vs. faithful functor, 10.59
  vs. finally dense subcategory, 10.69
  vs. indiscrete object, 10.42
  vs. initial subsource, in a fibre-small topological category, 21.36
  vs. mono-source, 17.13
  vs. subsource, 10.46
  w.r.t. a functor, 10.57
Initial subobject, 8.6
  vs. indiscrete object, 21.35
  vs. monotopological construct, 21.42
  vs. topological category, 21.35
Initial T-algebra, 20I
Injection morphism, 10.63
Injective, enough, 9.9, 15H
Injective automata, 9I
Injective extension, 9.11
  types of, 9.20
Injective hull, 9.16
  w.r.t. a class of morphisms, 9.22
  existence, 9C
  for finitary varieties, 16J
  uniqueness, 9.19
  vs. enough injectives, 9D
Injective morphism, vs. initial morphism, 8B
Injective object, 9.1
  is an absolute retract, 9.7
  in **CAT**, vs. topological category, 21.21
  characterization, 9B
  w.r.t. a class of morphisms, 9.22
  preserved by product, 10.40
  terminal object is, 9.4
  in topological construct, 28A
  vs. essential extension, 9.15
  vs. maximal essential extension, 9F
  vs. reflective subcategory, 9.25
  vs. retract, 9.5
Injectivity class, 9.26
Insertion of generators, 19.4
Intersection, of a family of sets, 2.1
Intersection of subobjects, 11.23
  closure under the formation of, 11.26
  existence, 12.1
  as pullback, 11F
  vs. (E, M)-structured category, 14.15
  vs. (Epi, ExtrMono)-structured category, 14.19
  vs. isomorphism, 11.25
  vs. limit, 11.25
  vs. regular subobject & section, 11K
Inverse functor, unique determination, 3.25
Inverse image, of a subject, 11.19
Inverse of an isomorphism, notation, 3.12
Inverse limit, 11.4
Inverse of a morphism, 3.8
  uniqueness, 3.11
"is equivalent to", is equivalence relation on categories, 3.34
"is isomorphic to" is an equivalence relation on categories, 3.25
Isomorphic, "is isomorphic to" is an equivalence relation on categories, 3.25
  concretely, 5.12
  naturally, 6.5
Isomorphic categories, 3.24, 3.25
  vs. equivalent categories, 3.35

INDEX

Isomorphic objects, 3.15
    yields equivalence relation on the object class, 3.16
Isomorphic structured arrows, 8.19, 8.34
Isomorphism, 3.8 ff, 3F
    in a category, vs. isomorphic objects, 3.15
    composite of, is an isomorphism, 3.14
    concrete, 5N
    creation, 13.35, 13M
    functor is isomorphism iff full, faithful, and bijective on objects, 3.28
    functors need not reflect, 3.22
    Galois, 6.26
    inverse is an isomorphism, 3.14
    natural, 6.5
    non-concrete, 5B
    is 1-product, 10.20
    preserved and reflected by equivalence functor, 7.47
    preserved by functors, 3.21
    preserved by products, 10.35
    reflection of, 13.35, 13F
    vs. coequalizer, 7.70
    vs. $(E, \mathbf{M})$-category, 15.5
    vs. epimorphism, 7.43, 7.66
    vs. equalizer, 7.54
    vs. extremal monomorphism, 7.66
    vs. final morphism, 8.14
    vs. identity morphism, 3.13
    vs. initial morphism, 8.14
    vs. intersection, 11.25
    vs. monomorphism, 7.36
    vs. product source, 10.26
    vs. retraction, 7.36
    vs. section, 7.43
Isomorphism-closed (full) subcategory, 4.9, 4B
Isomorphism-closed full subcategory, vs. limit-closed, 13.27
Isomorphism-dense full subcategory, 4.9, 4.10, 4.12

Isomorphism-dense functor, 3.33
Isomorphism functor, 3.24
    closed under composition, 3.30
    is isomorphism in **CAT**, 3.51
Iso-transformation, 6.5

Kernel, 7C
    vs. pullback, 11E
Kleisli category, 20.39, 20B

Large category, 3.44
Large class, 2.2
Largest essential extension, 9.20
Left adjoint, = co-adjoint, 19.11
Left inverse of a morphism, vs. right inverse, 3.10
Legitimate conglomerate, 2.3
Legitimate quasicategory, 6.16
Lifting of an adjoint situation, 21.26
    vs. topological category, 21.28
Lifting of colimits, 13.17
Lifting of $(E, \mathbf{M})$-factorizations, uniquely, 20.23
    vs. E-monadic functor, 20.24
Lifting of (Extremal Epi, Mono-Source)-factorizations uniquely, vs. algebraic category, 23.31
Lifting of limits, 13.17
    uniquely, and reflect limits, vs. create limits, 13.25
    vs. amnesticity, 13.21
    vs. create limits, 13.20
    vs. detect limits, 13.34
    vs. preserve small limits, 13.19
    vs. reflect limits, 13.23
    vs. transportability, 13E
Limit-closed subcategory, 13.26, 13I, 13J
    vs. isomorphism-closed subcategory, 13.27
Limit-dense subcategory, rarity, 12E

504   INDEX

Limit (of a diagram), 11.3 ff
  concrete, = concrete limit, 13.12
  construction via large colimits, 12.7
  creation, 13.17
  detection, = detection of limit, 13.22
  is extremal mono-source, 11.6
  inductive, 11.28
  inverse, 11.4
  lifting of, 13.17, 13D, 13E
  — vs. topological functor, 21.15
  preservation of, = preservation of
    limit, 13.1 ff, 13B
  — vs. preservation of monos, 13.5
  — vs. preservation of strong limits,
    13.1
  — vs. topological functor, 21.15
  projective, 11.4, 11B
  reflection of, 13.22, 13G, 17.13
  uniqueness, 11.7
  vs. factorization, 14.16
  vs. hom-functors, 13H
  vs. initial object, 11A
  vs. intersection, 11.25
  vs. pullback square, 11.9
  vs. reflective subcategory, 13.28
Locally presentable category, 20H

Mac Neille completion, 21H
  vs. solid category, 25G
M-action, category of all M-actions and
  action homomorphisms, 3.26
Map, quotient, = quotient map, 9.27
Matching condition, *see* Object-free
  category
Maximal essential extension, 9.20
  vs. injective object, 9F
Minimal injective extension, 9.20
  vs. smallest injective extension, 9G
$\mathbf{M}$-initially closed subcategory, vs.
  concretely E-reflective subcategory,
  22.9
Modification, amnestic, = amnestic

modification, 5.6, 5.33, 5.34, 5F
Modifications of structure, vs.
  (co)reflective subcategory, 4.17, 4.26
Monad, 20.1 ff
  associated with an adjoint situation,
    20.3, 20A
  finitary, 24.4
  gives rise to an adjoint situation, 20.7
  idempotent, 20F
  on **Set**, is finitary if and only if the
    associated construct is finitary, 24.6
  — is regular, 20.22
  regular, 20.21
  trivial, 20.2
  with rank, 20G
Monadic concrete category, 20.8 ff
  cocompleteness of, 20D
  $\cong$ its associated category of algebras,
    20.40
  properties of, 20.12
  vs. adjoint functor, 20.46
  vs. concrete functor, 20E
  vs. co-wellpoweredness, 20.29
  vs. (Epi,Mono-Source)-factorization,
    20.49
  vs. essentially algebraic category,
    23.16
  vs. extremal co-wellpoweredness,
    20.48
  vs. factorization structure, 20.28
  vs. final morphism, 20.28
  vs. finitary variety, 24.7
  vs. monadic subcategory, 20.19
  vs. reflective subcategory, 20.18
  vs. varietor, 20.56, 20.57, 20.58
Monadic construct, 20.34 ff
  bounded, vs. bounded variety, 24.11
  characterization theorem for, 20.35
  is complete, cocomplete, wellpowered,
    co-wellpowered, and has regular
    factorizations, 20.34
  is regularly algebraic, 23.37
  vs. algebraic construct, 23.41

vs. coequalizer of congruence relation, 20.35
vs. creation of finite limits, 20.35
Monadic functor, 20.8 ff
  characterization theorem for, 20.17
  composition of, need not be monadic, 20.45
  creates absolute colimits, 20.16
  deficiencies of, 20.45 ff
  need not detect colimits, 20.47
  need not reflect regular epimorphisms, 20.52
  properties of, 20.12
  regularly, = regularly monadic functor, 20.21 ff
  vs. colimit, 20.13
  vs. essentially algebraic functor, 23C
  vs. extremal monomorphism, 20O
  vs. order preserving map, 20N
  vs. preservation of extremal epimorphisms, 20.50
  vs. preservation of regular epimorphisms, 20.50
  vs. regularly monadic functor, 20.32, 20.35
  vs. solid functor, 25F
Monadic subcategory, vs. monadic category, 20.18
Monadic towers, 20J
Monad morphism, 20.55
Monoid (as a category), 3.3(4)
  M-actions as functors from a monoid M to **Set**, 3.20(13)
Monomorphism, 7.32 ff
  closed under composition, 7.34
  extremal, = extremal monomorphism, 7.61 ff
  normal, 7C
  preserved by products, 10.35
  preserved and reflected by equivalence functor, 7.47
  preserved by representable functor, 7.37

is pullback stable, 11.18
reflected by faithful functor, 7.37
regular, = regular monomorphism, 7.56 ff
as retraction means isomorphism, 7.36
split, = section, 7.19 ff
strict, = strict monomorphism, 7.76, 7D
types of, 7.76
vs. coequalizer, 7.70
vs. congruence relation, 11.20, 11R
vs. embedding, 8.7
vs. (E, M)-structured category, 14.9
vs. free object, 8.28, 8.29
vs. mono-source, 10.26
vs. preservation of limits, 13.5
vs. pullback square, 11.15, 11.16
vs. regular monomorphism, 7O
vs. section, 7.35, 7P
Monoreflective subcategory, 16.1
  is bireflective, 16.3
Mono-source, 10.5 ff
  characterization, 10A
  extremal, = extremal mono-source, 10.11
  is extremal in balanced category, 10.12
  is initial, vs. essentially algebraic functor, 23.2
  — vs. monadic category, 20.12
  — vs. reflection of limits, 17.13, 17.14
  preservation, by representable functors, 10.7
  — implied by (Generating, −)-factoriza- tion for 2-sources, 17.12
  preserved by composition, 10.9
  preserved by first factor, 10.9
  reflection, by faithful functors, 10.7
  vs. balanced category, 10.12
  vs. $(E, \mathbf{M})$-category, 15.6–15.9
  vs. initial source, 17.13
  vs. monomorphism, 10.26
  vs. point-separating source, 10.8, 10T

vs. preservation of limits, 13.5
vs. pushout, 11P
vs. subsource, 10.10

Monotopological category, 21.38 ff

Monotopological construct,
  characterization of fibre-small, 22.10
  vs. concrete products, 21.42
  vs. initial subobjects, 21.42
  vs. subconstruct of a functor-structured construct, 22.10

Mono-transformation, 6.5, 7R

Morphism, 3.1 ff
  (co)domain of, 3.2
  coessential, 9.27
  composite of, see Composition of morphisms
  constant, 7A, 10W
  diagonal, 10V, 10W
  epi-, see Epimorphism
  evaluation, 27.2
  exponential, 27.2
  final, = final morphism, 8.10 ff
  initial, = initial morphism, 8.6 ff
  isomorphism, 3.8
  mono-, see Monomorphism
  in an object-free category, 3.53
  partial, 28.1
  product of, 10.34
  quotient, = quotient morphism, 8.10
  retraction, 7.24 ff
  section, 7.19 ff
  zero, 7A

Morphism class of a category, 3.2
  as corresponding object-free category, 3.54
  of small category is a set, 3.45

M-reducible, 28.13

M-sources, closed under the formation of, 16.7

M-subobject, = singleton **M**-source, 16.8

M-topological category, 21.38 ff
  characterization theorem, 21.40
  is E-reflective in some topological category, 21.40
  fibre-small, characterization, 22.9
  vs. subcategory of a functor-structured category, 22.9

M-topological functor, 21K

M-topological structure theorem, 22.9

M-topos, 28.7

M-transformation, 6.5

Multiple equalizer, 11.4

Multiple pullback, 11L–11N, 12C
  preservation, 13A

Natural equivalence, = natural isomorphism, 6.5

Natural isomorphism, 6.5

Natural numbers, 2.1

Naturality condition, 6.1

Naturally isomorphic, 6.5

Natural sink, 11.27

Natural source, 11.3
  is natural transformation, 11.5

Natural transformation, 6.1 ff
  cardinality of, 6B
  composition, 6.13, 6A
  concrete, 6.23, 6.24
  identity-carried, = concrete natural transformation, 6.23
  M-transformation, 6.5
  vs. composition of functors, 6.3
  vs. opposite functor, 6.3
  vs. **Set**-valued functor, 6.18

Neighborhood space, 5N

Normal monomorphism, 7C

nth power functor, 3.20(10)

n-tuple of sets or classes, 2.1, 2.2

Object, 3.1 ff
  copower of, 10.63
  discrete, = discrete object, 8.1

equivalent to another, in a concrete
 category, 5.4
free, = free object, 8.22
indiscrete, = indiscrete object, 8.3
initial, = initial object, 7.1
injective, = injective object, 9.1
isomorphic, 3.15, 3.16
power, 27.2
power of, 10.37
projective, = projective object, 9.27
quotient, = quotient object, 7.84 ff
terminal, = terminal object, 7.4
zero, 7.7, 27A

Object class of a category, 3.1, 3.2
 vs. identity morphisms, 3.19

Object-free category, 3.53
 corresponding to a category, 3.54, 3.55

Object-free functor, 3.55

Opposite category, 3.5
 dually equivalent means opposite
  category is equivalent, 3.38
 vs. contravariant hom-functor, 3.20(5)
 vs. contravariant power-set functor,
  3.20(9)
 vs. dual functor for vector spaces,
  3.20(12)
 vs. Stone-functor, 3.20(11)

Opposite functor, 3.41
 vs. Galois connection, 6.27 (2)
 vs. natural transformation, 6.3

Ordered pair of sets, 2.1

Order on concrete functors, 5.18
 vs. concretely reflective subcategory,
  5.26

Order on objects, in a concrete category,
 5.4, 5.5

Order preserving map, = morphism in
 Pos, 4.3
 vs. monadic functor, 20N

Partial binary algebra, 3.52
Partial binary operation, 3.52

Partial morphism, 28.1
Pointed category, 3B, 7B
Point-separating source, 10.5
 vs. mono-source, 10.8, 10T
Posets, category of, 4.3
 down-directed & up-directed, 11.4
Power object, 27.2
Power-set, = set of all subsets, 2.1
Power-set functor, 3.20(8), 3.20(9)
 representability of, 6F
 vs. adjoint functor, 18E
Power-set monad, 20.2
Power-set-morphism, 28.11
Power (of an object), 10.37
 with discrete exponent is product, vs.
  cartesian closed topological
  construct, 27.24
 comparison of, 10K
 vs. coseparator, 10.38
Preimage-morphism, 28.11
Preordered class, as a category, 3.3(4)
Preorder relation on the fibres, in a
 concrete category, 5.5
Preservation of coequalizers, 13.1
Preservation of colimits, 13.1
Preservation of coproducts, 13.1
Preservation of equalizers, 13.1, 13.3
Preservation of extremal epimorphisms,
 vs. monadic functor, 20.50
Preservation of initial sources, 10.47
 vs. finally dense subcategory, 10.71
Preservation of limits & colimits, §13
Preservation of limits,
 adjoint functor does, 18.9
 vs. adjoint functor, 18.12, 18.14,
  18.17, 18.19
 vs. embedding, 13.11
 vs. hom-functor, 13.7
 vs. monomorphism & mono-source,
  13.5

vs. representable functor, 13.9
vs. solid functor, 25.14
Preservation of mono-sources, implied by (Generating, −)-factorization for 2-sources, 17.12
Preservation of products, 13.1, 13.3
Preservation of pullbacks, 13.3
Preservation and reflection of mono-sources, vs. free object, 18.10
vs. monadic functor, 20.12
Preservation of regular epimorphisms, vs. (regularly) monadic functor, 20.32, 20.50
Preservation of small limits, vs. lift limits, 13.19
Preservation of strong limits, vs. (ExtrGen, Mono-Source) functor, 17.11, 17H
Preservation of terminal object, 13.3
Pretopological space, 5N
Product (of morphisms), 10.34
of epimorphisms, vs. cartesian closed category, 27.8
Product (of objects), 10.19 ff
in abelian groups, 10G
for Banach spaces, 10J
characterization, 10Q
composition of, 10.25
concrete, = concrete product, 10.52, 13.12
with discrete factors are coproducts, vs. cartesian closed topological construct, 27.24
existence, 10.29, 10I, 12.1
is extremal mono-source, 10.21
notation for, 10.23
of pairs, 10.30
preservation of, 13.1
preserved by first mono-factor, 10.56
and pullback square, vs. equalizer, 11.11, 11.14
uniqueness, 10.22

vs. empty source, 10.20
vs. (E, M)-structured category, 14.15
vs. equalizer, 10.36
vs. finite product, 11B
vs. first factor, 10.25
vs. hom-functor, 10E, 10F
vs. isomorphism, 10.20
vs. projective limit, 11B
vs. pullback square, 11.13, 11C, 11D
vs. retraction, 10.28
vs. terminal object, 10.20, 10.30, 10H
Product category, 3.3(4)
Product source, vs. isomorphism, 10.26
Projection morphism, 10.23
vs. retraction, 10.27
Projective cover, 9.27
Projective hull, w.r.t. a class of morphisms, 9.27
Projective limit, 11.4
vs. (finite) product, 11B
Projective object, 9.27
w.r.t. a class of morphisms, 9.27
extremal, in algebraic category, 23.28, 23.29
regular, 9E
vs. free object, 9.29, 9.30
Proper class, 2.2 *See also* Large category
Proper quasicategory, 3.50, 3.51
quasicategory of all categories is, 3.51
Property, dual, 3.7
universal, 4.16, 4.25
Property of objects, as (isomorphism-closed) full subcategory, 4.8
Pulation square, 11.32, 11Q
vs. congruence relation, 11.33
Pullback, 11.8 ff. *See also* Pullback square
closure under the formation of, 11.17
existence, 12.1
of a morphism, 11.8
multiple, 11L–11N

INDEX

of a sink along a morphism, 28.13
of a 2-sink, 11.8
vs. (E, M)-structured category, 14.15
vs. epi-sink, 11I
vs. equalizer, 11S
vs. kernel, 11E
vs. product, 11C, 11D
vs. section, 11H
vs. strict monomorphism, 11H

Pullback square, 11.8 ff. *See also* Pullback
cancellation of, 11.10, 11.15
composition of, 11.10
and product, vs. equalizer, 11.14
vs. equalizer and product, 11.11
vs. extremal mono-source, 11.9
vs. limit, 11.9
vs. monomorphism, 11.15, 11.16
vs. product, 11.13
vs. terminal object, 11.13

Pullback stable, 11.17, 28.13

Pushout, 11.30 ff
existence, 12.1
of a 2-source, 11.30
vs. $(E, \mathbf{M})$-category, 15.14, 15.15, 15.16
vs. mono-source, 11P

Pushout square, 11.30 ff
vs. coequalizer, 11.33

Quasicategory, 3.49 ff
of all categories, is not a category, 3.5, 3.51
of all concrete categories over a given base category, 5.14, 5.15
of all object-free categories, 3.55
of all quasicategories, yields Russell-like paradox, 3.51, 3L
category is, 3.51
functor quasicategory, 6.15, 6H
(il)legitimate, 6.16
proper, 3.50, 3.51
vs. hom-functor, 3.51

Quasiconstruct of quasitopological spaces, 5.6

Quasitopos, 28.7
is (Epi, RegMono)-structured, 28.10

Quasivariety, = algebraic construct, 24.12
finitary, = finitary quasivariety, 16.12

Quotient map, 9.27
coessential, 9.27

Quotient morphism, 8.10
composition & first factor of, 8.13
vs. extremal epimorphism, 23G
vs. (regular) epimorphism & retraction, 8.12

Quotient object, 7.84 ff
extremal, = extremal quotient object, 7.84
order on, 7.85
regular, = regular quotient object, 7.84

Rational & real numbers, 2.1

Realization = full, concrete embedding, 5O

Reflection arrow, 4.16 ff
uniqueness, 4.19
vs. colimit, 13.30

Reflection, *See also* Reflector
concrete, 5.22, 5E
Galois, 6.26

Reflection of colimits, 13.22

Reflection of equalizers, 17.13
vs. essentially algebraic functor, 23.2
vs. faithful functor, 13.24, 17.13, 17.14
vs. reflection of limits, 17.13, 17.14

Reflection of (extremal) epimorphisms, 17.13
vs. adjoint situation, 19.14
vs. essentially algebraic functor, 23.2
vs. faithful functor, 19.14
vs. monadic functor, 20.12

Reflection of identities, vs. creation of isomorphisms, 13.36

Reflection of isomorphisms, 13.35
  functors need not, 3.22
  vs. creation of isomorphisms, 13.36
  vs. creation of limits, 13.25
  vs. essentially algebraic functor, 23.2
  vs. (Generating, Mono-Source)-factoriza- tions of 2-sources, 17.13
  vs. reflection of limits, 17.13, 17.14

Reflection of limits, 13.22, 13G, 17.13
  vs. essentially algebraic functor, 23.2
  vs. lifting of limits, 13.23
  vs. mono-sources are initial, 17.13, 17.14
  vs. reflection of equalizers, 17.13, 17.14
  vs. reflection of isomorphisms, 17.13, 17.14

Reflection of products, vs. faithful functor, 10.60

Reflection of regular epimorphisms, vs. initiality of mono-sources, 19.14

Reflective embedding, misbehaved, 13K

Reflective modification, of a concrete category, 5.22, 5K

Reflective subcategory, 4.16, 16D, 16E
  characterization of fullness, 4.20
  embedding of, preserves and reflects mono-sources, 18.7
  full, 4.20, 7F
  of a functor-structured category, vs. cocomplete category that has free objects, 26.8
  intersections, 4F
  nonfull, with reflection arrows isomorphisms, example, 4.21, 13K
  reflectors are naturally isomorphic, 6.7
  of special categories, 4D
  vs. adjoint situation, 19.4
  vs. cartesian closed subcategory, 27.9
  vs. cocomplete subcategory, 12K
  vs. colimit-closed subcategory, 13.29
  vs. detection of colimits, 13.32

  vs. E-monadic subcategory, 20.25
  vs. injective objects, 9.25
  vs. limits, 13.28
  vs. monadic category, 20.18

Reflector (for a reflective subcategory), 4.23, 4H
  as composite of epireflectors, 16.24
  concrete, 5.22, 5E
  existence, 4.22
  naturally isomorphic to others, 6.7
  uniqueness 4.24, 6.7
  vs. concretely reflective subcategory, 5.26, 5.27, 5.31, 5.32

(RegEpi, Mono-Source)-category, characterization, 15.25
  vs. regular factorization, 15.13

(RegEpi, Mono)-structured category, 14.22, 14D

Regular category, 14E

Regular co-wellpowered category, 7.87
  vs. category with separator or coseparator, 7.89
  vs. construct, 7.88

Regular epimorphism, 7.71 ff
  closed under composition, vs. (RegEpi, Mono-Source)-category, 15.25
  composition of, 14.22
  is extremal epimorphism, 7.75
  preservation and reflection, vs. regularly algebraic category, 23.39
  vs. $(E, \mathbf{M})$-category, 15.7, 15.8
  vs. $(E,$ Mono$)$-structured category, 14.14
  vs. (extremal) epimorphism, 7O, 21.13
  vs. final morphism, 8.11(4), 8O, 20.51
  vs. forgetful functor, 7.72(5), 7.73
  vs. monadic functor, 20.51, 20.52
  vs. quotient morphism, 8.12
  vs. retraction, 7.75

Regular epireflective subcategory, 16.1

Regular epireflective subconstruct, vs. (regular) equational subconstruct,

INDEX 511

16.18

Regular equation, 16.16

Regular equational subconstruct, vs. regular epireflective subconstruct, 16.18

Regular factorization, 15.12, 20.32
   vs. algebraic functor, 24.2
   vs. (RegEpi,Mono-Source)-category, 15.13
   vs. regularly algebraic category, 23.38, 23.39
   vs. regularly monadic category, 20.30

Regular monomorphism, 7.56 ff
   composition of, 7J, 10M, 14I, 28.6
   is (extremal) monomorphism, 7.59, 7.63
   first factor, 14I
   in functionally Hausdorff spaces, 7J
   preserved by products, 10.35
   is pullback stable, 11.18
   in semigroups, 14I
   vs. embedding, 8.7, 8A
   vs. equalizer, 13.6
   vs. extremal monomorphism, 7.62, 7.63, 7.65, 7J, 12B, 14.20, 14I, 21.13
   vs. monomorphism, 7O
   vs. section, 7.59
   vs. strict monomorphism, 12A
   vs. topos, 28F

Regular projective object, 9E

Regular quotient object, 7.84

Regular subobject, 7.77
   vs. intersection, 11K

Regular wellpowered category, 7.82
   vs. category with separator or co-separator, 7.89
   vs. construct, 7.88
   vs. wellpowered category, 7.83

Regularly algebraic category, 23.35 ff
   characterization theorem for, 23.38
   vs. adjoint functor, 23.38
   vs. regular factorization, 23.39

vs. uniquely transportable functor, 23.38

Regularly algebraic functor, 23.35 ff, 23E, 23M
   decomposition theorem for, 24.2
   implies associated monad is regular, 24.1
   implies comparison functor is a regular epireflective embedding, 24.1

Regularly monadic category, vs. regular factorization, 20.30

Regularly monadic functor, 20.21 ff, 20L
   characterization theorem for, 20.32
   composition of, 24.2
   detects colimits, 20.33
   detects extremal co-wellpoweredness, 20.31
   is regularly algebraic, 23.36
   vs. monadic functor, 20.35
   vs. solid functor, 26.4

Relation, congruence, = congruence relation, 11.20

Relational category, 28H

Representable extremal partial morphisms, 28.19

Representable functor, 6.9, 6K
   preserves limits, 13.9
   preserves monomorphisms, 7.37
   preserves mono-sources, 10.7
   vs. adjoint functor, 18C
   vs. duality, 10N
   vs. free object, 8.23

Representable M-partial morphisms, vs. topological category, 28.12

Representable partial morphisms,
   vs. final sinks are reducible, 28.15
   vs. topological category, 28.15

Represent M-partial morphisms, 28.1

Retract, 7.24
   absolute, 9.6
   in an isomorphism-closed full reflective subcategory, 7.31

vs. injective object, 9.5
Retraction, 7.24 ff
   is epimorphism, 7.42
   as monomorphism means isomorphism, 7.36
   preserved by all functors, 7.28
   preserved by products, 10.35
   preserved and reflected by equivalence functor, 7.47
   is pullback stable, 11.18
   reflected by full, faithful functor, 7.29
   is regular epimorphism, 7.75
   vs. (E, M)-structured category, 14.9
   vs. product, 10.28
   vs. projection morphism, 10.27
   vs. quotient morphism, 8.12
Right adjoint, = adjoint, 19.11
Right inverse of a morphism, vs. left inverse, 3.10
Russell's paradox, 2.2, 2.3, 3.51, 3L

Satisfaction of an implication, = e-injective, 16.11, 16.12
Satisfaction of topological (co-)axiom, 22.1, 22.6
Scheme, = domain of a diagram, 11.1
Section, 7.19 ff, 7E
   as epimorphism is isomorphism, 7.43
   preserved by all functors, 7.22
   preserved by products, 10.35
   reflected by full, faithful functor, 7.23
   is (regular) monomorphism, 7.35, 7.59
   vs. embedding, 8.7
   vs. equivalence functor, 7.47
   vs. intersection, 11K
   vs. monomorphism, 7P
   vs. pullback, 11H
Self-adjoint endofunctor, 19G
Self-dual concept, 3.7 *See also* Pulation square
Self-dual properties for functors, 3.43
Semifinal arrow, for a structured sink, 25.8, 25C
Semifinal solution, for a structured sink, 25.7, 25C
Separating set, 7.14
   vs. concretizable category, 7Q
Separator, 7.10 ff
   category with, is regular (co-)wellpowered, 7.89
   extremal, 10.63
   vs. bicoreflective subcategory, 16.4
   vs. (co)wellpoweredness and (co)complete- ness, 12.13
   vs. faithful hom-functor, 7.12
   vs. topological category, 21.16, 21.17
Set,
   constructions that can be performed with sets, 2.1
   morphism class of small category is a set, 3.45
   separating, = separating set, 7.14
   is a (small) class, 2.2
   underlying, = underlying set, *see* Forgetful functor
Set-indexed family of sets, 2.2
Set-indexed source, 10.2
Set-valued functor, vs. hom-functor, 6.18
   vs. natural transformation, 6.18
Singleton-morphism, 28.11
Sink, 10.62 ff
   costructured, 17.4
   final, 10.63, 28.15
   natural, 11.27
Situation, adjoint, = adjoint situation, 19.3 ff
Skeleton of a category, 4.12–4.15, 4I
Small category, 3.44. *See also* Fibre-small
   is a set, 3.45
Small class, 2.2
Small conglomerate, 2.3
Smallest containing E-reflective

subcategory, = E-reflective hull, 16.20, 16.21

Smallest injective extension, 9.20
vs. minimal injective extension, 9G

Smallness of hom-class condition, *see* Object-free category

Small object-free category, 3.55

Small source, 10.2

Solid concrete category, 25.10 ff
is almost topologically algebraic, 25.18, 25.19
characterization theorems, 25.19, 25I
structure theorem, 26.7
vs. (co)complete category, 25.15, 25.16
vs. Mac Neille completion, 25G
vs. subcategory of topological category, 26.2
vs. topological category, 25J

Solid construct, need not be topologically algebraic nor strongly cocomplete, 25.17

Solid functor, 25.10 ff
is almost topologically algebraic, 25.18, 25.19
composite of topological and essentially algebraic functors, 26.3
composite of topologically algebraic functors, 26.1
composite of topological and regularly monadic functors, 26.3
composition of, 25.13
detects colimits and preserves and detects limits, 25.14
is faithful and adjoint, 25.12
preserves extremal monomorphisms, 25.21
topologically algebraic functor is, 25.11
vs. adjoint functor, 25.12, 25.19
vs. detection and preservation of limits, 25.18, 25.19
vs. essentially algebraic functor, 26.3
vs. faithful functor, 25.12

vs. full reflective embedding, 26.1
vs. monadic functor, 25F
vs. regularly monadic functor, 26.4
vs. topologically algebraic functor, 25D, 25E, 26.1
vs. topological functor, 26.1, 26.3, 26.4

Solution set condition, 18.12

Source, 10.1 ff
codomain of, 10.1
composite of, 10.3
domain of, 10.1
empty, 10.2
initial, = initial source, 10.41, 10.57
natural, 11.3
notation for, 10.2, 10.4
set-indexed, 10.2
small, 10.2
structured, w.r.t. a functor, 17.1

Split coequalizer, 20.14

Split epimorphism, = retraction, 7.24 ff

Split fork, 20.14

Split monomorphism, = section, 7.19 ff

Square,
commutes, 3.4
pulation, 11.32, 11.33, 11Q
pullback, = pullback square, 11.8 ff

Stability under pullbacks, 11.17, 28.13

Stable epimorphism, 11J
vs. adjoint functor, 18I

Statement, dual, = dual statement, 3.7

Stone-functor, 3.20(11)

Strict monomorphism, 7D
vs. other types of monomorphisms, 7.76
vs. pullback, 11H
vs. regular monomorphism, 12A

Strong fibre-smallness, 26D

Strongly cocomplete category, 12.2
is (Epi,ExtrMono-Source)-category, 15.17
vs. (co)complete category, 12.5, 12I,

12N, 12O
Strongly complete category, 12.2
  is (E, M)-structured, 14.21
  vs. (co)complete category, 12.5, 12I, 12N, 12O
  vs. $(E, \mathbf{M})$-category, 15.25, 15.26
  vs. free monad & varietor, 20.59
Strongly fibre-small concrete category, 26.5
(Strongly Generating, Initial Source)-functor, is topologically algebraic functor, 25.6
Strongly generating structured arrow, 25.4
  is generating, 25.5
Strongly n-generating object, 20H
Strongly limit-closed subcategory, vs. epireflective subcategory, 16L
Strong monomorphism, 7.76, 14C
Structure, factorization:
  — for morphisms, 14.1 ff
  — for sources, 15.1 ff
  — w.r.t. a functor, 17.3 ff
Structured arrow, 8.15
  equivalence for, 26.5
  (extremally) generating, vs. (extremal) epimorphism, 8.36
  generating, = generating structured arrow, 8.15
  isomorphic, 8.19, 8.34
  strongly generating, 25.4, 25.5
  universal, 8.30
  w.r.t. a functor, 8.30
Structured source,
  w.r.t. a functor, 17.1
  notation for, 17.2
  self-indexed, 17.2
Structured 2-source, factorizations of, 17F
Subcategory, 4.1 ff
  algebraic, = algebraic subcategory, 23.32
  bireflective, 16.1, 16.3

  colimit-closed, 13I
  colimit-dense, = colimit-dense subcategory, 12.10
  concrete, 5.21 ff
  coreflective, = coreflective subcategory, 4.25, 16.1 ff
  definable by topological (co-)axioms, 22.1, 22.3, 22.4, 22.6
  dense, 12D
  E-equational, 16.16
  epireflective, 16.1
  E-reflective, 16.1
  extremally epireflective, 16.1
  finally closed, 21.29
  finally dense, 10.69, 10.69
  full, 4.1(2) ff
  of a functor-costructured category, vs. topological category, 22.8
  of a functor-structured category, vs. ($\mathbf{M}$-) topological category, 22.3, 22.4, 22.9
  implicational, 16.12
  initially closed, 21.29
  initially dense, 10.69
  isomorphism-dense, 4.9
  limit-closed, 13.26, 13.27, 13I, 13J
  limit-dense, 12E
  $\mathbf{M}$-initially closed, vs. concretely E-reflec- tive subcategory, 22.9
  monadic, conditions for, 20.19
  monoreflective, 16.1
  nonfull, 4.3(3)
  reflective, = reflective subcategory, 4.16, 16.1 ff
  regular epireflective, 16.1
  simultaneously reflective and coreflective, 4E
  of subcategories, 4G
  topological, = topological subcategory, 21.29 ff
  vs. identities, 4A
Subconstruct, 5.21
  of a functor-structured construct, vs. monotopological construct, 22.10

INDEX

of **Set** vs. concretely cartesian closed construct, 27.16
Subobject, 7.77 ff
    extremal, = extremal subobject, 7.77
    initial, = initial subobject, 8.6
    intersection of, 11.23
    non-isomorphic, 7K
    order on, 7.79
    regular, = regular subobject, 7.77
    vs. **M**-subobject, 16.8
Subset, 2.1. *See also* Power set
Swell epimorphism, 7.76, 15A
Swell separator, 19K

T-algebra, 5.37
T-algebra, category of, 20.4, 5G
    — closed under the formation of mono-sources, 20.11
    — three-step construction, 20.6
Taut lift theorem, 21.28
Terminal, weakly terminal set of objects, 12F
Terminal category, 3.3(4)
Terminal object, 7.4
    uniqueness, 7.6
    vs. product, 10.20, 10.30, 10H
    vs. pullback square, 11.13
    vs. weak terminal object, 12.9
Thin category, 3.26
    vs. faithful functor, 3.29, 3G
Thin quasicategory, 3.51
T-homomorphism, 5.37
Topological axiom, 22.1
Topological characterization theorems,
    external, 21.21
    internal, 21.18
Topological (co-)axiom, 22.6
Topological concrete category, 21.7 ff
    characterization, 21C
    duality for, 21.9
    is fibre-complete, 21.11

fibre-small, 21.34 ff
— characterization of, 22.3, 22.4, 22.8
— vs. closure under the formation of indiscrete objects, initial subobjects, and products, 21.37
fibre-small reflective subcategory, vs. solid strongly fibre-small category, 26.7
subcategory of, vs. solid category, 26.2
vs. (co)completeness, 21.16, 21.17
vs. concretely cartesian closed subcategory, 27.15
vs. concretely reflective subcategory, 21.32
vs. coseparator, 21.16, 21.17
vs. co-wellpoweredness, 21.16, 21.17
vs. discrete object, 21.11
vs. $(E, \mathbf{M})$-category, 21.14
vs. $(E, \mathbf{M})$-factorization, 21.16, 21.17
vs. extremal morphism, 21.13
vs. final lift, 21.34
vs. finally dense subcategory, 21.32
vs. has function spaces, 27.22
vs. indiscrete object, 21.11, 21.35
vs. initial lift, 21.34
vs. initially closed subcategory, 21.30, 21.31
vs. initial subobjects, 21.35
vs. injective object, 21.21
vs. lift of an adjoint situation, 21.28
vs. representable $(M\text{-})$partial morphisms, 28.12, 28.15
vs. separator, 21.16, 21.17
vs. solid category, 25J
vs. subcategory of a functor-(co)structured category, 22.3, 22.4, 22.8
vs. wellpoweredness, 21.16, 21.17
Topological construct, fibre-small,
    characterization of, 22.4
    vs. forgetful functor, 21L
Topological functor, 21.1 ff
    is (co-)adjoint, 21.12
    composition of, 21.6

is faithful, 21.3
lifts limits and colimits, 21.15
preserves limits and colimits, 21.15
preserves and reflects mono-sources and epi-sinks, 21.13
is topologically algebraic, 25.2
vs. amnestic concrete category, 21.5
vs. bimorphism, 21M
vs. discrete and indiscrete functors, 21.12
vs. indiscrete structures, 21.18, 21.19, 21.20
vs. solid functor, 26.1, 26.3, 26.4
vs. topologically algebraic functor, 25B
vs. unique lifting of limits, 21.18, 21.20
vs. uniquely transportable concrete category, 21.5

Topologically algebraic category, 25.1 ff
characterization theorem, 25.6
vs. universal initial completion, 25H

Topologically algebraic functor, 25.1 ff
almost is solid, 25.18, 25.19
is composite of other types, 26B
composites yield solid functor, 26.1
is faithful and adjoint, 25.3
is solid, 25.11
vs. adjoint functor, 25.3, 25.6, 25.19
vs. detection and preservation of limits, 25.18, 25.19
vs. solid functor, 25D, 25E
vs. topological functor, 25B

Topological structure theorem, 22.3

Topological subcategory, 21.29 ff
vs. concretely (co)reflective subcategory, 21.33

Topological theory, 22B

Topological universe, 28.21
equivalent conditions, 28.22

Topology, three approaches, 5N

Topos, 28.7
is balanced quasitopos, 28.8

characterization, 28E
vs. regular monomorphism, 28F

Total category, 6I

Transformation, natural, = natural transformation, 6.1 ff

Transportable concrete category, 5.28
vs. amnestic concrete category, 5.29, 5.30
vs. concrete category, 5.35, 5.36

Transportable functor, monadic functor is, 20.12
vs. lifting of limits, 13E

Triangle, commutes, 3.4

Trivial factorization structure, 15.3

Trivial monad, 20.2

T-space, 5.40, 5H

Underlying function, *see* Forgetful functor

Underlying functor, = forgetful functor, 3.20(3), 5.1 ff

Union, of a family of sets, 2.1

Unique diagonalization property, 14.1, 15.1

Unique lifting of limits, vs. topological functor, 21.18, 21.20

Uniquely transportable concrete category, 5.28
vs. topological functor, 21.5

Uniquely transportable functor,
vs. algebraic functor, 23.30
vs. regularly algebraic category, 23.38

Unit of an adjunction, 19.3

Unit ball functor, 5.2

Unit existence condition, *see* Object-free category

Unit of a partial binary algebra, 3.52, 19J

Universal arrow, 8.22
must be extremally generating, 8.33
uniqueness, 8.25, 8.35

vs. concrete generation, 8.24
vs. co-universal arrow, 19.1, 19.2
Universal category, 4J
Universal completion, 12M
Universal initial completion, 21I
    vs. topologically algebraic category, 25H
Universally initial morphism, 10P
Universally topological category, 28.16
    as injective object, 28G
    vs. concrete quasitopos, 28.18
Universal property, 4.16, 4.25, 8.22, 8.30
Universal structured arrow, 8.30
Universe, 2.2
Up-directed poset, 11.4
Upper semicontinuity, as adjointness, 18F

Varietal category & functor, 23F
Varietor, 20.53
    vs. colimits of ω-chains, 20P
    vs. free monad & monadic category, 20.56–20.58
Variety, = monadic construct, 24.12
    bounded, vs. bounded monadic construct, 24.11
    finitary, = finitary variety, 16.16, 20.20
    vs. equational subquasiconstruct of some $\mathbf{Alg}(\Omega)$, 24.11

Weakly algebraic functor, 23A
Weakly terminal set of objects, 12F
Weak terminal object, vs. terminal object, 12.9
Well-fibred construct, 27.20
Well-fibred topological construct, vs. cartesian closed construct, 27.22
Wellpowered category, 7.82 ff
    and complete category, is strongly complete, 12.5
    is regular wellpowered and extremally wellpowered, 7.83
    vs. adjoint functor, 18.19
    vs. co-wellpoweredness, 12.13
    vs. essentially algebraic category, 23.12, 23.13
    vs. free object, 18.10
    vs. topological category, 21.16, 21.17
Word-monad, 20.2

Yoneda embedding, 6J
Yoneda Lemma, 6.19

Zero morphism, 7A
Zero object, 7.7
    vs. Cartesian closed category, 27A

A CATALOG OF SELECTED
DOVER BOOKS
IN SCIENCE AND MATHEMATICS

Astronomy

CHARIOTS FOR APOLLO: The NASA History of Manned Lunar Spacecraft to 1969, Courtney G. Brooks, James M. Grimwood, and Loyd S. Swenson, Jr. This illustrated history by a trio of experts is the definitive reference on the Apollo spacecraft and lunar modules. It traces the vehicles' design, development, and operation in space. More than 100 photographs and illustrations. 576pp. 6 3/4 x 9 1/4. 0-486-46756-2

EXPLORING THE MOON THROUGH BINOCULARS AND SMALL TELESCOPES, Ernest H. Cherrington, Jr. Informative, profusely illustrated guide to locating and identifying craters, rills, seas, mountains, other lunar features. Newly revised and updated with special section of new photos. Over 100 photos and diagrams. 240pp. 8 1/4 x 11. 0-486-24491-1

WHERE NO MAN HAS GONE BEFORE: A History of NASA's Apollo Lunar Expeditions, William David Compton. Introduction by Paul Dickson. This official NASA history traces behind-the-scenes conflicts and cooperation between scientists and engineers. The first half concerns preparations for the Moon landings, and the second half documents the flights that followed Apollo 11. 1989 edition. 432pp. 7 x 10. 0-486-47888-2

APOLLO EXPEDITIONS TO THE MOON: The NASA History, Edited by Edgar M. Cortright. Official NASA publication marks the 40th anniversary of the first lunar landing and features essays by project participants recalling engineering and administrative challenges. Accessible, jargon-free accounts, highlighted by numerous illustrations. 336pp. 8 3/8 x 10 7/8. 0-486-47175-6

ON MARS: Exploration of the Red Planet, 1958-1978--The NASA History, Edward Clinton Ezell and Linda Neuman Ezell. NASA's official history chronicles the start of our explorations of our planetary neighbor. It recounts cooperation among government, industry, and academia, and it features dozens of photos from Viking cameras. 560pp. 6 3/4 x 9 1/4. 0-486-46757-0

ARISTARCHUS OF SAMOS: The Ancient Copernicus, Sir Thomas Heath. Heath's history of astronomy ranges from Homer and Hesiod to Aristarchus and includes quotes from numerous thinkers, compilers, and scholasticists from Thales and Anaximander through Pythagoras, Plato, Aristotle, and Heraclides. 34 figures. 448pp. 5 3/8 x 8 1/2. 0-486-43886-4

AN INTRODUCTION TO CELESTIAL MECHANICS, Forest Ray Moulton. Classic text still unsurpassed in presentation of fundamental principles. Covers rectilinear motion, central forces, problems of two and three bodies, much more. Includes over 200 problems, some with answers. 437pp. 5 3/8 x 8 1/2. 0-486-64687-4

BEYOND THE ATMOSPHERE: Early Years of Space Science, Homer E. Newell. This exciting survey is the work of a top NASA administrator who chronicles technological advances, the relationship of space science to general science, and the space program's social, political, and economic contexts. 528pp. 6 3/4 x 9 1/4. 0-486-47464-X

STAR LORE: Myths, Legends, and Facts, William Tyler Olcott. Captivating retellings of the origins and histories of ancient star groups include Pegasus, Ursa Major, Pleiades, signs of the zodiac, and other constellations. "Classic." – *Sky & Telescope*. 58 illustrations. 544pp. 5 3/8 x 8 1/2. 0-486-43581-4

A COMPLETE MANUAL OF AMATEUR ASTRONOMY: Tools and Techniques for Astronomical Observations, P. Clay Sherrod with Thomas L. Koed. Concise, highly readable book discusses the selection, set-up, and maintenance of a telescope; amateur studies of the sun; lunar topography and occultations; and more. 124 figures. 26 halftones. 37 tables. 335pp. 6 1/2 x 9 1/4. 0-486-42820-6

Browse over 9,000 books at www.doverpublications.com

CATALOG OF DOVER BOOKS

Chemistry

MOLECULAR COLLISION THEORY, M. S. Child. This high-level monograph offers an analytical treatment of classical scattering by a central force, quantum scattering by a central force, elastic scattering phase shifts, and semi-classical elastic scattering. 1974 edition. 310pp. 5 3/8 x 8 1/2. 0-486-69437-2

HANDBOOK OF COMPUTATIONAL QUANTUM CHEMISTRY, David B. Cook. This comprehensive text provides upper-level undergraduates and graduate students with an accessible introduction to the implementation of quantum ideas in molecular modeling, exploring practical applications alongside theoretical explanations. 1998 edition. 832pp. 5 3/8 x 8 1/2. 0-486-44307-8

RADIOACTIVE SUBSTANCES, Marie Curie. The celebrated scientist's thesis, which directly preceded her 1903 Nobel Prize, discusses establishing atomic character of radioactivity; extraction from pitchblende of polonium and radium; isolation of pure radium chloride; more. 96pp. 5 3/8 x 8 1/2. 0-486-42550-9

CHEMICAL MAGIC, Leonard A. Ford. Classic guide provides intriguing entertainment while elucidating sound scientific principles, with more than 100 unusual stunts: cold fire, dust explosions, a nylon rope trick, a disappearing beaker, much more. 128pp. 5 3/8 x 8 1/2. 0-486-67628-5

ALCHEMY, E. J. Holmyard. Classic study by noted authority covers 2,000 years of alchemical history: religious, mystical overtones; apparatus; signs, symbols, and secret terms; advent of scientific method, much more. Illustrated. 320pp. 5 3/8 x 8 1/2.
0-486-26298-7

CHEMICAL KINETICS AND REACTION DYNAMICS, Paul L. Houston. This text teaches the principles underlying modern chemical kinetics in a clear, direct fashion, using several examples to enhance basic understanding. Solutions to selected problems. 2001 edition. 352pp. 8 3/8 x 11. 0-486-45334-0

PROBLEMS AND SOLUTIONS IN QUANTUM CHEMISTRY AND PHYSICS, Charles S. Johnson and Lee G. Pedersen. Unusually varied problems, with detailed solutions, cover of quantum mechanics, wave mechanics, angular momentum, molecular spectroscopy, scattering theory, more. 280 problems, plus 139 supplementary exercises. 430pp. 6 1/2 x 9 1/4. 0-486-65236-X

ELEMENTS OF CHEMISTRY, Antoine Lavoisier. Monumental classic by the founder of modern chemistry features first explicit statement of law of conservation of matter in chemical change, and more. Facsimile reprint of original (1790) Kerr translation. 539pp. 5 3/8 x 8 1/2. 0-486-64624-6

MAGNETISM AND TRANSITION METAL COMPLEXES, F. E. Mabbs and D. J. Machin. A detailed view of the calculation methods involved in the magnetic properties of transition metal complexes, this volume offers sufficient background for original work in the field. 1973 edition. 240pp. 5 3/8 x 8 1/2. 0-486-46284-6

GENERAL CHEMISTRY, Linus Pauling. Revised third edition of classic first-year text by Nobel laureate. Atomic and molecular structure, quantum mechanics, statistical mechanics, thermodynamics correlated with descriptive chemistry. Problems. 992pp. 5 3/8 x 8 1/2. 0-486-65622-5

ELECTROLYTE SOLUTIONS: Second Revised Edition, R. A. Robinson and R. H. Stokes. Classic text deals primarily with measurement, interpretation of conductance, chemical potential, and diffusion in electrolyte solutions. Detailed theoretical interpretations, plus extensive tables of thermodynamic and transport properties. 1970 edition. 590pp. 5 3/8 x 8 1/2. 0-486-42225-9

Browse over 9,000 books at www.doverpublications.com

CATALOG OF DOVER BOOKS

Engineering

FUNDAMENTALS OF ASTRODYNAMICS, Roger R. Bate, Donald D. Mueller, and Jerry E. White. Teaching text developed by U.S. Air Force Academy develops the basic two-body and n-body equations of motion; orbit determination; classical orbital elements, coordinate transformations; differential correction; more. 1971 edition. 455pp. 5 3/8 x 8 1/2. 0-486-60061-0

INTRODUCTION TO CONTINUUM MECHANICS FOR ENGINEERS: Revised Edition, Ray M. Bowen. This self-contained text introduces classical continuum models within a modern framework. Its numerous exercises illustrate the governing principles, linearizations, and other approximations that constitute classical continuum models. 2007 edition. 320pp. 6 1/8 x 9 1/4. 0-486-47460-7

ENGINEERING MECHANICS FOR STRUCTURES, Louis L. Bucciarelli. This text explores the mechanics of solids and statics as well as the strength of materials and elasticity theory. Its many design exercises encourage creative initiative and systems thinking. 2009 edition. 320pp. 6 1/8 x 9 1/4. 0-486-46855-0

FEEDBACK CONTROL THEORY, John C. Doyle, Bruce A. Francis and Allen R. Tannenbaum. This excellent introduction to feedback control system design offers a theoretical approach that captures the essential issues and can be applied to a wide range of practical problems. 1992 edition. 224pp. 6 1/2 x 9 1/4. 0-486-46933-6

THE FORCES OF MATTER, Michael Faraday. These lectures by a famous inventor offer an easy-to-understand introduction to the interactions of the universe's physical forces. Six essays explore gravitation, cohesion, chemical affinity, heat, magnetism, and electricity. 1993 edition. 96pp. 5 3/8 x 8 1/2. 0-486-47482-8

DYNAMICS, Lawrence E. Goodman and William H. Warner. Beginning engineering text introduces calculus of vectors, particle motion, dynamics of particle systems and plane rigid bodies, technical applications in plane motions, and more. Exercises and answers in every chapter. 619pp. 5 3/8 x 8 1/2. 0-486-42006-X

ADAPTIVE FILTERING PREDICTION AND CONTROL, Graham C. Goodwin and Kwai Sang Sin. This unified survey focuses on linear discrete-time systems and explores natural extensions to nonlinear systems. It emphasizes discrete-time systems, summarizing theoretical and practical aspects of a large class of adaptive algorithms. 1984 edition. 560pp. 6 1/2 x 9 1/4. 0-486-46932-8

INDUCTANCE CALCULATIONS, Frederick W. Grover. This authoritative reference enables the design of virtually every type of inductor. It features a single simple formula for each type of inductor, together with tables containing essential numerical factors. 1946 edition. 304pp. 5 3/8 x 8 1/2. 0-486-47440-2

THERMODYNAMICS: Foundations and Applications, Elias P. Gyftopoulos and Gian Paolo Beretta. Designed by two MIT professors, this authoritative text discusses basic concepts and applications in detail, emphasizing generality, definitions, and logical consistency. More than 300 solved problems cover realistic energy systems and processes. 800pp. 6 1/8 x 9 1/4. 0-486-43932-1

THE FINITE ELEMENT METHOD: Linear Static and Dynamic Finite Element Analysis, Thomas J. R. Hughes. Text for students without in-depth mathematical training, this text includes a comprehensive presentation and analysis of algorithms of time-dependent phenomena plus beam, plate, and shell theories. Solution guide available upon request. 672pp. 6 1/2 x 9 1/4. 0-486-41181-8

Browse over 9,000 books at www.doverpublications.com

CATALOG OF DOVER BOOKS

HELICOPTER THEORY, Wayne Johnson. Monumental engineering text covers vertical flight, forward flight, performance, mathematics of rotating systems, rotary wing dynamics and aerodynamics, aeroelasticity, stability and control, stall, noise, and more. 189 illustrations. 1980 edition. 1089pp. 5 5/8 x 8 1/4. 0-486-68230-7

MATHEMATICAL HANDBOOK FOR SCIENTISTS AND ENGINEERS: Definitions, Theorems, and Formulas for Reference and Review, Granino A. Korn and Theresa M. Korn. Convenient access to information from every area of mathematics: Fourier transforms, Z transforms, linear and nonlinear programming, calculus of variations, random-process theory, special functions, combinatorial analysis, game theory, much more. 1152pp. 5 3/8 x 8 1/2. 0-486-41147-8

A HEAT TRANSFER TEXTBOOK: Fourth Edition, John H. Lienhard V and John H. Lienhard IV. This introduction to heat and mass transfer for engineering students features worked examples and end-of-chapter exercises. Worked examples and end-of-chapter exercises appear throughout the book, along with well-drawn, illuminating figures. 768pp. 7 x 9 1/4. 0-486-47931-5

BASIC ELECTRICITY, U.S. Bureau of Naval Personnel. Originally a training course; best nontechnical coverage. Topics include batteries, circuits, conductors, AC and DC, inductance and capacitance, generators, motors, transformers, amplifiers, etc. Many questions with answers. 349 illustrations. 1969 edition. 448pp. 6 1/2 x 9 1/4. 0-486-20973-3

BASIC ELECTRONICS, U.S. Bureau of Naval Personnel. Clear, well-illustrated introduction to electronic equipment covers numerous essential topics: electron tubes, semiconductors, electronic power supplies, tuned circuits, amplifiers, receivers, ranging and navigation systems, computers, antennas, more. 560 illustrations. 567pp. 6 1/2 x 9 1/4. 0-486-21076-6

BASIC WING AND AIRFOIL THEORY, Alan Pope. This self-contained treatment by a pioneer in the study of wind effects covers flow functions, airfoil construction and pressure distribution, finite and monoplane wings, and many other subjects. 1951 edition. 320pp. 5 3/8 x 8 1/2. 0-486-47188-8

SYNTHETIC FUELS, Ronald F. Probstein and R. Edwin Hicks. This unified presentation examines the methods and processes for converting coal, oil, shale, tar sands, and various forms of biomass into liquid, gaseous, and clean solid fuels. 1982 edition. 512pp. 6 1/8 x 9 1/4. 0-486-44977-7

THEORY OF ELASTIC STABILITY, Stephen P. Timoshenko and James M. Gere. Written by world-renowned authorities on mechanics, this classic ranges from theoretical explanations of 2- and 3-D stress and strain to practical applications such as torsion, bending, and thermal stress. 1961 edition. 560pp. 5 3/8 x 8 1/2. 0-486-47207-8

PRINCIPLES OF DIGITAL COMMUNICATION AND CODING, Andrew J. Viterbi and Jim K. Omura. This classic by two digital communications experts is geared toward students of communications theory and to designers of channels, links, terminals, modems, or networks used to transmit and receive digital messages. 1979 edition. 576pp. 6 1/8 x 9 1/4. 0-486-46901-8

LINEAR SYSTEM THEORY: The State Space Approach, Lotfi A. Zadeh and Charles A. Desoer. Written by two pioneers in the field, this exploration of the state space approach focuses on problems of stability and control, plus connections between this approach and classical techniques. 1963 edition. 656pp. 6 1/8 x 9 1/4. 0-486-46663-9

Browse over 9,000 books at www.doverpublications.com

CATALOG OF DOVER BOOKS

Mathematics–Bestsellers

HANDBOOK OF MATHEMATICAL FUNCTIONS: with Formulas, Graphs, and Mathematical Tables, Edited by Milton Abramowitz and Irene A. Stegun. A classic resource for working with special functions, standard trig, and exponential logarithmic definitions and extensions, it features 29 sets of tables, some to as high as 20 places. 1046pp. 8 x 10 1/2. 0-486-61272-4

ABSTRACT AND CONCRETE CATEGORIES: The Joy of Cats, Jiri Adamek, Horst Herrlich, and George E. Strecker. This up-to-date introductory treatment employs category theory to explore the theory of structures. Its unique approach stresses concrete categories and presents a systematic view of factorization structures. Numerous examples. 1990 edition, updated 2004. 528pp. 6 1/8 x 9 1/4. 0-486-46934-4

MATHEMATICS: Its Content, Methods and Meaning, A. D. Aleksandrov, A. N. Kolmogorov, and M. A. Lavrent'ev. Major survey offers comprehensive, coherent discussions of analytic geometry, algebra, differential equations, calculus of variations, functions of a complex variable, prime numbers, linear and non-Euclidean geometry, topology, functional analysis, more. 1963 edition. 1120pp. 5 3/8 x 8 1/2. 0-486-40916-3

INTRODUCTION TO VECTORS AND TENSORS: Second Edition--Two Volumes Bound as One, Ray M. Bowen and C.-C. Wang. Convenient single-volume compilation of two texts offers both introduction and in-depth survey. Geared toward engineering and science students rather than mathematicians, it focuses on physics and engineering applications. 1976 edition. 560pp. 6 1/2 x 9 1/4. 0-486-46914-X

AN INTRODUCTION TO ORTHOGONAL POLYNOMIALS, Theodore S. Chihara. Concise introduction covers general elementary theory, including the representation theorem and distribution functions, continued fractions and chain sequences, the recurrence formula, special functions, and some specific systems. 1978 edition. 272pp. 5 3/8 x 8 1/2. 0-486-47929-3

ADVANCED MATHEMATICS FOR ENGINEERS AND SCIENTISTS, Paul DuChateau. This primary text and supplemental reference focuses on linear algebra, calculus, and ordinary differential equations. Additional topics include partial differential equations and approximation methods. Includes solved problems. 1992 edition. 400pp. 7 1/2 x 9 1/4. 0-486-47930-7

PARTIAL DIFFERENTIAL EQUATIONS FOR SCIENTISTS AND ENGINEERS, Stanley J. Farlow. Practical text shows how to formulate and solve partial differential equations. Coverage of diffusion-type problems, hyperbolic-type problems, elliptic-type problems, numerical and approximate methods. Solution guide available upon request. 1982 edition. 414pp. 6 1/8 x 9 1/4. 0-486-67620-X

VARIATIONAL PRINCIPLES AND FREE-BOUNDARY PROBLEMS, Avner Friedman. Advanced graduate-level text examines variational methods in partial differential equations and illustrates their applications to free-boundary problems. Features detailed statements of standard theory of elliptic and parabolic operators. 1982 edition. 720pp. 6 1/8 x 9 1/4. 0-486-47853-X

LINEAR ANALYSIS AND REPRESENTATION THEORY, Steven A. Gaal. Unified treatment covers topics from the theory of operators and operator algebras on Hilbert spaces; integration and representation theory for topological groups; and the theory of Lie algebras, Lie groups, and transform groups. 1973 edition. 704pp. 6 1/8 x 9 1/4. 0-486-47851-3

Browse over 9,000 books at www.doverpublications.com